Lecture Notes in Bioinformatics 12314

Subseries of Lecture Notes in Computer Science

Alessandro Abate · Tatjana Petrov ·
Verena Wolf (Eds.)

Computational Methods in Systems Biology

18th International Conference, CMSB 2020
Konstanz, Germany, September 23–25, 2020
Proceedings

 Springer

Editors
Alessandro Abate 🆔
Department of Computer Science
University of Oxford
Oxford, UK

Verena Wolf 🆔
Department of Computer Science
Saarland University
Saarbrücken, Germany

Tatjana Petrov 🆔
Department of Computer
and Information Science
University of Konstanz
Konstanz, Germany

ISSN 0302-9743 ISSN 1611-3349 (electronic)
Lecture Notes in Bioinformatics
ISBN 978-3-030-60326-7 ISBN 978-3-030-60327-4 (eBook)
https://doi.org/10.1007/978-3-030-60327-4

LNCS Sublibrary: SL8 – Bioinformatics

This Springer imprint is published by the registered company Springer Nature Switzerland AG
The registered company address is: Gewerbestrasse 11, 6330 Cham, Switzerland

Preface

The 18th conference on Computational Methods in Systems Biology (CMSB 2020) was held during September 23–25, 2020. The conference was originally planned to be hosted by the University of Konstanz, but due to the COVID-19 pandemic, CMSB 2020 took place fully online. Recent editions of the CMSB conference series were organized by the University of Trieste (Italy), Masaryk University in Brno (Czech Republic), and TU Darmstad (Germany).

The scope of CMSB covers the analysis of biological systems, networks, data, and corresponding application domains. The conference brings together computer scientists, biologists, mathematicians, engineers, and physicists interested in a system-level understanding of biological processes. CMSB 2020 has retained the emphasis on system-level understanding of biological processes – by no means restricted to a narrow class of mathematical models – and has in particular stressed the importance of integrating the techniques developed separately in different areas. CMSB 2020 especially encouraged the presentation of new work concerning integration of machine learning techniques into modeling and analysis frameworks.

Topics featured in the workshop included formalisms for modeling biological processes, models and their biological applications, frameworks for model verification, validation, analysis, and simulation of biological systems, methods for synthetic biology and biomolecular computing, multi-scale modeling and analysis methods, collective behavior, high-performance computational systems biology, parallel implementations, and a new emphasis on machine learning for systems biology, model inference from experimental data, and model integration from biological databases.

CMSB 2020 was a three-day event, featuring three invited talks, two tutorials, single-track regular sessions, as well as an interactive session with tool presentations. The 30-minute presentations of regular papers and 20-minute presentations of tool papers were of high quality and the participation was lively, interactive, and stimulating.

45 Program Committee members helped to provide at least three reviews for each of the 30 submitted contributions, out of which 17 high-quality articles were accepted to be presented during the single-track sessions. Moreover, five tool paper submissions, each receiving three reviews for paper presentations, and two reviews from the Tool Evaluation Committee, were selected for presentation. All accepted contributions appear (a few after further feedback from a shepherding process by the Program Committee) as full papers in these proceedings. The articles were bundled into four thematic sessions, which is reflected in the organization of these proceedings: modeling and analysis; Boolean networks; identification and inference; and tools. We also hosted a session on tool papers, with five interactive tool demonstrations.

A highlight of CMSB 2020 was the presence of three high-profile invited speakers whom we also selected in view of the breadth of the event, covering formal and control theory methods, theoretical work and laboratory experiments, and modeling for systems biology, and for collective animal behavior. Iain Couzin, Director of the Max Planck Institute of Animal Behavior, Department of Collective Behaviour and the Chair of Biodiversity and Collective Behaviour at the University of Konstanz, Germany, gave a seminar titled "Employing Immersive Virtual Reality to Reveal Common Geometric Principles of Individual and Collective Decision-Making." Grégory Batt, a research scientist at Inria and at Institut Pasteur, France, gave a seminar titled "Methods and Tools for the Quantitative Characterization of Engineered Biomolecular Systems," and Domitilla Del Vecchio, Professor of Mechanical Engineering at MIT, USA, gave a talk titled "Context dependence of biological circuits: Predictive models and engineering solutions."

Further details on CMSB 2020 are featured on the website: https://cmsb2020.uni-saarland.de.

Finally, we would like to thank the local organizing team: Jacob Davidson, Matej Hajnal, Huy Phung, Stefano Tognazzi (University of Konstanz) for their supportive, can-do attitude. We thank to Stefano Tognazzi, for the technical and logistic solutions regarding streaming the conference online, as well as for organising the virtual tour of the city of Konstanz. We further thank to Jennifer Durai and Aamir Rizwani, for their engagement towards a smooth virtual experience for all presenters and participants. Thanks to Springer for continuing to host the CMSB proceedings in its Lecture Notes series. Thanks to the generous support of the Centre for the Advanced Study of Collective Behaviour (CASCB), it was possible to offer a free registration for the digital edition of CMSB 2020. Thanks to Ezio Bartocci, François Fages, and Jérôme Feret from the Steering Committee of CMSB for support and encouragement, to all the Program Committee members and additional reviewers for their great work in ensuring the quality of the contributions to CMSB 2020, as well as to all the participants for contributing to this memorable event. Thanks to Saarland University for hosting the web-domain of CMSB 2020 and to Whova Virtual Conference Platform for their services.

September 2020

Alessandro Abate
Tatjana Petrov
Verena Wolf

Organization

Program Committee Chairs

Alessandro Abate	University of Oxford, UK
Tatjana Petrov	University of Konstanz, Germany
Verena Wolf	Saarland University, Germany

Tool Evaluation Committee Chair

David Šafránek	Masaryk University, Czech Republic

Local Organization Chair

Tatjana Petrov	University of Konstanz, Germany

Local Organization Committee

Jacob Davidson	University of Konstanz, Germany
Matej Hajnal	Masaryk University, Czech Republic
Stefano Tognazzi	University of Konstanz, Germany

Program Committee

Nicos Angelopoulos	University of Essex, UK
Paolo Ballarini	CentraleSupélec, France
Ezio Bartocci	Vienna University of Technology, Austria
Pavol Bokes	Comenius University, Slovakia
Luca Bortolussi	University of Trieste, Italy
Pierre Boutillier	Harvard University, USA
Luca Cardelli	Microsoft Research, UK
Milan Česká	Brno University of Technology, Czech Republic
Eugenio Cinquemani	Inria, France
Neil Dalchau	Microsoft Research, UK
François Fages	Inria, France
Jerome Feret	Inria, France
Christoph Flamm	University of Vienna, Austria
Anastasis Georgoulas	University College London, UK
Ashutosh Gupta	TIFR, India
Matej Hajnal	Masaryk University, Czech Republic
Jan Hasenauer	University of Bonn, Germany
Monika Heiner	Brandenburg University of Technology, Germany
Jane Hillston	The University of Edinburgh, UK

Katsumi Inoue	NII, Japan
Jan Kretinsky	Technical University of Munich, Germany
Jean Krivine	CNRS, France
Hillel Kugler	Microsoft Research, UK
Morgan Magnin	CNRS, France
Dimitrios Milios	The University of Edinburgh, UK
Chris Myers	The University of Utah, USA
Laura Nenzi	University of Trieste, Italy
Joachim Niehren	Inria, France
Loïc Paulevé	CNRS, LaBRI, France
Carla Piazza	University of Udine, Italy
Ovidiu Radulescu	University of Montpellier, France
Andre Riberio	Tampere University, Finland
Olivier Roux	École Centrale de Nantes, France
David Šafránek	Masaryk University, Czech Republic
Guido Sanguinetti	The University of Edinburgh, UK
Heike Siebert	Freie Universität Berlin, Germany
Simone Silvetti	University of Udine, Italy
Abhyudai Singh	University of Delaware, USA
Scott Smolka	Stony Brook University, USA
Carlo Spaccasassi	Microsoft Research, UK
Carolyn Talcott	SRI International, USA
Mirco Tribastone	IMT School for Advanced Studies Lucca, Italy
Adelinde Uhrmacher	University of Rostock, Germany
Andrea Vandin	Sant'Anna School of Advanced Studies, Italy
Boyan Yordanov	Microsoft Research, UK

Additional Reviewers

Jacek Chodak	Brandenburg University of Technology, Germany
Pedro Fontanarossa	The University of Utah, USA
Shouvik Roy	Stony Brook University, USA
Jean-Paul Comet	Université Côte d'Azur, France
Aurélien Naldi	Inria, France
Maxime Folschette	École Centrale de Lille, France
Sylvain Soliman	Inria, France
Jeanet Mante	The University of Utah, USA
Usama Mehmood	Stony Brook University, USA
George Assaf	Brandenburg University of Technology, Germany

Sponsors

Universität
Konstanz

Centre for the Advanced Study
of Collective Behaviour

Invited Talks

Context Dependence of Biological Circuits: Predictive Models and Engineering Solutions

Domitilla del Vecchio

MIT, USA

ddv@mit.edu

Abstract. Engineering biology has tremendous potential to impact applications, from energy, to environment, to health. As the sophistication of engineered biological circuits increases, the ability to predict system behavior becomes more limited. In fact, while a system's component may be well characterized in isolation, its salient properties often change in surprising ways once it interacts with other systems in the cell. This context-dependence of biological circuits makes it difficult to perform rational design and leads to lengthy, combinatorial, design procedures where each component is re-designed ad hoc when other parts are added to a system. In this talk, I will overview some causes of context-dependence. I will then focus on problems of resource loading and describe a design-oriented mathematical model that accounts for it. I will introduce a general engineering framework, grounded on control theoretic concepts, that can serve as a basis for creating devices that are "insulated" from context. Example devices will be introduced for both bacterial and mammalian genetic circuits. These solutions support rational and modular design of sophisticated genetic circuits and can serve for engineering biological circuits that are more reliable and predictable.

Methods and Tools for the Quantitative Characterization of Engineered Biomolecular Systems

Gregory Batt[1,2]

[1] Inria Paris, 75012 Paris, France
[2] Institut Pasteur, USR 3756 IP CNRS, 75015 Paris, France
gregory.batt@inria.fr

Abstract. Despite many years of research, no standard approach has emerged to rationally design novel genetic circuits implementing non-trivial functions. Synthetic biology still largely relies on tinkering. This comes notably from our limited capacities to quantitatively predict the behavior of biological systems in different cellular contexts. Iterative approaches, employing design-built-test-and-learn (DBTL) strategies, have the potential to circumvent this problem.

In this presentation, I will describe some of our efforts to develop an integrated framework supporting DBTL approaches. Firstly, I will present experimental platforms that we have developed to run experiments in an automated manner. Automation increases throughput, and also, more importantly in fact, improves standardization and reproducibility. Secondly, I will present recent results we obtained on the characterization and modeling of several natural and engineered microbial systems, together with applications to real-time control and treatment optimization. Finally, I will conclude with recent results on the optimal design of parallel experiments.

Keywords: Bioreactor and cytometry automation · Microfluidics and microscopy automation · Plate reader automation · Optogenetics · Deterministic and stochastic modeling · Model predictive control · Process and treatment optimization · Optimal experimental design · Cybergenetics · Antimicrobial resistance · protein bioproduction in yeast

Acknowledgements. This work has been done in collaboration with all the members of the InBio group at Inria and Institut Pasteur.

Employing Immersive Virtual Reality to Reveal Common Geometric Principles of Individual and Collective Decision-Making

Iain Couzin[1,2,3]

[1] University of Konstanz
[2] Centre for the Advanced Study of Collective Behaviour
[3] Max-Planck Institute of Animal Behaviour, Germany
icouzin@ab.mpg.de

Abstract. Understanding how social influence shapes biological processes is a central challenge in contemporary science, essential for achieving progress in a variety of fields ranging from the organization and evolution of coordinated collective action among cells, or animals, to the dynamics of information exchange in human societies. Using an integrated experimental and theoretical approach, involving automated tracking, immersive virtual reality (VR), and computational visual field reconstruction, I will discuss the discovery of universal geometric principles of perceptual decision-making across vast scales of biological organization, from neural collectives within the brains of individual invertebrates and vertebrates, to the collective movement decisions made by fish schools and primate societies.

Contents

Inference and Identification

Tools

Tutorials

Modelling and Analysis

Rate Equations for Graphs

Vincent Danos[1], Tobias Heindel[2], Ricardo Honorato-Zimmer[3],
and Sandro Stucki[4(✉)]

[1] CNRS, ENS-PSL, INRIA, Paris, France
vincent.danos@ens.fr
[2] Institute of Commercial Information Technology and Quantitative Methods,
Technische Universität Berlin, Berlin, Germany
heindel@tu-berlin.de
[3] Centro Interdisciplinario de Neurociencia de Valparaíso, Universidad de Valparaíso,
Valparaiso, Chile
ricardo.honorato@cinv.cl
[4] Department of Computer Science and Engineering, University of Gothenburg,
Gothenburg, Sweden
sandro.stucki@gu.se

Abstract. In this paper, we combine ideas from two different scientific traditions: 1) graph transformation systems (GTSs) stemming from the theory of formal languages and concurrency, and 2) mean field approximations (MFAs), a collection of approximation techniques ubiquitous in the study of complex dynamics. Using existing tools from algebraic graph rewriting, as well as new ones, we build a framework which generates rate equations for stochastic GTSs and from which one can derive MFAs of any order (no longer limited to the humanly computable). The procedure for deriving rate equations and their approximations can be automated. An implementation and example models are available online at https://rhz.github.io/fragger. We apply our techniques and tools to derive an expression for the mean velocity of a two-legged walker protein on DNA.

Keywords: Mean field approximations · Graph transformation systems · Algebraic graph rewriting · Rule-based modelling

1 Introduction

Mean field approximations (MFAs) are used in the study of complex systems to obtain simplified and revealing descriptions of their dynamics. MFAs are used in many disparate contexts such as Chemical Reaction Networks (CRNs) and their derivatives [13,23,34], walkers on bio-polymers [24,44], models of epidemic spreading [27], and the evolution of social networks [20]. These examples witness both the power and universality of MFA techniques, and the fact that they are pursued in a seemingly ad hoc, case-by-case fashion.

RH-Z was supported by ANID FONDECYT/POSTDOCTORADO/No3200543.

© Springer Nature Switzerland AG 2020
A. Abate et al. (Eds.): CMSB 2020, LNBI 12314, pp. 3–26, 2020.
https://doi.org/10.1007/978-3-030-60327-4_1

Fig. 1. Stukalin model of a walking DNA bimotor.

The case of CRNs is particularly interesting because they provide a human-readable, declarative language for a common class of complex systems. The stochastic semantics of a CRN is given by a continuous-time Markov chain (CTMC) which gives rise to the so-called *master equation* (ME). The ME is a system of differential equations describing the time evolution of the probability of finding the CRN in any given state. Various tools have been developed to automate the generation and solution of the ME from a given CRN, liberating modellers from the daunting task of working with the ME directly (e.g. [25,36,45]).

Its high dimensionality often precludes exact solutions of the ME. This is where MFA techniques become effective. The generally countably infinite ME is replaced by a finite system of differential equations, called the *rate equations* (RE) [28,34], which describe the time evolution of the average occurrence count of individual species. Here, we extend this idea to the case of graphs and, in fact, the resulting framework subsumes all the examples mentioned above (including CRNs). The main finding is summarised in a single Eq. (15) which we call the *generalised rate equations for graphs* (GREG). We have published in previous work a solution to this problem for the subclass of reversible graph rewriting systems [17,18]. The solution presented here is valid for *any* such system, reversible or not. The added mathematical difficulty is substantial and concentrates in the backward modularity Lemma 2. As in Ref. [18], the somewhat informal approach of Ref. [17] is replaced with precise category-theoretical language with which the backward modularity Lemma finds a concise and natural formulation.

As the reader will notice, Eq. (15) is entirely combinatorial and can be readily implemented. Our implementation can be played with at https://rhz.github.io/fragger. Its source can be found at https://github.com/rhz/fragger.

1.1 Two-Legged DNA Walker

Let us start with an example from biophysics [44]. The model describes a protein complex walking on DNA. The walker contains two special proteins – the *legs* – each binding a different DNA strand. The legs are able to move along the strands independently but can be at most m DNA segments apart.

Following Stukalin et al. [44], we are interested in computing the velocity at which a two-legged walker moves on DNA with $m = 1$. In this case, and assuming the two legs are symmetric, there are only two configurations a walker can be in: either extended (E) or compressed (C). Therefore all possible transitions can be compactly represented by the four rules shown in Fig. 1, where the grey node represents the walker and white nodes are DNA segments. The polarisation of the DNA double helix is represented by the direction of the edge that binds two consecutive DNA segments. Rules are labelled by two subscripts: the first tells us if the leg that changes position is moving *forward* (F) or *backward* (B), while the second states whether the rule extends or compresses the current configuration.

The *mean velocity* V of a single walker in the system can be computed from the rates at which they move forward and backward and their expected number of occurrences $\mathbb{E}[G_i]$, where G_i is in either of the three possible configurations depicted in Fig. 1, and $[G_i]$ is short for $[G_i](X(t))$, the integer-valued random variable that tracks the number of occurrences of G_i in the (random) state of the system $X(t)$ at time t. We call any real- or integer-valued function on $X(t)$ an *observable*.

$$V = \frac{1}{2}\left(k_{F,E}\mathbb{E}\left[\,\vcenter{\hbox{\includegraphics{g1}}}\,\right] + k_{F,C}\mathbb{E}\left[\,\vcenter{\hbox{\includegraphics{g2}}}\,\right] - k_{B,E}\mathbb{E}\left[\,\vcenter{\hbox{\includegraphics{g3}}}\,\right] - k_{B,C}\mathbb{E}\left[\,\vcenter{\hbox{\includegraphics{g4}}}\,\right]\right)$$

In the case there is only a single motor in the system, the observables $[G_i]$ are Bernoulli-distributed random variables, and the expectations $\mathbb{E}[G_i]$ correspond to the probabilities of finding the motor in the configuration G_i at any given time. Thus by constructing the ODEs for these observables, we can compute the mean velocity of a single motor in the system. That is, we must compute the rate equations for these graphs.

Intuitively, to compute rate equations we must find all ways in which the rules can create or destroy an occurrence of an observable of interest. When, and only when, a rule application and an occurrence of the given observable overlap, can this occurrence be created or destroyed. A systematic inventory of all such overlaps can be obtained by enumerating the so-called *minimal gluings* (MGs) of the graph underlying the given observable and the left- and right-hand sides of each rule in the system. MGs show how two graphs can overlap (full definition in the next section). Such an enumeration of MGs is shown in Fig. 2, where the two graphs used to compute the MGs are the extended walker motif – the middle graph in Fig. 1 – and the left-hand side of the forward-extension rule. The MGs are related and partially ordered by graph morphisms between them.

In theory, since we are gluing with the left-hand side of a rule each one of the MGs represents a configuration in which the application of the rule might destroy an occurrence of the observable. However, if we suppose that walkers initially have two legs, then 13 of the 21 MGs in Fig. 2 are impossible to produce by the rules, because no rule can create additional legs. Therefore those configurations will never be reached by the system and we can disregard them. If we further suppose the DNA backbone to be simple and non-branching, we eliminate three more gluings. Finally, if there is only one motor, the remaining four non-trivial gluings are eliminated. In this way, invariants can considerably reduce the number of gluings that have to be considered. Removing terms corresponding to observables which, under the assumptions above, are identically zero, we get the following series of ODEs. For readability, only a subset of the terms is shown, and we write G instead of the proper $\mathbb{E}[G]$ in ODEs.

$$\frac{d}{dt}\;[\text{graph}] \;=\; k_{F,E}\,[\text{graph}] \;-\;k_{B,C}\,[\text{graph}] \;-\;k_{F,C}\,[\text{graph}] \;+\;k_{B,E}\,[\text{graph}]$$

$$\frac{d}{dt}\;[\text{graph}] \;=\; -k_{F,E}\,[\text{graph}] \;+\;k_{B,C}\,[\text{graph}] \;+\;k_{F,C}\,[\text{graph}] \;-\cdots$$

$$\frac{d}{dt}\;[\text{graph}] \;=\; k_{F,E}\,[\text{graph}] \;-\;k_{B,C}\,[\text{graph}] \;-\;k_{F,C}\,[\text{graph}] \;+\cdots$$

$$\frac{d}{dt}\;[\text{graph}] \;=\; -k_{F,E}\,[\text{graph}] \;+\;k_{B,C}\,[\text{graph}] \;+\;k_{F,C}\,[\text{graph}] \;-\cdots$$

$$\frac{d}{dt}\;[\text{graph}] \;=\; \cdots$$

Notice how only graphs with extra white nodes to the right are obtained when computing the ODE for the left graph in Fig. 1. The opposite is true for the right graph in Fig. 1. This infinite expansion can be further simplified if we assume the DNA chain to be infinite or circular. In this case we can avoid boundary conditions and replace the left- and right-hand observables below by the simpler middle observable:

$$\mathbb{E}\Big[\,[\text{graph}]\,\Big] = \mathbb{E}\Big[\,[\text{graph}]\,\Big] = \mathbb{E}\Big[\,[\text{graph}]\,\Big]$$

The infinite expansion above now boils down to a simple finite ODE system.

$$\frac{d}{dt}\;[\text{graph}] \;=\; k_{F,E}\,[\text{graph}] \;-\;k_{B,C}\,[\text{graph}] \;-\;k_{F,C}\,[\text{graph}] \;+\;k_{B,E}\,[\text{graph}]$$

$$\frac{d}{dt}\;[\text{graph}] \;=\; -k_{F,E}\,[\text{graph}] \;+\;k_{B,C}\,[\text{graph}] \;+\;k_{F,C}\,[\text{graph}] \;-\;k_{B,E}\,[\text{graph}]$$

From the above ODEs and assumptions, we get the steady state equation.

$$(k_{F,E} + k_{B,E})\mathbb{E}\Big[\,[\text{graph}]\,\Big] = (k_{F,C} + k_{B,C})\mathbb{E}\Big[\,[\text{graph}]\,\Big]$$

Since we have only one motor,

$$\mathbb{E}\Big[\,[\text{graph}]\,\Big] + \mathbb{E}\Big[\,[\text{graph}]\,\Big] = 1$$

Using this, we can derive the steady state value for the mean velocity:

$$V = \frac{1}{2}\Big((k_{F,E} - k_{B,E})\mathbb{E}\Big[\,[\text{graph}]\,\Big] + (k_{F,C} - k_{B,C})\mathbb{E}\Big[\,[\text{graph}]\,\Big] \Big)$$

$$= \frac{(k_{F,C} + k_{B,C})(k_{F,E} - k_{B,E}) + (k_{F,E} + k_{B,E})(k_{F,C} - k_{B,C})}{2(k_{F,E} + k_{B,E} + k_{F,C} + k_{B,C})}$$

This exact equation is derived in Ref. [44]. We obtain it as a particular case of the general notion of rate equations for graph explained below. It is worth noting that, despite the simplicity of the equation, it is not easily derivable by hand. This and other examples are available to play with in our web app at https://rhz.github.io/fragger/. The example models include

- the DNA walker model described above;
- a population model tracking parent-child and sibling relationships;
- the voter model from Ref. [17];
- the preferential attachment model from Ref. [18].

The DNA walker model presented in this introduction is small and reversible. It requires no approximation to obtain a finite expansion. By contrast, the population model and the preferential attachment model are irreversible; the population and the voter model require an approximation to obtain a finite expansion.

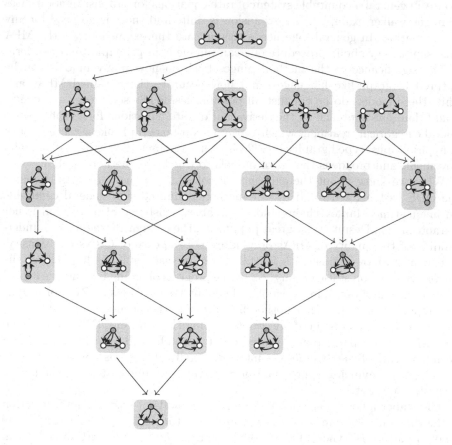

Fig. 2. The poset of minimal gluings of G_2 and G_1. The disjoint sum is at the top. Gluings are layered by the number of node and edge identifications or, equivalently, by the size of their intersection.

1.2   Discussion

The reasoning and derivation done above in the DNA walker can in fact made completely general. Given a *graph observable* $[F]$, meaning a function counting the number of embeddings of the graph F in the state X, one can build an ODE which describes the rate at which the mean occurrence count $\mathbb{E}([F](X(t)))$ changes over time.

Because the underlying Markov process $X(t)$ is generated by graph-rewriting rules, the one combinatorial ingredient to build that equation is the notion of *minimal gluings* (MGs) of a pair of graphs. Terms in the ODE for F are derived from the set of MGs of F with the left and right sides of the rules which generate $X(t)$. Besides, each term in F's ODE depends on the current state *only* via expressions of the form $\mathbb{E}([G])$ for G a graph defining a new observable. Thus each fresh observable $[G]$ can then be submitted to the same treatment, and one obtains in general a countable system of rate equations for graphs. In good cases (as in the walker example), the expansion is finite and there is no need for any approximation. In general, one needs to truncate the expansion. As the MFA expansion is a symbolic procedure one can pursue it in principle to any order.

The significance of the method hinges both on how many models can be captured in graph-like languages, and how accurate the obtained MFAs are. While these models do no exhaust all possibilities, GTSs seem very expressive. In addition, our approach to the derivation of rate equations for graphs uses a general categorical treatment which subsumes various graph-like structures such as: hyper-graphs, typed graphs, etc. [2, 35]. This abstract view is mathematically convenient, and broadens the set of models to which the method applies.

What we know about the existence of solutions to the (in general) countable ODE systems generated by our method is limited. For general countable continuous-time Markov chains and observables, existence of a solution is not guaranteed [43]. Despite their great popularity, the current mathematical understanding of the quality of MFAs only addresses the case of CRNs and density-dependent Markov chains, with Kurtz' theory of scalings [23, Chap. 11], or the case of dynamics on static graphs [27]. Some progress on going beyond the formal point of view and obtaining existence theorems for solutions of REs for graphs were reported in Ref. [16]. Another limitation is the accuracy of MFAs once truncated (as they must be if one wants to plug them in an ODE solver). Even if an MFA can be built to any desired order, it might still fall short of giving a sensible picture of the dynamics of interest. Finally, it may also be that the cost of running a convincing approximation is about the same as that of simulating the system upfront.

This paper follows ideas on applying the methods of abstract interpretation to the differential semantics of *site graph* rewriting [15, 26, 30]. Another more remote influence is Lynch's finite-model theoretic approach to MFAs [39]. From the GTS side, the theory of site graph rewriting had long been thought to be a lucky anomaly until a recent series of work showed that most of its ingredients could be made sense of, and given a much larger basis of applications, through the use of algebraic graph-rewriting techniques [3, 31, 32]. These latter investigations motivated us to try to address MFA-related questions at a higher level of generality.

1.3   Relation to the rule-algebraic approach

Another approach to the same broad set of questions started with Behr et al. introducing ideas from representation theory commonly used in statistical physics [5]. They show that one can study the algebra of rules and their composition law in a way that is decoupled from what is actually being re-written. The "rule-algebraic" theory allows one to derive Kolmogorov equations for observables based on a systematic use of rule commutators (recently implemented in [12]). Interestingly, novel notions of graph rewriting appear [11]. Partial differential equations describing the generating function of such observables can be derived systematically [7]. As the theory can handle adhesive categories in general and sesqui-pushout rewriting [10], it offers an treatment of irreversible rewrites alternative to the one presented in this paper. (The rule-algebraic approach can also handle application conditions [9]). It will need further ork to precisely pinpoint how these two threads of work articulate both at the theoretical and at the implementation levels.

Outline. The paper is organised as follows: Sect. 2 collects preliminaries on graph-rewriting and establishes the key *forward* and *backward modularity* lemmas; Sect. 3 derives our main result namely a concrete formula for the action of a generator associated to a set of graph-rewriting rules as specified in Sect. 2. From this formula, the rate equation for graphs follows easily. Basic category-theoretical definitions needed in the main text are given in App. A; axiomatic proofs in App. B.

2   Stochastic Graph Rewriting

We turn now to the graphical framework within which we will carry out the derivation of our generalised rate equation (GREG) in Sect. 3. We use a categorical approach know as algebraic graph rewriting, specifically the *single pushout (SPO)* approach [22,37]. The reasons for this choice are twofold: first, we benefit from a solid body of preexisting work; second, it allows for a succinct and 'axiomatic' presentation abstracting over the details of the graph-like structures that are being rewritten. Working at this high level of abstraction allows us to identify a set of generic properties necessary for the derivation of the GREG without getting bogged down in the details of the objects being rewritten. Indeed, while we only treat the case of directed multigraphs (graphs with an arbitrary number of directed edges between any two nodes) in this section, the proofs of all lemmas are set in the more general context of *adhesive categories* [35] in App. B. This extends the applicability of our technique to rewrite systems over typed graphs and hypergraphs, among others.

For the convenience of the reader, we reproduce from Ref. [18] our basic definitions for the category **Grph** of *directed multigraphs*. Next, we briefly summarise the SPO approach and its *stochastic semantics* [33]. We conclude with the *modularity lemmas*, which are key to the derivation of the GREG in the next section.

2.1 The Category of Directed Multigraphs

A *directed multigraph* G consists of a finite set of *nodes* V_G, a finite set of *edges* E_G, and *source* and *target* maps $s_G, t_G \colon E_G \to V_G$. A *graph morphism* $f \colon G \to H$ between graphs G and H is a pair of maps $f_E \colon E_G \to E_H$, $f_V \colon V_G \to V_H$ which preserve the graph structure, i.e. such that for all $e \in E_G$,

$$s_H(f_E(e)) = f_V(s_G(e)) \qquad \text{and} \qquad t_H(f_E(e)) = f_V(t_G(e)).$$

The graphs G and H are called the *domain* and *codomain* of f. A graph morphism $f \colon G \to H$ is a *monomorphism*, or simply a *mono*, if f_V and f_E are injective; it is a *graph inclusion* if both f_V and f_E are inclusion maps, in which case G is a *subgraph* of H and we write $G \subseteq H$. Every morphism $f \colon G \to H$ induces a subgraph $f(G) \subseteq H$ called the *direct image* (or just the *image*) of f in H, such that $V_{f(G)} = f_V(V_G)$ and $E_{f(G)} = f_E(E_G)$. Figure 3 illustrates a graph and a graph morphism.

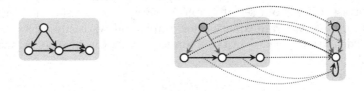

Fig. 3. Examples of a) a directed multigraph, b) a graph morphism.

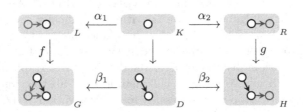

Fig. 4. A derivation. (Color figure online)

A *partial* graph morphism $p \colon G \rightharpoonup H$ is a pair of partial maps $p_V \colon V_G \rightharpoonup V_H$ and $p_E \colon E_G \rightharpoonup E_H$ that preserve the graph structure. Equivalently, p can be represented as a *span* of (total) graph morphisms, that is, a pair of morphisms

$p_1\colon K \to G$, $p_2\colon K \to H$ with common domain K, where p_1 is mono and K is the *domain of definition* of p. We will use whichever representation is more appropriate for the task at hand. Graphs and graph morphisms form the category **Grph** (with the obvious notion of composition and identity) while graphs and partial graph morphisms form the category **Grph**$_*$.

Graph morphisms provide us with a notion of pattern matching on graphs while partial graph morphisms provide the accompanying notion of rewrite rule. We restrict pattern matching to monos: a *match* of a pattern L in a graph G is a monomorphism $f\colon L \to G$. We write $[L, G]$ for the set of matches of L in G. We also restrict rules: a *rule* is a partial graph morphism $\alpha\colon L \rightharpoonup R = (\alpha_1\colon K \to L, \alpha_2\colon K \to R)$ where both α_1 and α_2 are monos. We say that L and R are α's left and right hand side (LHS and RHS). Rules are special cases of partial graph morphisms and compose as such. Given a rule $\alpha\colon L \rightharpoonup R = (\alpha_1, \alpha_2)$, we define the *reverse* rule $\alpha^\dagger\colon R \rightharpoonup L$ as the pair $\alpha^\dagger := (\alpha_2, \alpha_1)$, not to be confused with the inverse of α (which does not exist in general). Note that $-^\dagger$ is an involution, that is, $(\alpha^\dagger)^\dagger = \alpha$.

2.2   Graph Rewriting

The basic rewrite steps of a GTS are called *derivations*. We first describe them informally. Figure 4 shows a commutative square, with a match $f\colon L \to G$ on the left and a rule $\alpha\colon L \rightharpoonup R$, on top. The match f identifies the subgraph in G that is to be modified, while the rule α describes how to carry out the modification. In order to obtain the *comatch* $g\colon R \to H$ on the right, one starts by removing nodes and edges from $f(L)$ which do not have a preimage under $f \circ \alpha_1$, as well as any edges left dangling (coloured red in the figure). To complete the derivation, one extends the resulting match by adjoining to D the nodes and edges in R that do not have a preimage under α_2 (coloured green in the figure).

Derivations constructed in this way have the defining property of *pushout squares* (PO) in **Grph**$_*$, hence the name SPO for the approach. Alternatively, one can describe a derivation through the properties of its inner squares: the left square is the *final pullback complement* (FPBC) of α_1 and f, while the right one is a PO in **Grph** [14]. (Definitions and basic properties of POs and FPBCs are given in App. A.)

Definition 1. *A* derivation *of a comatch* $g\colon R \to H$ *from a match* $f\colon L \to G$ *by a rule* $\alpha = (\alpha_1\colon K \to L, \alpha_2\colon K \to R)$ *is a diagram in* **Grph** *such as* (1), *where the left square is an FPBC of* f *and* α_1 *and the right square is a PO,*

$$
\begin{array}{ccc}
L \xleftarrow{\alpha_1} K \xrightarrow{\alpha_2} R \\
f\downarrow \quad \ \downarrow h \quad \ \downarrow g \quad (1) \\
G \xleftarrow{\beta_1} D \xrightarrow{\beta_2} H
\end{array}
\qquad
\begin{array}{ccc}
L \xrightarrow{\alpha} R \\
f\downarrow \quad \ \downarrow g \quad (2) \\
G \xrightarrow{\beta} H
\end{array}
$$

with h, g *matches and* $\beta = (\beta_1, \beta_2)$ *a rule, called the* corule *of the derivation.*

Equivalently, a derivation of g from f by α is a PO in **Grph**$_*$ as in (2), with corule β. We will mostly use this second characterisation of derivations.

Write $f \Rightarrow_\alpha g$ if there is a derivation of g from f by α. Since derivations are POs of partial morphisms and $\mathbf{Grph}_*$ has all such POs [37], the relation $\Rightarrow_\alpha$ is *total*, that is, for any match f and rule α (with common domain), we can find a comatch g. However, the converse is not true: not every match g having the RHS of α as its domain is a comatch of a derivation by α. Which is to say, there might not exist f such that $f \Rightarrow_\alpha g$ (the relation $\Rightarrow_\alpha$ is not surjective). When there is such an f, we say g is *derivable* by α. Consider the example in Fig. 5. Here, g is α-derivable (as witnessed by f) but h is not: no match of the LHS could contain a "preimage" of the extra (red) edge e in the codomain of h because the target node of e has not yet been created.

We say a derivation $f \Rightarrow_\alpha g$ (with corule β) is *reversible* if $g \Rightarrow_{\alpha^\dagger} f$ (with corule $\beta^\dagger$), and *irreversible* otherwise. Clearly, derivations are not reversible in general, otherwise $\Rightarrow_\alpha$ would be surjective. Consider the derivation shown in Fig. 4. The derivation removes two (red) edges from the codomain of f; the removal of the lower edge is specified in the LHS of α, whereas the removal of the upper edge is a *side effect* of removing the red node to which the edge is connected (graphs cannot contain dangling edges). Applying the reverse rule $\alpha^\dagger$ to the comatch g restores the red node and the lower red edge, but not the upper red edge. In other words, f is not $\alpha^\dagger$-derivable, hence the derivation in Fig. 4 is irreversible. In previous work, we have shown how to derive rate equations for graph transformation systems with only reversible derivations [15,17,18]. In Sect. 3, we overcome this limitation, giving a procedure that extends to the irreversible case.

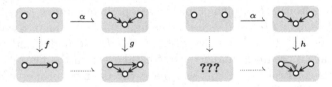

Fig. 5. The match g is α-derivable, while h is not. (Color figure online)

Since POs are unique (up to unique isomorphism), $\Rightarrow_\alpha$ is also *functional* (up to isomorphism). The fact that derivations are only defined up to isomorphism is convenient as it allows us to manipulate them without paying attention to the concrete naming of nodes and edges. Without this flexibility, stating and proving properties such as Lemma 2 and 3 below would be exceedingly cumbersome. On the other hand, when defining the stochastic semantics of our rewrite systems, it is more convenient to restrict $\Rightarrow_\alpha$ to a properly functional relation. To this end, we fix once and for all, for any given match $f: L \to G$ and rule $\alpha: L \rightharpoonup R$, a *representative* $f \Rightarrow_\alpha \alpha(f)$ from the corresponding isomorphism class of derivations, with (unique) comatch $\alpha(f): R \to H$, and (unique) corule $f(\alpha): G \rightharpoonup H$.

A set of rules $\mathcal{R}$ thus defines a *labelled transition system (LTS)* over graphs, with corules as transitions, labelled by the associated pair (f, α). Given a rule

$\alpha\colon L \rightharpoonup R$, we define a stochastic rate matrix $Q_\alpha := (q_{GH}^\alpha)$ over graphs as follows.

$$q_{GH}^\alpha := |\{f \in [L, G] \mid \alpha(f) \in [R, H]\}| \qquad \text{for } G \neq H,$$
$$q_{GG}^\alpha := \sum_{H \neq G} -q_{GH}^\alpha \qquad\qquad \text{otherwise.} \qquad (3)$$

Given a *model*, that is to say a finite set of rules $\mathcal{R}$ and a *rate map* $k\colon \mathcal{R} \to \mathbb{R}^+$, we define the model rate matrix $Q(\mathcal{R}, k)$ as

$$Q(\mathcal{R}, k) := \sum_{\alpha \in \mathcal{R}} k(\alpha) Q_\alpha \qquad (4)$$

Thus a model defines a CTMC over $\mathbf{Grph}_*$. As $\mathcal{R}$ is finite, $Q(\mathcal{R}, k)$ is row-finite.

2.3 Composition and Modularity of Derivations

By the well-known Pushout Lemma, derivations can be composed horizontally (rule composition) and vertically (rule specialisation) in the sense that if inner squares below are derivations, so are the outer ones:

$$
\begin{array}{ccc}
L & \xrightarrow{\alpha_1} R_1 & \xrightarrow{\alpha_2} R_2 \\
f\downarrow & \quad\downarrow g_1 & \quad\downarrow g_2 \\
G & \longrightarrow H_1 & \longrightarrow H_2
\end{array}
\qquad
\begin{array}{cc}
L & \xrightarrow{\alpha_1} R \\
f_1\downarrow & \quad\downarrow g_1 \\
G_1 & \xrightarrow{\alpha_2} H_1 \\
f_2\downarrow & \quad\downarrow g_2 \\
G_2 & \longrightarrow H_2
\end{array}
$$

Derivations can also be decomposed vertically. First, one has a forward decomposition (which follows immediately from pasting of POs in $\mathbf{Grph}_*$):

Lemma 1 (Forward modularity). *Let α, β, γ be rules and f_1, f_2, g, g_1 matches such that diagrams (5) and (6) are derivations. Then there is a unique match g_2 such that diagram (7) commutes (in $\mathbf{Grph}_*$) and is a vertical composition of derivations.*

$$
\begin{array}{cc}
L & \xrightarrow{\alpha} R \\
f_1\downarrow & \\
S & \quad\downarrow g \;(5) \\
f_2\downarrow & \\
G & \xrightarrow{\beta} H
\end{array}
\qquad
\begin{array}{cc}
L & \xrightarrow{\alpha} R \\
f_1\downarrow & \quad\downarrow g_1 \;(6) \\
S & \xrightarrow{\gamma} T
\end{array}
\qquad
\begin{array}{cc}
L & \xrightarrow{\alpha} R \\
f_1\downarrow & \quad\downarrow g_1 \\
S & \xrightarrow{\gamma} T \;\; g \;(7) \\
f_2\downarrow & \quad\downarrow g_2 \\
G & \xrightarrow{\beta} H
\end{array}
$$

A novel observation, which will play a central role in the next section, is that one also has a backward decomposition:

Lemma 2 (Backward modularity). *Let α, β, γ be rules and f, f_1, g_1, g_2 matches such that diagrams (8) and (9) are derivations. Then there is a unique match f_2 such that diagram (10) commutes (in $\mathbf{Grph}_*$) and is a vertical composition of derivations.*

$$
\begin{array}{ccc}
L & \xrightarrow{\alpha} & R \\
\downarrow{f} & \downarrow{g_1} & \\
 & T & \quad(8) \\
 & \downarrow{g_2} & \\
G & \xrightarrow{\beta} & H
\end{array}
\qquad
\begin{array}{ccc}
L & \xleftarrow{\alpha^\dagger} & R \\
\downarrow{f_1} & & \downarrow{g_1} \\
S & \xleftarrow{\gamma^\dagger} & T
\end{array}
\quad(9)
\qquad
\begin{array}{ccc}
L & \xrightarrow{\alpha} & R \\
\downarrow{f_1} & & \downarrow{g_1} \\
S & \xrightarrow{\gamma} & T \\
\downarrow{f_2} & & \downarrow{g_2} \\
G & \xrightarrow{\beta} & H
\end{array}
\quad(10)
$$

Forward and backward modularity look deceptively similar, but while Lemma 1 is a standard property of POs, Lemma 2 is decidedly non-standard. Remember that derivations are generally irreversible. It is therefore not at all obvious that one should be able to transport factorisations of comatches backwards along a rule, let alone in a unique fashion. Nor is it obvious that the top half of the resulting decomposition should be reversible. The crucial ingredient that makes backward modularity possible is that both matches and rules are monos. Because rules are (partial) monos, we can reverse α and β in (8), and the resulting diagram still commutes (though it is no longer a derivation in general). The existence and uniqueness of f_2 is then a direct consequence of the universal property of (9), seen as a PO. The fact that (9) is reversible relies on matches also being monos, but in a more subtle way. Intuitively, the graph T cannot contain any superfluous edges of the sort that render the derivation in Fig. 4 irreversible because, g_2 being a mono, such edges would appear in H as subgraphs, contradicting the α-derivability of $g_2 \circ g_1$. Together, the factorisation of f and the reversibility of (9) then induce the decomposition in (10) by Lemma 1. A full, axiomatic proof of Lemma 2 is given in App. B.3.

Among other things, Lemma 2 allows one to relate *derivability* of matches to *reversibility* of derivations:

Lemma 3. *A match* $g\colon R \to H$ *is derivable by a rule* $\alpha\colon L \rightharpoonup R$ *if and only if the derivation* $g \Rightarrow_{\alpha^\dagger} f$ *is reversible.*

2.4   Gluings

Given $G_1 \subseteq H$ and $G_2 \subseteq H$, the *union* of G_1 and G_2 in H is the unique subgraph $G_1 \cup G_2$ of H, such that $V_{(G_1 \cup G_2)} = V_{G_1} \cup V_{G_2}$ and $E_{(G_1 \cup G_2)} = E_{G_1} \cup E_{G_2}$. The *intersection* $(G_1 \cap G_2) \subseteq H$ is defined analogously. The subgraphs of H form a complete distributive lattice with $\cup$ and $\cap$ as the join and meet operations. One can *glue* arbitrary graphs as follows:

Definition 2. *A gluing of graphs* G_1, G_2 *is a pair of matches* $i_1\colon G_1 \to U$, $i_2\colon G_2 \to U$ *with common codomain* U; *if in addition* $U = i_1(G_1) \cup i_2(G_2)$, *one says the gluing is* minimal.

Two gluings $i_1\colon G_1 \to U$, $i_2\colon G_2 \to U$ and $j_1\colon G_1 \to V$, $j_2\colon G_2 \to V$ are said to be *isomorphic* if there is an isomorphism $u\colon U \to V$, such that $j_1 = u \circ i_1$ and $j_2 = u \circ i_2$. We write $G_1 *_{\simeq} G_2$ for the set of isomorphism classes of minimal gluings (MG) of G_1 and G_2, and $G_1 * G_2$ for an arbitrary choice of representatives from each class in $G_1 *_{\simeq} G_2$. Given a gluing $\mu\colon G_1 \to H \leftarrow G_2$, denote by $\hat{\mu}$ its "tip", i.e. the common codomain $\hat{\mu} = H$ of μ.

It is easy to see the following (see App. B for an axiomatic proof):

Lemma 4. *Let G_1, G_2 be graphs, then $G_1 * G_2$ is finite, and for every gluing $f_1: G_1 \to H$, $f_2: G_2 \to H$, there is a unique MG $i_1: G_1 \to U$, $i_2: G_2 \to U$ in $G_1 * G_2$ and match $u: U \to H$ such that $f_1 = u \circ i_1$ and $f_2 = u \circ i_2$.*

See Fig. 2 in Sect. 1 for an example of a set of MGs.

3 Graph-Based GREs

To derive the GRE for graphs (GREG) we follow the development in our previous work [17,18] with the important difference that we do not assume derivations to be reversible. The key technical innovation that allows us to avoid the assumption of reversibility is the backward modularity lemma (Lemma 2).

As sketched in Sect. 1.2, our GRE for graphs is defined in terms of graph observables, which we now define formally. Fix S to be the countable (up to iso) set of finite graphs, and let $F \in S$ be a graph. The *graph observable* $[F]: S \to \mathbb{N}$ is the integer-valued function $[F](G) := \|F, G\|$ counting the number of occurrences (i.e. matches) of F in a given graph G. Graph observables are elements of the vector space $\mathbb{R}^S$ of real-valued functions on S.

The stochastic rate matrix Q_α for a rule $\alpha: L \rightharpoonup R$ defined in (3) is a linear map on $\mathbb{R}^S$. Its action on an observable $[F]$ is given by

$$(Q_\alpha [F])(G) := \sum_H q^\alpha_{GH}([F](G) - [F](H)) \qquad \text{for } G, H \in S. \tag{11}$$

Since the sum above is finite, $Q_\alpha [F]$ is indeed a well-defined element of $\mathbb{R}^S$. We call $Q_\alpha [F]$ the *jump* of $[F]$ relative to Q_α. Intuitively, $(Q_\alpha [F])(G)$ is the expected rate of change in $[F]$ given that the CTMC sits at G.

To obtain the GREG as sketched in Sect. 1, we want to express the jump as a finite linear combination of graph observables. We start by substituting the definition of Q_α in (11).

$$
\begin{aligned}
(Q_\alpha [F])(G) &= \sum_H q^\alpha_{GH}([F](H) - [F](G)) \\
&= \sum_H \sum_{f \in [L,G] \text{ s.t. } \alpha(f) \in [R,H]} (\|F,H\| - \|F,G\|) \\
&= \sum_{f \in [L,G]} (\|F, \mathrm{cod}(\alpha(f))\| - \|F,G\|).
\end{aligned}
$$

where the simplification in the last step is justified by the fact that f and α uniquely determine $\alpha(f)$. The last line suggests a decomposition of $Q_\alpha [F]$ as $Q_\alpha [F] = Q^+_\alpha [F] - Q^-_\alpha [F]$, where Q^+_α produces new instances of F while Q^-_α consumes existing ones.

By Lemma 4, we can factor the action of the consumption term Q^-_α through the MGs $L * F$ of L and F to obtain

$$(Q^-_\alpha [F])(G) = \sum_{f \in [L,G]} \|F,G\| = \|L,G\| \cdot \|F,G\| = \sum_{\mu \in L * F} \|\hat{\mu}, G\|.$$

The resulting sum is a linear combination of a finite number of graph observables, which is exactly what we are looking for.

Simplifying the production term requires a bit more work. Applying the same factorisation Lemma 4, we arrive at

$$(Q_\alpha^+[F])(G) = \sum_{f\in[L,G]} |[F, \hat{\alpha}(f)]|$$
$$= \sum_{f\in[L,G]} \sum_{(\mu_1,\mu_2)\in R*F} |\{g \in [\hat{\mu}, \hat{\alpha}(f)] \mid g \circ \mu_1 = \alpha(f)\}|.$$

where $\hat{\alpha}(f) = \text{cod}(\alpha(f))$ denotes the codomain of the comatch of f. To simplify this expression further, we use the properties of derivations introduced in Sect. 2.3. First, we observe that μ_1 must be derivable by α for the set of g's in the above expression to be nonempty

Lemma 5. *Let* $\alpha: L \rightharpoonup R$ *be a rule and* $f: L \to G$, $g: R \to H$, $g_1: R \to T$ *matches such that* $f \Rightarrow_\alpha g$, *but* g_1 *is not derivable by* α. *Then there is no match* $g_2: T \to H$ *such that* $g_2 \circ g_1 = g$.

Proof. By the contrapositive of backward modularity. Any such g_2 would induce, by Lemma 2, a match $f_1: L \to S$ and a derivation $f_1 \Rightarrow_\alpha g_1$.   □

We may therefore restrict the set $R * F$ of right-hand MGs under consideration to the subset $\alpha *_R F := \{(\mu_1,\mu_2) \in R * F \mid \exists h. h \Rightarrow_\alpha \mu_1\}$ of MGs with a first projection derivable by α. Next, we observe that the modularity Lemma 1 and 2 establish a *one-to-one correspondence* between the set of factorisations of the comatches $\alpha(f)$ (through the MGs in $\alpha *_R F$) and a set of factorisations of the corresponding matches f.

Lemma 6 (correspondence of matches). *Let* α, β, γ, f, f_1, g, g_1 *such that diagrams* (12) *and* (13) *are derivations and* g_1 *is derivable by* α. *Then the set* $M_L = \{f_2 \in [S,G] \mid f_2 \circ f_1 = f\}$ *is in one-to-one correspondence with the set* $M_R = \{g_2 \in [T,H] \mid g_2 \circ g_1 = g\}$.

$$
\begin{array}{ccc}
L & \xrightarrow{\alpha} & R \\
f\downarrow & & \downarrow g \quad (12) \\
G & \xrightarrow{\beta} & H
\end{array}
\qquad
\begin{array}{ccc}
L & \xleftarrow{\alpha^\dagger} & R \\
f_1\downarrow & & \downarrow g_1 \quad (13) \\
S & \xleftarrow{\gamma^\dagger} & T
\end{array}
$$

Proof. Since g_1 is α-derivable, the diagram (13) is reversible, that is, $f_1 \Rightarrow_\alpha g_1$, with corule γ (by Lemma 3). Hence, if we are given a match f_2 in M_L, we can forward-decompose (12) vertically along the factorisation $f_2 \circ f_1 = f$, resulting in the diagram below (by forward modularity, Lemma 1). Furthermore, the comatch g_2 is unique with respect to this decomposition, thus defining a function $\phi: M_L \to M_R$ that maps any f_2 in M_L to the corresponding comatch $\phi(f_2) = g_2$ in M_R. We want to show that ϕ is a bijection. By backward modularity (Lemma 2), there is a match $f_2 \in M_L$ for any match $g_2 \in M_R$ such that $\phi(f_2) = g_2$ (surjectivity), and furthermore, f_2 is the unique match for which $\phi(f_2) = g_2$ (injectivity).   □

$$
\left(
\begin{array}{ccc}
L & \xrightarrow{\alpha} & R \\
f_1\downarrow & & \downarrow g_1 \\
S & \xrightarrow{\gamma} & T \\
f_2\downarrow & & \downarrow g_2 \\
G & \xrightarrow{\beta} & H
\end{array}
\right) g
$$

Using Lemma 5 and 6, we can simplify Q_α^+ as follows:

$$(Q_\alpha^+[F])(G) = \sum_{f \in [L,G]} \sum_{\mu \in \alpha *_R F} |\{g_2 \in [\hat{\mu}, \hat{\alpha}(f)] \mid g_2 \circ \mu_1 = \alpha(f)\}|$$

$$= \sum_{\mu \in \alpha *_R F} \sum_{f \in [L,G]} |\{f_2 \in [\hat{\alpha}^\dagger(\mu_1), G] \mid f_2 \circ \alpha^\dagger(\mu_1) = f\}|$$

$$= \sum_{\mu \in \alpha *_R F} |[\hat{\alpha}^\dagger(\mu_1), G]|$$

If we set $\alpha *_L F := L * F$ to symmetrise notation, we obtain

$$Q_\alpha([F]) = \sum_{\mu \in \alpha *_R F} [\hat{\alpha}^\dagger(\mu_1)] - \sum_{\mu \in \alpha *_L F} [\hat{\mu}] \qquad (14)$$

Now, in general for a CTMC on a countable state space S, the Markov-generated and time-dependent probability p on S follows the master equation [1,40]: $\frac{d}{dt} p^T = p^T Q$. Given an abstract observable f in $\mathbb{R}^S$, and writing $\mathbb{E}_p(f) := p^T f$ for the expected value of f according to p, we can then derive the formal[1] Kolmogorov equation for f:

$$\frac{d}{dt} \mathbb{E}_p(f) = \frac{d}{dt} p^T f = p^T Q f = \mathbb{E}_p(Qf),$$

giving us an equation for the rate of change of the mean of $f(X(t))$. Following this general recipe gives us the GRE for graphs immediately from (14).

$$\frac{d}{dt} \mathbb{E}_p([F]) = -\sum_{\alpha \in \mathcal{R}} k(\alpha) \sum_{\mu \in \alpha *_L F} \mathbb{E}_p[\hat{\mu}] + \sum_{\alpha \in \mathcal{R}} k(\alpha) \sum_{\mu \in \alpha *_R F} \mathbb{E}_p[\hat{\alpha}^\dagger(\mu_1)]. \qquad (15)$$

Remember that μ_1 denotes the left injection of the MG $\mu = (\mu_1, \mu_2)$ while $\hat{\mu}$ denotes its codomain, and that $\hat{\alpha}^\dagger(f) = \text{cod}(\alpha^\dagger(f))$.

Unsurprisingly, the derivation of (15) was more technically challenging than that of the GRE for reversible graph rewrite systems (cf. [18, Theorem 2]). Yet the resulting GREs look almost identical (cf. [18, Eq. (7)]). The crucial difference is in the production term Q_α^+, where we no longer sum over the full set of right-hand MGs $R * F$ but only over the subset $\alpha *_R F$ of MGs that are α-derivable. This extra condition is the price we pay for dealing with irreversibility: irreversible rules can consume all MGs, but only produce some.

Note that the number of terms in (15) depends on the size of the relevant sets of left and right-hand MGs, which is worst-case exponential in the size of the graphs involved, due to the combinatorial nature of MGs. (See Fig. 2 in Sect. 1 for an example.) In practice, one often finds many pairs of *irrelevant* MGs, the terms of which cancel out exactly. This reduces the effective size of the equations but not the overall complexity of generating the GREG.

---

[1] In the present paper, we elide the subtle issues of ensuring that the system of interest actually satisfies this equation. See the work of Spieksma [43] for the underlying mathematics or our previous work [16], which additionally considers computability of the solutions to arbitrary precision.

Finally, as said in Sect. 1.2, the repeated application of (15) will lead to an infinite expansion in general. In practice, the system of ODEs needs to be truncated. For certain models, one can identify invariants in the underlying rewrite system via static analysis, which result in a finite closure even though the set of reachable components is demonstrably infinite [19]. We have seen an example in Sect. 1.

4   Conclusion

We have developed a computer supported method for mean field approximations (MFA) for stochastic systems with graph-like states that are described by rules of SPO rewriting. The underlying theory unifies a large and seemingly unstructured collection of MFA approaches which share a graphical "air de famille". Based on the categorical frameworks of graph transformation systems (GTS), we have developed MFA-specific techniques, in particular concerning the combinatorics of minimal gluings. The main technical hurdle consisted in showing that the set of subgraph observables is closed under the action of the rate matrix (a.k.a. the infinitesimal generator) of the continuous-time Markov chain generated by an irreversible GTS. The proof is constructive and gives us an explicit term for the derivative of the mean of any observable of interest.

Mean field approximation and moment-closure methods are of wide use in applications, as typical probabilistic systems tend to have state spaces which defy more direct approaches. To reach their full potential, MFAs need to be combined with reachability and invariant analysis (as illustrated in Sect. 1).

We have worked the construction at the general axiomatic level of SPO-rewriting with matches and rules restricted to monomorphisms. One interesting extension is to include nested application conditions (NACs) [29,41] where the application of a rule can be modulated locally by the context of the match. NACs are useful in practice, and bring aboard the expressive power of first order logic in the description of transformation rules. We plan to investigate the extension of our approach to NACs, and, in particular, whether it is possible to incorporate them axiomatically, and what additional complexity cost they might incur.

Another direction of future work is to improve on the method of truncation. In the literature, one often finds graphical MFAs used in combination with conditional independence assumptions to control the size of connected observables, as e.g. the so-called pair approximation [20,27]. As these methods are known to improve the accuracy of naive truncation, we wish to understand if and how they can be brought inside our formal approach.

A   Pushout and pull-back complements

Algebraic graph rewriting relies on certain category-theoretical *limits* and *colimits* [4]. We give definitions of the relevant (co-)limits here along with some of their basic properties. Among these, pullback complements are the least known. We refer the interested reader to Ref. [14,21] for a thorough treatment.

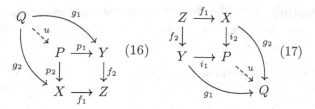

Let $\mathcal{C}$ be a category.

Definition 3 (Pullback). *A* pullback *of a cospan of morphisms* $X \xrightarrow{f_1} Z \xleftarrow{f_2} X$ *in* $\mathcal{C}$ *is a span* $X \xleftarrow{p_1} P \xrightarrow{p_2} Y$ *making the bottom-right square in* (16) *commute, and such that for any other span* $X \xleftarrow{g_1} Q \xrightarrow{g_2} Y$ *for which the outer square commutes, there is a unique morphism* $u\colon Q \to P$ *making the diagram commute.*

Definition 4 (Pushout). *A* pushout *of a span of morphisms* $X \xrightarrow{f_1} Z \xleftarrow{f_2} Y$ *in* $\mathcal{C}$ *is a cospan* $X \xrightarrow{i_1} P \xleftarrow{i_2} Y$ *making the top-left square in* (17) *commute, and such that for any other cospan* $X \xrightarrow{g_1} Q \xleftarrow{g_2} Y$ *for which the outer square commutes, there is a unique morphism* $u\colon P \to Q$ *making the diagram commute.*

$$
\begin{array}{ccc}
P & & \\
& \searrow^{\,p} & \\
g_1' \downarrow & X \xrightarrow{f_1} Y & \\
& g_1 \downarrow \quad \downarrow f_2 & (18) \\
& W \xrightarrow{g_2} Z & \\
& \nearrow^{\,u} & \\
Q & \xrightarrow{g_2'} &
\end{array}
$$

Definition 5 (Final pullback complement). *A* final pullback complement *(FPBC)* *(or simply* pullback complement*) of a pair of composable morphisms* $X \xrightarrow{f_1} Y \xrightarrow{f_2} Z$ *in some category* $\mathcal{C}$ *is a pair of composable morphisms* $X \xrightarrow{g_1} W \xrightarrow{g_2} Z$ *making the right inner square in* (18) *a pullback, such that for any other pullback* $P \xrightarrow{f_1'} Y \xrightarrow{f_2} Z \xleftarrow{g_2'} Q \xleftarrow{g_1'} P$ *and morphism* $p\colon P \to X$ *for which the diagram commutes, there is a unique morphism* $u\colon Q \to W$ *that makes the diagram commute.*

The following lemmas, pertaining to the composition of pullbacks, pushouts and FPBCs, respectively, are used throughout the proofs in App. B. The first two are dual versions of the well-known "pasting" lemma for pullbacks and pushouts, and we leave their proofs as an exercise to the reader. A proof of the third lemma can be found in [38, Proposition 5].

$$
\begin{array}{ccccc}
A & \xrightarrow{f_1} & B & \xrightarrow{f_2} & C \\
g_1 \downarrow & & \downarrow g_2 & & \downarrow g_3 \qquad (19) \\
D & \xrightarrow{h_1} & E & \xrightarrow{h_2} & F
\end{array}
$$

Lemma 7 (Pasting of pullbacks). *Suppose the right inner square in* (19) *is a pullback in some category* C. *Then the left inner square is a pullback if and only if the outer square is.*

Lemma 8 (Pasting of pushouts). *Suppose the left inner square in* (19) *is a pushout in some category* C. *Then the right inner square is a pushout if and only if the outer square is.*

Lemma 9 (Composition of FPBCs). *Consider again diagram* (19) *in some category* C,

- *(horizontal composition) if* $A \xrightarrow{g_1} D \xrightarrow{h_1} E$ *and* $B \xrightarrow{g_2} E \xrightarrow{h_2} F$ *are the FPBCs of* $A \xrightarrow{f_1} B \xrightarrow{g_2} E$ *and* $B \xrightarrow{f_2} C \xrightarrow{g_3} F$, *respectively, then* $A \xrightarrow{g_1} D \xrightarrow{h_2 \circ h_1} F$ *is the FPBC of* $A \xrightarrow{f_2 \circ f_1} C \xrightarrow{g_3} F$;
- *(vertical composition) if* $A \xrightarrow{f_1} B \xrightarrow{g_2} E$ *and* $B \xrightarrow{f_2} C \xrightarrow{g_3} F$ *are the FPBCs of* $A \xrightarrow{g_1} D \xrightarrow{h_1} E$ *and* $B \xrightarrow{g_2} E \xrightarrow{h_2} F$, *respectively, then* $A \xrightarrow{f_2 \circ f_1} C \xrightarrow{g_3} F$ *is the FPBC of* $A \xrightarrow{g_1} D \xrightarrow{h_2 \circ h_1} F$.

B   Generalised proofs of lemmas

This section contains detailed proofs of the various lemmas introduced in previous sections. We will present the proofs in a slightly more general setting, namely that of *sesqui-pushout (SqPO) rewriting* [14] in arbitrary *adhesive categories* [35]. To be precise, we assume an ambient category G, such that

- G is adhesive (among other things, this implies that G has all pullbacks as well as all pushouts along monomorphisms, that monomorphism are stable under pushout, and that all such pushouts are also pullbacks, cf. [35]),
- G has all final pullback complements (FPBCs) above monomorphisms.

Both these assumptions hold in **Grph**. Within G, we define derivations as in Definition 1, taking matches and rules to be monomorphisms and spans thereof, respectively.

Alternatively, rules can be seen as partial maps [42] in the category G_*, generalising the interpretation of rules as partial graph morphisms in **Grph**$_*$. Derivations can then be shown to correspond exactly to pushouts of rules along monomorphisms in G_* [2, Proposition 2.10], and composition of derivations corresponds to pushout composition in G_*.

B.1   Proof of Lemma 4 (minimal gluings)

Let G_1 and G_2 be graphs, then

1. the set $G_1 * G_2$ of MGs of G_1 and G_1 is finite, and

2. for every cospan $G_1 \xrightarrow{f_1} H \xleftarrow{f_2} G_2$ of matches, there is a unique MG $(G_1 \xrightarrow{i_1} U \xleftarrow{i_2} G_2) \in G_1 * G_2$ and match $u\colon U \to H$ such that $f_1 = u \circ i_1$ and $f_2 = u \circ i_2$.

Proof. For this proof we will make two additional assumptions on $\mathcal{G}$, namely that $\mathcal{G}$ has all binary products, and that the objects of $\mathcal{G}$ are finitely powered, that is, any object A in $\mathcal{G}$ has a finite number of subobjects. Both these assumptions hold in **Grph**.

Recall that the subobjects of any object A in $\mathcal{G}$ form a poset category $\mathbf{Sub}(A)$ with *subobject intersections* as products and *subobject unions* as coproducts. By stability of monomorphisms under pullback, products (intersections) in $\mathbf{Sub}(A)$ are given by pullbacks in $\mathcal{G}$, and since $\mathcal{G}$ is adhesive, coproducts (unions) in $\mathbf{Sub}(A)$ are given by pushouts of pullbacks in $\mathcal{G}$. See [35, Theorem 5.1] for more details.

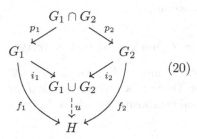

$$(20)$$

We will start by showing that any cospan $G_1 \xrightarrow{f_1} H \xleftarrow{f_2} G_2$ of matches in $\mathcal{G}$ factorises uniquely through an element of $G_1 * G_2$. Given such a cospan, let $u\colon G_1 \cup G_2 \to H$ be a representative in $\mathcal{G}$ of the subject union of f_1 and f_2 in $\mathbf{Sub}(H)$, with coproduct injections $i_1\colon G_1 \to G_1 \cup G_2$ and $i_2\colon G_2 \to G_1 \cup G_2$ as in (20). Since u is the mediating morphism of a pullback, it is unique up to isomorphism of $G_1 \cup G_2$. It remains to show that $G_1 \xrightarrow{i_1} G_1 \cup G_2 \xleftarrow{i_2} G_2$ is a MG. By adhesiveness of $\mathcal{G}$, the pushout square at the top of (20) is also a pullback, and hence an intersection of i_1 and i_2 in $\mathbf{Sub}(G_1 \cup G_2)$. It follows that $\mathrm{id}_{G_1 \cup G_2}$ represents the subobject union of i_2 and i_2 in $\mathbf{Sub}(G_1 \cup G_2)$ and hence $G_1 \xrightarrow{i_1} G_1 \cup G_2 \xleftarrow{i_2} G_2$ is indeed a MG.

The finiteness of $G_1 * G_2$ follows from a similar argument. First, note that $|G_1 * G_2| = |G_1 *_{\simeq} G_2|$, so it is sufficient to show that $G_1 *_{\simeq} G_2$ is finite. Being a subobject union, every MG is the pushout of a span $G_1 \xleftarrow{p_1} G_1 \cap G_2 \xrightarrow{p_2} G_2$ of matches as in (20). Since isomorphic spans have isomorphic pushouts, there can be at most as many isomorphism classes of MGs of G_1 and G_2 as there are isomorphism classes of spans over G_1 and G_2. Furthermore, the spans $G_1 \xleftarrow{p_1} X \xrightarrow{p_2} G_2$ are in one-to-one correspondence with the pairings $\langle p_1, p_2 \rangle \colon X \to G_1 \times G_2$ in $\mathcal{G}$, which represent subobjects in $\mathbf{Sub}(G_1 \times G_2)$ (with isomorphic spans corresponding to identical subobjects). Since $G_1 \times G_2$ is finitely powered, there are only a finite number of such subobjects, and hence there can only be a finite number of isomorphism classes of spans over G_1 and G_2, which concludes the proof. $\square$

B.2   Proof of Lemma 1 (forward modularity)

Let f_1, f_2, g, g_1 be matches, and α, β, γ rules, such that the diagrams (5) and (6) are derivations. Then there is a unique match g_2, such that diagram (7) commutes and is a vertical composition of derivations.

$$
\begin{array}{ccc}
L \xrightarrow{\alpha} R & & L \xrightarrow{\alpha} R \\
f_1\downarrow \quad\quad & L \xrightarrow{\alpha} L & f_1\downarrow \quad g_1\downarrow \\
S \quad\quad g \ (5) & f_1\downarrow \quad\downarrow g_1 \ (6) & S \xrightarrow{\gamma} T \ \Big) g \ (7) \\
f_2\downarrow \quad\quad & S \xrightarrow{\gamma} T & f_2\downarrow \quad g_2\downarrow \\
G \xrightarrow{\beta} H & & G \xrightarrow{\beta} H
\end{array}
$$

Proof. Using the universal property of the pushout (6), we obtain the mediating morphism g_2 and apply the Pasting Lemma for pushouts to conclude that the lower square in (7) is a pushout.   □

B.3   Proof of Lemma 2 (backward modularity)

Let f, f_1, g_1, g_2 be matches, and α, β, γ rules, such that the diagrams (8) and (9) are derivations. Then there is a unique match f_2, such that diagram (10) commutes and is a vertical composition of derivations.

$$
\begin{array}{ccc}
L \xrightarrow{\alpha} R & & L \xrightarrow{\alpha} R \\
\downarrow \quad\downarrow g_1 & L \xrightarrow{\alpha^\dagger} R & \downarrow f_1 \quad\downarrow g_1 \\
f\downarrow \ T \ (8) & f_1\downarrow \quad\downarrow g_1 \ (9) & f\Big(S \xrightarrow{\gamma} T \ (10) \\
\downarrow g_2 & S \xleftarrow{\gamma^\dagger} T & f_2\downarrow \quad\downarrow g_2 \\
G \xrightarrow{\beta} H & & G \xrightarrow{\beta} H
\end{array}
$$

Proof. The proof is in three steps: we first construct f_1 and f_2 in $\mathcal{G}$, then we show that diagram (10) is indeed a composition of derivations, and finally we verify the uniqueness of f_2 for this property.

Consider diagram (21) below, which is the underlying diagram in $\mathcal{G}$ of derivation (8) from the lemma:

$$
\begin{array}{ccc}
L \xleftarrow{\alpha_1} K \xrightarrow{\alpha_2} R & & L \xleftarrow{\alpha_1} K \xrightarrow{\alpha_2} R \\
\quad\quad\downarrow g_1 & & \downarrow f_1 \quad\downarrow h_1 \quad\downarrow g_1 \\
f\downarrow \ h\downarrow \ T \ (21) & & f\Big(S \xleftarrow{\gamma_1} E \xrightarrow{\gamma_2} T \ (22) \\
\quad\quad\downarrow g_2 & & \downarrow f_2 \quad\downarrow h_2 \quad\downarrow g_2 \\
G \xleftarrow{\beta_1} D \xrightarrow{\beta_2} H & & G \xleftarrow{\beta_1} D \xrightarrow{\beta_2} H
\end{array}
$$

The right-hand square is a pushout along monomorphisms, and hence it is also a pullback in $\mathcal{G}$, and we can decompose it along g_1 and g_2 to obtain the upper and lower right squares of diagram (22). By stability of pushouts in $\mathcal{G}$ (see [35, Lemma 4.7]), both these squares are also pushouts. To complete diagram (22), let its upper-left square be a pushout, and f_2 the unique mediating morphism such that the right-hand side of the diagram commutes.

Note that all morphisms in (22), except possibly f_2, are monic. The composites $\gamma_1 \circ h_1$ and $\gamma_2 \circ h_1$ are pushout complements of $f_1 \circ \alpha_1$ and $g_1 \circ \alpha_2$, respectively, and hence by [14, proposition 12], they are also FPBCs. It follow that the upper half of (22) is indeed the underlying diagram in $\mathcal{G}$ of both the derivations in (9) and the upper half of (10). For the lower half of (22) to also be a derivation, f_2 must be a match, so we need to show that it is monic. To show this, let $S \xrightarrow{i_1} G' \xleftarrow{i_2} D$ be a pushout of $S \xleftarrow{\gamma_1} E \xrightarrow{h_2} D$, and let $u \colon G' \to G$ be its mediating morphism with respect to the lower-left square of (22). Since h_2 is monic, so is i_1 (by adhesiveness of $\mathcal{G}$). By pasting of pushouts, u is also the mediating morphism of the pushout $L \xrightarrow{i_1 \circ f_1} G' \xleftarrow{i_2} D$ with respect to the left-hand square in (21), which in turn, is also a pullback square. In fact, the composite pushout is the union of the subobjects represented by f and β_1, and hence by [35, Theorem 5.1], u is a monomorphism. It then follows that the composite $f_2 = u \circ i_1$ is also a monomorphism.

$$
f\left(\begin{array}{ccc}
L \xleftarrow{\alpha_1} K \xrightarrow{\alpha_2} R \\
\downarrow f_1 \quad \downarrow h_1 \quad g_1 \downarrow \\
S \xleftarrow{\gamma_1} E \xrightarrow{\gamma_2} T \\
\downarrow f_2 \quad \downarrow h_2' \quad g_2' \downarrow \\
G \xleftarrow{\beta_1} D \xrightarrow{\beta_2} H
\end{array}\right) g \quad (23)
\qquad
f\left(\begin{array}{ccc}
L \xleftarrow{\alpha_1} K \xrightarrow{\alpha_2} R \\
\downarrow f_1 \quad \downarrow h_1 \quad g_1 \downarrow \\
S \xleftarrow{\gamma_1} E \xrightarrow{\gamma_2} T \\
\downarrow f_2' \quad \downarrow h_2'' \quad g_2 \downarrow \\
G \xleftarrow{\beta_1} D \xrightarrow{\beta_2} H
\end{array}\right) g \quad (24)
$$

Now let h_2' and g_2' be matches such that diagram (23) commutes and is a composition of tiles as per Lemma 1. Then we have $\beta_1 \circ h_2 = f_2 \circ \gamma_1 = \beta_1 \circ h_2'$, and hence $h_2 = h_2'$ because β_1 is monic. Furthermore, the top-right square of (23) is a pushout, and hence g_2' is the unique mediating morphism such that $\beta_2 \circ h_2 = g_2' \circ \gamma_2$ and $g = g_2' \circ g_1$. But from diagram (22) we know that $\beta_2 \circ h_2 = g_2 \circ \gamma_2$ and $g = g_2 \circ g_1$, and hence $g_2' = g_2$. It follows that the bottom half of (22) is indeed a derivation.

Finally, let f_2' and h_2'' be any matches such that diagram (24) commutes and is a composition of tiles. Then $h_2'' = h_2$ (because $\beta_2 \circ h_2'' = g_2 \circ \gamma_2 = \beta_2 \circ h_2$ and β_2 is monic) and $f_2' = f_2$ (because it is the unique mediating morphism of the top-left pushout-square such that $\beta_1 \circ h_2 = f_2' \circ \gamma_1$ and $f = f_2' \circ f_1$), which concludes the proof. $\square$

B.4   Proof of Lemma 3 (derivability)

A match $g \colon R \to H$ is derivable by a rule $\alpha \colon L \rightharpoonup R$ if and only if $g \Rightarrow_{\alpha \circ \alpha^\dagger} g$. Equivalently, g is derivable from f by α if and only if the derivation $g \Rightarrow_{\alpha^\dagger} f$ is reversible.

Proof. This is a direct consequence of Lemma 2. First, assume that $g \colon R \to H$ is derivable by $\alpha \colon L \rightharpoonup R$ from some match $f \colon L \to G$, and let $h \colon L \to E$ be the comatch of some derivation $g \Rightarrow_{\alpha^\dagger} h$. By Lemma 2 (setting $g_1 = g$ and $f_1 = h$), the derivation $h \Rightarrow_\alpha g$ exists, and so does $g \Rightarrow_{\alpha \circ \alpha^\dagger} g$ (by horizontal composition of derivations).

Now assume that we are given the derivation $g \Rightarrow_{\alpha \circ \alpha^\dagger} g$ instead, and let $f' \colon L \to G'$ and $h' \colon R \to E'$ be the comatches of some derivations $g \Rightarrow_{\alpha^\dagger} f'$ and $f' \Rightarrow_\alpha h'$. By horizontal composition and uniqueness of derivations up to isomorphism, we have $g \Rightarrow_{\alpha \circ \alpha^\dagger} h'$ and $g = u \circ h'$ for some (unique) isomorphism $u \colon E' \xrightarrow{\simeq} H$. Hence there is a derivation $f' \Rightarrow_\alpha g$. □

References

1. Anderson, W.J.: Continuous-Time Markov Chains: An Applications-Oriented Approach. Springer, New York (2012). https://doi.org/10.1007/978-1-4612-3038-0
2. Baldan, P., Corradini, A., Heindel, T., König, B., Sobocinski, P.: Processes and unfoldings: concurrent computations in adhesive categories. Math. Struct. Comput. Sci. **24**, 56–103 (2014)
3. Bapodra, M., Heckel, R.: From graph transformations to differential equations. ECEASST **30**, 21 (2010). https://doi.org/10.14279/tuj.eceasst.30.431.405
4. Barr, M., Wells, C.: Category theory for computing science, 2 ed., Prentice Hall International Series in Computer Science, Prentice Hall (1995)
5. Behr, N., Danos, V., Garnier, I.: Stochastic mechanics of graph rewriting. In: Proceedings 31st Annual ACM/IEEE Symposium on Logic in Computer Science, LICS 2016, New York, pp. 46–55. ACM (2016). https://doi.org/10.1145/2933575.2934537. ISBN 9781450343916
6. Behr, N., Sobocinski, P.: Rule algebras for adhesive categories. Log. Methods Comput. Sci. **16**(3) (2020). https://lmcs.episciences.org/6628
7. Behr, N., Danos, V., Garnier, I.: Combinatorial conversion and moment bisimulation for stochastic rewriting systems. Log. Methods Comput. Sci. **16**(3) (2020). https://lmcs.episciences.org/6628
8. Behr, N., Krivine, J.: Compositionality of rewriting rules with conditions. CoRR arXiv:1904.09322 (2019)
9. Behr, N., Krivine, J.: Rewriting theory for the life sciences: a unifying framework for CTMC semantics. In: Gadducci, F., Kehrer, T. (eds.) Proceedings Graph Transformation, 13th International Conference, ICGT 2020. LNCS, vol. 12150. Springer (2020). https://doi.org/10.1007/978-3-030-51372-6_11
10. Behr, N.: Sesqui-pushout rewriting: concurrency, associativity and rule algebra framework. Electron. Proc. Theoret. Comput. Sci. **309**, 23–52 (2019). https://doi.org/10.4204/eptcs.309.2. ISSN 2075-2180
11. Behr, N., Danos, V., Garnier, I., Heindel, T.: The algebras of graph rewriting. CoRR arXiv:1612.06240 (2016)
12. Behr, N., Saadat, M.G., Heckel, R.: Commutators for stochastic rewriting systems: theory and implementation in Z3. CoRR arXiv:2003.11010 (2020)
13. Bortolussi, L., Hillston, J., Latella, D., Massink, M.: Continuous approximation of collective system behaviour: a tutorial. Performance Eval. **70**(5), 317–349 (2013)
14. Corradini, A., Heindel, T., Hermann, F., König, B.: Sesqui-pushout rewriting. In: Corradini, A., Ehrig, H., Montanari, U., Ribeiro, L., Rozenberg, G. (eds.) ICGT 2006. LNCS, vol. 4178, pp. 30–45. Springer, Heidelberg (2006). https://doi.org/10.1007/11841883_4
15. Danos, V., Harmer, R., Honorato-Zimmer, R., Stucki, S.: Deriving rate equations for site graph rewriting systems. In: Proceedings 4th International Workshop on Static Analysis and Systems Biology (SASB 2013). Seattle, WA, USA (2013), (to appear)

16. Danos, V., Heindel, T., Garnier, I., Simonsen, J.G.: Computing continuous-time markov chains as transformers of unbounded observables. In: Esparza, J., Murawski, A.S. (eds.) FoSSaCS 2017. LNCS, vol. 10203, pp. 338–354. Springer, Heidelberg (2017). https://doi.org/10.1007/978-3-662-54458-7_20

17. Danos, V., Heindel, T., Honorato-Zimmer, R., Stucki, S.: Approximations for stochastic graph rewriting. In: Merz, S., Pang, J. (eds.) ICFEM 2014. LNCS, vol. 8829, pp. 1–10. Springer, Cham (2014). https://doi.org/10.1007/978-3-319-11737-9_1

18. Danos, V., Heindel, T., Honorato-Zimmer, R., Stucki, S.: Moment semantics for reversible rule-based systems. In: Krivine, J., Stefani, J.-B. (eds.) RC 2015. LNCS, vol. 9138, pp. 3–26. Springer, Cham (2015). https://doi.org/10.1007/978-3-319-20860-2_1

19. Danos, V., Honorato-Zimmer, R., Jaramillo-Riveri, S., Stucki, S.: Coarse-graining the dynamics of ideal branched polymers. In: Proceedings 3rd International Workshop on Static Analysis and Systems Biology (SASB 2012). ENTCS, vol. 313, pp. 47–64 (2015)

20. Durrett, R., et al.: Graph fission in an evolving voter model. Proc. Nat. Acad. Sci. **109**(10), 3682–3687 (2012)

21. Dyckhoff, R., Tholen, W.: Exponentiable morphisms, partial products and pullback complements. J. Pure Appl. Algebra **49**(1–2), 103–116 (1987)

22. Ehrig, H., et al.: Algebraic approaches to graph transformation. Part II: Single pushout approach and comparison with double pushout approach. In: Rozenberg, G. (ed.) Handbook of Graph Grammars and Computing by Graph Transformation, pp. 247–312. World Scientific, River Edge, NJ, USA (1997)

23. Ethier, S.N., Kurtz, T.G.: Markov Processes: Characterization and Convergence. Wiley (1986)

24. Evans, M.R., Ferrari, P.A., Mallick, K.: Matrix representation of the stationary measure for the multispecies TASEP. J. Stat. Phys. **135**(2), 217–239 (2009)

25. Fages, F., Soliman, S.: Formal cell biology in biocham. In: Bernardo, M., Degano, P., Zavattaro, G. (eds.) SFM 2008. LNCS, vol. 5016, pp. 54–80. Springer, Heidelberg (2008). https://doi.org/10.1007/978-3-540-68894-5_3

26. Feret, J., Danos, V., Harmer, R., Krivine, J., Fontana, W.: Internal coarse-graining of molecular systems. PNAS **106**(16), 6453–8 (2009)

27. Gleeson, J.P.: High-accuracy approximation of binary-state dynamics on networks. Phys. Rev. Lett. **107**(6), 068701 (2011)

28. Grima, R., Thomas, P., Straube, A.V.: How accurate are the nonlinear chemical Fokker-Planck and chemical Langevin equations? J. Chem. Phys. **135**(8), 084103 (2011)

29. Habel, A., Pennemann, K.H.: Correctness of high-level transformation systems relative to nested conditions. Math. Struct. Comput. Sci. **19**(2), 245–296 (2009)

30. Harmer, R., Danos, V., Feret, J., Krivine, J., Fontana, W.: Intrinsic information carriers in combinatorial dynamical systems. Chaos **20**(3), 037108-1–037108-16 (2010). https://doi.org/10.1063/1.3491100

31. Hayman, J., Heindel, T.: Pattern graphs and rule-based models: the semantics of kappa. In: Pfenning, F. (ed.) FoSSaCS 2013. LNCS, vol. 7794, pp. 1–16. Springer, Heidelberg (2013). https://doi.org/10.1007/978-3-642-37075-5_1

32. Heckel, R.: DPO transformation with open maps. In: Ehrig, H., Engels, G., Kreowski, H.-J., Rozenberg, G. (eds.) ICGT 2012. LNCS, vol. 7562, pp. 203–217. Springer, Heidelberg (2012). https://doi.org/10.1007/978-3-642-33654-6_14

33. Heckel, R., Lajios, G., Menge, S.: Stochastic graph transformation systems. Fundam. Inform. **74**(1), 63–84 (2006)

34. van Kampen, N.: Stochastic Processes in Physics and Chemistry. North-Holland, 3rd edition (2007)
35. Lack, S., Sobociński, P.: Adhesive and quasiadhesive categories. ITA **39**(3), 511–545 (2005)
36. Lopez, C.F., Muhlich, J.L., Bachman, J.A., Sorger, P.K.: Programming biological models in Python using PySB. Molecular Syst. Biol. **9**(1), 602–625 (2013)
37. Löwe, M.: Algebraic approach to single-pushout graph transformation. Theor. Comput. Sci. **109**(1&2), 181–224 (1993)
38. Löwe, M.: Graph rewriting in span-categories. In: Ehrig, H., Rensink, A., Rozenberg, G., Schürr, A. (eds.) ICGT 2010. LNCS, vol. 6372, pp. 218–233. Springer, Heidelberg (2010). https://doi.org/10.1007/978-3-642-15928-2_15
39. Lynch, J.F.: A logical characterization of individual-based models. In: Logic in Computer Science, 2008. In: 23rd Annual IEEE Symposium on LICS 2008, pp. 379–390. IEEE (2008)
40. Kulik, R., Soulier, P.: Markov chains. Heavy-Tailed Time Series. SSORFE, pp. 373–423. Springer, New York (2020). https://doi.org/10.1007/978-1-0716-0737-4_14
41. Rensink, A.: Representing first-order logic using graphs. In: Ehrig, H., Engels, G., Parisi-Presicce, F., Rozenberg, G. (eds.) ICGT 2004. LNCS, vol. 3256, pp. 319–335. Springer, Heidelberg (2004). https://doi.org/10.1007/978-3-540-30203-2_23
42. Robinson, E., Rosolini, G.: Categories of partial maps. Inf. Comput. **79**(2), 95–130 (1988)
43. Spieksma, F.M.: Kolmogorov forward equation and explosiveness in countable state Markov processes. Ann. Oper. Res. **2012**, 3–22 (2012). https://doi.org/10.1007/s10479-012-1262-7
44. Stukalin, E.B., Phillips III, H., Kolomeisky, A.B.: Coupling of two motor proteins: a new motor can move faster. Phys. Rev. Lett. **94**(23), 238101 (2005)
45. Thomas, P., Matuschek, H., Grima, R.: Intrinsic noise analyzer: a software package for the exploration of stochastic biochemical kinetics using the system size expansion. PLoS ONE **7**(6), e38518 (2012)

Stationary Distributions and Metastable Behaviour for Self-regulating Proteins with General Lifetime Distributions

Candan Çelik[1], Pavol Bokes[1,2(✉)], and Abhyudai Singh[3]

[1] Department of Applied Mathematics and Statistics, Comenius University,
84248 Bratislava, Slovakia
{candan.celik,pavol.bokes}@fmph.uniba.sk
[2] Mathematical Institute, Slovak Academy of Sciences, 81473 Bratislava, Slovakia
[3] Department of Electrical and Computer Engineering, University of Delaware,
Newark, Delaware 19716, USA
absingh@udel.edu

Abstract. Regulatory molecules such as transcription factors are often present at relatively small copy numbers in living cells. The copy number of a particular molecule fluctuates in time due to the random occurrence of production and degradation reactions. Here we consider a stochastic model for a self-regulating transcription factor whose lifespan (or time till degradation) follows a general distribution modelled as per a multi-dimensional phase-type process. We show that at steady state the protein copy-number distribution is the same as in a one-dimensional model with exponentially distributed lifetimes. This invariance result holds only if molecules are produced one at a time: we provide explicit counterexamples in the bursty production regime. Additionally, we consider the case of a bistable genetic switch constituted by a positively autoregulating transcription factor. The switch alternately resides in states of up- and downregulation and generates bimodal protein distributions. In the context of our invariance result, we investigate how the choice of lifetime distribution affects the rates of metastable transitions between the two modes of the distribution. The phase-type model, being non-linear and multi-dimensional whilst possessing an explicit stationary distribution, provides a valuable test example for exploring dynamics in complex biological systems.

Keywords: Stochastic gene expression · Master equation · Stationary distribution · Metastable systems

CÇ is supported by the Comenius University grant for doctoral students Nos. UK/201/2019 and UK/106/2020. PB is supported by the Slovak Research and Development Agency under the contract No. APVV-18-0308, by the VEGA grant 1/0347/18, and the EraCoSysMed project 4D-Healing. AS is supported by the National Science Foundation grant ECCS-1711548 and ARO W911NF-19-1-0243.

A. Abate et al. (Eds.): CMSB 2020, LNBI 12314, pp. 27–43, 2020.
https://doi.org/10.1007/978-3-030-60327-4_2

1   Introduction

Biochemical processes at the single-cell level involve molecules such as transcription factors that are present at low copy numbers [6,50]. The dynamics of these processes is therefore well described by stochastic Markov processes in continuous time with discrete state space [16,24,45]. While few-component or linear-kinetics systems [17] allow for exact analysis, in more complex system one often uses approximative methods [13], such as moment closure [4], linear-noise approximation [3,10], hybrid formulations [27,28,35], and multi-scale techniques [41,42].

In simplest Markovian formulations, the lifetime of a regulatory molecule is memoryless, i.e. exponentially distributed [11,51]. However, non-exponential decay patterns have been observed experimentally for both mRNA transcripts and proteins [19,39]. Therefore, in this paper we shall consider lifetime distributions that can assume far more complex forms than the simple exponential. Previous studies of gene-expression models with delayed degradation also provide examples of non-exponential lifetime distributions [14,40].

In Sect. 2, we formulate, both in the deterministic and stochastic settings, a one-dimensional model for the abundance of a transcription factor with a memoryless lifetime. Since many transcription factors regulate their own gene expression [2], we allow the production rate to vary with the copy number. We show that the deterministic solutions tend to the fixed points of the feedback response function; in the stochastic framework, we provide the stationary distribution of the protein copy number.

In Sect. 3, we proceed to characterise the steady-state behaviour of a structured model that accounts for complex lifetime pathways. The model is multidimensional, each dimension corresponding to a different class and stage of a molecule's lifetime; the chosen structure accounts for a wide class of phase-type lifetime distributions [36,49]. We demonstrate that the deterministic fixed points and the stochastic stationary distribution that were found for the one-dimensional framework remain valid for the total protein amount in the multidimensional setting.

We emphasise that the distribution invariance result rests on the assumption of non-bursty production of protein. The case of bursty production is briefly discussed in Sect. 4, where explicit counter-examples are constructed by means of referring to explicit mean and variance formulae available from literature for systems without feedback [30,38].

In the final Sect. 5, we approximate the stochastic protein distribution by a mixture of Gaussians with means at deterministic fixed points and variances given by the linear-noise approximation [9,32]. Additionally, we study the rates of metastable transitions [43,47] between the Gaussian modes in the one-dimensional and structured settings.

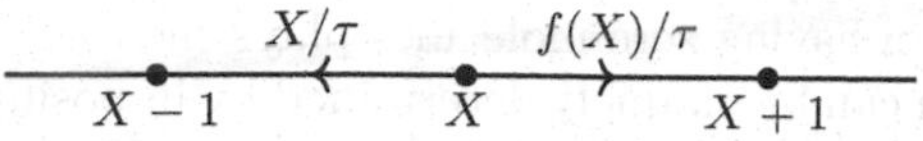

Fig. 1. A diagram of the one-dimensional model. The number of molecules X can decrease by one or increase by one. The stochastic rates (or propensities) of these transitions are indicated above the transition edges.

2 One-Dimensional Model

Deterministic framework. The dynamics of the abundance of protein X at time t can be modelled deterministically by an ordinary differential equation

$$\frac{dX}{dt} = \tau^{-1}\left(f(X) - X\right), \tag{1}$$

which states that the rate of change in X is equal to the difference of production and decay rates. The decay rate is proportional to X; the factor of proportionality is the reciprocal of the expected lifetime τ. The rate of production per unit protein lifetime is denoted by $f(X)$ in (1); the dependence of the production rate on the protein amount X implements the feedback in the model. Equating the right-hand side of (1) to zero yields

$$f(X) = X, \tag{2}$$

meaning that steady states of (1) are given by the fixed point of the production response function $f(X)$.

Stochastic framework. The stochastic counterpart of (1) is the Markov process with discrete states $X \in \mathbb{N}_0$ in continuous time with transitions $X \to X - 1$ or $X \to X+1$, occurring with rates X/τ and $f(X)/\tau$ respectively (see the schematic in Fig. 1). Note that in case of a constant production rate, i.e. $f(X) \equiv \lambda$, the model turns into the immigration-and-death process [34]; in queueing theory this is also known as $M/M/\infty$ queue [23]. The stationary distribution of the immigration–death process is known to be Poissonian with mean equal to λ [34].

For a system with feedback, the probability $P(X, t)$ of having X molecules at time t satisfies the master equation

$$\frac{dP(X,t)}{dt} = \tau^{-1}\left(\mathbb{E}^{-1} - 1\right) f(X)P(X,t) + \tau^{-1}\left(\mathbb{E} - 1\right) X P(X,t), \tag{3}$$

in which $\mathbb{E}$ is the van-Kampen step operator [32]. Inserting $P(X,t) = \pi(X)$ into (3) and solving the resulting difference equation, one finds a steady-state distribution in the explicit form

$$\pi(X) = \pi(0)\frac{\prod_{k=0}^{X-1} f(k)}{X!}. \tag{4}$$

The probability $\pi(0)$ of having zero molecules plays the role of the normalisation constant in (4), which can be uniquely determined by imposing the normalisation condition $\pi(0) + \pi(1) + \ldots = 1$. Note that inserting $f(X) \equiv \lambda$ into (4) results in the aforementioned Poissonian distribution with $\pi(0) = e^{-\lambda}$.

3   Multiclass–multistage Model

In this section, we introduce a structured multiclass–multistage model which is an extension of one-dimensional model introduced in the previous section. The fundamentals of the multidimensional model are as shown in Fig. 2. A newly produced molecule is assigned into one of K distinct classes. Which class is selected is chosen randomly according to a discrete distribution $p_1, \ldots, p_K$. The lifetime of a molecule in the i-th class consists of S_i stages. The holding time in any of these stages is memoryless (exponential), and parametrised by its mean τ_{ij}, where i indicates which class and j indicates which stage. Note that

$$\tau = \sum_{i=1}^{K} \sum_{j=1}^{S_i} p_i \tau_{ij} \tag{5}$$

gives the expected lifetime of a newly produced molecule. After the last (S_i-th) stage, the molecule is degraded. The total distribution of a molecule lifetime is a mixture, with weights p_i, of the lifetime distributions of the individual classes, each of which is a convolution of exponential distributions of the durations of the individual stages; such distributions are referred to as phase-type distribution and provide a wide family of distribution to approximate practically any distribution of a positive random variable [49].

We denote by X_{ij} the number of molecules in the i-th class and the j-th stage of their lifetime, by

$$X = (X_{11}, \ldots, X_{1S_1}, X_{21}, \ldots, X_{2S_2}, \ldots, X_{K1}, \ldots, X_{KS_K})$$

the $\sum_{i=1}^{K} S_i$-dimensional copy-number vector, and by

$$\|X\| = \sum_{i=1}^{K} \sum_{j=1}^{S_i} X_{ij} \tag{6}$$

the total number of molecules across all classes and stages.

Deterministic framework. The deterministic description of the structured model is given by a system of coupled ordinary differential equations

$$\frac{dX_{i1}}{dt} = \frac{p_i f(\|X\|)}{\tau} - \frac{X_{i1}}{\tau_{i1}}, \quad i = 1, \ldots, K, \tag{7}$$

$$\frac{dX_{ij}}{dt} = \frac{X_{ij-1}}{\tau_{ij-1}} - \frac{X_{ij}}{\tau_{ij}}, \quad i = 1, \ldots, K \text{ and } j = 2, \ldots, S_i. \tag{8}$$

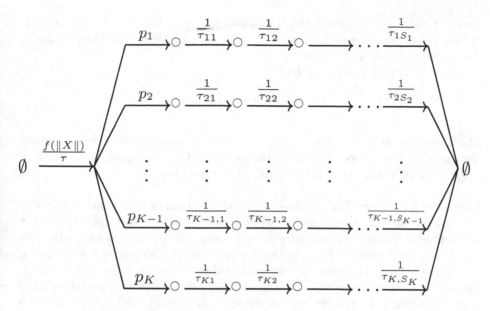

Fig. 2. A schematic representation of multiclass–multistage model. A newly produced molecule is randomly assigned, according to a prescribed distribution $p_1, \ldots, p_K$, into one of K distinct classes. The lifetime of a molecule in the i-th class consists of S_i consecutive memoryless stages, and ends in the degradation of the molecule. The expected holding time in the j-th stage of the i-th class is τ_{ij}. The production rate is a function of the total number $\|X\|$ of molecules across all stages and classes.

The right-hand sides of (7)–(8) are each equal to the difference of appropriate arrival and departure rates at/from a particular compartment of the structured model. The departure rates are proportional to the number of molecules in the compartment, with the reciprocal of the holding time giving the factor of proportionality. The arrival rate takes a different form for the first stages (7) and for the other stages (8). For the first stage, the arrival is obtained by the product of the production rate $f(\|X\|)/\tau$ and the probability p_i of selecting the i-th class. For the latter stages, the arrival rate is equal to the departure rate of the previous stage.

Equating (7)–(8) to zero, we find that

$$\frac{p_i f\left(\|X\|\right)}{\tau} = \frac{X_{i1}}{\tau_{i1}} = \frac{X_{i2}}{\tau_{i2}} = \cdots = \frac{X_{ij}}{\tau_{ij}} \tag{9}$$

hold at steady state, from which it follows that

$$X_{ij} = \frac{p_i \tau_{ij} f\left(\|X\|\right)}{\tau}. \tag{10}$$

Summing (10) over $i = 1, \ldots, K$ and $j = 2, \ldots, S_i$, and using (5) and (6), yield

$$\|X\| = f\left(\|X\|\right) \tag{11}$$

for the total protein amount (6). Thus, the protein amount at steady state is obtained, like in the one-dimensional model, by calculating the fixed points of the feedback response function.

Combining (11) and (9) we find

$$X_{ij} = \frac{p_i \tau_{ij} \|X\|}{\tau}, \tag{12}$$

which means that at steady state the total protein amount is distributed among the compartments proportionally to the product of class assignment probability and the mean holding time of the particular compartment.

Stochastic framework. Having shown that the stationary behaviour of the one-dimensional and the structured multi-dimensional models are the same in the deterministic framework, we next aim to demonstrate that the same is also true in the stochastic context. Prior to turning our attention to the feedback system, it is again instructive to discuss the case without regulation, i.e. $f(\|X\|) \equiv \lambda$; the new molecule arrivals are then exponentially distributed. In the language of queueing theory, the process can be reinterpreted as the $M/G/\infty$ queue with exponential arrivals of customers, a general phase-type distribution of service times, and an infinite number of servers. It is well known that the steady-state distribution of an $M/G/\infty$ queue is Poisson with mean equal to λ [44]. Thus, without feedback, we obtain the very same Poisson(λ) distribution that applies in the one-dimensional case.

In the feedback case, the probability $P(X, t)$ of having $X = (X_{11}, \ldots, X_{K,S_K})$ copy numbers in the individual compartments at any time t satisfies the master equation

$$\frac{\mathrm{d}P(X,t)}{\mathrm{d}t} = \tau^{-1} \sum_{i=1}^{K} p_i \left(\mathbb{E}_{i1}^{-1} - 1\right) f(\|X\|) P(X,t) \tag{13}$$

$$+ \sum_{i=1}^{K} \sum_{j=1}^{S_i - 1} \tau_{ij}^{-1} \left(\mathbb{E}_{ij} \mathbb{E}_{ij+1}^{-1} - 1\right) X_{ij} P(X,t) \tag{14}$$

$$+ \sum_{i=1}^{K} \tau_{iS_i}^{-1} \left(\mathbb{E}_{iS_i} - 1\right) X_{iS_i} P(X,t). \tag{15}$$

The right-hand-side terms (13), (14), and (15) stand for the change in probability mass function due to the production, moving to next stage, and decay reactions, respectively. Note that $\mathbb{E}_{ij}$ is a step operator which increases the copy number of molecules in the i-th class at the j-th stage by one [32]. Likewise, $\mathbb{E}_{ij}^{-1}$ decreases the same copy number by one. Rearrangement of terms in the master equation yields

$$\frac{\mathrm{d}P(X,t)}{\mathrm{d}t} = \sum_{i=1}^{K} \left(\tau^{-1} p_i \mathbb{E}_{i1}^{-1} f(\|X\|) P(X,t) - \tau_{i1}^{-1} X_{i1} P(X,t) \right)$$

$$+ \sum_{i=1}^{K} \sum_{j=1}^{S_i-1} \left(\tau_{ij}^{-1} \mathbb{E}_{ij} \mathbb{E}_{ij+1}^{-1} X_{ij} P(X,t) - \tau_{ij+1}^{-1} X_{ij+1} P(X,t) \right)$$

$$+ \sum_{i=1}^{K} \tau_{iS_i}^{-1} \mathbb{E}_{iS_i} X_{iS_i} P(X,t) \; - \; \tau^{-1} f(\|X\|) P(X,t).$$

Equating the derivative to zero, we derive for the stationary distribution $\pi(X)$ an algebraic system

$$0 = \sum_{i=1}^{K} \left(\tau^{-1} p_i \mathbb{E}_{i1}^{-1} f(\|X\|) \pi(X) - \tau_{i1}^{-1} X_{i1} \pi(X) \right)$$

$$+ \sum_{i=1}^{K} \sum_{j=1}^{S_i-1} \left(\tau_{ij}^{-1} \mathbb{E}_{ij} \mathbb{E}_{ij+1}^{-1} X_{ij} \pi(X) - \tau_{ij+1}^{-1} X_{ij+1} \pi(X) \right) \qquad (16)$$

$$+ \sum_{i=1}^{K} \tau_{iS_i}^{-1} \mathbb{E}_{iS_i} X_{iS_i} \pi(X) \; - \; \tau^{-1} f(\|X\|) \pi(X).$$

Clearly, it is sufficient that

$$\begin{aligned}
\tau^{-1} p_i \mathbb{E}_{i1}^{-1} f(\|X\|) \pi(X) &= \tau_{i1}^{-1} X_{i1} \pi(X), \\
\tau_{ij}^{-1} \mathbb{E}_{ij} \mathbb{E}_{ij+1}^{-1} X_{ij} \pi(X) &= \tau_{ij+1}^{-1} X_{ij+1} \pi(X), \\
\sum_{i=1}^{K} \tau_{iS_i}^{-1} \mathbb{E}_{iS_i} X_{iS_i} \pi(X) &= \tau^{-1} f(\|X\|) \pi(X)
\end{aligned} \qquad (17)$$

hold for $\pi(X)$ in order that (16) be satisfied. One checks by direct substitution that

$$\pi(X) \propto \prod_{k=0}^{\|X\|-1} f(k) \times \prod_{i=1}^{K} \prod_{j=1}^{S_i} \frac{(p_i \tau_{ij}/\tau)^{X_{ij}}}{X_{ij}!} \qquad (18)$$

satisfies (17); therefore, (18) represents the stationary distribution of the structured model. In order to interpret (18), we condition the joint distribution on the total protein copy number, writing

$$\pi(X) = \pi_{\mathrm{cond}}(X \mid \|X\|) \pi_{\mathrm{tot}}(\|X\|), \qquad (19)$$

in which the conditional distribution is recognised as the multinomial [31]

$$\pi_{\mathrm{cond}}(X \mid \|X\|) = \binom{\|X\|}{X} \prod_{i=1}^{K} \prod_{j=1}^{S_i} (p_i \tau_{ij}/\tau)^{X_{ij}}, \qquad (20)$$

and the total copy number distribution is given by

$$\pi_{\text{tot}}(\|X\|) = \pi_{\text{tot}}(0)\frac{\prod_{k=0}^{\|X\|-1} f(k)}{\|X\|!}. \tag{21}$$

By (20), the conditional means of X_{ij} coincide with the deterministic partitioning of the total copy number (12). Importantly, comparing (21) to (4), we conclude that the one-dimensional and multi-dimensional models generate the same (total) copy number distributions.

4 Bursting

The independence of stationary distribution on the lifetime distribution relies on the assumption of non-bursty production of protein that has implicitly been made in our model. In this section, we allow for the synthesis of protein in bursts of multiple molecules at a single time [15,18]. Referring to previously published results [30,38], we provide a counterexample that demonstrates that in the bursty case different protein lifetime distributions can lead to different stationary copy-number distributions. The counterexample can be found even in the absence of feedback.

Bursty production means that the number of molecules can increase within an infinitesimally small time interval of length dt from X to $X+j$, where $j \geq 1$, with probability $\lambda\tau^{-1}b_j dt$, in which λ is the burst frequency (a constant in the absence of feedback), τ is the mean protein lifetime, and $b_j = \text{Prob}[B = j]$ is the probability mass function of the burst size B. Protein molecules degrade independently of one another. The distribution of their lifetime T can in general be described by the survival function $G(t) = \text{Prob}[T > t]$; the mean lifetime thereby satisfies

$$\tau = -\int_0^\infty tG'(t)\mathrm{d}t = \int_0^\infty G(t)\mathrm{d}t. \tag{22}$$

The copy protein number X at a given time is given by the number of products that have been produced in a past burst and survived until the given time; this defines a random process, cf. [30], whose steady-state moments are provided below. In queueing theory, bursty increases in the state variable are referred to as batch customer arrivals. Specifically, a bursty gene-expression model without feedback and with general lifetime distribution corresponds to the $M^X/G/\infty$ queue with memoryless (exponential) batch arrivals, general service distribution, and an infinite number of servers.

Previous analyses [30,38] show that the steady-state protein mean $\langle X \rangle$ and the Fano factor $F = \text{Var}(X)/\langle X \rangle$ are given by

$$\langle X \rangle = \lambda\langle B \rangle, \quad F = 1 + K_s\left(\frac{\langle B^2 \rangle}{\langle B \rangle} - 1\right), \tag{23}$$

where

$$K_s = \frac{\int_0^\infty G^2(t)\mathrm{d}t}{\tau} \tag{24}$$

is referred to as the senescence factor. Elementary calculation shows that $K_s = 1/2$ if the lifetime distribution is exponential with survival function $G(t) = e^{-t/\tau}$ and that $K_s = 1$ if the lifetime distribution is deterministic with survival function $G(t) = 1$ for $t < \tau$ and $G(t) = 0$ for $t \geq \tau$. Thus, although two lifetime distributions result in the same value of the stationary mean protein copy number, they give a different value of the noise (the Fano factor); therefore the copy-number distributions are different.

5   Metastable Transitioning

Transcription factors that self-sustain their gene expression by means of a positive feedback loop can act as a simple genetic switch [5,22]. A positive-feedback switch can be in two states, one in which the gene is fully activated through its feedback loop, while in the other the gene is expressed at a basal level. The switch serves as a basic memory unit, retaining the information on its initial state on long timescales, and very slowly relaxing towards an equilibrium distribution. It is therefore important to investigate not only the stationary, but also transient distributions, which are generated by a positively autoregulating transcription factor.

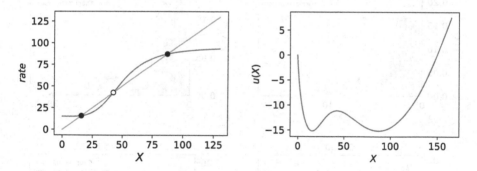

Fig. 3. *Left:* A sigmoid feedback response function (blue curve) intersects the diagonal (orange line) in multiple fixed points. Ones that are stable to the rate Eq. (1) (full circles) are interspersed by unstable ones (empty circle). *Right:* The potential $u(x)$, defined by (33), is a Lyapunov function of the rate Eq. (1). The local minima, or the troughs/wells, of the potential are situated at its stable fixed points; the local maximum, or the barrier, of the potential coincides with the unstable fixed point. *Parameter values for both panels:* We use the Hill-type response (25) with $a_0 = 0.3$, $a_1 = 1.6$, $H = 4$, $\Omega = 50$.(Color figure online)

Following previous studies [7,12,21], we model positive feedback by the Hill function response curve

$$f(X) = \Omega \left(a_0 + \frac{a_1 X^H}{\Omega^H + X^H} \right), \tag{25}$$

in which a_0 and a_1 represent the basal and regulable production rates, H is the cooperativity coefficient, and Ω gives the critical amount of protein required for half-stimulation of feedback. Provided that $H > 1$, one can find a_0 and a_1 such that (25) possesses three distinct fixed points $X_- < X_0 < X_+$, of which the central is unstable and the other two are stable (Fig. 3, left). The two stable fixed points provide alternative large-time outcomes of the deterministic models (1) and (7)–(8).

Bistability of deterministic models translates into bimodal distributions in the stochastic framework. For large values of Ω, the bimodal protein distribution can be approximated by a mixture of Gaussian modes which are located at the stable fixed points $X_\pm$ (see Fig. 4), cf. [9,32],

$$P(X,t) \sim p_-(t)\frac{e^{-\frac{(X-X_-)^2}{2\sigma_-^2}}}{\sqrt{2\pi}\sigma_-} + p_+(t)\frac{e^{-\frac{(X-X_+)^2}{2\sigma_+^2}}}{\sqrt{2\pi}\sigma_+}. \tag{26}$$

The mixture approximation (26) is determined not only by the locations $X_\pm$, but also on the variances $\sigma_\pm^2$ and the weights $p_\pm(t)$ of the two modes (which

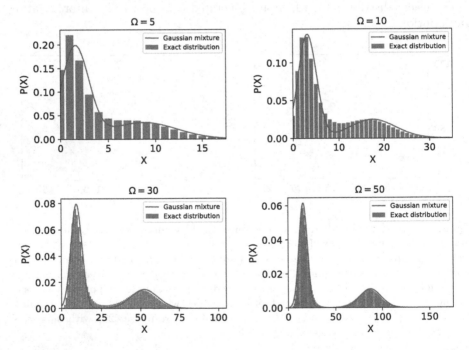

Fig. 4. Exact stationary protein distribution (4) and the Gaussian-mixture approximation (26) in varying system-size conditions. The means of the Gaussians are given by the stable fixed points of $f(X)$; the variances are given by linear-noise approximation (27). The mixture weights are given by $p_+(\infty) = T_+/(T_+ + T_-)$, $p_-(\infty) = T_-/(T_+ + T_-)$, where the residence times are given by the Arrhenius-type formula (32). We use a Hill-type response (25) with $a_0 = 0.3$, $a_1 = 1.6$, $H = 4$, and Ω shown in panel captions.

are given below). The weights in (26) are allowed to vary with time in order to account for the slow, metastable transitions that occur between the distribution modes.

The invariance result for stationary distributions derived in the preceding sections implies that, in the limit of $t \to \infty$, the protein distribution (26) becomes independent of the choice of the protein lifetime distribution. In particular, the same variances $\sigma_{\pm}^2$ and the same limit values $p_{\pm}(\infty)$ of the weights will apply for exponentially distributed and phase-type decay processes. In what follows, we first consult literature to provide results $\sigma_{\pm}^2$ and $p_{\pm}(t)$ that apply for the one-dimensional model with exponential decay. Next, we use stochastic simulation to investigate the effect of phase-type lifespan distributions on the relaxation rate of $p_{\pm}(t)$ to the stationary values.

The variances of the modes are obtained by the linear-noise approximation [37,48] of the master Eq. (3), which yields

$$\sigma_{\pm}^2 = \frac{X_{\pm}}{1 - f'(X_{\pm})};$$ (27)

the right-hand side of (27) is equal to the ratio of a fluctuation term (equal to the number of molecules) to a dissipation term (obtained by linearising the rate Eq. (1) around a stable fixed point).

The metastable transitions between the distribution modes can be described by a random telegraph process (cf. Fig. 5, left), cf. [46],

$$\ominus \; \underset{1/T_+}{\overset{1/T_-}{\rightleftharpoons}} \; \oplus,$$ (28)

in which the lumped states $\ominus$ and $\oplus$ correspond to the basins of attractions of the two stable fixed points; T_- and T_+ are the respective residence times. The mixture weights $p_-(t)$ and $p_+(t)$ in (26) are identified with the probabilities of the lumped states in (28); these satisfy the Chapman–Kolmogorov equations [8]

$$\frac{dp_-}{dt} = -\frac{p_-}{T_-} + \frac{p_+}{T_+}, \quad \frac{dp_+}{dt} = \frac{p_-}{T_-} - \frac{p_+}{T_+},$$ (29)

which admit an explicit solution

$$p_+(t) = \frac{T_+}{T_+ + T_-} + \left(p_+(0) - \frac{T_+}{T_+ + T_-} \right) \exp\left(-\left(\frac{1}{T_+} + \frac{1}{T_-} \right) t \right),$$ (30)

$$p_-(t) = \frac{T_-}{T_+ + T_-} + \left(p_-(0) - \frac{T_-}{T_+ + T_-} \right) \exp\left(-\left(\frac{1}{T_+} + \frac{1}{T_-} \right) t \right).$$ (31)

The initial probability $p_+(0) = 1 - p_-(0)$ is set to one or zero in (30)–(31) depending on whether the model is initialised in the neighbourhood of the upper or the lower stable fixed point.

With (30)–(31) at hand, the problem of determining the mixture weights in (26) is reduced to that of determining the residence times $T_{\pm}$. Previous large-deviation and WKB analyses of the one-dimensional model [20,25,26] provide an Arrhenius-type formula

$$T_\pm = 2\pi\tau X_\pm^{-1}\sigma_\pm\sqrt{-\sigma_0^2}\exp(u(X_0) - u(X_\pm)). \tag{32}$$

Formula (32) features, on top of the familiar symbols (the mean lifetime τ, fixed points $X_\pm$ and X_0, linearised variances $\sigma_\pm$, and the Ludolph-van-Ceulen constant π), two new symbols: a value σ_0^2 and a function $u(X)$. The value σ_0^2 is readily calculated by inserting 0 instead of $\pm$ into the fluctuation–dissipation relation (27); note that for the unstable fixed point X_0, the denominator in (27) is negative (cf. Fig. 3, left), which renders the whole fraction also negative.

In analogy with the Arrhenius law, the function $u(X)$ represents an "energy" of state X, and is given here explicitly by an indefinite integral [20,25,26]

$$u(X) = \int \ln\left(\frac{X}{f(X)}\right) dX. \tag{33}$$

Note that the derivative of (33),

$$u'(X) = \ln\left(\frac{X}{f(X)}\right), \tag{34}$$

is zero if $f(X) = X$, i.e. at the fixed points of the feedback response function, is negative if $f(X) > X$ and positive if $f(X) < X$. Substituting into (33) the

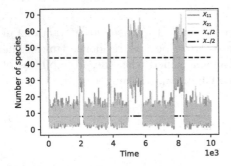
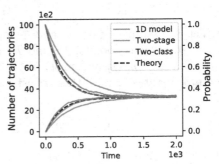

Fig. 5. *Left:* Large-time stochastic trajectories of a structured two-class model with parameters as given below. The horizontal lines represent deterministic fixed points as given by (11)–(12). *Right:* The number of trajectories, out of 10^4 simulation repeats, that reside in the basin of attraction of the upper stable fixed point as function of time. Simulation is initiated at the upper stable fixed point (the decreasing function) or at the lower stable fixed point (the increasing function). The dashed black curve gives the theoretical probability (30) with initial condition $p_+(0) = 1$ (the decreasing solution) or $p_+(0) = 0$ (the increasing solution). *Parameter values:* The Hill-function parameters are: $\Omega = 50$, $H = 4$, $a_0 = 0.3$, $a_1 = 1.6$. The mean lifetime is $\tau = 1$. The two-stage model parameters are: $K = 1$, $p_1 = 1$, $S_1 = 2$, $\tau_{11} = \tau_{12} = 0.5$. The two-class model parameters are: $K = 2$, $p_1 = 1/6$, $p_2 = 5/6$, $S_1 = S_2 = 1$, $\tau_{11} = 3$, $\tau_{21} = 3/5$.

solution $X = X(t)$ to the deterministic rate Eq. (1) and evaluating the time derivative, we find

$$\frac{du(X(t))}{dt} = u'(X(t))\frac{dX(t)}{dt} = \tau^{-1}(f(X) - X)\ln\left(\frac{X}{f(X)}\right)\Bigg|_{X=X(t)} \leq 0, \quad (35)$$

with equality in (35) holding if and only if X is a fixed point of the feedback response function $f(X)$. Therefore, the energy function $u(X)$ is a Lyapunov function of the ordinary differential Eq. (1) (Fig. 3, right). The exponentiation in (32) dramatically amplifies the potential difference between the stable and the unstable fixed points. For example, a moderately large potential barrier, say 5 (which is about the height of the potential barrier in Fig. 3, right), introduces a large factor $e^5 \approx 150$ in (32). This confirms an intuition that metastable transitions between the distribution modes are very (exponentially) slow.

The random telegraph solution (30) is compared in Fig. 5 to the residence of stochastically generated trajectories in the basin of attraction of the upper fixed point. The agreement is close for simulations of the one-dimensional model (with an exponential lifetime) and for a structured model with one class and two stages (with an Erlangian lifetime). For a two-class model (with an exponential mixture lifetime), the transitioning also occurs on the exponentially slow timescale, but is perceptibly slower. Sample trajectories were generated in Python's package for stochastic simulation of biochemical systems GillesPy2 [1]. The one-dimensional model was initiated with $\lfloor X_+ \rfloor$ molecules. The two species in the two-stage and two-class models were initiated to S and $\lfloor X_+ \rfloor - S$, where S was drawn from the binomial distribution $\mathrm{Binom}(\lfloor X_+ \rfloor, 0.5)$.

6 Discussion

In this paper we studied a stochastic chemical reaction system for a self-regulating protein molecule with exponential and phase-type lifetimes. We demonstrated that the exponential and phase-type models support the same stationary distribution of the protein copy number. While stationary distributions of similar forms have previously been formulated in the context of queueing theory [23, 29, 33], our paper provides a self-contained and concise treatment of the one-dimensional model and the multi-dimensional structured model that is specifically tailored for applications in systems biology.

We showed that the invariance result rests on the assumption of non-bursty production of protein. We demonstrated that, in the presence of bursts, exponential and deterministic lifetimes generate stationary protein-level distributions with different variances.

Deterministic modelling approaches are used in systems biology as widely as stochastic ones. Therefore, we complemented the stationary analysis of the stochastic Markov-chain models by a fixed-point analysis of deterministic models based on differential equations. The result is that, irrespective of lifetime distribution, the deterministic protein level is attracted, for large times, to the stable fixed points of the feedback response function. Connecting the stochastic and deterministic frameworks, we demonstrated that the stationary distribution of the Markovian model is sharply peaked around the fixed points of the deterministic equation. We showed that the distribution can be approximated by a mixture of Gaussian modes with means given by the deterministic fixed points and variances that are consistent with the traditional linear-noise analysis results.

Next, we focused on the transitions between the distribution modes. These occur rarely with rates that are exponentially small. We compared an asymptotic result, derived in previous literature for the one-dimensional model, to stochastic simulation results of the one-dimensional model and two specific structured models: we chose a model with one class and two stages and a model with two classes each with one stage. The simulation results of the one-dimensional and two-stage models agreed closely to the theoretical prediction; intriguingly, the agreement with theory was closer for the two-stage model. On the other hand, a two-class model showed slower transitioning rates. The theoretical asymptotic results have been derived in [20,25,26] only for the one-dimensional model. Large deviations in multi-dimensional models are much harder to quantify than one-variable ones. We believe that the current model, being multi-dimensional while possessing a tractable steady-state distribution, provides a convenient framework on which such methodologies can be developed.

In summary, our study provides an invariance-on-lifetime-distribution result in the deterministic and stochastic contexts for a non-bursty regulatory protein. While the main results concern the stationary behaviour, our study also performs simulation, and opens avenue for future enquiries, into the transient transitioning dynamics.

References

1. Abel, J.H., Drawert, B., Hellander, A., Petzold, L.R.: GillesPy: a python package for stochastic model building and simulation. IEEE Life Sci. Lett. **2**(3), 35–38 (2017)
2. Alon, U.: An Introduction to Systems Biology: Design Principles of Biological Circuits. Chapman & Hall/CRC (2007)
3. Andreychenko, A., Bortolussi, L., Grima, R., Thomas, P., Wolf, V.: Distribution approximations for the chemical master equation: comparison of the method of moments and the system size expansion. In: Graw, F., Matthäus, F., Pahle, J. (eds.) Modeling Cellular Systems. CMCS, vol. 11, pp. 39–66. Springer, Cham (2017). https://doi.org/10.1007/978-3-319-45833-5_2
4. Backenköhler, M., Bortolussi, L., Wolf, V.: Control variates for stochastic simulation of chemical reaction networks. In: Bortolussi, L., Sanguinetti, G. (eds.) CMSB 2019. LNCS, vol. 11773, pp. 42–59. Springer, Cham (2019). https://doi.org/10.1007/978-3-030-31304-3_3
5. Becskei, A., Séraphin, B., Serrano, L.: Positive feedback in eukaryotic gene networks: cell differentiation by graded to binary response conversion. EMBO J. **20**, 2528–2535 (2001)

6. Blake, W., Kaern, M., Cantor, C., Collins, J.: Noise in eukaryotic gene expression. Nature **422**, 633–637 (2003)
7. Bokes, P., Lin, Y., Singh, A.: High cooperativity in negative feedback can amplify noisy gene expression. Bull. Math. Biol. **80**, 1871–1899 (2018)
8. Bokes, P.: Postponing production exponentially enhances the molecular memory of a stochastic switch. BioRxiv (2020). https://doi.org/10.1101/2020.06.19.160754
9. Bokes, P., Borri, A., Palumbo, P., Singh, A.: Mixture distributions in a stochastic gene expression model with delayed feedback: a WKB approximation approach. J. Math. Biol. **81**, 343–367 (2020). https://doi.org/10.1007/s00285-020-01512-y
10. Bokes, P., Hojcka, M., Singh, A.: Buffering gene expression noise by microRNA based feedforward regulation. In: Češka, M., Šafránek, D. (eds.) CMSB 2018. LNCS, vol. 11095, pp. 129–145. Springer, Cham (2018). https://doi.org/10.1007/978-3-319-99429-1_8
11. Bokes, P., King, J.R., Wood, A.T., Loose, M.: Exact and approximate distributions of protein and mRNA levels in the low-copy regime of gene expression. J. Math. Biol. **64**, 829–854 (2012). https://doi.org/10.1007/s00285-011-0433-5
12. Bokes, P., Singh, A.: Controlling noisy expression through auto regulation of burst frequency and protein stability. In: Češka, M., Paoletti, N. (eds.) HSB 2019. LNCS, vol. 11705, pp. 80–97. Springer, Cham (2019). https://doi.org/10.1007/978-3-030-28042-0_6
13. Bortolussi, L., Lanciani, R., Nenzi, L.: Model checking Markov population models by stochastic approximations. Inf. Comput. **262**, 189–220 (2018)
14. Bratsun, D., Volfson, D., Tsimring, L.S., Hasty, J.: Delay-induced stochastic oscillations in gene regulation. Proc. Natl. Acad. Sci. U.S.A. **102**(41), 14593–14598 (2005)
15. Cai, L., Friedman, N., Xie, X.: Stochastic protein expression in individual cells at the single molecule level. Nature **440**, 358–362 (2006)
16. Cinquemani, E.: Identifiability and reconstruction of biochemical reaction networks from population snapshot data. Processes **6**(9), 136 (2018)
17. Cinquemani, E.: Stochastic reaction networks with input processes: analysis and application to gene expression inference. Automatica **101**, 150–156 (2019)
18. Dar, R.D., Razooky, B.S., Singh, A., Trimeloni, T.V., McCollum, J.M., Cox, C.D., Simpson, M.L., Weinberger, L.S.: Transcriptional burst frequency and burst size are equally modulated across the human genome. Proc. Natl. Acad. Sci. U.S.A. **109**, 17454–17459 (2012)
19. Deneke, C., Lipowsky, R., Valleriani, A.: Complex degradation processes lead to non-exponential decay patterns and age-dependent decay rates of messenger RNA. PLoS ONE **8**(2), e55442 (2013)
20. Escudero, C., Kamenev, A.: Switching rates of multistep reactions. Phys. Rev. E **79**(4), 041149 (2009)
21. Friedman, N., Cai, L., Xie, X.: Linking stochastic dynamics to population distribution: an analytical framework of gene expression. Phys. Rev. Lett. **97**, 168302 (2006)
22. Griffith, J.: Mathematics of cellular control processes II. Positive feedback to one gene. J. Theor. Biol. **20**(2), 209–216 (1968)
23. Gross, D.: Fundamentals of Queueing Theory. Wiley, Hoboken (2008)
24. Guet, C., Henzinger, T.A., Igler, C., Petrov, T., Sezgin, A.: Transient memory in gene regulation. In: Bortolussi, L., Sanguinetti, G. (eds.) CMSB 2019. LNCS, vol. 11773, pp. 155–187. Springer, Cham (2019). https://doi.org/10.1007/978-3-030-31304-3_9

25. Hanggi, P., Grabert, H., Talkner, P., Thomas, H.: Bistable systems: master equation versus Fokker-Planck modeling. Phys. Rev. A **29**(1), 371 (1984)
26. Hinch, R., Chapman, S.J.: Exponentially slow transitions on a Markov chain: the frequency of calcium sparks. Eur. J. Appl. Math. **16**(04), 427–446 (2005)
27. Innocentini, G.C.P., Antoneli, F., Hodgkinson, A., Radulescu, O.: Effective computational methods for hybrid stochastic gene networks. In: Bortolussi, L., Sanguinetti, G. (eds.) CMSB 2019. LNCS, vol. 11773, pp. 60–77. Springer, Cham (2019). https://doi.org/10.1007/978-3-030-31304-3_4
28. Innocentini, G.C., Hodgkinson, A., Radulescu, O.: Time dependent stochastic mRNA and protein synthesis in piecewise-deterministic models of gene networks. Front. Phys. **6**, 46 (2018)
29. Jackson, J.R.: Jobshop-like queueing systems. Manage. Sci. **10**(1), 131–142 (1963)
30. Jia, T., Kulkarni, R.: Intrinsic noise in stochastic models of gene expression with molecular memory and bursting. Phys. Rev. Lett. **106**(5), 58102 (2011)
31. Johnson, N., Kotz, S., Kemp, A.: Univariate Discrete Distributions, 3rd edn. Wiley, Hoboken (2005)
32. van Kampen, N.: Stochastic Processes in Physics and Chemistry. Elsevier, Amsterdam (2006)
33. Kelly, F.P.: Reversibility and Stochastic Networks. Cambridge University Press, Cambridge (2011)
34. Kendall, D.: Stochastic processes and population growth. J. Roy. Stat. Soc. B **11**, 230–282 (1949)
35. Kurasov, P., Lück, A., Mugnolo, D., Wolf, V.: Stochastic hybrid models of gene regulatory networks – a PDE approach. Math. Biosci. **305**, 170–177 (2018)
36. Lagershausen, S.: Performance Analysis of Closed Queueing Networks, vol. 663. Springer, Heidelberg (2012). https://doi.org/10.1007/978-3-642-32214-3
37. Lestas, I., Paulsson, J., Ross, N., Vinnicombe, G.: Noise in gene regulatory networks. IEEE Trans. Circ. I **53**(1), 189–200 (2008)
38. Liu, L., Kashyap, B., Templeton, J.: On the $GI^X/G/\infty$ system. J. Appl. Probab. **27**(3), 671–683 (1990)
39. McShane, E., Sin, C., Zauber, H., Wells, J.N., Donnelly, N., Wang, X., Hou, J., Chen, W., Storchova, Z., Marsh, J.A., et al.: Kinetic analysis of protein stability reveals age-dependent degradation. Cell **167**(3), 803–815 (2016)
40. Miękisz, J., Poleszczuk, J., Bodnar, M., Foryś, U.: Stochastic models of gene expression with delayed degradation. Bull. Math. Biol. **73**(9), 2231–2247 (2011)
41. Michaelides, M., Hillston, J., Sanguinetti, G.: Geometric fluid approximation for general continuous-time Markov chains. Proc. Roy. Soc. A **475**(2229), 20190100 (2019)
42. Michaelides, M., Hillston, J., Sanguinetti, G.: Statistical abstraction for multi-scale spatio-temporal systems. ACM Trans. Model. Comput. Simul. **29**(4), 1–29 (2019)
43. Newby, J., Chapman, S.J.: Metastable behavior in Markov processes with internal states. J. Math. Biol. **69**(4), 941–976 (2014)
44. Norris, J.R.: Markov Chains. Cambridge Univ Press, Cambridge (1998)
45. Prajapat, M.K., Ribeiro, A.S.: Added value of autoregulation and multi-step kinetics of transcription initiation. R. Soc. Open Sci. **5**(11), 181170 (2018)
46. Ross, S.M.: Introduction to probability models. Academic Press, Cambridge (2014)
47. Samal, S.S., Krishnan, J., Esfahani, A.H., Lüders, C., Weber, A., Radulescu, O.: Metastable regimes and tipping points of biochemical networks with potential applications in precision medicine. In: Liò, P., Zuliani, P. (eds.) Automated Reasoning for Systems Biology and Medicine. CB, vol. 30, pp. 269–295. Springer, Cham (2019). https://doi.org/10.1007/978-3-030-17297-8_10

48. Schnoerr, D., Sanguinetti, G., Grima, R.: Approximation and inference methods for stochastic biochemical kinetics–a tutorial review. J. Phys. A: Math. Theor. **50**(9), 093001 (2017)

49. Soltani, M., Vargas-Garcia, C.A., Antunes, D., Singh, A.: Intercellular variability in protein levels from stochastic expression and noisy cell cycle processes. PLoS Comput. Biol. **12**(8), e1004972 (2016)

50. Taniguchi, Y., Choi, P., Li, G., Chen, H., Babu, M., Hearn, J., Emili, A., Xie, X.: Quantifying E. coli proteome and transcriptome with single-molecule sensitivity in single cells. Science **329**, 533–538 (2010)

51. Thattai, M., van Oudenaarden, A.: Intrinsic noise in gene regulatory networks. Proc. Natl. Acad. Sci. U.S.A. **98**(15), 8614–8619 (2001)

Accelerating Reactions at the DNA Can Slow Down Transient Gene Expression

Pavol Bokes[3], Julia Klein[1], and Tatjana Petrov[1,2(✉)]

[1] Department of Computer and Information Sciences, University of Konstanz,
Konstanz, Germany
tatjana.petrov@gmail.com
[2] Centre for the Advanced Study of Collective Behaviour, University of Konstanz,
Konstanz, Germany
[3] Department of Applied Mathematics and Statistics, Comenius University,
Bratislava, Slovakia

Abstract. The expression of a gene is characterised by the upstream transcription factors and the biochemical reactions at the DNA processing them. Transient profile of gene expression then depends on the amount of involved transcription factors, and the scale of kinetic rates of regulatory reactions at the DNA. Due to the combinatorial explosion of the number of possible DNA configurations and uncertainty about the rates, a detailed mechanistic model is often difficult to analyse and even to write down. For this reason, modelling practice often abstracts away details such as the relative speed of rates of different reactions at the DNA, and how these reactions connect to one another. In this paper, we investigate how the transient gene expression depends on the topology and scale of the rates of reactions involving the DNA. We consider a generic example where a single protein is regulated through a number of arbitrarily connected DNA configurations, without feedback. In our first result, we analytically show that, if all switching rates are uniformly speeded up, then, as expected, the protein transient is faster and the noise is smaller. Our second result finds that, counter-intuitively, if all rates are fast but some more than others (two orders of magnitude vs. one order of magnitude), the opposite effect may emerge: time to equilibration is slower and protein noise increases. In particular, focusing on the case of a mechanism with four DNA states, we first illustrate the phenomenon numerically over concrete parameter instances. Then, we use singular perturbation analysis to systematically show that, in general, the fast chain with some rates even faster, reduces to a slow-switching chain. Our analysis has wide implications for quantitative modelling of

TP's research is supported by the Ministry of Science, Research and the Arts of the state of Baden-Württemberg, and the DFG Centre of Excellence 2117 'Centre for the Advanced Study of Collective Behaviour' (ID: 422037984), JK's research is supported by Committee on Research of Univ. of Konstanz (AFF), 2020/2021. PB is supported by the Slovak Research and Development Agency under the contract No. APVV-18-0308 and by the VEGA grant 1/0347/18. All authors would like to acknowledge Jacob Davidson and Stefano Tognazzi for useful discussions and feedback.

© Springer Nature Switzerland AG 2020
A. Abate et al. (Eds.): CMSB 2020, LNBI 12314, pp. 44–60, 2020.
https://doi.org/10.1007/978-3-030-60327-4_3

gene regulation: it emphasises the importance of accounting for the network topology of regulation among DNA states, and the importance of accounting for different magnitudes of respective reaction rates. We conclude the paper by discussing the results in context of modelling general collective behaviour.

1   Introduction

Gene regulation is one of the most fundamental processes in living systems. The experimental systems of *lac* operon in bacteria *E. coli* and the genetic switch of bacteriophage lambda virus allowed to unravel the basic molecular mechanisms of how a gene is turned on and off. These were followed by a molecular-level explanation of stochastic switching between lysis and lysogeny of phage [22], all the way to more complex logic gate formalisms that attempt to abstract more complex biological behaviour [6,12,21]. To date, synthetic biology has demonstrated remarkable success in engineering simple genetic circuits that are encoded in DNA and perform their function *in vivo*. However, significant conceptual challenges remain, related to the still unsatisfactory quantitative but also qualitative understanding of the underlying processes [19,30]. Partly, this is due to the unknown or unspecified interactions in experiments *in vivo* (crosstalk, host-circuit interactions, loading effect). Another major challenge towards rational and rigorous design of synthetic circuits is computational modelling: gene regulation has a combinatorial number of functional entities, it is inherently stochastic, exhibits multiple time-scales, and experimentally measuring kinetic parameters/rates is often difficult, imprecise or impossible. In such context, predicting the transient profile - how gene expression in a population of cells evolves over time - becomes a computationally expensive task. However, predicting how the transient phenotype emerges from the mechanistic, molecular interactions, is crucial both for engineering purposes of synthetic biology (e.g. when composing synthetic systems), as well as for addressing fundamental biological and evolutionary questions (e.g. for understanding whether the cell aims to create variability by modulating timing).

Mechanistically, the transcription of a single gene is initiated whenever a subunit of RNA polymerase binds to that gene's promoter region at the DNA [35]. While such binding can occur spontaneously, it is typically promoted or inhibited through other species involved in regulation, such as proteins and transcription factors (TFs). Consequently, the number of possible molecular configurations of the DNA grows combinatorially with the number of operator sites regulating the gene in question. For instance, one hypothesised mechanism in lambda-phage, containing only three left and three right operators, leads to 1200 different DNA configurations [31]. The combinatorial explosion of the number of possible configurations makes the model tedious to even write down, let alone execute and make predictions about it. The induced stochastic process enumerates states which couple the configuration of the DNA, with the copy number of the protein, and possibly other species involved in regulation, such as mRNA and transcription factors. In order to faithfully predict the stochastic evolution of the gene product (protein) over time, the modelled system can be solved numerically, by integrating the Master equation of the stochastic process. This is often prohibitive in practice, due to large dimensionality and a combinatorial number of reachable states. For this reason, modelling practice often abstracts away details and adds assumptions. One popular approach is simulating the system by Gillespie simulation [7] and statistically inferring the protein expression profile, hence trading off accuracy and precision. Other approaches are based on mean-field approximations (e.g. deterministic limit [17] and linear-noise approximation [5]), significantly reducing the computational effort. However, mean-field models do not capture the inherent stochasticity, which is especially prominent in gene regulation. Further model reduction ideas exploit multi-scaleness of the system: fast subsystems are identified (possibly dynamically), and assumed to be reaching an equilibrium fast, relative to the observable dynamics [2,13,29,34]. A special class of reductions based on steady-state assumption is the experimentalists' favourite approach of *statistical thermodynamics limit*. This widely and successfully used method (e.g. [3,24,33]) estimates the probability of being in any of the possible DNA binding configurations from their relative binding energies (Boltzmann weights) and the protein concentrations, both of which can often be experimentally accessed. The statistical thermodynamics limit model is rooted in the argument that, when the switching rates among DNA configurations are fast, the probability distribution over the configurations is rapidly arriving at its stationary distribution. While this model takes into account the stochasticity inherent to the DNA binding configurations, it neglects the transient probabilities in the DNA switching, before the equilibrium is reached. It abstracts away the relative speed of rates of different reactions at the DNA, and how they connect to one another. The question arises: how does the transient gene expression - its shape and duration - depend on the topology and scale of the rates of reactions involving the DNA? Is it justifiable, in this context, to consider sufficiently fast propensities as an argument for applying a (quasi-)steady-state assumption?

In this paper, we investigate how the transient gene expression depends on the topology and scale of the rates of reactions involving the DNA. In Sect. 2, we introduce reaction networks, a stochastic process assigned to it, and the equations for the transient dynamics. In Sect. 3, we describe a generic example where a single protein is regulated through a number of arbitrarily connected DNA configurations, without feedback. This means that any transition between two states of the network is possible. In our first result, we analytically show that, if all switching rates are uniformly speeded up, then, as expected, the protein transient is faster and the noise is smaller. In Sect. 3.1, we introduce concrete parameter instances to illustrate the phenomenon numerically. Then, in Sect. 4, we present our main result: counter-intuitively, if all rates are fast but some more than others (two orders of magnitude vs. one order of magnitude), the opposite effect may emerge: time to equilibration is slower and protein noise increases. We use singular perturbation analysis to systematically show that, in general, the fast chain with some rates even faster, exactly reduces to a slow-switching chain. We conclude the paper by discussing the implications of our results in Sect. 5.

1.1   Related Works

Timing aspects of gene regulation are gaining increasing attention, such as explicitly modelling delays in gene expression [28], showcasing dramatic phenotypic consequences of small delays in the arrival of different TFs [11], resolving the temporal dynamics of gene regulatory networks from time-series data [10], as well as the study of transient hysteresis and inherent stochasticity in gene regulatory networks [27]. Following the early works on examining the relation between topology and relaxation to steady states of reaction networks [9], stochastic gene expression from a promoter model has been studied for multiple states [14]. Singular perturbation analysis has been used for lumping states of Markov chains arising in biological applications [4,34]. To the best of our knowledge, none of these works showcases the phenomenon of obtaining slower dynamics through faster rates, or, more specifically, slowing down gene expression by speeding up the reactions at the DNA.

2   Preliminaries

The default rate of gene expression, also referred to as the basal rate, can be modified by the presence of transcriptional activators and repressors. *Activators* are transcription factors (TFs) that bind to specific locations on the DNA, or to other TFs, and enhance the expression of a gene by promoting the binding of RNAP. *Repressors* reduce the expression of gene g, by directly blocking the binding of RNAP, or indirectly, by inhibiting the activators, or promoting direct repressors. The mechanism of how and at which rates the molecular species are interacting is transparently written in a list of reactions. Reactions are equipped with the stochastic semantics which is valid under mild assumptions [7]. In the following, we will model gene regulatory mechanisms with the standard Chemical Reaction Network formalism (CRN).

Definition 1. A reaction system is a pair (S, R), such that $\mathsf{S} = \{S_1, \ldots, S_s\}$ is a finite set of species, and $\mathsf{R} = \{r_1, \ldots, r_r\}$ is a finite set of reactions. The state of a system can be represented as a multi-set of species, denoted by $x = (x_1, \ldots, x_s) \in \mathbb{N}^s$. Each reaction is a triple $r_j \equiv (a_j, \nu_j, c_j) \in \mathbb{N}^s \times \mathbb{N}^s \times \mathbb{R}_{\geq 0}$, written down in the following form:

$$a_{1j}S_1, \ldots, a_{sj}S_s \xrightarrow{c_j} a'_{1j}S_1, \ldots, a'_{sj}S_s, \text{ such that } \forall i. a'_{ij} = a_{ij} + \nu_{ij}.$$

The vectors a_j and a'_j are often called respectively the *consumption* and *production* vectors due to jth reaction, and c_j is the respective *kinetic rate*. If the jth reaction occurs, after being in state x, the next state will be $x' = x + \nu_j$. This will be possible only if $x_i \geq a_{ij}$ for $i = 1, \ldots, s$.

Stochastic Semantics. The species' multiplicities follow a continuous-time Markov chain (CTMC) $\{X(t)\}_{t\geq0}$, defined over the state space $S = \{x \mid x$ is reachable from x_0 by a finite sequence of reactions from $\{r_1,\ldots,r_r\}\}$. In other words, the probability of moving to the state $x + \nu_j$ from x after time Δ is

$$\mathsf{P}(X(t+\Delta) = x + \nu_j \mid X(t) = x) = \lambda_j(x)\Delta + o(\Delta),$$

with λ_j the propensity of jth reaction, assumed to follow the principle of mass-action: $\lambda_j(x) = c_j \prod_{i=1}^{s} \binom{x_i}{a_{ij}}$. The binomial coefficient $\binom{x_i}{a_{ij}}$ reflects the probability of choosing a_{ij} molecules of species S_i out of x_i available ones.

Computing the Transient. Using the vector notation $\mathbf{X}(t) \in \mathbb{N}^n$ for the marginal of process $\{X(t)\}_{t\geq0}$ at time t, we can compute this transient distribution by integrating the *chemical master equation* (CME). Denoting by $p_x(t) := \mathsf{P}(\mathbf{X}(t) = x)$, the CME for state $x \in \mathbb{N}^s$ reads

$$\frac{\mathrm{d}}{\mathrm{d}t}p_x(t) = \sum_{j=1,x-\nu_j \in S}^{r} \lambda_j(x - \nu_j)p_{(x-\nu_j)}(t) - \sum_{j=1}^{r} \lambda_j(x)p_x(t). \tag{1}$$

The solution may be obtained by solving the system of differential equations, but, due to its high (possibly infinite) dimensionality, it is often statistically estimated by simulating the traces of $\{X_t\}$, known as the stochastic simulation algorithm (SSA) in chemical literature [7]. As the statistical estimation often remains computationally expensive for desired accuracy, for the case when the deterministic model is unsatisfactory due to the low multiplicities of many molecular species [18], different further approximation methods have been proposed, major challenge to which remains the quantification of approximation accuracy (see [32] and references therein for a thorough review on the subject).

3   Moment Calculations

We consider a generic example with m different DNA states regulating a single protein, without feedback. The configurations of the DNA are indexed by $1, 2 \ldots m$, and we denote the transition rates between them (reaction propensities) by q_{ij} (we additionally define $q_{ii} = -\sum_{j=1}^{m} q_{ij}$). We assume that the gene chain is irreducible, justified by the reversibility of all reactions at the DNA. The dynamics of the protein copy number is modelled as usually by a birth–death process (with gene-state-dependent birth rate k_i and linear death rate δ per protein). The respective reaction system is schemed in Table 1, left.

The underlying stochastic process $\{X(t)\}$ takes values in the state space $S \subseteq \mathbb{N}^{m+1}$, such that the first m components represent the DNA states, and the last one is the protein count.

In the following, we will use notation $X_{1:m}(t) \in \{0,1\}^m$, to denote the projection of the marginal process at time t, to the DNA-regulatory elements, and, for better readability, we introduce $N(t) := X_{m+1}(t)$ to denote the protein count at time t.

In total, since there is exactly one copy of the DNA, any state in S can be seen as a gene state coupled with the protein copy number, i.e. $S \cong \{1, 2, \ldots, m\} \times \mathbb{N}$. We introduce short-hand notation $s_{(i,n)}$ for state $\boldsymbol{x} = (\underbrace{0, \ldots, 1, \ldots, 0}_{i}, n) \in S$.

Allowable transitions and their rates are summarised in Table 1, right.

Table 1. Two equivalent formulations of a multi-state gene expression model. *Left*: a reaction system with $m + 1$ reaction species S_1, ..., S_m (gene states) and $S_{m+1} = P$ (protein) with copy numbers $X_1(t)$, ..., $X_m(t)$ and $X_{m+1}(t) = N(t)$, whereby $X_1(t) + \ldots + X_m(t) = 1$ holds initially (and throughout time). *Right*: a two-component Markov chain in which the first component indexes the gene state and the second component gives the protein copy number.

Reaction	Rate	Reset map
$S_i \to S_j$	$q_{ij} X_i$	$X_i \to X_i - 1$ $X_j \to X_j + 1$
$S_i \to S_i + P$	$k_i X_i$	$N \to N + 1$
$P \to \emptyset$	δN	$N \to N - 1$

Transition	Rate
$(i, n) \to (j, n)$	q_{ij}
$(i, n) \to (i, n+1)$	k_i
$(i, n) \to (i, n-1)$	δn

We arrange the probabilities $p_{n,i}(t) := P(\mathbf{X}(t) = s_{(i,n)})$ of being in gene state i and having n protein into a column vector

$$\boldsymbol{p}_n(t) = (p_{n,1}(t), \ldots, p_{n,m}(t))^T.$$

The probability vector satisfies a system of difference–differential equations

$$\frac{\mathrm{d}\boldsymbol{p}_n}{\mathrm{d}t} = \boldsymbol{A}\boldsymbol{p}_n + \boldsymbol{\Lambda}_k(\boldsymbol{p}_{n-1} - \boldsymbol{p}_n) + \delta((n+1)\boldsymbol{p}_{n+1} - n\boldsymbol{p}_n), \tag{2}$$

where $\boldsymbol{A} = \boldsymbol{Q}^\mathsf{T}$ is the Markovian generator matrix and $\boldsymbol{\Lambda}_k$ is a diagonal square matrix with the elements of the vector $\boldsymbol{k} = (k_1, \ldots, k_m)^T$ placed on the main diagonal. We study (2) subject to the initial condition

$$\boldsymbol{p}_n(0) = \delta_{n,n_0}\boldsymbol{e}_{j_0} \tag{3}$$

in which n_0 is the initial protein copy number, j_0 is the initial gene state, δ_{n,n_0} represents the Kronecker delta, $\boldsymbol{e}_{j_0}$ is the j_0-th element of the standard basis in the m-dimensional Euclidean space.
Let us introduce the variables

$$\boldsymbol{p}(t) = \sum_{n=0}^{\infty} \boldsymbol{p}_n(t), \quad \langle N(t) \rangle = \sum_{n=0}^{\infty} n\mathbf{1}^T \boldsymbol{p}_n(t), \tag{4}$$

$$\boldsymbol{f}(t) = \left(\sum_{n=0}^{\infty} n\boldsymbol{p}_n(t)\right) - \langle N(t) \rangle \boldsymbol{p}(t), \tag{5}$$

$$\sigma^2(t) = \left(\sum_{n=0}^{\infty} n^2 \mathbf{1}^T \boldsymbol{p}_n(t)\right) - \langle N(t) \rangle^2. \tag{6}$$

Note that $\mathbf{1}^T\boldsymbol{p}_n(t) = p_{n,1}(t) + \ldots + p_{n,m}(t)$, where $\mathbf{1}^{\mathsf{T}} = (1,\ldots,1)$ is the m-dimensional row vector of ones, gives the marginal protein probability mass function. It is instructive to interpret the variables $\boldsymbol{p}(t)$ and $\boldsymbol{f}(t)$ from the standpoint of the reaction-network formulation of the model (Table 1, left). The elements of the copy-number vector $\boldsymbol{X}_{1:m}(t)^{\mathsf{T}}$ of gene states can be zero or one, with exactly one of them being equal to one; the gene-state copy-number statistics can be expressed in terms of (4)–(6) as

$$\langle \boldsymbol{X}_{1:m}(t) \rangle = \boldsymbol{p}(t),$$

$$\langle N(t)\boldsymbol{X}_{1:m}(t) \rangle - \langle N(t) \rangle \langle \boldsymbol{X}_{1:m}(t) \rangle = \boldsymbol{f}(t),$$

$$\langle \boldsymbol{X}_{1:m}(t)\boldsymbol{X}_{1:m}(t)^{\mathsf{T}} \rangle - \langle \boldsymbol{X}_{1:m}(t) \rangle \langle \boldsymbol{X}_{1:m}(t) \rangle^{\mathsf{T}} = \boldsymbol{\Sigma}(t) = \boldsymbol{\Lambda}_{\boldsymbol{p}(t)} - \boldsymbol{p}(t)\boldsymbol{p}(t)^T.$$

The variables (4)–(6) thus fully describe the mean and covariance of the reaction system in Tabl 1, right. In particular, $\boldsymbol{f}(t)$ is the covariance between the gene state and protein copy number; $\boldsymbol{\Sigma}(t)$ is the covariance matrix of the gene state (with itself).

The variables (4)–(6) satisfy a system of differential equations (see Appendix B for derivation)

$$\frac{d\boldsymbol{p}}{dt} = \boldsymbol{A}\boldsymbol{p}, \quad \frac{d\langle N(t) \rangle}{dt} = \boldsymbol{k}^T\boldsymbol{p} - \delta\langle N(t) \rangle, \tag{7}$$

$$\frac{d\boldsymbol{f}}{dt} = \boldsymbol{A}\boldsymbol{f} - \delta\boldsymbol{f} + \boldsymbol{\Sigma}\boldsymbol{k}, \quad \text{where} \quad \boldsymbol{\Sigma} = \boldsymbol{\Lambda}_{\boldsymbol{p}} - \boldsymbol{p}\boldsymbol{p}^T, \tag{8}$$

$$\frac{d\sigma^2}{dt} = 2\boldsymbol{k}^T\boldsymbol{f} + \boldsymbol{k}^T\boldsymbol{p} + \delta\langle N(t) \rangle - 2\delta\sigma^2 \tag{9}$$

subject to initial conditions

$$\boldsymbol{p}(0) = \boldsymbol{e}_{j_0}, \quad \langle N(0) \rangle = n_0, \quad \boldsymbol{f}(0) = \boldsymbol{0}, \quad \sigma^2(0) = 0, \tag{10}$$

where $\boldsymbol{e}_{j_0}$ is the j_0-th element of the standard basis in m-dimensional Euclidean space.

Equating the derivatives in (7)–(9) to zero, we obtain steady state protein mean and Fano factor in the form

$$\langle N(t) \rangle = \frac{\boldsymbol{k}^T\bar{\boldsymbol{p}}}{\delta}, \quad \frac{\sigma^2}{\langle N(t) \rangle} = 1 + \frac{\boldsymbol{k}^T\bar{\boldsymbol{f}}}{\boldsymbol{k}^T\bar{\boldsymbol{p}}}, \tag{11}$$

where $\bar{\boldsymbol{p}}$ and $\bar{\boldsymbol{f}}$ satisfy algebraic equations

$$\boldsymbol{A}\bar{\boldsymbol{p}} = 0, \quad (\boldsymbol{A} - \delta)\bar{\boldsymbol{f}} + \boldsymbol{\Sigma}\boldsymbol{k} = 0. \tag{12}$$

We note that the solution $\bar{\boldsymbol{f}}$ to (12) can be represented as

$$\bar{\boldsymbol{f}} = \int_0^\infty e^{-\delta t} \left(e^{t\boldsymbol{A}} - \bar{\boldsymbol{p}}\mathbf{1}^T \right) dt\,\boldsymbol{\Sigma}\boldsymbol{k}. \tag{13}$$

Equation (13) connects the magnitude of $\bar{\boldsymbol{f}}$ to the equilibration timescale of the gene-state Markov chain (note that $\bar{\boldsymbol{p}}\mathbf{1}^{\mathsf{T}} = \lim_{t\to\infty} e^{t\boldsymbol{A}}$). Specifically, if $\boldsymbol{A} = \tilde{\boldsymbol{A}}/\varepsilon$, where $\varepsilon \ll 1$, i.e. the gene transition rates are $O(1/\varepsilon)$ large, then substituting $t = \varepsilon s$ into (13) implies that

$$\bar{f} = \varepsilon \int_0^\infty \left(e^{s\tilde{A}} - \bar{p}\mathbf{1}^T \right) \mathrm{d}s \, \Sigma k + O(\varepsilon^2),$$

which is $O(\varepsilon)$ small. Correspondingly, the (steady-state) protein Fano factor (11) will differ from the Poissonian value of 1 by an $O(\varepsilon)$ quantity. This concludes the argument that, in agreement with intuition, if all rates are faster by an order of magnitude (ε^{-1}), then, as expected, the magnitude of equilibration time-scale of the whole chain scales down with the same factor. The fast fluctuations of the gene chain are thereby averaged out at the downstream level of the protein.

However, as will be demonstrated in the next section with a specific example of a four state chain, the largeness of transition rates does not guarantee, on its own, fast equilibration, and the connectivity of the chain can play a crucial role. Indeed, we will show that one can slow down equilibriation (and hence increase protein noise) by increasing some of the transition rates.

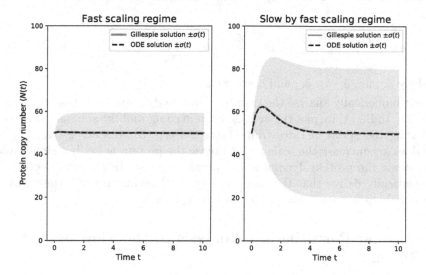

Fig. 1. Dependence of $\langle N(t) \rangle \pm \sigma(t)$ on t for a value of $\varepsilon = 0.01$, using two regimes. Left: "fast" scaling regime where all transition rates are $O(1/\varepsilon)$. Right: "slow by fast" scaling regime, where the backward rates are speeded up to $O(1/\varepsilon^2)$. ODE results (dashed line) are cross-validated by Gillespie simulations (solid line). The chain is initially at state $j_0 = 2$, and the amount of protein is set to $n_0 = 50$. The transition matrix parameters are set to $\tilde{a}_g = \tilde{b}_g = \tilde{a}_r = \tilde{b}_r = \tilde{\tilde{a}}_b = \tilde{\tilde{b}}_b = 1$.

3.1   Four-State Chain

We specifically focus on a case with four gene states, with transition matrix

$$
A = \begin{pmatrix}
-a_r & a_b & 0 & 0 \\
a_r & -a_b - a_g & b_g & 0 \\
0 & a_g & -b_g - b_b & b_r \\
0 & 0 & b_b & -b_r
\end{pmatrix}
\tag{14}
$$

(recall that the matrix A is shown as a transpose of the graph of connections, the respective graph is depicted in Fig. 2, left). We investigate two alternative, different scaling regimes with respect to a small dimensionless parameter ε:

Fast. We assume that all transition rates are $O(1/\varepsilon)$, i.e.

$$
a_g = \frac{\tilde{a}_g}{\varepsilon}, \quad a_b = \frac{\tilde{a}_b}{\varepsilon}, \quad a_r = \frac{\tilde{a}_r}{\varepsilon}, \quad b_g = \frac{\tilde{b}_g}{\varepsilon}, \quad b_b = \frac{\tilde{b}_b}{\varepsilon}, \quad b_r = \frac{\tilde{b}_r}{\varepsilon},
$$

where $\tilde{a}_g, \tilde{a}_b, \tilde{a}_r, \tilde{b}_g, \tilde{b}_b$, and $\tilde{b}_r$ are $O(1)$.
Slow by fast. We speed up the backward rates by making them $O(1/\varepsilon^2)$, i.e.

$$
a_g = \frac{\tilde{a}_g}{\varepsilon}, \quad a_b = \frac{\tilde{a}_b}{\varepsilon^2}, \quad a_r = \frac{\tilde{a}_r}{\varepsilon}, \quad b_g = \frac{\tilde{b}_g}{\varepsilon}, \quad b_b = \frac{\tilde{b}_b}{\varepsilon^2}, \quad b_r = \frac{\tilde{b}_r}{\varepsilon}.
\tag{15}
$$

where $\tilde{a}_g, \tilde{\tilde{a}}_b, \tilde{a}_r, \tilde{b}_g, \tilde{\tilde{b}}_b$, and $\tilde{b}_r$ are $O(1)$.
We first numerically analyse the transient protein dynamics for these two scaling scenarios. In Fig. 1, we plot the average protein count and the standard deviation. Increasing the speed of rates from states 2 (resp. 3) to state 1 (resp. 4) not only does not increase the scale and decrease the protein noise, but significantly slows down the protein dynamics and increases noise. In the next section, we systematically derive that the case 'slow by fast' regime is approximated by a slow-switching 2-state chain (shown in Fig. 2, right).

4   Singular-Perturbation Analysis of the Slow-by-fast Regime

The probability dynamics generated by the transition matrix (14) in the slow-by-fast scaling regime (15) is given by a system of four differential equations

$$
\varepsilon^2 \frac{dp_1}{dt} = \tilde{\tilde{a}}_b p_2 - \varepsilon \tilde{a}_r p_1,
\tag{16}
$$

$$
\varepsilon^2 \frac{dp_2}{dt} = \varepsilon \tilde{a}_r p_1 - \varepsilon \tilde{a}_g p_2 - \tilde{\tilde{a}}_b p_2 + \varepsilon \tilde{b}_g p_3,
\tag{17}
$$

$$
\varepsilon^2 \frac{dp_3}{dt} = \varepsilon \tilde{a}_g p_2 - \varepsilon \tilde{b}_g p_3 - \tilde{\tilde{b}}_b p_3 + \varepsilon \tilde{b}_r p_4,
\tag{18}
$$

$$
\varepsilon^2 \frac{dp_4}{dt} = \tilde{\tilde{b}}_b p_3 - \varepsilon \tilde{b}_r p_4.
\tag{19}
$$

Equations such as (16)–(19) whose right-hand sides depend on a small parameter ε are referred to as perturbation problems. Additionally, problems in which, like in (16)–(19), the small parameter multiplies one or more derivatives on the left-hand side, are classified as singularly perturbed [16,23]. We study solutions to system (16)–(19) that satisfy an intial condition

$$p_i(0) = p_i^{\text{init}}, \quad i = 1, 2, 3, 4, \tag{20}$$

where the right-hand side of (20) is a prescribed probability distribution. The aim of what follows is to characterise the behaviour as $\varepsilon \to 0$ of the solution to (16)–(19) subject to (20).

We look for a solution to (16)–(19) in the form of a regular power series

$$p_i(t; \varepsilon) = p_i^{(0)}(t) + \varepsilon p_i^{(1)}(t) + O(\varepsilon^2), \quad i = 1, 2, 3, 4. \tag{21}$$

Inserting (21) into (16) and (19) and collecting terms of same order yields

$$O(1): \quad p_2^{(0)} = p_3^{(0)} = 0, \tag{22}$$

$$O(\varepsilon): \quad \tilde{a}_{\text{b}} p_2^{(1)} = \tilde{a}_{\text{r}} p_1^{(0)}, \tag{23}$$

$$\tilde{b}_{\text{b}} p_3^{(1)} = \tilde{b}_{\text{r}} p_4^{(0)}. \tag{24}$$

Equations (22) imply that the probability of states 2 or 3 is $O(\varepsilon)$-small. Equation (23) means that, at the leading order, the probability of state 2 is proportional to that of state 1; equation (24) establishes the analogous for states 3 and 4. Adding (16) to (17), and (18) to (19), yield

$$\varepsilon \frac{d}{dt}(p_1 + p_2) = -\tilde{a}_{\text{g}} p_2 + \tilde{b}_{\text{g}} p_3, \tag{25}$$

$$\varepsilon \frac{d}{dt}(p_3 + p_4) = \tilde{a}_{\text{g}} p_2 - \tilde{b}_{\text{g}} p_3. \tag{26}$$

Inserting (21) into (25)–(26) and collecting $O(\varepsilon)$ terms gives

$$\frac{d}{dt}(p_1^{(0)} + p_2^{(0)}) = -\tilde{a}_{\text{g}} p_2^{(1)} + \tilde{b}_{\text{g}} p_3^{(1)}, \tag{27}$$

$$\frac{d}{dt}(p_3^{(0)} + p_4^{(0)}) = \tilde{a}_{\text{g}} p_2^{(1)} - \tilde{b}_{\text{g}} p_3^{(1)}. \tag{28}$$

Inserting (22)–(24) into (27)–(28) yields

$$\frac{dp_1^{(0)}}{dt} = -\tilde{a}_{\text{g}} \tilde{a}_{\text{r}} \tilde{a}_{\text{b}}^{-1} p_1^{(0)} + \tilde{b}_{\text{g}} \tilde{b}_{\text{r}} \tilde{b}_{\text{b}}^{-1} p_4^{(0)}, \tag{29}$$

$$\frac{dp_4^{(0)}}{dt} = \tilde{a}_{\text{g}} \tilde{a}_{\text{r}} \tilde{a}_{\text{b}}^{-1} p_1^{(0)} - \tilde{b}_{\text{g}} \tilde{b}_{\text{r}} \tilde{b}_{\text{b}}^{-1} p_4^{(0)}. \tag{30}$$

Equations (29)–(30) describe the probability dynamics of a two-state (or random-telegraph) chain with states 1 and 4 and transition rates $\tilde{a}_{\text{g}} \tilde{a}_{\text{r}}/\tilde{a}_{\text{b}}$ and $\tilde{b}_{\text{g}} \tilde{b}_{\text{r}}/\tilde{b}_{\text{b}}$ between them. Intriguingly, the emergent dynamics of (29)–(30) occurs on the $t = O(1)$ scale although the original system (16)–(19) featured only $O(1/\varepsilon)$ rates (or faster).

4.1   Inner Solution and Matching

In singular-perturbation studies, the leading-order term of a regular solution (21) is referred to as the outer solution [15,23]. As is typical in singularly perturbed problems, the outer solution satisfies a system, here (29)–(30), that is lower-dimensional than the original system (16)–(19); the remaining components of the outer solution are trivially given by (22). Therefore, the initial condition (20) cannot be immediately imposed on the outer solution. In order to formulate an appropriate initial condition for (29)–(30), we need to study (16)–(19) on the fast timescale, find the so-called inner solution, and use an asymptotic matching principle [15,23] to connect the two asymptotic solutions together.

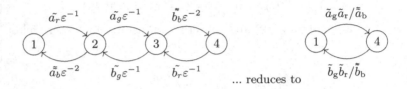

Fig. 2. (left) Four-state chain at the DNA: all rates are faster than $O(1)$: some by order or magnitude ε^{-1}, and some even faster, by order of magnitude ε^{-2}. (left) The emergent dynamics is approximated by a two-state chain. Intriguingly, the emergent dynamics of occurs on the $t = O(1)$ scale although the original system featured only $O(1/\varepsilon)$ rates and faster.

In order to construct the inner solution, we focus on the fast dynamics of system (16)–(19) by means of a transformation

$$t = \varepsilon^2\, T, \quad p_i(t) = P_i(T). \tag{31}$$

Inserting (31) into (16)–(19) yields a time-rescaled system

$$\frac{\mathrm{d}P_1}{\mathrm{d}T} = \tilde{\tilde{a}}_b P_2 - \varepsilon\tilde{a}_r P_1, \tag{32}$$

$$\frac{\mathrm{d}P_2}{\mathrm{d}T} = \varepsilon\tilde{a}_r P_1 - \varepsilon\tilde{a}_g P_2 - \tilde{\tilde{a}}_b P_2 + \varepsilon\tilde{b}_g P_3, \tag{33}$$

$$\frac{\mathrm{d}P_3}{\mathrm{d}T} = \varepsilon\tilde{a}_g P_2 - \varepsilon\tilde{b}_g P_3 - \tilde{\tilde{b}}_b P_3 + \varepsilon\tilde{b}_r P_4, \tag{34}$$

$$\frac{\mathrm{d}P_4}{\mathrm{d}T} = \tilde{\tilde{b}}_b P_3 - \varepsilon\tilde{b}_r P_4. \tag{35}$$

Note that in the time-rescaled system (32)–(35), the time derivative is no longer multiplied by a small parameter.

$$P_i(T;\varepsilon) = P_i^{(0)}(T) + \varepsilon P_i^{(1)}(T) + O(\varepsilon^2), \quad i = 1,2,3,4. \tag{36}$$

into (32)–(35) and collecting the $O(1)$ terms yield

$$\frac{\mathrm{d}P_1^{(0)}}{\mathrm{d}t} = -\frac{\mathrm{d}P_2^{(0)}}{\mathrm{d}t} = \tilde{\tilde{a}}_b P_2^{(0)}, \quad \frac{\mathrm{d}P_3^{(0)}}{\mathrm{d}t} = -\frac{\mathrm{d}P_4^{(0)}}{\mathrm{d}t} = -\tilde{\tilde{b}}_b P_3^{(0)}. \tag{37}$$

Since the reduced problem (37) retains the dimensionality of the original problem (32)–(35), we can solve it subject to the same initial condition

$$P_i^{(0)}(0) = p_i^{\text{init}}, \tag{38}$$

which yields

$$P_1^{(0)}(T) = p_1^{\text{init}} + p_2^{\text{init}}(1 - e^{-\tilde{a}_b T}), \quad P_2^{(0)}(T) = p_2^{\text{init}} e^{-\tilde{a}_b T}, \tag{39}$$

$$P_3^{(0)}(T) = p_3^{\text{init}} e^{-\tilde{b}_b T}, \quad P_4^{(0)}(T) = p_4^{\text{init}} + p_3^{\text{init}}(1 - e^{-\tilde{b}_b T}). \tag{40}$$

Thus, on the inner timescale, there occurs a fast transfer of probability mass from the states 2 and 3 into the states 1 and 4, respectively.

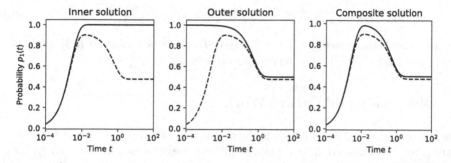

Fig. 3. The inner, outer, and composite approximations (solid curves) to the first component of the exact solution (dashed curve) to (16)–(20) (time is shown at logarithmic scale). The timescale separation parameter is set to $\varepsilon = 0.01$. The chain is initially at state 2, i.e. $p_2^{\text{init}} = 1$, $p_1^{\text{init}} = p_3^{\text{init}} = p_4^{\text{init}} = 0$. The transition matrix parameters are set to $\tilde{a}_g = \tilde{b}_g = \tilde{a}_r = \tilde{b}_r = \tilde{a}_b = \tilde{b}_b = 1$.

According to the asymptotic matching principle [15,23], the large-T behaviour of the inner and the small-t behaviour of the outer solution overlap, i.e.

$$p_1^{(0)}(0) = P_1^{(0)}(\infty) = p_1^{\text{init}} + p_2^{\text{init}}, \tag{41}$$

$$p_4^{(0)}(0) = P_4^{(0)}(\infty) = p_3^{\text{init}} + p_4^{\text{init}}. \tag{42}$$

Equations (41)–(42) establish the relationship between the original initial condition (20) and the initial condition that needs to be imposed for the outer solution; solving (29)–(30) subject to (41)–(42) yields

$$p_1^{(0)}(t) = \frac{\dfrac{\tilde{b}_g \tilde{b}_r}{\tilde{b}_b}}{\dfrac{\tilde{a}_g \tilde{a}_r}{a_b} + \dfrac{\tilde{b}_g \tilde{b}_r}{b_b}} + \left(p_1^{\text{init}} + p_2^{\text{init}} - \frac{\dfrac{\tilde{b}_g \tilde{b}_r}{\tilde{b}_b}}{\dfrac{\tilde{a}_g \tilde{a}_r}{a_b} + \dfrac{\tilde{b}_g \tilde{b}_r}{b_b}} \right) e^{-\left(\frac{\tilde{a}_g \tilde{a}_r}{a_b} + \frac{\tilde{b}_g \tilde{b}_r}{b_b}\right) t}, \tag{43}$$

$$p_4^{(0)}(t) = \frac{\frac{\tilde{a}_g \tilde{a}_r}{a_b}}{\frac{\tilde{a}_g \tilde{a}_r}{a_b} + \frac{\tilde{b}_g \tilde{b}_r}{b_b}} + \left(p_3^{\text{init}} + p_4^{\text{init}} - \frac{\frac{\tilde{a}_g \tilde{a}_r}{a_b}}{\frac{\tilde{a}_g \tilde{a}_r}{a_b} + \frac{\tilde{b}_g \tilde{b}_r}{b_b}} \right) e^{-\left(\frac{\tilde{a}_g \tilde{a}_r}{a_b} + \frac{\tilde{b}_g \tilde{b}_r}{b_b} \right) t}. \tag{44}$$

We note that the second and third components of the outer solution are trivially given by $p_2^{(0)}(t) = p_3^{(0)}(t) = 0$ by (22). The outer solution (43)–(44) provides a close approximation to the original solution for $t = O(1)$ but fails to capture the behaviour of the initial transient; the inner solution (39)–(40) provides a close approximation for $T = O(1)$, i.e. $t = O(\varepsilon^2)$, but disregards the outer dynamics. A uniformly valid composite solution can be constructed by adding the inner and outer solutions up, and subtracting the matched value, i.e.

$$p_1^{\text{comp}}(t) = p_1^{(0)}(t) - p_2^{\text{init}} e^{-\tilde{a}_b t/\varepsilon^2}, \quad p_2^{\text{comp}}(t) = p_2^{\text{init}} e^{-\tilde{a}_b t/\varepsilon^2}, \tag{45}$$

$$p_3^{\text{comp}}(t) = p_3^{\text{init}} e^{-\tilde{b}_b t/\varepsilon^2}, \quad p_4^{\text{comp}}(t) = p_4^{(0)}(t) - p_3^{\text{init}} t^{-\tilde{b}_b t/\varepsilon^2}. \tag{46}$$

Figure 3 shows the exact solution to (16)–(20), the inner solution (39)–(40), the outer solution (43)–(44), and the composite solution (45)–(46).

5   Discussion and Future Work

The key ingredient of our analysis is the separation of temporal scales at the level of the gene state chain. If all gene transition rates are of the same order, say $O(1/\varepsilon)$, then the chain equilibrates on a short $O(\varepsilon)$ timescale (the $O(1)$ timescale is assumed to be that of protein turnover). This situation has been widely considered in literature, e.g. [25, 34]. In the example on which we focused in our analysis, however, some transition rates are of a larger order, $O(1/\varepsilon^2)$. These faster rates generate an $O(\varepsilon^2)$ short timescale in our model. Importantly (and counterintuitively), the acceleration of these rates drives an emergent slow transitioning dynamics on the slow $O(1)$ timescale. This means, in particular, that the transient behaviour, as well as stochastic noise, is not averaged out but retained at the downstream level of protein dynamics. We expect that more general networks of gene states can generate more than two timescales (fast and slow). Results from other works can be used to compute approximations for multiple timescales [9]. In particular, we note that although our example retains some $O(1/\varepsilon)$ transition rates, no distinguished dynamics occurs on the corresponding $O(\varepsilon)$ timescale. We expect, however, the intermediate $O(\varepsilon)$ timescale can play a distinguished role in more complex systems.

The possibility of realistic GRNs implementing slow gene expression dynamics by accelerating reactions at the DNA, opens up fundamental biological questions related to their regulatory and evolutionary roles. For modelling, the uncertainty about even the magnitude of biochemical reaction rates pressures us to account for the potentially emerging slow-by-fast phenomenon: approximations resting on the argument that all rates are 'sufficiently fast', while not accounting for the topology of interactions at the DNA, can lead to wrong conclusions. The key feature of the 4-state example presented in this paper are very fast rates towards two different states which are poorly connected and consequently hard to leave. This situation will likely be seen in larger, realistic gene regulatory networks, because the rate of forming larger functional complexes typically depends on the order of TF's binding at the DNA. For instance, in a biologically realisable gene regulatory circuit shown in [11], a pair of activators and a pair of repressors compete to bind the DNA, so to rapidly transition to highly stable conformational change at the DNA. One of the interesting directions for future work is automatising the derivation of singular perturbation reduction shown in Sect. 4. Such a procedure would allow us to systematically explore reductions for larger gene regulatory networks. Additionally, we want to examine different topologies and sizes of networks to generalize our results. This could reveal if the backward reactions are always the crucial factor in causing the slow-by-fast phenomenon.

Slow-by-fast phenomena we show here, could appear in application domains beyond gene regulation, i.e., wherever nodes over a weighted network regulate a collective response over time. For instance, in network models used to predict the spread of information or spread of disease, among coupled agents [26,36], or in network-models for studying the role of communication in wisdom of the crowds (known to be enhanced by interaction, but at the same time hindered by information exchange [1,20]). Finally, networks of neurons are known to have different intrinsic time-scales, in addition to the time-scales that arise from network connections [8].

Appendix A: Mechanism of Gene Regulation - Examples

Example 1 (basal gene expression). Basal gene expression with RNAP binding can be modelled with four reactions, where the first reversible reaction models binding between the promoter site at the DNA and the polymerase, and the second two reactions model the protein production and degradation, respectively:

$$\text{DNA, RNAP} \leftrightarrow \text{DNA.RNAP at rates } k, k^-$$
$$\text{DNA.RNAP} \rightarrow \text{DNA.RNAP} + \text{P at rate } \alpha$$
$$\text{P} \rightarrow \emptyset \text{ at rate } \beta.$$

The state space of the underlying CTMC $S \cong \{0,1\} \times \{0,1,2,\ldots\}$, such that $s_{(1,x)} \in S$ denotes an active configuration (where the RNAP is bound to the DNA) with $x \in \mathbb{N}$ protein copy number.

Example 2 (adding repression). Repressor blocking the polymerase binding can be modelled by adding a reaction

$$\text{DNA}, R \leftrightarrow \text{DNA.}R$$

In this case, there are three possible promoter configurations, that is, $S \cong \{\text{DNA, DNA.RNAP, DNA.}R\} \times \{0,1,2,\ldots\}$ (states DNA and DNA.$R\}$ are inactive promoter states).

Appendix B: Derivation of Moment Equations

Multiplying the master equation (2) by $n(n-1)\ldots(n-j+1)$ and summing over all $n \geq 0$ yields differential equations [37]

$$\frac{\mathrm{d}\boldsymbol{\nu}_j}{\mathrm{d}t} = \boldsymbol{A}\boldsymbol{\nu}_j + j\left(\boldsymbol{\Lambda}_k\boldsymbol{\nu}_{j-1} - \delta\boldsymbol{\nu}_j\right) \tag{B1}$$

for the factorial moments

$$\boldsymbol{\nu}_j(t) = \sum_{n=0}^{\infty} n(n-1)\ldots(n-j+1)\boldsymbol{p}_n(t). \tag{B2}$$

The quantities (4)–(6) can be expressed in terms of the factorial moments as

$$\boldsymbol{p} = \boldsymbol{\nu}_0, \quad \langle n \rangle = \boldsymbol{1}^{\mathsf{T}}\boldsymbol{\nu}_1, \quad \boldsymbol{f} = \boldsymbol{\nu}_1 - (\boldsymbol{1}^{\mathsf{T}}\boldsymbol{\nu}_1)\boldsymbol{\nu}_0, \quad \sigma^2 = \boldsymbol{1}^{\mathsf{T}}\boldsymbol{\nu}_2 + \boldsymbol{1}^{\mathsf{T}}\boldsymbol{\nu}_1 - (\boldsymbol{1}^{\mathsf{T}}\boldsymbol{\nu}_1)^2. \tag{B3}$$

Differentiating (B3) with respect to t and using (B1), one recovers equations (7)–(9).

References

1. Becker, J., Brackbill, D., Centola, D.: Network dynamics of social influence in the wisdom of crowds. Proc. Natl. Acad. Sci. **114**(26), E5070–E5076 (2017)
2. Beica, A., Guet, C.C., Petrov, T.: Efficient reduction of kappa models by static inspection of the rule-set. In: Abate, A., Šafránek, D. (eds.) HSB 2015. LNCS, vol. 9271, pp. 173–191. Springer, Cham (2015). https://doi.org/10.1007/978-3-319-26916-0_10
3. Bintu, L.: Transcriptional regulation by the numbers: applications. Curr. Opin. Genet. Dev. **15**(2), 125–135 (2005)
4. Bo, S., Celani, A.: Multiple-scale stochastic processes: decimation, averaging and beyond. Phys. Rep. **670**, 1–59 (2017)
5. Cardelli, L., Kwiatkowska, M., Laurenti, L.: Stochastic analysis of chemical reaction networks using linear noise approximation. Biosystems **149**, 26–33 (2016)
6. Gardner, T.S., Cantor, C.R., Collins, J.J.: Construction of a genetic toggle switch in Escherichia coli. Nature **403**(6767), 339 (2000)
7. Gillespie, D.T.: Exact stochastic simulation of coupled chemical reactions. J. Phys. Chem. **81**, 2340–2361 (1977)
8. Gjorgjieva, J., Drion, G., Marder, E.: Computational implications of biophysical diversity and multiple timescales in neurons and synapses for circuit performance. Curr. Opin. Neurobiol. **37**, 44–52 (2016)
9. Goban, A.N., Radulescu, O.: Dynamic and static limitation in multiscale reaction networks, revisited. Adv. Chem. Eng. **34**, 103–107 (2008)
10. Greenham, K., McClung, C.R.: Time to build on good design: resolving the temporal dynamics of gene regulatory networks. Proc. Natl. Acad. Sci. **115**(25), 6325–6327 (2018)

11. Guet, C., Henzinger, T.A., Igler, C., Petrov, T., Sezgin, A.: Transient memory in gene regulation. In: Bortolussi, L., Sanguinetti, G. (eds.) CMSB 2019. LNCS, vol. 11773, pp. 155–187. Springer, Cham (2019). https://doi.org/10.1007/978-3-030-31304-3_9

12. Guet, C.C., Elowitz, M.B., Hsing, W., Leibler, S.: Combinatorial synthesis of genetic networks. Science **296**(5572), 1466–1470 (2002)

13. Gunawardena, J.: Time-scale separation-Michaelis and Menten's old idea, still bearing fruit. FEBS J. **281**(2), 473–488 (2014)

14. da Costa Pereira Innocentini, G., Forger, M., Ramos, A.F., Radulescu, O., Hornos, J.E.M.: Multimodality and flexibility of stochastic gene expression. Bull. Math. Biol. **75**(12), 2360–2600 (2013)

15. Kevorkian, J., Cole, J.D.: Perturbation Methods in Applied Mathematics. Springer, New York (1981). https://doi.org/10.1007/978-1-4757-4213-8

16. Kevorkian, J., Cole, J.D., Nayfeh, A.H.: Perturbation methods in applied mathematics. Bull. Am. Math. Soc. **7**, 414–420 (1982)

17. Kurtz, T.G.: Solutions of ordinary differential equations as limits of pure jump Markov processes. J. Appl. Prob. **7**(1), 49–58 (1970)

18. Kurtz, T.G.: Limit theorems for sequences of jump Markov processes approximating ordinary differential processes. J. Appl. Prob. **8**(2), 344–356 (1971)

19. Kwok, R.: Five hard truths for synthetic biology. Nature **463**(7279), 288–290 (2010)

20. Lorenz, J., Rauhut, H., Schweitzer, F., Helbing, D.: How social influence can undermine the wisdom of crowd effect. Proc. Natl. Acad. Sci. **108**(22), 9020–9025 (2011)

21. Marchisio, M.A., Stelling, J.: Automatic design of digital synthetic gene circuits. PLoS Comput. Biol. **7**(2), e1001083 (2011)

22. McAdams, H.H., Arkin, A.: It's a noisy business! genetic regulation at the Nanomolar scale. Trends Genet. **15**(2), 65–69 (1999)

23. Murray, J.D.: Mathematical Biology: I. Springer, Introduction (2003)

24. Myers, C.J.: Engineering Genetic Circuits. CRC Press, Boca Raton (2009)

25. Newby, J., Chapman, J.: Metastable behavior in Markov processes with internal states. J. Math. Biol. **69**(4), 941–976 (2013). https://doi.org/10.1007/s00285-013-0723-1

26. Pagliara, R., Leonard, N.E.: Adaptive susceptibility and heterogeneity in contagion models on networks. IEEE Trans. Automatic Control (2020)

27. Pájaro, M., Otero-Muras, I., Vázquez, C., Alonso, A.A.: Transient hysteresis and inherent stochasticity in gene regulatory networks. Nat. Commun. **10**(1), 1–7 (2019)

28. Parmar, K., Blyuss, K.B., Kyrychko, Y.N., Hogan., S.J.: Time-delayed models of gene regulatory networks. In: Computational and Mathematical Methods in Medicine (2015)

29. Peleš, S., Munsky, B., Khammash, M.: Reduction and solution of the chemical master equation using time scale separation and finite state projection. J. Chem. Phys. **125**(20), 204104 (2006)

30. Rothenberg, E.V.: Causal gene regulatory network modeling and genomics: second-generation challenges. J. Comput. Biol. **26**(7), 703–718 (2019)

31. Santillán, M., Mackey, M.C.: Why the lysogenic state of phage λ is so stable: a mathematical modeling approach. Biophys. J. **86**(1), 75–84 (2004)

32. Schnoerr, D., Sanguinetti, G., Grima, R.: Approximation and inference methods for stochastic biochemical kinetics–a tutorial review. J. Phys. A: Math. Theor. **50**(9), 093001 (2017)

33. Segal, E., Widom, J.: From DNA sequence to transcriptional behaviour: a quantitative approach. Nat. Rev. Genet. **10**(7), 443–456 (2009)

34. Srivastava, R., Haseltine, E.L., Mastny, E., Rawlings, J.B.: The stochastic quasi-steady-state assumption: reducing the model but not the noise. J. Chem. Phys. **134**(15), 154109 (2011)
35. Trofimenkoff, E.A.M., Roussel, M.R.: Small binding-site clearance delays are not negligible in gene expression modeling. Math. Biosci. 108376 (2020)
36. Zhong, Y.D., Leonard, N.E.: A continuous threshold model of cascade dynamics. arXiv preprint arXiv:1909.11852 (2019)
37. Zhou, T., Liu, T.: Quantitative analysis of gene expression systems. Quant. Biol. **3**(4), 168–181 (2015). https://doi.org/10.1007/s40484-015-0056-8

Graphical Conditions for Rate Independence in Chemical Reaction Networks

Élisabeth Degrand, François Fages[(✉)], and Sylvain Soliman

Inria Saclay-Île de France, Palaiseau, France
Francois.Fages@inria.fr

Abstract. Chemical Reaction Networks (CRNs) provide a useful abstraction of molecular interaction networks in which molecular structures as well as mass conservation principles are abstracted away to focus on the main dynamical properties of the network structure. In their interpretation by ordinary differential equations, we say that a CRN with distinguished input and output species computes a positive real function $f : \mathbb{R}_+ \to \mathbb{R}_+$, if for any initial concentration x of the input species, the concentration of the output molecular species stabilizes at concentration $f(x)$. The Turing-completeness of that notion of chemical analog computation has been established by proving that any computable real function can be computed by a CRN over a finite set of molecular species. Rate-independent CRNs form a restricted class of CRNs of high practical value since they enjoy a form of absolute robustness in the sense that the result is completely independent of the reaction rates and depends solely on the input concentrations. The functions computed by rate-independent CRNs have been characterized mathematically as the set of piecewise linear functions from input species. However, this does not provide a mean to decide whether a given CRN is rate-independent. In this paper, we provide graphical conditions on the Petri Net structure of a CRN which entail the rate-independence property either for all species or for some output species. We show that in the curated part of the Biomodels repository, among the 590 reaction models tested, 2 reaction graphs were found to satisfy our rate-independence conditions for all species, 94 for some output species, among which 29 for some non-trivial output species. Our graphical conditions are based on a non-standard use of the Petri net notions of place-invariants and siphons which are computed by constraint programming techniques for efficiency reasons.

1 Introduction

Chemical Reaction Networks (CRNs) are one fundamental formalism widely used in chemistry, biochemistry, and more recently computational systems biology and synthetic biology. CRNs provide an abstraction of molecular interaction networks in which molecular structures as well as mass conservation principles are abstracted away. They come with a hierarchy of dynamic Boolean, discrete,

© Springer Nature Switzerland AG 2020
A. Abate et al. (Eds.): CMSB 2020, LNBI 12314, pp. 61–78, 2020.
https://doi.org/10.1007/978-3-030-60327-4_4

stochastic and differential interpretations [17] which is at the basis of a rich theory for the analysis of their qualitative dynamical properties [3,14,19], of their computational power [8,11,15], and on their relevance as a design method for implementing high-level functions in synthetic biology, using either DNA [10,29] or DNA-free enzymatic reactions [12,32].

In their interpretation by ordinary differential equations, we say that a CRN with distinguished input and output species computes a positive real function $f : \mathbb{R}_+ \to \mathbb{R}_+$, if for any initial concentration x of the input species, the concentration of the output molecular species stabilizes at concentration $f(x)$. The Turing-completeness of that notion of chemical analog computation has been shown by proving that any computable real function can be computed by a CRN over a finite set of molecular species [15].

In the perspective of biochemical implementations with real enzymes however, the strong property of rate independence, i.e. independence of the computed result of the rates of the reactions [33], is a desirable property that greatly eases their concrete realization, and guarantees a form of absolute robustness of the CRN. The set of input/output functions computed by a rate-independent CRNs has been characterized mathematically in [4,9] as the set of piecewise linear functions. However, this does not give any mean to decide whether a given CRN is rate-independent or not.

In this paper, we provide purely graphical conditions on the CRN structure which entail the rate-independence property either for all molecular species or for some output species. These conditions can be checked statically on the reaction hypergraph of the CRN, i.e. on its Petri net structure, or can be used as structural constraints in rate-independent CRN design problems.

Example 1. For instance, the reaction a+b=>c computes at steady state the minimum of a and b, i.e. $c^* = \min(a(0), b(0)) + c(0)$, $a^* = \max(0, a(0) - b(0))$, $b^* = \max(0, b(0) - a(0))$ whatever the reaction rate is. Our graphical condition for rate independence on all species assumes that there is no synthesis reaction, no fork and no loop in the reaction hypergraph (Theorem 2 below). This is trivially the case in this CRN and suffices to prove rate-independence for all species in this example.

Example 2. Similarly, the CRN

```
a   => x+c
b   => y+c
x+y => z
c+z => r
```

assuming $x(0) = y(0) = c(0) = z(0) = r(0) = 0$, computes at steady state the maximum of a and b: $c^* = \max(a(0), b(0))$ (as $a(0)+b(0)-\min(a(0),b(0))$), $x^* = \max(0, a(0) - b(0))$, $y^* = \max(0, b(0) - a(0))$, $z^* = 0$, $r^* = \min(a(0), b(0))$, $a^* = 0$, $b^* = 0$, independently of the reaction rates. Figure 1 shows some trajectories obtained with different values for the mass action law kinetics constants k_1, k_2, k_3, k_4 of the four reactions above, with initial concentrations $a(0) = 3, b(0) = 1$ and 0 for the other species. Here again, our graphical condition is trivially satisfied and demonstrates the rate-independence property of that CRN for all species, by Theorem 2.

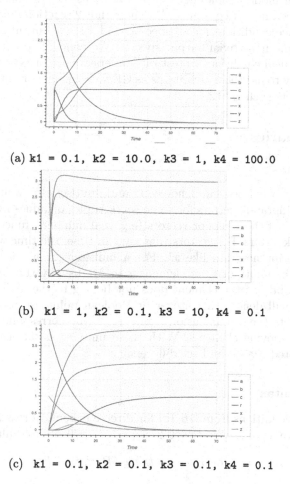

(a) k1 = 0.1, k2 = 10.0, k3 = 1, k4 = 100.0

(b) k1 = 1, k2 = 0.1, k3 = 10, k4 = 0.1

(c) k1 = 0.1, k2 = 0.1, k3 = 0.1, k4 = 0.1

Fig. 1. Computation of max(a, b) with the rate-independent CRN of Example 2 with mass action law kinetics with different reaction rate constants.

The rest of this paper is organized as follows. In Sect. 3, we first give a sufficient condition for the rate independence of output species of the CRN. That condition tests the existence of particular P-invariants and siphons in the Petri net structure of the CRN. This test is modelled as a constraint satisfaction problem, and implemented using constraint programming techniques in order to avoid the enumeration of all P-invariants and siphons that can be in exponential number. Then in Sect. 4, we give another sufficient condition that entails the existence of a unique steady state, and ensures that the computed functions for all species of a CRN are rate-independent. None of these conditions are necessary conditions but we show with examples that they cover a large class of rate-independent CRNs. In Sect. 5, we evaluate our conditions on the curated part of the repository of models BioModels [7] by taking as output species the species that are produced and not consumed. We show that 2 reaction graphs satisfy our rate-independence conditions for all species, 94 for some output species, among which 29 for some non-trivial output species. We conclude on the efficiency of our purely graphical conditions to test rate-independence of existing CRNs, and on the possibility to use those conditions as CRN design constraints for synthetic biology constructs such as [12].

2   Preliminaries

2.1   Notations

Unless explicitly noted, we will denote sets and multisets by capital letters (e.g. S, also using calligraphic letters for some sets), tuples of values by vectors (e.g., $\boldsymbol{x}$), and elements of those sets or vectors (e.g. real numbers, functions) by small Roman or Greek letters. For vectors that vary in time, the time will be denoted using a superscript notation like $\boldsymbol{x}^t$. For a multiset (or a set) $M : S \to \mathbb{N}$, $M(x)$ denotes the multiplicity of element x in M (usually the stoichiometry in the following), and 0 if the element does not belong to the multiset. By abuse of notation, $\geq$ will denote the integer or Boolean pointwise order on vectors, multisets and sets (i.e. set inclusion), and $+$, $-$ the corresponding operations for adding or removing elements. With these unifying notations, set inclusion may thus be noted $S \leq S'$ and set difference $S - S'$.

2.2   CRN Syntax

We recall here definitions from [16,18] for directed chemical reactions networks. In this paper, we assume a finite set $S = \{x_1, \ldots, x_n\}$ of molecular species.

Definition 1. *A reaction over S is a triple (R, P, f), where*

- *R is a multiset of reactants in S,*
- *P a multiset of products in S,*
- *and $f : \mathbb{R}^n \to \mathbb{R}$ is a rate function over molecular concentrations or numbers.*

A chemical reaction network (CRN) $\mathcal{C}$ is a finite set of reactions.

It is worth noting that a molecular species in a reaction can be both a reactant and a product, i.e. a *catalyst*. Those mathematical definitions are mainly compatible with SBML [22], however there are some differences. Unlike SBML, we find it useful to consider only directed reactions (reversible reactions being represented here by two reactions).

Furthermore, we enforce the following compatibility conditions between the rate function and the structure of a reaction:

Definition 2 ([16,18]). *A reaction (R, P, f) over S is* well-formed *if the following conditions hold:*

1. *f is a non-negative partially differentiable function,*
2. *$x_i \in R$ iff $\partial f / \partial x_i(\boldsymbol{x}) > 0$ for some value $\boldsymbol{x} \in \mathbb{R}_+^n$,*
3. *$f(x_1, \ldots, x_n) = 0$ iff there exists $x_i \in R$ such that $x_i = 0$.*

A CRN is well-formed *if all its reactions are well-formed.*

Those compatibility conditions are necessary to perform structural analyses of CRN dynamics. They ensure that the reactants contribute positively to the rate of the reaction at least in some region of the concentration space (condition 2), that the system remains positive (Proposition 2.8 in [16]) and that a reaction stops only when one of the reactant has been entirely consumed, whatever the rate function is.

To analyse the notion of function computed by a CRN, we will study the steady states of the ODE system, i.e., states where $\frac{dx}{dt} = 0$, the flux f_i of each reaction of $\mathcal{C}$ at steady state will be called its *steady flux*.

A directed weighted bipartite graph $\mathcal{G}_\mathcal{C}$ can be naturally associated to a chemical reaction network $\mathcal{C}$, with species and reactions as vertices, and stoichiometric coefficients, i.e. multiplicity in the multisets R and P, as weights for the incoming/outgoing edges.

Example 3. Figure 2 shows the bipartite graph $\mathcal{G}_\mathcal{C}$ of the Example 2 of the introduction. For this graph, the weights are all 1 and are not written for that reason.

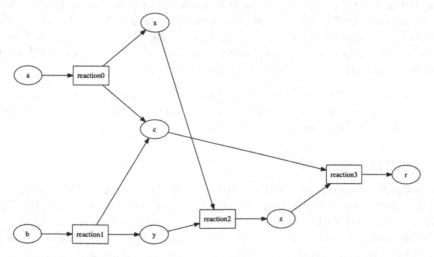

Fig. 2. Bipartite graph $\mathcal{G}_\mathcal{C}$ associated to the CRN given in Example 2. The weights are all equal to 1 and not displayed.

2.3 CRN Semantics

As detailed in [17], a CRN can be interpreted in a hierarchy of semantics with different formalisms that can be formally related by abstraction relationships in the framework of abstract interpretation [13]. In this article, we consider the *differential semantics* which associates with a CRN $\mathcal{C}$ the system ODE($\mathcal{C}$) of Ordinary Differential Equations

$$\frac{dx_j}{dt} = \sum_{(R_i, P_i, f_i) \in \mathcal{C}} (P_i(j) - R_i(j)) \cdot f_i$$

Example 4. Assuming mass action law kinetics for the CRN of Example 2, the ODEs are:

$$da/dt = -k_1 \cdot a \tag{1}$$
$$db/dt = -k_2 \cdot b \tag{2}$$
$$dc/dt = k_1 \cdot a + k_2 \cdot b - k_4 \cdot c \cdot z \tag{3}$$
$$dr/dt = k_4 \cdot c \cdot z \tag{4}$$
$$dx/dt = k_1 \cdot a - k_3 \cdot x \cdot y \tag{5}$$
$$dy/dt = k_2 \cdot b - k_3 \cdot x \cdot y \tag{6}$$
$$dz/dt = k_3 \cdot x \cdot y - k_4 \cdot c \cdot z \tag{7}$$

Definition 3 [15]. *The function of time computed by a CRN $\mathcal{C}$ from initial state $x \in \mathbb{R}^n$ is, if it exists, the solution of the ODE associated to $\mathcal{C}$ with initial conditions $x \in \mathbb{R}^n$*

Definition 4 [15]. *The input/output function computed by a CRN with n species, on an output species z, a set of m input species $y \in \mathbb{R}^m$ and a fixed initial state for the other species $x \in \mathbb{R}^{n-m}$ is, if it exists, the function $f : \mathbb{R}^m \to \mathbb{R}$ for which the ODEs associated to $\mathcal{C}$ have a solution which moreover stabilizes on some value $f(x, y)$ on the z species component.*

Definition 5. *A CRN is rate-independent on an output species z if the input/output function computed on z with all species considered as input does not depend of the rate functions of the reactions.*

2.4 Petri Net Structure

The bipartite graph $\mathcal{G}_\mathcal{C}$ of a CRN can be naturally seen as a Petri-net graph [5,6,30], here used with the continuous Petri-net semantics [20,21,31]. The species correspond to places and the reactions to transitions of the Petri-net. We recall here some classical Petri-net concepts [24,28] used in the next section, since they may have various names depending on the community.

Definition 6. *A minimal semi-positive P-invariant is a vector of $\mathbb{N}^n$ that is in the left-kernel of the stoichiometric matrix. Equivalently it is a weighted sum over places concentrations that remains constant by any transition.*

A P-surinvariant is a weighted sum that only increases.

The support *of a P-invariant or P-surinvariant is the set of places with non-zero value. Those places will be said to be* covered *by the P-invariant or P-surinvariant.*

Intuitively a P-invariant is a conservation law of the CRN. The notion of P-surinvariant will be used to identify the output species of a CRN.

Definition 7. *A siphon is a set of places such that for each edge from a transition to any place of the siphon, there is an edge from a place of the siphon to that transition.*

Intuitively a siphon is a set of places that once empty remains empty, i.e., a set of species that cannot be produced again once they have been completely consumed. Our first condition for rate independence will be based on the following.

Definition 8. *A critical siphon is a siphon that does not contain the support of any P-invariant.*

A siphon that is not critical contains the support of a P-invariant, therefore it cannot ever get empty. A critical siphon on the other hand is thus a set of species that might disappear completely and then always remain absent.

3 Rate Independence Condition for Persistent Outputs

The persistence concept has been introduced to identify Petri nets for which places remain non-zero [1]. Here we establish a link between this notion of persistence and the rate-independence property of the input/output function computed on some output species.

3.1 Sufficient Graphical Condition

As in [2], we are interested by the persistence not of the whole CRN but of some species. We will say that a species is an output of a CRN if it is produced and not consumed (and thus can only increase), i.e. if the stoichiometry of that species in the product part of any reaction is greater or equal to the reactant part, or equivalently:

Definition 9. *A species is an* output *of a CRN if it is the singleton support of a P-surinvariant.*

Example 5. In the CRN of Example 2, the *max* is computed on a non-output node, c. The *min* is computed on an output node r. The rate independence of that CRN on r follows from Theorem 1 below.

Definition 10. *A species is* structurally persistent *if it is covered by a P-invariant and does not belong to any critical siphon.*

Such species' concentrations will not reach zero for well-formed CRNs as proved in [1,2], but this section shows that if they are also output species they converge to a value that is independent of the rates of reactions. Note that such species might still belong to some non-critical siphons, for instance siphons that cover the whole P-invariant it is part of.

Theorem 1. *If a species p of a well-formed CRN is a structurally persistent output, then that CRN is rate-independent on p.*

Proof. Since p is structurally persistent, it is covered by some P-invariant and therefore bounded. Since p is an *output species* and the CRN is well-formed, $\frac{dp}{dt} \geq 0$. Hence its concentration converges to some value p^*.

When p reaches that steady state, all incoming reactions that modify it have null flux, hence by well-formedness one of their reactants has 0 concentration. If there are only incoming reactions that do not affect p then it is trivially constant and therefore rate-independent. Otherwise there are some such incoming reactions with a null reactant.

Now, notice that Prop. 1 of [1] states, albeit with completely different notations, that if one species of a well-formed CRN reaches 0 then all the species of a siphon reach 0. Therefore there exists a whole siphon S containing that reactant and with 0 concentration (intuitively, this reactant also has its input fluxes null, and one can thus build recursively a whole siphon).

By construction, $S' = \{p\} \cup S$ is also a siphon, and since p is persistent, S' is not critical. S' therefore covers some P-invariant P and all concentrations are null except that of p in S'. Now P necessarily covers p since otherwise its conservation would be violated by having all 0 concentrations.

Note that by definition, for each P-invariant V containing p we have at any time t with state vector $\boldsymbol{x}^t$ that $V \cdot \boldsymbol{x}^t = V \cdot \boldsymbol{x}^0$. Hence:

$$p(t) = \boldsymbol{x}^t(p) = \frac{V^0 - V \cdot \boldsymbol{x}^t_{p \to 0}}{V(p)} \leq \frac{V^0}{V(p)}$$

where $\boldsymbol{x}^t_{p \to 0}$ is the state vector except for the concentration of p replaced by 0, and V^0 is a shorthand for $V \cdot \boldsymbol{x}^0$.

At steady state we get $p^* = \frac{P^0}{P(p)}$ since we proved that all concentrations other than that of p are null. Hence, we have:

$$p^* = \min_W \frac{V^0}{V(p)}$$

where $W = \{V \mid V$ is a P-invariant covering $p\}$, and which is obviously rate-independent. $\qquad \square$

3.2 Constraint-Based Programming

It is well-known that there may be an exponential number of P-invariants and siphons in a Petri net. Therefore, it is important to combine the constraints of both structural conditions for the computation of the minimal P-invariants and the union of critical siphons, without computing all siphons and P-invariants. This is the essence of constraint programming and of constraint-based modeling of such a decision problem as a constraint satisfaction problem. Furthermore, deciding the existence of a minimal siphon containing a given place is an NP-complete problem for which constraint programming has already shown its practical efficiency for enumerating all minimal siphons in BioModels, see [26].

We have thus developed a *constraint program* dedicated to the computation of structurally persistent species. For the minimal P-invariants, the constraint solving problem is the same as in [34] and is quite efficient on CRNs. For the second part about critical siphons, we use a similar approach but with Boolean variables to represent our siphons as in [26]. However, we enumerate *maximal* siphons here. This amounts to enumerate values 1 before 0, and to add in the branch-and-bound procedure for optimization that each new siphon must include at least one new place. Furthermore, we add the constraint that they are critical: for each P-invariant P, one of the species of its support must be absent (0). We get the flexibility of our constraint-based approach to add this kind of supplementary constraint while keeping some of the efficiency already demonstrated before.

In Sect. 5, this constraint program is used to compute the set of outputs and check if they are structurally persistent for many models of the `biomodels.net` repository. There are however a few models on which our constraint program is quite slow. An alternative constraint solving technique to solve those hard instances could be to use a SAT solver, at least for the enumeration of critical siphons, as shown in [26].

4   Global Rate Independence Condition

Theorem 1 above can be used to prove the rate-independence property on some output species of a CRN, like r in Example 2 for computing the max, but not on some intermediate species, like c for computing min. In this section we provide a sufficient condition for proving the rate-independence of a CRN on all species.

4.1   Sufficient Graphical Condition

Definition 11. *A chemical reaction network C is* synthesis-free *if for all reactions (R_i, P_i, f_i) of C we have $R_i \not\leq P_i$.*

In other words any reaction need to consume something to produce something.

Definition 12. *A chemical reaction network C is* loop-free *if there is no circuit in its associated graph $\mathcal{G}_C$.*

Definition 13. *A chemical reaction network C is* fork-free *if for all species $x \in S$ there is at most one reaction (R_i, P_i, f_i) such that $R_i(x) > 0$.*

This is equivalent to saying that the out-degree of species vertices is at most one in $\mathcal{G}_C$.

Definition 14. *A funnel CRN is a CRN that is:*

1. *synthesis-free*
2. *loop-free*
3. *fork-free*

In Example 2 for computing the maximum concentration of two input species, A and B, one can easily check that the CRN satisfies the funnel condition (see Fig. 2). More generally, we can prove that any well-formed funnel CRN has a single stable state and that this state does not depend on the precise values of the parameters of the rate functions f_i.

Lemma 1. *The structure of the bipartite graph $\mathcal{G}_C$ of a funnel CRN C is a DAG with leaves that are only species.*

Proof. Since C is loop-free, $\mathcal{G}_C$ is acyclic. Since C is synthesis-free, leaves cannot be reactions. □

Lemma 2. *All steady fluxes of a funnel CRN C are equal to 0.*

Proof. Let us prove the lemma by induction on the topological order of reactions in $\mathcal{G}_C$, this is enough thanks to Lemma 1.

For the base case (smallest reaction in the order), at least one of the species x such that $R_i(x) > P_i(x)$ is a leaf (synthesis-freeness), then notice that at steady state, $\frac{dx}{dt} = 0 = (P_i(x) - R_i(x))f_i$ since there is no production of x as it is a leaf, and no other consumption as C is fork-free. Hence $f_i = 0$.

For the induction case, consider a reactant x s.t. $R_i(x) > P_i(x)$ of our reaction. By induction hypothesis, at steady state we have $\frac{dx}{dt} = 0 = (P_i(x) - R_i(x))f_i$ since all productions of x are lower in the topological order, and there is no other consumption of x as C is fork-free. Hence $f_i = 0$. □

Definition 15. *We shall denote x_i^+ the total amount of species x_i available in an execution of the corresponding ODE system.*

$$x_i^+ = x_i^0 + \int_0^{+\infty} \frac{dx_i^+}{dt} = x_i^0 + \int_0^{+\infty} \sum_{P_j(x_i) > R_j(x_i)} (P_j(x_i) - R_j(x_i))f_j$$

Lemma 3. *Let C be a well-formed funnel CRN, then for each initial state $\boldsymbol{x}^0$, if C reaches a steady state $\boldsymbol{x}^*$, then the total amount x_i^+ of any species x_i can be computed and is independent from the kinetic functions f_j of C.*

Proof. Let us proceed by induction on the topological order of species x_i in $\mathcal{G}_C$.

If x_i is a leaf, then since nothing produces it $x_i^+ = x_i^0$.

Now let us look at the induction case for x_i, and consider the set J of reactions producing x_i (i.e., such that $P_j(x_i) > R_j(x_i)$).

From Lemma 2 we know that for all these reactions $f_j = 0$ at stable state, and since C is well-formed, it means that there exists at least one species x_{j_0} such that $x_{j_0}^* = 0$. As C is fork-free and well-formed x_{j_0} has only been consumed by reaction r_j, which led to precisely producing an amount of x_i equal to $x_{j_0}^+(P_j(x_i) - R_j(x_i))/(R_j(x_{j_0}) - P_j(x_{j_0}))$, where $x_{j_0}^+$ is available via induction hypothesis. Note also that $j_0 = \mathrm{argmin}_{x_k | R_j(x_k) > P_j(x_k)} x_k^+(R_j(x_k) - P_j(x_k))$ since the reaction will stop as soon as it has depleted one of its inputs.

Hence $x_i^+ = x_i^0 + \sum_J x_{j_0}^+(P_j(x_i) - R_j(x_i))/(R_j(x_{j_0}) - P_j(x_{j_0}))$, which only depends on the initial state and the stoichiometry. □

Theorem 2. *Let C be a well-formed funnel CRN, then the ODE system associated to C has a single steady state x^* that does not depend on the kinetic functions f_i of C.*

Proof. From proof of Lemma 3 one notices that either x_i is not consumed at all and we have $x_i^* = x_i^+$ or if j is the only reaction consuming x_i, its total consumption is given by $x_{j_0}(R_j(x_{j_0}) - P_j(x_{j_0}))$, with j_0 defined as in the proof of Lemma 3.

These x_i^* do not depend on the kinetic functions f_i of C.

We prove now that every x_i is convergent. It can be first noticed that

$$x_i(t) = x_{i_0} + F_e(t) - F_s(t)$$

where $F_e(t) = \int_0^t \sum_{P_j(x_i) > R_j(x_i)} (P_j(x_i) - R_j(x_i)) f_j$ is the incoming flux and $F_s(t) = \int_0^t (P_k(x_i) - R_k(x_i)) f_k$ is the outgoing flux.

F_e is the integral of a positive quantity, it is then an increasing function. Moreover, as $F_e(t) \leq x_i^+$, this function is bounded and then converges to a real number limit.

Similarly, F_s is increasing and, as $x_i(t) \geq 0$, we have $F_s(t) \leq x_i^+$ then it is bounded and converges.

To conclude, x_i is a difference of two convergent functions, hence it converges to a real number.

Corollary 1. *Any well-formed funnel CRN is rate-independent for any output species.*

We have thus given here a sufficient condition for a very strong notion of *rate-independence* in which all the species of the CRN have a steady state independent of the reaction rates, as in Example 2.

4.2   Necessary Condition

Our sufficient condition is not a necessary condition for global rate independence. Basically, forks that join and circuits that leak do not prevent rate independence:

Example 6. The CRN

```
a=>b.
b=>a.
b=>c.
```

is not a funnel CRN as it has both a loop (formed by a and b) and a fork (b is a reactant in two distinct reactions). Nevertheless, this CRN is rate-independent on all species. The circuit formed by a and b has a leak with the third reaction. Every molecule of a and b will thus be finally transformed into c whatever the reaction kinetics are. At the steady state, the concentration of a and b will be null, and the concentration of c will be the sum of all the initial concentrations.

Nevertheless, we can show that any function computable by a rate independent CRN can be computed by a funnel CRN. We first show that funnel CRNs are composable under certain conditions for rate independent CRNs, similarly to the composability conditions given in [4].

Definition 16. *Two CRNs C_1 and C_2 are composable if*

$$(\bigcup_{(R,P,f)\in C_1} R\cup P)\cap(\bigcup_{(R',P',f')\in C_2} R'\cup P') = \{x\}$$

i.e., there is a single species appearing in both sets of reactions.

The composition of C_1 and C_2 is the union of their sets of reactions. The species x is called the link between both CRNs.

Lemma 4. *The composition of two funnel CRNs is a funnel CRN if the composition does not create forks on their link.*

Proof. As the reaction rates of the two original CRNs are well-formed, the reaction rates of the resultant CRN are well-formed too. No synthesis and no loop can be created by the union of two CRNs as all species are different except for the link. Therefore, the condition to create no fork by composition is sufficient to ensure that the resultant CRN is a funnel CRN.

Corollary 2. *The composition of two funnel CRNs is a funnel CRN if the link x is reactant in at most one of the CRNs.*

Proof. Since both CRNs are funnel, x is a reactant in at most one reaction in each. Now from our hypothesis it is not reactant at all in one of the CRNs, hence it appears as reactant in at most one reaction and therefore in no fork. By Lemma 4, the resulting CRN is a funnel CRN.

Theorem 3. *Any function computable by a rate independent CRN is computable by a funnel CRN.*

Proof. Using the same theorem from Ovchinnikov [27] as in [9] we note that any such function f with components $f_j, 1 \le j \le p$ can be written as $f(x) = \max_{1\le i\le q} \min_{j\in S_i} f_j(x)$ for some family $S_i \le \{1,\ldots,p\}$.

Each f_j is rational linear, so this function can be written: $f_j(x) = \sum_1^n \frac{\alpha_{j,i}}{n_j} x_i$. To compute this linear sum, the following reactions are needed: For every x_i, we add the reactions $x_i => \alpha_{j,i} \cdot w_j$ which compute $w = \sum_1^n \frac{\alpha_{j,i}}{x} i$.

Then we add the reaction $n_j \cdot w_j => y_j$ which compute $y_j = \frac{1}{n_j} w_j$.

The output of the CRN that computes a linear function is a funnel CRN. Both max and min can be written with a funnel CRN (see respectively Examples 2 and 2) and min can be composed by max as the output of min is not a reactant in the CRN that computes min. From Corollary 2, the conclusion is immediate.

5 Evaluation on Biomodels

In this section, we evaluate our sufficient condition for rate-independence on the reaction graphs of the curated part of the repository of models BioModels [7]. These models are numbered from BIOMD0000000001 to BIOMD0000000705. After excluding the empty models (i.e. models with no reactions or species), 590 models have been tested in total. As already noted in [16] however, many models in the curated of BioModels come from ODE models that have not been transcribed in SBML with well-formed reactions. Basically, some species appearing in the kinetics are missing as reactants or modifiers in the reactions, or some kinetics are negative. In this section, we test our graphical conditions for rate independence on the reaction graphs given for those models, without rewriting the structure of the reactions when they were not well-formed. Therefore, the actual rate-independence of the models that satisfy our sufficient criteria is conditioned to the well-formedness of the CRN.

The evaluation has been performed using Biocham[1] with a timeout of 240 s. The computer used for the evaluation has a quad-processor Intel X3.07 GHz with 8 Gb of RAM.

5.1 Computation of Rate-Independent Output Species

Following Definition 9, we tested the species that constitute the singleton support of a P-surinvariant. Among the 590 models tested, 340, i.e. 57.6% of them, were found to have no output species. 94 models, i.e. 15.9% of the models, were found to have at least one rate-independent output. 27 models, i.e. 4.5%, have both one rate-independent output and one undecided output, i.e. an output not satisfying our sufficient condition. 86 models, i.e. 14.5%, have at least one undecided output.

It is worth noting however that the species that are never modified by a reaction, i.e. that are only catalysts, remain always constant and thus constitute trivial rate-independent outputs. Amongst the 94 models with at least one rate-independent output found during evaluation, 29 have at least one non-trivial rate-independent output. Table 1 gives some details on the size and computation time for those 29 models.

---

[1] All our experiments are available on https://lifeware.inria.fr/wiki/Main/Software# CMSB20b.

Table 1. Model numbers in Biomodels containing non-trivial structurally persistent output species which are thus rate-independent by Theorem 1. For each model, we indicate the numbers of species, reactions, rate-independent species, non-trivial rate-independent species and total computation time in seconds.

Biomodel#	#species	#reactions	#outputs	#RI	#NTRI	NTRI-species	Time (s)
037	12	12	2	2	2	Yi, Pi	0.950
104	6	2	3	3	1	species_4	0.074
105	39	94	11	3	1	AggP_Proteasome	63.366
143	20	20	4	1	1	MLTH_c	3.333
178	6	4	1	1	1	lytic	0.139
227	60	57	2	1	1	s194	17.299
259	17	29	1	1	1	s10	2.308
260	17	29	1	1	1	s10	2.310
261	17	29	1	1	1	s10	2.297
267	4	3	1	1	1	lytic	0.086
283	4	3	1	1	1	Q	0.053
293	136	316	14	4	3	aggE3, aggParkin, AggP_Proteasome	>240
313	16	16	4	2	1	IL13_DecoyR	2.071
336	18	26	1	1	1	IIa	4.148
344	54	80	7	2	1	AggP_Proteasome	>240
357	9	12	1	1	1	T	0.561
358	12	9	4	2	1	Xa_ATIII	0.892
363	4	4	1	1	1	IIa	0.067
366	12	9	4	2	1	Xa_ATIII	0.901
415	10	5	7	7	7	s10, s11, s12, s13, s14, s9, s15	0.894
437	61	40	22	8	1	T	16.109
464	14	10	6	3	1	s12	2.282
465	16	14	5	5	1	s23	59.554
525	18	19	8	3	1	p18inactive	33.479
526	18	19	8	3	1	p18inactive	33.858
540	22	11	12	11	8	s14, s15, s16, s17, s18, s19, s20, s21	56.134
541	37	32	13	9	7	s14, s15, s16, s17, s18, s19, s21	31.573
559	90	136	18	2	2	s493, s502	150.954
575	76	58	9	1	1	DA_GSH	66.806

Now, evaluating by simulation the actual rate-independence property of those models, and thereby the empirical completeness of our purely graphical criterion in this benchmark, would raise a number of difficulties. First, as said above, many SBML models coming from ODE models have not been properly transcribed with well-formed reactions and would need to be rewritten [16]. Second, some models may contain additional events or assignment rules which are not reflected in the CRN reaction graph. Third, the relevant time horizon to consider for simulation is not specified in the SBML file. In the curated part of BioModels, this time horizon can range from 20 s to 1 000 000 s.

Nevertheless, we performed some manual testing on 9 models from Table 1, namely models 37, 104, 105, 143, 178 and 227, which have at least one non-trivial rate-independent output, and models 50, 52 and 54, which have only undecided outputs. For each model, numerical simulations were done with two different sets of initial concentrations and two different sets of parameters. Even when it was not the case in the original models, all the parameters were set to positive values. All outputs in models 37 and 104 were found rate-independent which was confirmed by numerical simulation. For model 105, 3 outputs among the 11 outputs of this model were found rate-independent by our algorithm which seemed again to be confirmed by numerical simulation. Models 143 and 227 are not well-formed which explains why the species satisfying our graphical criterion were shown not be rate-independent by numerical simulation. For models with only undecided outputs, i.e. models 50, 52 and 54, numerical simulations show that none of their outputs is rate-independent. For these 3 models, 11 undecided outputs were tested in total. In this manual testing, we did not find any output that was left undecided by the algorithm and was found rate-independent by numerical simulation.

5.2   Test of Global Rate-Independence

In this section, we test the criterion given in Definition 14 that ensures the rate-independence of all the species of a given CRN.

On the 590 reaction models tested, 20 models have reached the time-out limit of 240 s and were therefore not evaluated. Two models were found to be rate-independent on all species, namely models BIOMD0000000178 and BIOMD0000000267. These models constitute a chain of respectively 4 and 3 species. At steady state, all species have a null concentration, except the last one. The steady state value of the last species is equal to the sum of all the initial concentrations. These models simulate the onset of paralysis of skeletal muscles induced by botulinum neurotoxin serotype A. They are used in particular to get an upper time limit for inhibitors to have an effect [25].

These two models were also found to have rate-independent outputs during the evaluation of the previous criterion for outputs. The global criterion here shows that not only the output species of the chain are rate-independent, but also all the inner species of the chain.

6   Conclusion

We have given two graphical conditions for verifying the rate-independence property of a chemical reaction network. First, the absence of synthesis, circuit and fork in the reaction graph, ensures the existence of a single steady state that does not depend on the reaction rates, thereby ensuring the existence of a computed input/output function for all species of the CRN and their independence of the rate of the reactions. Second, the covering of a given output species by one P-invariant and no critical siphon, provides a criterion to ensure the rate-independence property of the computed function on that output species.

These graphical conditions are sufficient but none of them is necessary. Evaluation in BioModels suggests however that they are already quite powerful since among the 590 models of the curated part of BioModels tested, 94 reaction graphs were found rate-independent for some output species, 29 for non-trivial output species, and 2 for all species which was confirmed for well-formed models.

It is worth noting that our second condition uses the classical Petri net notions of P-invariant and siphons in a non-standard way for continuous systems. A similar use has already been done for instance in [1] for the study of persistence and monotone systems, and interestingly in [23], where the authors remark the discrepancy there is on the Petri net property of trap between the standard discrete interpretation, under which a non empty trap remains non empty, and the continuous interpretation under which a non empty trap may become empty. This shows the remarkable power of Petri net notions and tools for the study of continuous dynamical systems, thus beyond standard discrete Petri nets and outside Petri net theory properly speaking.

As already remarked in previous work [26,34], modeling the computation of Petri net invariants, siphons and other structural properties as a constraint satisfaction problem provides efficient implementations using general purpose constraint solvers, often showing better efficiency than with dedicated algorithms. This was illustrated here by the use of a constraint logic program to implement our condition on P-invariants and critical siphons by constraining the search to those sets of places that satisfy the condition, without having to actually compute the sets of all P-invariants and critical siphons.

Finally, it is also worth noting that beyond verifying the rate-independence property of a CRN and identifying the output species for which the computed function is rate-independent, our graphical conditions may also be considered as structural constraints to satisfy for the design of rate-independent CRNs in synthetic biology [12]. They should thus play an important role in CRN design systems in the future.

Acknowledgement. This work was jointly supported by ANR-MOST *BIOPSY Biochemical Programming System* grant ANR-16-CE18-0029 and ANR-DFG *SYMBIONT Symbolic Methods for Biological Networks* grant ANR-17-CE40-0036.

References

1. Angeli, D., Leenheer, P.D., Sontag, E.D.: A Petri net approach to persistence analysis in chemical reaction networks. In: Queinnec, I., Tarbouriech, S., Garcia, G., Niculescu, S.I. (eds.) Biology and Control Theory: Current Challenges. LNCIS, vol. 357, pp. 181–216. Springer, Heidelberg (2007). https://doi.org/10.1007/978-3-540-71988-5_9
2. Angeli, D., Leenheer, P.D., Sontag, E.D.: Persistence results for chemical reaction networks with time-dependent kinetics and no global conservation laws. In: Proceedings of the 48h IEEE Conference on Decision and Control (CDC), pp. 4559–4564. IEEE (2009)

3. Baudier, A., Fages, F., Soliman, S.: Graphical requirements for multistationarity in reaction networks and their verification in biomodels. J. Theor. Biol. **459**, 79–89 (2018). https://hal.archives-ouvertes.fr/hal-01879735

4. Chalk, C., Kornerup, N., Reeves, W., Soloveichik, D.: Composable rate-independent computation in continuous chemical reaction networks. In: Češka, M., Šafránek, D. (eds.) CMSB 2018. LNCS, vol. 11095, pp. 256–273. Springer, Cham (2018). https://doi.org/10.1007/978-3-319-99429-1_15

5. Chaouiya, C.: Petri net modelling of biological networks. Brief. Bioinform. **8**(4), 210–219 (2007)

6. Chaouiya, C., Remy, E., Thieffry, D.: Petri net modelling of biological regulatory networks. J. Discret. Algorithms **6**(2), 165–177 (2008)

7. Chelliah, V., Laibe, C., Novère, N.: Biomodels database: a repository of mathematical models of biological processes. In: Schneider, M.V. (ed.) In Silico Systems Biology, Methods in Molecular Biology, vol. 1021, pp. 189–199. Humana Press (2013)

8. Chen, H.L., Doty, D., Soloveichik, D.: Deterministic function computation with chemical reaction networks. Nat. Comput. **7433**, 25–42 (2012)

9. Chen, H.L., Doty, D., Soloveichik, D.: Rate-independent computation in continuous chemical reaction networks. In: Proceedings of the 5th Conference on Innovations in Theoretical Computer Science, ITCS 2014, pp. 313–326. ACM, New York (2014)

10. Chen, Y., et al.: Programmable chemical controllers made from DNA. Nat. Nanotechnol. **8**, 755–762 (2013)

11. Cook, M., Soloveichik, D., Winfree, E., Bruck, J.: Programmability of chemical reaction networks. In: Condon, A., Harel, D., Kok, J.N., Salomaa, A., Winfree, E. (eds.) Algorithmic Bioprocesses, pp. 543–584. Springer, Heidelberg (2009). https://doi.org/10.1007/978-3-540-88869-7_27

12. Courbet, A., Amar, P., Fages, F., Renard, E., Molina, F.: Computer-aided biochemical programming of synthetic microreactors as diagnostic devices. Mol. Syst. Biol. **14**(4), e7845 (2018)

13. Cousot, P., Cousot, R.: Abstract interpretation: a unified lattice model for static analysis of programs by construction or approximation of fixpoints. In: POPL 1977: Proceedings of the 6th ACM Symposium on Principles of Programming Languages, pp. 238–252. ACM Press, New York, Los Angeles (1977)

14. Craciun, G., Feinberg, M.: Multiple equilibria in complex chemical reaction networks: II. The species-reaction graph. SIAM J. Appl. Math. **66**(4), 1321–1338 (2006)

15. Fages, F., Le Guludec, G., Bournez, O., Pouly, A.: Strong turing completeness of continuous chemical reaction networks and compilation of mixed analog-digital programs. In: Feret, J., Koeppl, H. (eds.) CMSB 2017. LNCS, vol. 10545, pp. 108–127. Springer, Cham (2017). https://doi.org/10.1007/978-3-319-67471-1_7

16. Fages, F., Gay, S., Soliman, S.: Inferring reaction systems from ordinary differential equations. Theor. Comput. Sci. **599**, 64–78 (2015)

17. Fages, F., Soliman, S.: Abstract interpretation and types for systems biology. Theor. Comput. Sci. **403**(1), 52–70 (2008)

18. Fages, F., Soliman, S.: From reaction models to influence graphs and back: a theorem. In: Fisher, J. (ed.) FMSB 2008. LNCS, vol. 5054, pp. 90–102. Springer, Heidelberg (2008). https://doi.org/10.1007/978-3-540-68413-8_7

19. Feinberg, M.: Mathematical aspects of mass action kinetics. In: Lapidus, L., Amundson, N.R. (eds.) Chemical Reactor Theory: A Review, Chap. 1, pp. 1–78. Prentice-Hall (1977)

20. Gilbert, D., Heiner, M.: From petri nets to differential equations – an integrative approach for biochemical network analysis. In: Donatelli, S., Thiagarajan, P.S. (eds.) ICATPN 2006. LNCS, vol. 4024, pp. 181–200. Springer, Heidelberg (2006). https://doi.org/10.1007/11767589_11

21. Heiner, M., Gilbert, D., Donaldson, R.: Petri nets for systems and synthetic biology. In: Bernardo, M., Degano, P., Zavattaro, G. (eds.) SFM 2008. LNCS, vol. 5016, pp. 215–264. Springer, Heidelberg (2008). https://doi.org/10.1007/978-3-540-68894-5_7

22. Hucka, M., et al.: The systems biology markup language (SBML): a medium for representation and exchange of biochemical network models. Bioinformatics **19**(4), 524–531 (2003)

23. Johnston, M.D., Anderson, D.F., Craciun, G., Brijder, R.: Conditions for extinction events in chemical reaction networks with discrete state spaces. J. Math. Biol. **76**(6), 1535–1558 (2018)

24. von Kamp, A., Schuster, S.: Metatool 5.0: fast and flexible elementary modes analysis. Bioinformatics **22**(15), 1930–1931 (2006)

25. Lebeda, F.J., Adler, M., Erickson, K., Chushak, Y.: Onset dynamics of type A botulinum neurotoxin-induced paralysis. J. Pharmacok. Pharmacodyn. **35**(3), 251–267 (2008)

26. Nabli, F., Martinez, T., Fages, F., Soliman, S.: On enumerating minimal siphons in Petri nets using CLP and SAT solvers: theoretical and practical complexity. Constraints **21**(2), 251–276 (2016)

27. Ovchinnikov, S.: Max-min representation of piecewise linear functions. Contrib. Algebra Geom. **43**(1), 297–302 (2002)

28. Peterson, J.L.: Petri Net Theory and the Modeling of Systems. Prentice Hall, New Jersey (1981)

29. Qian, L., Soloveichik, D., Winfree, E.: Efficient turing-universal computation with DNA polymers. In: Sakakibara, Y., Mi, Y. (eds.) DNA 2010. LNCS, vol. 6518, pp. 123–140. Springer, Heidelberg (2011). https://doi.org/10.1007/978-3-642-18305-8_12

30. Reddy, V.N., Mavrovouniotis, M.L., Liebman, M.N.: Petri net representations in metabolic pathways. In: Hunter, L., Searls, D.B., Shavlik, J.W. (eds.) Proceedings of the 1st International Conference on Intelligent Systems for Molecular Biology (ISMB), pp. 328–336. AAAI Press (1993)

31. Sackmann, A., Heiner, M., Koch, I.: Application of Petri net based analysis techniques to signal transduction pathways. BMC Bioinform. **7**, 482 (2006)

32. Schneider, F.S., et al.: Biomachines for medical diagnosis. Adv. Mater. Lett. **11**(4), 1535–1558 (2020)

33. Senum, P., Riedel, M.: Rate-independent constructs for chemical computation. PLoS ONE **6**(6), e21414 (2011)

34. Soliman, S.: Invariants and other structural properties of biochemical models as a constraint satisfaction problem. Algorithms Mol. Biol. **7**, 15 (2012)

Interval Constraint Satisfaction and Optimization for Biological Homeostasis and Multistationarity

Aurélien Desoeuvres[1][(✉)], Gilles Trombettoni[2], and Ovidiu Radulescu[1]

[1] LPHI UMR CNRS 5235, University of Montpellier, Montpellier, France
{aurelien.desoeuvres,ovidiu.radulescu}@umontpellier.fr
[2] LIRMM, University of Montpellier, CNRS, Montpellier, France
gilles.trombettoni@lirmm.fr

Abstract. *Homeostasis* occurs in a biological system when some output variable remains approximately constant as one or several input parameters change over some intervals. When the variable is exactly constant, one talks about absolute concentration robustness (ACR). A dual and equally important property is *multistationarity*, which means that the system has multiple steady states and possible outputs, at constant parameters. We propose a new computational method based on interval techniques to find species in biochemical systems that verify homeostasis, and a similar method for testing multistationarity. We test homeostasis, ACR and multistationarity on a large collection of biochemical models from the Biomodels and DOCSS databases. The codes used in this paper are publicly available at: https://github.com/Glawal/IbexHomeo.

1 Introduction

The 19th century French physiologist Claude Bernard introduced the concept of homeostasis that plays a crucial role in understanding the functioning of living organisms. As he put it, homeostasis, defined as constancy, despite external changes, of the "milieu intérieur" that contains organs, tissues and cells, is a prerequisite of life. A simple example of homeostasis is the constancy of body temperature: our body temperature is maintained in a narrow range around 37° C despite large variation of the environment temperature. Another example is the concentration of many biochemical species (cell processes drivers and regulators such as glucose, ATP, calcium, potassium, cell surface receptors, transcription factors, etc.) whose steady state values are kept constant by tight control. Rather generally, homeostasis refers to constancy of the output w.r.t. variation of parameters or inputs [10]. Several other concepts such as robustness, resilience or viability are closely related to homeostasis and sometimes used with overlapping meaning. Robustness refers to the lack of sensitivity of temporal and static properties of systems w.r.t. parameters and/or initial conditions variation, thus encompassing homeostasis [4,11,26]. Resilience or viability has a more global, dynamical significance, meaning the capacity of systems to recover from perturbations via transient states that stay within bounds [3].

© Springer Nature Switzerland AG 2020
A. Abate et al. (Eds.): CMSB 2020, LNBI 12314, pp. 79–101, 2020.
https://doi.org/10.1007/978-3-030-60327-4_5

In [27], a special type of homeostasis is studied, named absolute concentration robustness (ACR), consisting in invariance of the steady state w.r.t. changes of initial conditions. For chemical reaction network (CRN) models, they proposed a sufficient graph-theoretical criterion for ACR. Replacing invariance by infinitesimal sensitivity, [10] presents a way to detect homeostasis in parametric systems of ordinary differential equations (ODEs). Their approach, based on the singularity theory, was applied to gene circuit models [1]. Our work builds on this reasoning, using a slightly different definition of homeostasis. Instead of looking at the infinitesimal variation, we choose to look at intervals. The steady states of ODE models are computed as solutions of algebraic equations. We are interested in finding intervals containing all the steady state values of model variables. If the variables values are contained inside sufficiently narrow intervals, those variables are stated homeostatic. Sensitivity-like calculations of the input-output relationship compute derivatives of the output w.r.t. the input and boil down to linearizations. In contrast, our method guarantees intervals containing the output w.r.t the initial system. This is crucial in applications, whenever parameters change on wide ranges and models are strongly nonlinear. The interval approach has thus larger applicability.

Our approach also provides a novel method for testing the multistationarity of CRNs, occurring when one or several variables can have several values at the steady state. In this case, we are interested in the number and the actual space position of all the steady states. Multistationarity is an important problem in mathematical biology and considerable effort has been devoted to its study, with a variety of methods: numerical, such as homotopy continuation [28] or symbolic, such as real triangularization and cylindrical algebraic decomposition [6]. However, as discussed in [6], numerical errors in homotopy based methods may lead to failure in the identification of the correct number of steady states, whereas symbolic methods have a double exponential complexity in the number of variables and parameters. As solving a system of algebraic equations is the same as finding intersections of manifolds, each manifold corresponding to an equation, our problem is equivalent to solving a system of constraints.

In this paper we use the interval constraint programming (ICP) and optimization solvers provided by the Ibex (Interval Based EXplorer) tool, for testing homeostasis and multistationarity. Although interval solvers are combinatorial in the worst-case, polynomial-time acceleration algorithms embedded in these solvers generally make them tractable for small or medium-sized systems. Interval constraint satisfaction methods offer an interesting compromise between good precision and low complexity calculations. ICP is an important field in computer science and its interaction with biology has cross-fertilization potential. Although interval methods have already been used in systems biology for coping with parametric models uncertainty [17,31], to the best of our knowledge, this is their first application to homeostasis and multistationarity. In an algorithmic point of view, this paper reports the first (portfolio) distributed variant of the IbexOpt Branch and Bound optimizer, where several variants of the solver are run on different threads and exchange information. Our approach has been tested on two databases, and this benchmarking represents the first systematic study of homeostasis, in particular ACR, on realistic CRN models.

Together, our tools can be used to address numerous problems in fundamental biology and medicine, whenever the stability and controlability of biochemical variables are concerned. In fundamental biology, both homeostasis and multistationarity are key concepts for understanding cell decision making in development and adaptation. In personalized medicine, our tests could be used not only for a better understanding of the loss of homeostasis, for instance in aging and degenerative disease, but also for diagnosis and for predicting the effect of therapy to bring back a normal functioning.

2   Settings and Definitions

Our definition of homeostasis is general and can be applied to any system of ODEs. For all the applications discussed in this paper, the ODEs systems result from chemical reactions whose rates are given either by mass action (as in Feinberg/Shinar's analysis of ACR) or by more general kinetic laws.

We consider thus the variables $x_1, ..., x_n$, representing species concentrations, the parameters $p_1, ..., p_r$, representing kinetic constants, and a set of differential equations :

$$\frac{dx_1}{dt} = f_1(x_1, ..., x_n, p_1, ..., p_r), ..., \frac{dx_n}{dt} = f_n(x_1, ..., x_n, p_1, ..., p_r), \quad (1)$$

where the functions $f_i, 1 \leq i \leq n$ are at least piecewise differentiable.

We are interested in systems that have steady states, $i.e.$ such that the system

$$f_1(x_1, ..., x_n, p_1, ..., p_r) = 0, ..., f_n(x_1, ..., x_n, p_1, ..., p_r) = 0 \quad (2)$$

admits real solutions for fixed parameters $p_1, ..., p_r$. Because $x_1, ..., x_n$ represent concentrations, we constrain our study to real positive solutions.

Generally, it is possible to have one or several steady states, or no steady state at all. The number of steady states can change at bifurcations. For practically all biochemical models, the functions $f_1, ..., f_n$ are rational, and at fixed parameters (2) defines an algebraic variety. The local dimension of this variety is given by the rank defect of the Jacobian matrix $\boldsymbol{J}$, of elements $J_{i,j} = \frac{\partial f_i}{\partial x_j}, 1 \leq i, j \leq n$.

When $\boldsymbol{J}$ has full rank, then by the implicit function theorem, the steady states are isolated points (zero dimensional variety) and all the species are locally expressible as functions of the parameters:

$$x_1 = \Phi_{x_1}(\boldsymbol{p}), ..., x_n = \Phi_{x_n}(\boldsymbol{p}). \quad (3)$$

The functions Φ_y were called input-output functions in [10], where the input is the parameters p_i and the output variable y is any of the variables $x_i, 1 \leq i \leq n$. Also in the full rank case, a system is called *multistationary* when, for fixed parameters there are multiple solutions of (2), $i.e.$ multiple steady states.

$\boldsymbol{J}$ has not full rank in two cases. The first case is at bifurcations, when the system output changes qualitatively and there is no homeostasis. The second case is when (1) has $l \leq n$ independent first integrals, $i.e.$ functions of x that are constant on any solution of the ODEs (1). In this case the Jacobian matrix has rank defect l everywhere and steady states form an l-dimensional variety. For instance, for many biochemical models, there is a full rank constant matrix C such that $\sum_{j=1}^{n} C_{ij} f_j(x_1, ..., x_n, p_1, ..., p_r) = 0$, for all $x_1, ..., x_n, p_1, ..., p_r, 1 \leq i \leq l$. In this case there are l linear conservation laws, *i.e.* $\sum_{i=1}^{n} C_{ij} x_j = k_i, 1 \leq i \leq l$ are constant on any solution. Here k_i depends only on the initial conditions, $k_i = \sum_{i=1}^{n} C_{ij} x_j(0)$. In biochemistry, linear conservation laws occur typically when certain molecules are only modified, or complexified, or translocated from one compartment to another one, but neither synthesized, nor degraded. The constant quantities k_i correspond to total amounts of such molecules, in various locations, in various complexes or with various modifications.

A biological system is characterized not only by its parameters but also by the initial conditions. For instance, in cellular biology, linear conservation laws represent total amounts of proteins of a given type and of their modifications, that are constant within a cell type, but may vary from one cell type to another. Therefore we are interested in the dependence of steady states on initial conditions, represented as values of conservation laws. Because conservation laws can couple many species, steady states are generically sensitive to their values. ACR represents a remarkable exception when steady states do not depend on conservation laws. In order to compute steady states at fixed initial conditions, we solve the extended system

$$f_1(x_1, ..., x_n, p_1, ..., p_r) = 0, \ldots, f_{n-l}(x_1, ..., x_n, p_1, ..., p_r) = 0,$$
$$C_{11}x_1 + ... + C_{1n}x_n = k_1, \ldots, C_{l1}x_1 + ... + C_{ln}x_n = k_l, \tag{4}$$

where k_i are considered as extra parameters, and $f_1, \ldots, f_{n-l}$ are linearly independent functions. In this case, excepting the degenerate steady states with zero concentrations discussed at the end of this section, the Jacobian of the extended system has full rank and one can define again input-output functions as unique solutions of (4). Similarly, when the biochemical models has non-linear conservation laws [9], an independent set of those can be added at the end of the system to obtain a full rank Jacobian. A system is *multistationary* if at fixed parameters there are multiple solutions of (4).

Homeostasis is defined using the input-output functions.

Definition 1. *We say that y is a k_{hom}-homeostatic variable if no bifurcations happen in P and if in the path of steady states given by Φ_y we get :*

$$\frac{\max_{\mathbf{p} \in P}(\Phi_y(\mathbf{p}))}{\min_{\mathbf{p} \in P}(\Phi_y(\mathbf{p}))} \leq k_{hom},$$

where $k_{hom} \geq 1$. We take $k_{hom} = 2$ in this paper, but a different k_{hom} can be used, depending on the tolerance of the biological system to variation. For instance, for human glucose homeostasis $k_{hom} \in [1.32, 1.95]$. P represents the space of parameters (P is compact in our examples), and $\mathbf{p}$ is a point inside P. So, we consider homeostasis of y for any change of parameters in P.

We exclude from our definition trivial solutions $x_i = 0$ obtained when

$$f_i(x_1, ..., x_n, p_1, ..., p_r) = x_i^{n_i} g_i(x_1, ..., x_n, p_1, ..., p_r),$$

where n_i are strictly positive integers, g_i are smooth functions with non-zero derivatives $\partial f_i / \partial x_i$ for $x_i = 0$. These solutions persist for all values of the parameters and are thus trivially robust. In this case we replace the problem $f_i = 0$ by the problem $g_i = 0$ that has only non-trivial solutions $x_i \neq 0$.

3   Interval Methods for Nonlinear Constraint Solving and Optimization

3.1   Intervals

Contrary to standard numerical analysis methods that work with single values, interval methods can manage sets of values enclosed in intervals. By these methods one can handle exhaustively the set of possible constraint systems solutions, with guarantees on the answer. Interval methods are therefore particularly useful for handling nonlinear, non-convex constraint systems.

Definition 2. *An interval* $[x_i] = [\underline{x_i}, \overline{x_i}]$ *defines the set of reals* x_i *such that* $\underline{x_i} \leq x_i \leq \overline{x_i}$. $\mathbb{IR}$ *denotes the set of all intervals. A* **box** $[x]$ *denotes a Cartesian product of intervals* $[x] = [x_1] \times ... \times [x_n]$. *The size or width of a box* $[x]$ *is given by* $w[x] = \max_i(w([x_i]))$ *where* $w([x_i]) = \overline{x_i} - \underline{x_i}$.

Interval arithmetic [22] has been defined to extend to $\mathbb{IR}$ the usual mathematical operators over $\mathbb{R}$ (such as $+$, $\cdot$, $/$, power, sqrt, exp, log, sine). For instance, the interval sum is defined by $[x_1] + [x_2] = [\underline{x_1} + \underline{x_2}, \overline{x_1} + \overline{x_2}]$. When a function f is a composition of elementary functions, an *extension* of f to intervals must be defined to ensure a conservative image computation.

Definition 3. (Extension of a function to $\mathbb{IR}$)
Consider a function $f : \mathbb{R}^n \to \mathbb{R}$.
$[f] : \mathbb{IR}^n \to \mathbb{IR}$ *is said to be an* **extension** *of f to intervals iff:*

$$\forall [x] \in \mathbb{IR}^n \quad [f]([x]) \supseteq \{f(y), \ y \in [x]\}$$
$$\forall x \ \in \mathbb{R}^n \quad f(x) = [f]([x, x])$$

The *natural extension* of a real function f corresponds to the mapping of f to intervals using interval arithmetic. More sophisticated interval extensions have been defined, based on interval Taylor forms or exploiting function monotonicity [13].

3.2   Interval Methods for Constraint Solving

Several interval methods have been designed to approximate *all* the real solutions of equality constraints ($h(x) = 0$) in a domain defined by an initial box $[x]$. These methods build a search tree that explores the search space exhaustively by subdividing $[x]$. The tree built contains a set of nodes, each of them corresponding to a sub-box of $[x]$. At each node, the Branch and Contract process achieves two main operations:

- **Bisection**: The current box is split into two sub-boxes along one variable interval.
- **Contraction**: Both sub-boxes are handled by *contraction* algorithms that can remove sub-intervals without solution at the bounds of the boxes.

At the end of this tree search, the "small" boxes of size less than a user-given precision ϵ contain all the solutions to the equation system. The process is combinatorial, but the contraction methods are polynomial-time acceleration algorithms that make generally the approach tractable for small or medium-sized systems. Without detailing, contraction methods are built upon interval arithmetic and can be divided into constraint programming (CP) [5, 23, 32] and convexification [21, 29] algorithms.

4   Multistationarity

The constraint solving strategy roughly described above is implemented by the `IbexSolve` strategy available in the `Ibex` C++ interval library. `IbexSolve` can find all the solutions of (4) with fixed parameters in a straightforward way.

This method is useful for small and medium systems, and sometimes for large systems, depending on the nature of the constraints and the efficiency of the contractors. Also, it provides as output each solution box. This output is easy to read, because (4) has always a finite set of solutions.

In case of large systems, it can be easier to answer the question: do we have zero, one, or several steady states? In this case, we can use another strategy, described in the next section, where the problem is reformulated in terms of $2n$ constrained global optimization problems: for every variable x_i, we call twice an optimization code that searches for the minimum and the maximum value of x_i while respecting the system (4).

- If the system admits at least two distinct solutions, the criterion used in Definition 1 (using k_{hom} close to 1) will fail for at least one species, *i.e.* we will find a species x_i whose minimum and maximum values are not close to each other.
- If the system admits no solution, the first call to the optimizer (*i.e.*, minimizing x_1) will assert it.
- And if we have only one solution, every species will respect the criterion.

Let us give a simple example given by the model 233 in the Biomodels database [15]. In this model we have two species x and y together with seven parameters (one for the volume of the compartment, four for kinetic rates, and two for assumed fixed species). The system of ODEs is given by:

$$\frac{dx}{dt} = \frac{2k_2k_6y - k_3x^2 - k_4xy - k_5x}{k_1}, \qquad \frac{dy}{dt} = \frac{-k_2k_6y + k_3x^2}{k_1}. \tag{5}$$

After replacing the symbolic parameters by their given values, the steady state equations read:

$$16y - x^2 - xy - \frac{3}{2}x = 0, \qquad -8y + x^2 = 0. \tag{6}$$

The system (6) has two non-zero solutions, given by (6,4.5) and (2,0.5). When the system (6) is tested by IbexHomeo (the dedicated strategy for homeostasis) on a strictly positive box (to avoid the trivial solution (0,0)), we find $x \in [2,6]$ and $y \in [0.5, 4.5]$. The homeostasy criterion fails at fixed parameters and we know that we have multistationarity.

5   IbexHomeo for Finding Homeostatic Species

The new interval solver dedicated to homeostasis proposed in this paper resorts to several calls to optimization processes. Let us first recall the principles behind interval Branch and Bound codes for constrained optimization.

5.1   Interval Branch and Bound Methods for Constrained Global Optimization

Constrained global optimization consists in finding a vector in the domain that satisfies the constraints while minimizing an *objective function*.

Definition 4. (Constrained Global Optimization)
Let $x = (x_1, ..., x_n)$ varying in a box $[x]$, and functions $f : \mathbb{R}^n \to \mathbb{R}$, $g : \mathbb{R}^n \to \mathbb{R}^m$, $h : \mathbb{R}^n \to \mathbb{R}^p$.

Given the system $S = (f, g, h, x, [x])$, the constrained global optimization problem consists in finding f^* :

$$f^* \equiv \min_{x \in [x]} f(x) \text{ subject to } g(x) \leq 0 \text{ and } h(x) = 0.$$

f denotes the objective function (Φ_y in Definition 1), f^* being the objective function value (or best "cost"), g and h are inequality and equality constraints respectively. x is said to be feasible if it satisfies the constraints.

Interval methods can handle constrained global optimization (minimization) problems having non-convex operators with a Branch and Bound strategy generalizing the Branch and Contract strategy described in the previous section. The Branch and Bound solver maintains two bounds lb and ub of f^*. The upper bound ub of f^* is the best (lowest) value of $f(x)$ satisfying the constraints found so far, and the lower bound lb of f^* is the highest value under which it does not exist any solution (feasible point). The strategy terminates when $ub - lb$ (or a relative distance) reaches a user-defined precision ϵ_f. To do so, a variable x_{obj} representing the objective function value and a constraint $x_{obj} = f(x)$ are first added to the system. Then a tree search is run that calls at each node a bisection procedure, a contraction procedure, but also an additional bounding procedure that aims at decreasing ub and increasing lb. Improving lb can be performed by contraction: it is given by the minimum value of x_{obj} over all the nodes in the search tree. Improving the upper bound is generally achieved by local numerical methods. Like any other Branch and Bound method, improving the upper bound ub allows the strategy to eliminate nodes of the tree for which $ub < x_{obj}$.

Remark. Interval Branch and Bound codes can solve the optimization problem defined in Definition 4, but they sometimes require a significant CPU time because of the guarantee on the equality constraints. A way to better tackle the problem in practice is to relax equalities $h(x) = 0$ by pairs of inequalities $-\epsilon_h \leq h(x)$ and $h(x) \leq +\epsilon_h$, where ϵ_h is a user-defined positive parameter. Therefore, in practice, interval Branch and Bound codes generally compute a feasible vector x satisfying the constraints $g(x) \leq 0$ and $-\epsilon_h \leq h(x) \leq +\epsilon_h$ such that $|f^* - f(x)| \leq \epsilon_f$.

The interval Branch and Bound strategy roughly described above is implemented by the `IbexOpt` strategy available in the `Ibex` C++ interval library [7]. `IbexOpt` is described in more details in [24, 30].

5.2   A Dedicated Solver for Homeostasis Based on IbexOpt

Remember that we consider a variable x_i to be homeostatic if it verifies Definition 1. For identifying homeostasis, we consider the system (4) in which the parameters p_i and k_i can vary.

Bi-optimization for a Given Species x_i

Since we want to compute the minimum and the maximum value of $x_i = \Phi_{x_i}(\mathbf{p})$, the homeostasis detection amounts to two optimization problems, one minimizing the simple objective function x_i, and one maximizing x_i, *i.e.* minimizing $-x_i$. The two values returned are finally compared to decide the x_i homeostasis. It is useful to consider that minimizing and maximizing x_i are somehow symmetric, allowing the strategy to transmit bounds of x_i from one optimization process to the dual one. These bounds can also be compared during optimization to stop both optimizations if they give enough information about homeostasis. Indeed, an optimizer minimizing x_i computes $[ln, un] \ni \min(x_i)$, where ln and un are lb and ub of the objective function x_i. An optimizer maximizing x_i computes $[ux, lx] \ni \max(x_i)$, where ux and lx are $-ub$ and $-lb$ of the objective function $-x_i$. Without detailing, lx/ln is an overestimate of the "distance" between any two feasible values of x_i, and a small value states that the species is homeostatic (see Definition 1). Conversely, ux/un is an underestimate of any two feasible values distance, and $ux/un > k_{hom}$ asserts that the species is not homeostatic. This `TestHom` decision procedure is implemented by Algorithm 1.

Algorithm `TestHom`$(un, ln, ux, lx, k_{hom})$
    if $lx/ln \leq k_{hom}$ **then**
         return 2 /* homeostatic variable */
    if $ux/un > k_{hom}$ **then**
         return 1 /* non homeostatic variable */
    else
         return 0 /* not enough information */

Algorithm 1: The `TestHom` decision procedure.

Improving Upper Bounding with Fixed Parameters

The bi-optimization described above runs on the system S corresponding to the system (4), where the equations $f_j(x, p) = 0$ are relaxed by inequalities $-\epsilon_h \leq f_j(x, p) \leq +\epsilon_h$; the parameters p can vary in a box $[p]$ and are added to the set of processed variables. As we are checking for homeostasis, it is important to notice that a steady state is expected for *every* parameter vector $p \in [p]$ (this is not valid, for instance, in the neighborhood of a saddle-node bifurcation, which should be avoided by re-defining $[p]$). We exploit this key point by also running minimization and maximization of x_i on a system S', corresponding to the system S where the parameters have been fixed to a random value $p \in [p]$, with the hope that reducing the parameter space allows a faster optimization. The computed values constitute feasible points for the initial problem (*i.e.*, with parameters that can vary) and can fasten the bi-optimization algorithm described above. Recall indeed that finding feasible points enables to improve the upper bound ub of f^* and to remove from the search tree the nodes with a greater cost.

Overall, homeostasis detection of species x_i is performed by Algorithm 2.

Algorithm Bi-Optimize$(x_i, t, S = (x \times p, [x] \times [p],$ *system* (4), $\epsilon_h),$
$P = (\epsilon_f, t), FP)$

> Execute in parallel until timeout t:
> $(un, ln, FP) \leftarrow$ Minimize$(x_i, S, \min_{x_i}(FP), P)$
> $(ux, lx, FP) \leftarrow$ Minimize$(-x_i, S, \max_{x_i}(FP), P)$
> **while** *true* **do**
>> $S' \leftarrow$ FixRandomParameters(S)
>> $FP \leftarrow$ Minimize$(x_i, S', +\infty, P)$
>
> **return** (un, ln, ux, lx, FP)

Algorithm 2: The double optimization process on a given species x_i. P is the set of solver parameters: ϵ_f is the user-defined precision on the objective function value, t is the timeout required.

All the optimization processes are run in parallel and exchange newly found feasible points stored in FP. Every call to Minimize on S can start with an initial upper bound initialized with the best feasible point found so far ($min_{x_i}(FP)$ or $max_{x_i}(FP)$).

The minimization processes on S' are generally fast so that several ones can be called in a loop (with different parameters fixed to random values) until the end of the main minimization processes on S.

A Portfolio Strategy for the Bi-optimization

It is important to understand that IbexSolve and IbexOpt are generic strategies. That is, different procedures can be selected for carrying out the choice of the next variable interval to bisect (called branching heuristic) or for selecting the next node to handle in the search tree. It is known that some heuristics in general useful can be sometimes bad for some specific problems. Therefore we propose a *portfolio* parallelization strategy where different processes (threads) run Branch and Bound algorithms using different branching heuristics (called *cutters* hereafter) or node selection heuristics (called *nodeSel*). These threads can communicate their bounds to each other, reducing the risks of an ineffective strategy. In practice, we should modify a call to `Minimize` as follows:

$$\text{Minimize}(x_i, S, P, cutters, nodeSel)$$

where *cutters* denotes a set of branching heuristics and *nodeSel* denotes a set of node selection heuristics. This routine calls $|cutters| \times |nodeSel|$ threads, each of them corresponding to one Branch and Bound using one branching heuristic in *cutters* and one node selection heuristic in *nodeSel*. These threads work in the same time on the same problem, but they build different search trees. Therefore one optimizer can compute an *lb* value better (greater) than the others. In this case, it sends it to the other threads.

Heuristics used to split a box are all the variants of the *smear* branching strategy described in [30] and [2]. Strategies used to select the next node to be handled are described in [24]. The cutting strategy *lsmear* [2] is generally more efficient than the others, and will be more often used.

The Main IbexHomeo Algorithm

Finally, because we want to determine all the homeostatic species, we run the double optimization n times, for every species x_i, as shown in Algorithm 3. After a first call to a `FirstContraction` procedure that contracts the domain $[x] \times [p]$, IbexHomeo calls two successive similar loops of different performance. The first loop iterates on every species x_i and calls on it the double optimization function `Bi-Optimize`. The optimization threads are all run using the *lsmear* branching heuristic and have a "short" timeout in order to not be blocked by a given species computation. If a bi-Optimization call on x_i reaches the timeout t without enough information about homeostasis, x_i is stored in L and the computation continues on subsequent species. Since the feasible region defined by S is the same for each optimization, the next iterations can learn (and store in FP) new feasible points than can be exploited by other optimization processes. Therefore the second loop is similar to the first one, but with a greater timeout and more threads in parallel running more various branching heuristics.

To summarize, the `IbexHomeo` algorithm creates communicating threads for:

- exploiting the duality min/max of the bi-optimization related to a given species homeostasis detection,
- finding feasible points more easily,
- running a portfolio of similar Branch and Bound algorithms using different heuristics.

Algorithm IbexHomeo *(S = (x × p, [x] × [p], system (4), ϵ_h), P = (ϵ_f, t, k_{hom}))*
   $cutters \leftarrow \{lsmear\}$
   $nodeSel \leftarrow \{double\_heap, cell\_beam\_search\}$
   $[x] \leftarrow FirstContraction([x], S)$
   $FP \leftarrow \emptyset, L \leftarrow \emptyset$
   foreach $x_i \in x$ **do**
      $(un, ln, ux, lx, FP) \leftarrow$ Bi-Optimize$(x_i, t, S, P, FP, cutters, nodeSel)$
      $[x_i] \leftarrow [un, ux]$
      if *timeout(t)* **and** TestHom*(un, ln, ux, lx, k_{hom})=0* **then**
         $L \leftarrow L \cup x_i$
   $t \leftarrow 10\,t$
   $cutters \leftarrow \{lsmear, smearSum, smearSumRel, smearMax, smearMaxRel\}$
   foreach $x_i \in L$ **do**
      $(un, ln, ux, lx, FP) \leftarrow$ Bi-Optimize$(x_i, t, S, P, FP, cutters, nodeSel)$
      $[x_i] \leftarrow [un, ux]$
   return HomeostaticSpecies *([x], x, k_{hom})*

Algorithm 3: Main frame of IbexHomeo. $k_{hom} \in [1, 2]$ is defined in Definition 1. Via the procedure HomeostaticSpecies, the algorithm returns the set of homeostatic variables.

6 Experimental Results

For benchmarking the multistationarity test we have used DOCSS (Database of Chemical Stability Space, http://docss.ncbs.res.in), a repository of multistationary biochemical circuits. DOCSS contains biochemical circuits with up to four species and up to five catalytic reactions. The catalytic reactions are decomposed into several mass action laws, elementary steps. In DOCSS, the models are specified as short strings of symbols coding for the catalytic reactions and as lists of numeric parameters. These specifications were first parsed to SBML files, then to systems of differential equations and conservation laws using tools developed in [16], and transformed into an input file for our algorithms. For the benchmarking we have selected all the 210 DOCSS circuits with 3 species (denoted a,b,c) and 3 catalytic reactions. The mass action models have up to 6 variables (*i.e.*, the species a,b,c, and several complexes resulting from the decomposition of catalytic reactions into mass action steps). The steady states of all models in DOCSS were numerically computed in [25] using a homotopy continuation method [28]. For all the 3 × 3 models both homotopy and interval IbexSolve methods find 3 or 4 steady states. Although the positions of most of the solutions are almost identical using the two methods (see Fig. 1), there are a few exceptions where the two solutions diverge. We have investigated each of these exceptions. The result is presented in Table 1.

The main reason of discrepancy is a different number of solutions computed by the two methods. For the models with discrepancies we have also computed symbolic steady state solutions using the Symbolic Math Toolbox of Matlab R2013b (MathWorks, Natick, USA), though this was not possible for all the models. The comparison to IbexSolve and homotopy solutions shows that IbexSolve always finds the right number of solutions in a fraction of a second

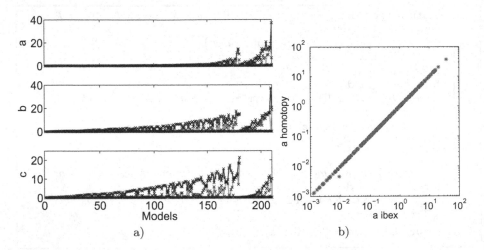

a)   b)

Fig. 1. Comparison between homotopy and `IbexSolve` steady states. All the tested models are multistationary. a) Models were partitioned into two classes, with 3 (appearing first) and 4 homotopy solutions, then sorted by the average of the steady state concentrations in the homotopy solutions. Homotopy and `IbexSolve` solutions are represented as lines (red, green and blue for models with 3 steady states, cyan for the fourth) and crosses, respectively. b) Values of the steady state concentration a computed by homotopy and `IbexSolve`. Each `IbexSolve` steady state was related to the closest homotopy state (red +), in the Euclidean distance sense; reciprocally, each homotopy state was related to the closest `IbexSolve` state (blue crosses) (Color figure online).

Table 1. Comparison of most divergent Ibex vs. homotopy solutions to symbolic solutions. n is the number of steady states. $dist$ is the distance between sets of steady states solutions computed by homotopy or IbexSolve and the symbolic solutions, computed as $\frac{1}{2n}(\sum_{i=1}^{n} \min_j d_{i,j}) + \frac{1}{2n_s}(\sum_{j=1}^{n_s} \min_i d_{i,j})$, where $d_{i,j}$ is the Euclidean distance between the numerical solution i and the symbolic solution j, n_s is either the number of solutions n_h found by the homotopy method or the number n_i found by IbexSolve.

Model	n sym	n_h homo	n_i Ibex	$dist$ homo	$dist$ Ibex
M338-2	3	4	3	0.020087	2.9101e−05
M464-1	3	4	3	0.012388	4.7049e−05
M464-2	3	4	3	0.029019	3.5893e−05
M488-1	3	4	3	0.011739	2.4117e−05
M488-2	3	4	3	0.010935	2.3473e−05
M488-3	3	4	3	0.017165	5.5298e−05
M506-1	4	3	4	0.10086	2.6487e−05
M506-2	4	3	4	0.0069613	4.3250e−06
M506-14	4	3	4	0.0092411	5.0559e−05
M95-1	4	3	4	0.0033133	1.1911e−05
M95-2	4	3	4	0.00063113	1.1230e−05

and computes their positions with better precision than the homotopy method. We conclude that discrepancies result from the failure of the homotopy method to identify the right number of solutions.

The homeostasis tests were benchmarked using the database Biomodels (https://www.ebi.ac.uk/biomodels/), a repository of mathematical models of biological and biomedical systems; parsed from SBML files to systems of differential equations and conservation laws using tools developed in [16], and transformed into an input file for our algorithms. Of the 297 models initially considered, 72 were selected. These models have a unique steady state where every species has a non null concentration. To select them we have considered several tests described in Table 2.

The selected models were partitioned into three categories depending on the possible tests: kinetics rates, conservation laws, and volume compartments. We also tested for ACR the three models described in [27] in which the parameters have been fixed to random values. These models were previously tested for ACR by the Shinar/Feinberg topological criterion, therefore should remain so for any parameter set. As expected, the three models respect the ACR condition.

The initial boxes/domains for the conservation laws and parameters values were determined from the nominal initial conditions and parameter values found in the SBML files. The initial intervals bounds were obtained by dividing and multiplying these nominal values by a factor 10 for total amount of conservation laws and for volume compartments, and by a factor 100 for kinetics parameters, respectively. Homeostasis was tested using Definition 1. ACR was tested using Definition 1 with $k \to 1$, *i.e.* almost zero width intervals, and where the varying input parameters are the conservation laws values.

The execution time statistics are given in Table 3.

Table 2. The methodology applied to select models to be tested for homeostasis from the initial set of models. A test using `IbexSolve` guarantees the existence and position of all steady states. Then, each model with a unique steady state having a non zero concentration is selected. To avoid false positive answer due to a given precision, a time course beginning from the steady state indicated by `IbexSolve` has been achieved using COPASI [12]. These false positive answers may occur in a limit cycle or in a focus since we use relaxed equalities.

Test	# tested	# passed
|steady states| ≥ 1 (`IbexSolve`)	297	191
unique steady state > 0	191	107
non-oscillatory steady state (COPASI)	107	72

We used Biomodels also for multistationarity tests. Among the 297 models tested for multistationarity using `IbexSolve`, 63 provide a timeout. For the solved models, 35 do not have steady-state, 153 have a unique steady state, 42 provide multistationarity, 4 have a continuum of steady states, see Appendix A2.

Table 3. Statistics on the homeostasis test using `IbexHomeo`. Selected biomodels have been classified for three tests. All of them have been tested w.r.t. the kinetics rates, and each model presenting at least one conservation law has been tested for ACR. Moreover, models with several compartments have been tested w.r.t. their volume. As we have many timeout (computed as 360 s per species, with an average dimension of 51.92 (15.48 for species, and 36.4 for parameters)), the time columns consider only models that passed the test. The last two columns indicate the models with ACR in the first test, or with a 2-homeostatic species in other cases (because we get data during computation it may occur that a timeout model gives us a homeostasis).

Test `IbexHomeo`	# models	# timeout	Time (s) (min/median/max)	Yes	No
ACR only	33	17	0.19/4.32/10887	3	14
Kinetics only	72	41	0.4/122/1923	4	29
Compartments only	14	9	0.4/70/189	3	3

7　Discussion and Conclusion

The results of the tests show that interval methods are valuable tools for studying multistationarity and homeostasis of biochemical models.

In multistationarity studies, our interval algorithms outperform homotopy continuation based numerical methods; they find the correct number of steady states, and with a high accuracy. In terms of complexity of calculations they behave better than symbolic methods. Indeed, `IbexSolve` solves all the models in the chosen database (DOCSS). The 210 DOCSS models correspond to 13 different symbolic systems of equations (the remaining differences concern numerical parameters). The symbolic solver did not find explicit solutions for 1 of these symbolic systems, reduced 3 other models to 4th degree equations in 3.5 to 47 s, and solved the 9 others in times from 2 to 20 s which should be compared to the fractions of a second needed for the `IbexSolve` calculations. We did not perform symbolic calculations on more complex models from Biomodels, that we expect out of reach of the Matlab symbolic solver. However, the multistationarity test using `IbexSolve` performed well on Biomodels with only 63 models out of the 297 tested producing a timeout. These results are very promising since multistationarity is a computationally hard problem with numerous applications to cell fate decision processes in development, cancer, tissue remodelling. As future work, we plan to test multistationarity of larger models using ICP.

In homeostasis studies, interval methods perform well for small and medium size models in the Biomodels database. Only 18% of the tested models have some form of homeostasis (see Appendix A3). When the size of compartments change, 3 models have homeostasis, with 2 of them presenting species independent from parameter changes. For the kinetical parameters change, 4 models present homeostasis and two of them are of small size (for details, see Appendix A1): BIOMD614 is a univariate model when steady-state happens only with a concentration equal to one, and BIOMD629 presents a buffering mechanism (a buffer is a molecule occurring in much larger amounts than its interactors and whose concentration is nearly constant). The other two models are BIOMD048 and BIOMD093, where a timeout happens. For the initial conditions change,

3 models presents ACR, and two others (BIOMD041 and BIOMD622) home-ostasis. Among them, BIOMD413 verifies the conditions of the Shinar-Feinberg theorem [27], but the other two models (BIOMD489 and BIOMD738) do not since the deficiency of their reaction network is different from 1. This confirms that these conditions are sufficient, but not necessary. Our new examples could be the starting point of research on more general conditions for ACR.

The low proportion of homeostasis could be explained by the possible incompleteness of the biochemical pathways models. Not only these models are not representing full cells or organisms, but they may also miss regulatory mechanisms required for homeostasis. Negative feed-back interaction is known to be the main cause of homeostasis (although feed-forward loops can also produce homeostasis) [8]. As well known in machine learning, it is notoriously difficult to infer feed-back interaction. For this reason, many of the models in the Biomodels database were built with interactions that are predominantly forward and have only few feed-back interactions. It is therefore not a surprise that models that were on purpose reinforced in negative feed-back to convey biological homeostasis, such as BIOMD041, a model of ATP homeostasis in the cardiac muscle, or BIOMD433, a model of MAPK signalling robustness, or BIOMD355, a model of calcium homeostasis, were tested positively for homeostatic species.

Interestingly, our approach emphasizes a duality relationship between home-ostasis and multistationarity; the former means constant output at variable input, whereas the latter means multiple output at constant input. In this paper we were mainly motivated by the formal aspects of this duality allowing to treat the two problems within the same formalism. Nevertheless, it would be intriguing to look for biological consequences of this formal duality.

For future work, several directions will be investigated. Homeostasis benchmarking was restricted to non-oscillating steady states to avoid a detection of a second steady state inside the oscillatory component, due to equality relaxation used in IbexOpt, that could break the homeostasis test. This includes limit cycles and foci. However, oscillations are ubiquitous in biology and it is worth extending our homeostasis definitions to these cases as well. Biological systems are characterized not only by their attractors, but also by the characteristic times of relaxation to attractors. Homeostasis of relaxation could be approached with our formalism because relaxation times are reciprocals to solutions of polynomial characteristic equations.

Other possible improvements concern the performances. As one could see, many models tested cannot be solved within the timeout. If we can explain this with a high dimension of the model, several improvements are possible. First, we know that some convexification contractors using affine arithmetic [19,20], or specific to quadratic forms, in particular bi-linear forms [18] (and the pattern $x_1 x_2$ is often present in chemical reaction networks) could improve the contraction part of our strategy.

Another promising improvement should be to exploit that these models come from ODEs. Indeed, if a steady state is attractive, it is possible to perform a simulation (with fixed parameters) starting from a random or chosen point of the box that can end near this steady state. Then, a hybrid strategy using this new simulation and the branch and bound strategy should be able to get a feasible point more easily, rendering the comparison for homeostasis more efficient. Moreover this simulation could be used to check if a steady state found is unstable or not. Also, adding as a stop criterion the homeostasis test could be another way to improve it, instead of checking it at a given time.

Acknowledgement. We thank M. Golubiski and F. Antoneli for presenting us the problem of homeostasy, U. Bhalla and N. Ramakrishnan for sharing their data, C. Lüders for parsing Biomodels, and F. Fages, G. Chabert and B. Neveu for useful discussions. The Ibex computations were performed on the high performance facility MESO@LR of the University of Montpellier. This work is supported by the PRCI ANR/DFG project Symbiont.

Appendix A1: Homeostasis of Low Complexity Models

BIOMD614 is a one species model, with equation:

$$\dot{x} = k_1 + k_2 k_3 x - k_1 x - k_2 k_3 x^2 \tag{7}$$

At steady state, this leads to:

$$k_1 + k_2 k_3 x = (k_1 + k_2 k_3 x)x \tag{8}$$

If $k_1 \neq 0$, the only solution to (8) is $x = 1$, which is the answer given by IbexHomeo. The model describes the irreversible reaction kinetics of the conformational transition of a human hormone, where x is the fraction of molecules having undergone the transition, which is inevitably equal to one at the steady state [14].

BIOMD629 has 2 reactions and 5 species, provides a 2-homeostasis for kinetics parameters with conserved total amounts fixed, provided by the SBML file. This model does not provide ACR, and the homeostasis found can be explained by the conserved total amounts, that lock species to a small interval. But if we change these total amounts and try again an homeostasis test, it should fail. Indeed this model is given by the equations :

$$\dot{x}_1 = -k_2 x_1 x_3 + k_3 x_2, \ \dot{x}_2 = k_2 x_1 x_3 - k_3 x_2 - k_4 x_2 x_4 + k_5 x_5,$$
$$\dot{x}_3 = -k_2 x_1 x_3 + k_3 x_2, \ \dot{x}_4 = -k_4 x_2 x_4 + k_5 x_5, \ \dot{x}_5 = k_4 x_2 x_4 - k_5 x_5,$$
$$x_4 + x_5 = k_6, \ x_2 + x_5 + x_3 = k_7, \ x_2 + x_5 + x_1 = k_8 \tag{9}$$

Here x_3 (receptor), and x_4 (coactivator) have been found homeostatic w.r.t. variations of the kinetics parameters. The total amounts are $k_6 = 30$, $k_7 = \frac{7}{2000}$, $k_8 = \frac{1}{2000}$. With these values, we get $x_5 \in \,]0, \frac{1}{2000}[$, which implies $x_4 \in \,]29.9995, 30.0005[$. In the same way we have $x_2 + x_5 \in \,]0, \frac{1}{2000}[$, which implies $x_3 \in \,]\frac{6}{2000}, \frac{7}{2000}[$. If k_6, k_7, k_8 were closer to each other, there would be no reason for homeostasis. This example corresponds to the homeostasis mechanism known in biochemistry as buffering: a buffer is a molecule in much larger amounts than its interactors and whose concentration is practically constant.

Appendix A2: Multistationarity Statistics

Among the 297 models tested for multistationarity using `IbexSolve`, 63 provide a timeout. For the solved models, 35 do not have steady-state, 153 have a unique steady state, 42 provide multistationarity, 4 have a continuum of steady states. Theses results are given by `IbexSolve`, but some multistationarity models, such as 003, have in reality an oscillatory behavior (Table 4, 5, 6, 7, 8, 9 and 10).

Table 4. Time statistics for multistationarity

# models	Size (min/max/median/average)	Median time (s)
63 (timeout)	3/194/23/37.26	10800
234 (solved)	1/166/8/13.89	0.009439

Multistationnary models :

- 10 steady states: 703.
- 7 steady states: 435.
- 4 steady states: 228, 249, 294, 517, 518, 663, 709.
- 3 steady states: 003, 008, 026, 027, 029, 069, 116, 166, 204, 233, 257, 296, 447, 519, 573, 625, 630, 687, 707, 708, 714, 729.
- 2 steady states: 079, 100, 156, 230, 315, 545, 552, 553, 688, 713, 716.

Appendix A3: Homeostasis Results Tables

Table 5. Models with less than 9 species tested for homeostasis w.r.t. the volume of compartments. Minimum and maximum volume of each compartment is set as value given by the SBML file divided by 10 and multiplied by 10, respectively. Kinetic rates and total amount of conservation laws stay fixed. In BIOMD355, all species are 2-homeostatic (and seems independent of the volume). In BIOMD433, the species MK_P is 2-homeostatic.

Model (compartments)	# varhom	Time (s)	# var	# param varying
BIOMD0000000115	0	120	2	3
BIOMD0000000191	0	0.4	2	2
BIOMD0000000355	9	189	9	4
BIOMD0000000432	0	70	8	2
BIOMD0000000433	1	13.8	8	2

Table 6. Models with less than 11 species tested for ACR, the second number indicate the number of species that verify 2-homeostasis (ACR included). The minimum and maximum value of each total amount of conservation laws is the value computed from the SBML file, divided by 10 and multiplied by 10, respectively. Kinetic rates and volume compartments stay fixed. FeinbergShinar models serve as tests to confirm that we detect ACR. In BIOMD041, ATP and ATPi are detected to be 2-homeostatic despite the timeout. In BIOMD622, R1B, R1Bubd, and Z are also detected to be 2-homeostatic despite the timeout. In BIOMD413, auxin verify ACR. In BIOMD738, FeDuo, FeRBC, FeSpleen, FeLiver, Hepcidin, FeRest, FeBM are ACR.

Model (for ACR)	# ACR-# hom	Time (s)	# var	# param varying
BIOMD0000000031	0–0	3.81	3	1
BIOMD0000000041	?–≥ 2	Timeout (2830)	10	3
BIOMD0000000057	0–0	0.56	6	1
BIOMD0000000060	0–0	0.19	4	1
BIOMD0000000084	0–0	2.39	8	4
BIOMD0000000150	0–0	1.13	4	2
BIOMD0000000213		Timeout (1987)	6	1
BIOMD0000000258	0–0	3.15	3	1
BIOMD0000000405	0–0	0.52	5	1
BIOMD0000000413	1–1	6.68	5	1
BIOMD0000000423		Timeout (3071)	9	3
BIOMD0000000432		Timeout (2711)	8	3
BIOMD0000000433		Timeout (2709)	8	3
BIOMD0000000454	0–0	0.3	3	1
BIOMD0000000622	?–≥ 3	Timeout (3606)	11	1
BIOMD0000000629	0–0	4.32	5	3
BIOMD0000000646	0–0	801	11	1
BIOMD0000000647	0–0	306	11	5
BIOMD0000000738	7–7	557	11	1
FeinbergShinar1	1–1	5.31	7	2
FeinbergShinar2	1–1	3.08	8	2
FeinbergShinar3	1–1	15.23	5	2

Table 7. Models with less than 9 species tested for homeostasis w.r.t. kinetics rates. The minimum and maximum value of each kinetic rate is given by the SBML file divided by 100 and multiplied by 100, respectively. In BIOMD614, the unique species is 2-homeostatic (moreover independent). In BIOMD629, receptor and coactivator are 2-homeostatic.

Models (kinetics)	# varhom	Time (s)	# var	# param varying
BIOMD0000000023	0	301	5	60
BIOMD0000000031	0	93.8	3	12
BIOMD0000000057	0	284	6	25
BIOMD0000000060	0	122	4	10

(continued)

Table 7. (*continued*)

Models (kinetics)	# varhom	Time (s)	# var	# param varying
BIOMD0000000065		Timeout (2885)	8	23
BIOMD0000000067		Timeout (2523)	7	20
BIOMD0000000076	0	60	1	20
BIOMD0000000084	0	22	8	17
BIOMD0000000115		Crashed	2	7
BIOMD0000000150	0	243	4	6
BIOMD0000000159	0	91	3	6
BIOMD0000000191	0	89	2	19
BIOMD0000000203	0	876	5	34
BIOMD0000000213		Timeout (2173)	6	46
BIOMD0000000219	0	540	9	74
BIOMD0000000221		Timeout (2902)	8	53
BIOMD0000000222		Timeout (2909)	8	53
BIOMD0000000228	0	1147	9	40
BIOMD0000000240		Timeout (2163)	6	20
BIOMD0000000249		Timeout (2883)	8	7
BIOMD0000000258	0	132	3	9
BIOMD0000000284	0	31	6	3
BIOMD0000000325	0	300	5	16
BIOMD0000000355		Timeout (2287)	9	22
BIOMD0000000405	0	801	5	5
BIOMD0000000413	0	481	5	12
BIOMD0000000414	0	3.7	1	4
BIOMD0000000417		Timeout (360)	1	12
BIOMD0000000423		Timeout (3132)	9	16
BIOMD0000000425	0	0.4	1	6
BIOMD0000000432	≤ 1	Timeout (1031)	8	26
BIOMD0000000433	≤ 3	Timeout (1671)	8	26
BIOMD0000000454		Timeout (1022)	3	11
BIOMD0000000456		Timeout (1375)	4	16
BIOMD0000000458	0	93	2	15
BIOMD0000000459	0	63	3	8
BIOMD0000000460	0	122	3	8
BIOMD0000000495	0	365	9	36
BIOMD0000000519		Timeout (1050)	3	8
BIOMD0000000530	0	1923	7	17
BIOMD0000000590	0	540	9	30
BIOMD0000000614	1	17	1	3
BIOMD0000000615	0	160	4	12
BIOMD0000000626		Timeout (632)	6	20
BIOMD0000000629	2	1	5	4
BIOMD0000000708	0	1770	5	13
BIOMD0000000728	0	1	2	4

Table 8. Models with more than 10 species tested for homeostasis w.r.t. the volume of compartments. Minimum and maximum volume of each compartment is set as value given by the SBML file divided by 10 and multiplied by 10, respectively. Kinetic rates and total amount of conservation laws stay fixed. In BIOMD738, we know that at least Hepcidin is 2-homeostatic (and independent) despite the timeout.

Model (compartments)	# varhom	Time (s)	# var	# param varying
BIOMD0000000041		Timeout (3759)	10	2
BIOMD0000000093		Timeout (12901)	34	2
BIOMD0000000123		Timeout (5681)	14	2
BIOMD0000000192		Timeout (3924)	13	2
BIOMD0000000482		Timeout (8308)	23	3
BIOMD0000000491		Timeout (20962)	57	3
BIOMD0000000492		Timeout (19066)	52	3
BIOMD0000000581		Timeout (5359)	27	2
BIOMD0000000738	≥ 1	Timeout (3676)	11	7

Table 9. Models with more than 12 species tested for ACR, the second number indicate the number of species that verify 2-homeostasis (ACR included). The minimum and maximum value of each total amount of conservation laws is the value computed from the SBML file, divided by 10 and multiplied by 10, respectively. Kinetic rates and volume compartments stay fixed. In BIOMD489, LPS:LBP:CD14: TLR4:TIRAP:MyD88:IRAK4, IkBb_mRNA, IkBe_mRNA, LPS:LBP:CD14:TLR4: RIP1:TRAM:TRIF:TBK/IKKe are detected ACR despite the timeout.

Model (for ACR)	# ACR-# hom	Time (s)	# var	# param varying
BIOMD0000000009		Timeout (7988)	22	7
BIOMD0000000011		Timeout (7804)	22	7
BIOMD0000000030		Timeout (6318)	18	3
BIOMD0000000038		Timeout (4400)	13	4
BIOMD0000000048		Timeout (8380)	23	6
BIOMD0000000093		Timeout (12115)	34	6
BIOMD0000000123	0–0	4810	14	3
BIOMD0000000192	0–0	1316	13	3
BIOMD0000000270	0–0	10887	32	9
BIOMD0000000431		Timeout (9808)	27	6
BIOMD0000000489	≥ 4 - ≥ 4	Timeout (11104)	52	4
BIOMD0000000491		Timeout (19323)	57	1
BIOMD0000000492		Timeout (17524)	52	1
BIOMD0000000581		Timeout (8485)	27	10

Table 10. Models with more than 10 species tested for homeostasis w.r.t. kinetics rates. The minimum and maximum value of each kinetic rate is given by the SBML file divided by 100 and multiplied by 100, respectively. In BIOMD048, EGF is detected 2-homeostatic despite the timeout. In BIOMD093, SHP2 is detected homeostatic despite the crash.

Models (kinetics)	# varhom	Time (s)	# var	# param varying
BIOMD0000000009		Timeout (8239)	22	32
BIOMD0000000011		Timeout (8009)	22	30
BIOMD0000000028		Timeout (5810)	16	27
BIOMD0000000030		Timeout (6571)	18	32
BIOMD0000000038		Timeout (4722)	13	24
BIOMD0000000041		Timeout (3952)	10	25
BIOMD0000000048	≥1	Timeout (9551)	23	50
BIOMD0000000093	≥1	Crashed	34	73
BIOMD0000000123		Timeout(5654)	14	22
BIOMD0000000192		Timeout (3923)	13	18
BIOMD0000000218	0	541	12	70
BIOMD0000000270		Timeout (12271)	32	30
BIOMD0000000294		Timeout (3608)	10	10
BIOMD0000000388		Timeout (1801)	11	19
BIOMD0000000431		Timeout (10028)	27	44
BIOMD0000000482		Timeout (8452)	23	56
BIOMD0000000489		Timeout (15513)	52	105
BIOMD0000000491		Timeout	57	172
BIOMD0000000492		Timeout	52	176
BIOMD0000000581		Timeout (9131)	27	35
BIOMD0000000622		Timeout (3618)	11	25
BIOMD0000000646		Timeout (3976)	11	33
BIOMD0000000647		Timeout (1173)	11	11
BIOMD0000000707		Timeout (1746)	5	10
BIOMD0000000738		Timeout (3986)	11	33

References

1. Antoneli, F., Golubitsky, M., Stewart, I.: Homeostasis in a feed forward loop gene regulatory motif. J. Theor. Biol. **445**, 103–109 (2018)
2. Araya, I., Neveu, B.: lsmear: a variable selection strategy for interval branch and bound solvers. J. Glob. Optim. **71**(3), 483–500 (2018)
3. Aubin, J.-P.: Viability Theory. SCFA. Birkhäuser, Boston (2009). https://doi.org/10.1007/978-0-8176-4910-4
4. Barr, K., Reinitz, J., Radulescu, O.: An in silico analysis of robust but fragile gene regulation links enhancer length to robustness. PLoS Comput. Biol. **15**(11), e1007497 (2019)

5. Benhamou, F., Goualard, F., Granvilliers, L., Puget, J.-F.: Revising hull and box consistency. In: Proceedings of ICLP, pp. 230–244 (1999)
6. Bradford, R., et al.: Identifying the parametric occurrence of multiple steady states for some biological networks. J. Symb. Comput. **98**, 84–119 (2020)
7. Chabert, G. (2020). http://www.ibex-lib.org
8. Cooper, S.J.: From Claude Bernard to Walter cannon. Emergence of the concept of homeostasis. Appetite **51**(3), 419–427 (2008)
9. Desoeuvres, A., Iosif, A., Radulescu, O., Seiß, M.: Approximated conservation laws of chemical reaction networks with multiple time scales. preprint, April 2020
10. Golubitsky, M., Stewart, I.: Homeostasis, singularities, and networks. J. Math. Biol. **74**(1–2), 387–407 (2017)
11. Gorban, A.N., Radulescu, O.: Dynamical robustness of biological networks with hierarchical distribution of time scales. IET Syst. Biol. **1**(4), 238–246 (2007)
12. Hoops, S., et al.: Copasi-a complex pathway simulator. Bioinformatics **22**(24), 3067–3074 (2006)
13. Jaulin, L., Kieffer, M., Didrit, O., Walter, E.: Applied Interval Analysis. Springer, London (2001). https://doi.org/10.1007/978-1-4471-0249-6
14. Kamihira, M., Naito, A., Tuzi, S., Nosaka, A.Y., Saito, H.: Conformational transitions and fibrillation mechanism of human calcitonin as studied by high-resolution solid-state 13 C NMR. Protein Sci. **9**(5), 867–877 (2000)
15. Novere, N., et al.: Biomodels database: a free, centralized database of curated, published, quantitative kinetic models of biochemical and cellular systems. Nucleic Acids Res. **34**(suppl_1), D689–D691 (2006)
16. Lüders, C., Radulescu, O., et al.: Computational algebra oriented CRN collection of models (2020, in preparation)
17. Markov, S.: Biomathematics and interval analysis: a prosperous marriage. In: AIP Conference Proceedings, vol. 1301, pp. 26–36. American Institute of Physics (2010)
18. McCormick, G.P.: Computability of global solutions to factorable nonconvex programs: part I - convex underestimating problems. Math. Program. **10**(1), 147–175 (1976)
19. Messine, F.: Extensions of affine arithmetic: application to unconstrained global optimization. J. Univ. Comput. Sci. **8**(11), 992–1015 (2002)
20. Messine, F., Touhami, A.: A general reliable quadratic form: an extension of affine arithmetic. Reliable Comput. **12**(3), 171–192 (2006)
21. Misener, R., Floudas, C.: ANTIGONE: algorithms for continuous/integer global optimization of nonlinear equations. J. Glob. Optim. (JOGO) **59**(2–3), 503–526 (2014)
22. Moore, R.E.: Interval Analysis, vol. 4. Prentice-Hall, Englewood Cliffs (1966)
23. Neveu, B., Trombettoni, G., Araya, I.: Adaptive constructive interval disjunction: algorithms and experiments. Constraints J. **20**(4), 452–467 (2015)
24. Neveu, B., Trombettoni, G., Araya, I.: Node selection strategies in interval branch and bound algorithms. J. Glob. Optim. **64**(2), 289–304 (2016)
25. Ramakrishnan, N., Bhalla, U.S.: Memory switches in chemical reaction space. PLoS Comput. Biol. **4**(7), e1000122 (2008)
26. Rizk, A., Batt, G., Fages, F., Soliman, S.: A general computational method for robustness analysis with applications to synthetic gene networks. Bioinformatics **25**(12), i169–i178 (2009)
27. Shinar, G., Feinberg, M.: Structural sources of robustness in biochemical reaction networks. Science **327**(5971), 1389–1391 (2010)
28. Sommese, A., Wampler, C.I.: The Numerical Solution of Systems of Polynomials Arising in Engineering and Science. World Scientific, Singapore (2005)
29. Tawarmalani, M., Sahinidis, N.V.: A polyhedral branch-and-cut approach to global optimization. Math. Program. **103**(2), 225–249 (2005)

30. Trombettoni, G., Araya, I., Neveu, B., Chabert, G.: Inner regions and interval linearizations for global optimization. In: Twenty-Fifth AAAI Conference on Artificial Intelligence (2011)
31. Tucker, W., Kutalik, Z., Moulton, V.: Estimating parameters for generalized mass action models using constraint propagation. Math. Biosci. **208**(2), 607–620 (2007)
32. Van Hentenryck, P., Michel, L., Deville, Y.: Numerica : A Modeling Language for Global Optimization. MIT Press, Cambridge (1997)

Growth Dependent Computation
of Chokepoints in Metabolic Networks

Alexandru Oarga[1]📧, Bridget Bannerman[2]📧, and Jorge Júlvez[1(✉)]📧

[1] Department of Computer Science and Systems Engineering,
University of Zaragoza, Zaragoza, Spain
{718123,julvez}@unizar.es
[2] Department of Medicine, University of Cambridge, Cambridge, UK
bpc28@cam.ac.uk

Abstract. Bacterial infections are among the major causes of mortality in the world. Despite the social and economical burden produced by bacteria, the number of new drugs to combat them increases very slowly due to the cost and time to develop them. Thus, innovative approaches to identify efficiently drug targets are required. In the absence of genetic information, chokepoint reactions represent appealing drug targets since their inhibition might involve an important metabolic damage. In contrast to the standard definition of chokepoints, which is purely structural, this paper makes use of the dynamical information of the model to compute chokepoints. This novel approach can provide a more realistic set of chokepoints. The dependence of the number of chokepoints on the growth rate is assessed on a number of metabolic networks. A software tool has been implemented to facilitate the computation of growth dependent chokepoints by the practitioners.

Keywords: Chokepoint reactions · Metabolic networks · Petri nets · Flux Balance Analysis

1 Introduction

Diseases caused by bacteria are one of the main causes of mortality in both developed and in-development countries. According to the World Health Organisation (WHO) in 2016, tuberculosis was the tenth cause of death worldwide which makes its pathogen, *Mycobacterium tuberculosis*, the infectious agent with the higher caused mortality. Moreover, upper respiratory system's diseases caused by microorganisms, like virus and bacteria, were the fourth cause of mortality [20]. In 2010 pneumonia was the leading cause of child mortality causing nearly 1.4 million deaths among children younger than 5 years of age [7].

This work was supported by the Spanish Ministry of Science, Innovation and Universities [ref. Medrese-RTI2018-098543-B-I00], and by the Medical Research Council, UK, MR/N501864/1.

ⓒ Springer Nature Switzerland AG 2020
A. Abate et al. (Eds.): CMSB 2020, LNBI 12314, pp. 102–119, 2020.
https://doi.org/10.1007/978-3-030-60327-4_6

Despite the high mortality caused by bacteria the development of new antibiotics is slow and challenging. Furthermore, bacteria have evolved complex mechanisms which make them difficult to fight. Thus, there is an urgent need to design novel methods for the development of new drugs. A promising possibility is to consider basic cellular processes as targets for antibiotic development [12].

Metabolism is the set of basic life processes that take place in the cell, and it is the means by which cells can maintain life and grow from their environment. The metabolism of a cell can be represented by a *metabolic network* that accounts for all the metabolic reactions that take place in the cell. A possible strategy for drug discovery is to find and damage critical vulnerabilities of the metabolic network that could stop the growth and replication of the bacteria.

Metabolism as a target has been proven to be an interesting approach in other areas like oncology [9] or viral diseases [10]. A number of methods have been proposed in order to find vulnerabilities in the metabolism that may lead to therapeutic results. Some of these methods consider topological properties of the metabolism with the purpose of finding possible critical spots, as for example: determine the importance of a metabolite based on the k-shortest paths between metabolites [15], or consider the inter-reactions dependence to find out how much influence a reaction has on metabolism [16]. Other methods focus on the genetic information associated with the metabolism, and compute, for instance, the set of genes that are essential for the survival of the cell [21].

Although a number of genome-scale models (GEMs) have been developed recently, most of them just account for the stoichiometry of their reactions and lack genetic information. This is usually the case in GEMs of bacteria. This dearth of data hampers the analysis of models, namely those based on gene essentiality, and calls for the design of computational methods that exploit as much as possible the available biological information. Here, we focus on the computation of chokepoints in metabolic networks [18], where a chokepoint is a reaction that is either the only producer or the only consumer of a given metabolite. Hence, the inhibition of a chokepoint would lead to the depletion or unlimited accumulation of metabolites, thus, potentially leading to an important disruption in the cellular metabolism. Chokepoints are, therefore, appealing drug targets of the bacterial metabolism.

The current approaches to compute chokepoints are based exclusively on the topology of the metabolic network and disregard the dynamic information that might be available. This dynamic information usually refers to the flux bounds of some metabolic reactions. As it will be shown, ignoring such an information can lead to the misidentification of chokepoints. The approach presented in this paper exploits the available flux bounds and computes, for a given growth rate of the cell, the set of chokepoints of the metabolic network. Such chokepoints are potential drug targets whose inhibition could involve a metabolic burden at the given growth rate.

The rest of the paper is organized as follows: Sect. 2 introduces the basic concepts and definition that will be used in the paper. Section 3 describes the computational method to obtain growth dependent chokepoints. Section 4 analyses the relationship between growth rate and number of chokepoints in the

GEM of *Mycobacterium leprae*. The main conclusions of the paper are drawn in Sect. 5. Finally, Appendix A introduces the software tool developed to compute growth dependent chokepoints, and Appendix B reports the number of chokepoints found in the GEMs of different microorganisms.

2   Preliminary Concepts and Definitions

2.1   Constraint-Based Models

A *constraint-based model* [13,19] is a tuple $\{\mathcal{R}, \mathcal{M}, \mathcal{S}, lb, ub\}$ where $\mathcal{R}$ is a set of reactions, $\mathcal{M}$ is a set of metabolites, $\mathcal{S} \in \mathbb{R}^{|\mathcal{M}| \times |\mathcal{R}|}$ is the stoichiometric matrix, and $lb, ub : \mathcal{R} \to \mathbb{R}$ are lower and upper flux bounds of the reactions.

Each reaction is associated with a set of reactant metabolites and a set of product metabolites (one of these sets can be empty). For instance, the reaction $r_1 : A \to 2B$ has one reactant, A, and one product, B. The number 2 expresses the stoichiometric weight, i.e. two units of B are produced per each unit of A that is consumed. The stoichiometric matrix $\mathcal{S}$ accounts for all the stoichiometric weights of the reactions, i.e. $S[m, r]$ is the stoichiometric weight of metabolite $m \in \mathcal{M}$ for reaction $r \in \mathcal{R}$. Thus, if $S[m, r] < 0$ then m is consumed when r occurs; if $S[m, r] > 0$ then m is produced when r occurs; and if $S[m, r] = 0$ then m is neither consumed nor produced when r occurs.

Constraint-based models can be represented graphically as Petri nets [5,11] where places, which are drawn as circles, are associated with metabolites, and transitions, which are drawn as rectangles, are associated with reactions. An arc from a place(transition) to a transition(place) means that the place is a reactant(product). The weights of the arcs of the Petri net correspond to the stoichiometric weights, in other words, the stoichimetric matrix of a constraint-base model and the incidence matrix of its corresponding Petri net coincide.

Example 1. The Petri net in Fig. 1 represents a simple contraint-based model that consists of 13 reactions and 9 metabolites. As an example, transition r_6 models the reaction $r_6 : m_a \to 2m_d$.

2.2   Topological Definitions.

Borrowing the usual Petri net notation (given a node x of a Petri net, $^{\bullet}x$ and $x^{\bullet}$ denote the sets of the input and output nodes of x respectively), we define the following sets for constraint-based models:

- Set of products of r: $r^{\bullet} = \{m \in \mathcal{M} | S(m, r) > 0\}$
- Set of reactants of r: $^{\bullet}r = \{m \in \mathcal{M} | S(m, r) < 0\}$
- Set of consumers of m: $m^{\bullet} = \{r \in \mathcal{R} | S(m, r) < 0\}$
- Set of producers of m: $^{\bullet}m = \{r \in \mathcal{R} | S(m, r) > 0\}$

A chokepoint is a reaction that is the only producer or the only consumer of a metabolite. More formally:

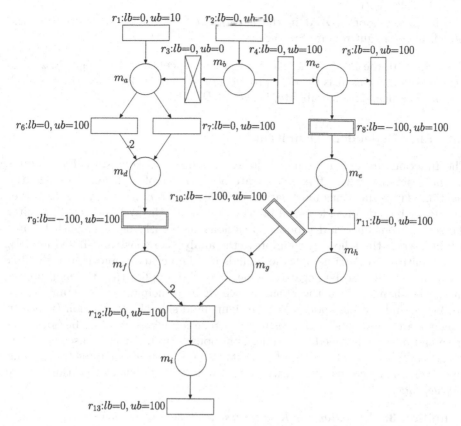

Fig. 1. Petri net modelling a constraint-based model. The values lb and ub are the lower and upper flux bounds of reactions. Non-reversible reactions are represented by simple rectangles, reversible reactions by double rectangles and dead-reactions by rectangles with a cross.

Definition 1. *A reaction $r \in \mathcal{R}$ is a chokepoint if there exists $m \in \mathcal{M}$ such that $m^\bullet = \{r\}$ or $^\bullet m = \{r\}$.*

The set of chokepoint reactions will be denoted as CP. Notice that the inhibition of the enzymes associated with a chokepoint will lead either to the depletion of metabolites (which might be essential for the cell) if the chokepoint is the only producer, or to the indefinite accumulation of metabolites (which will not be used as expected or might be toxic) if the chokepoint is the only consumer. Thus, a chokepoint is an attractive drug target because in both cases, essential functions of the cell can be affected by its inhibition [18].

A dead-end metabolite (DEM) is a metabolite that lacks either producing or consuming reactions:

Definition 2. *A metabolite $m \in \mathcal{M}$ is a dead-end metabolite (DEM) if $m^\bullet = \{\}$ or $^\bullet m = \{\}$.*

The presence of a DEM in the network reflects an incompleteness in the model, which might require further curation [8].

Example 2. In the Petri net in Fig. 1, r_1 is a producer of m_a, i.e. $r_1 \in {}^{\bullet}m_a$; r_4 is a chokepoint because it is the only producer of m_c, i.e. $\{r_4\} = {}^{\bullet}m_c$ and $r_4 \in CP$; and m_h is a dead-end metabolite, i.e. $m_h \in DEM$.

2.3 Flux Dependent Definitions

The functions lb and ub establish lower and upper steady state flux bounds on the reactions, where flux is the rate of turnover of molecules through the reaction. These functions must satisfy that $lb(r) \leq ub(r)$ for every $r \in \mathcal{R}$. Lower and upper bounds provide useful information about the system and might alter the sets of consumer and producer reactions previously defined. Such bounds will be used in the following to improve the analysis of constraint-based models.

In contrast to Petri nets, these bounds can be negative, and hence, the flux of a reaction can also be negative. A negative flux implies that the metabolites on the left-hand side of the reaction (which in principle are "reactants") are produced, and the metabolites on the right-hand side of the reaction (which in principle are "products") are consumed. A reaction whose flux can be both negative and positive is called reversible. Functions lb and ub will be used to define the sets of flux dependent reversible reactions (RR_d), dead reactions (DR_d), and non-reversible reactions (NR_d), where the subindex d indicates that the sets are flux *dependent*:

Definition 3. *A reaction $r \in \mathcal{R}$ is reversible if $lb(r) < 0 < ub(r)$.*

The set of reversible reactions is denoted RR_d, i.e. $RR_d = \{r \in \mathcal{R} \mid r$ is reversible$\}$.

Definition 4. *A reaction $r \in \mathcal{R}$ is dead if $lb(r) = ub(r) = 0$.*

The set of dead reactions is denoted DR_d, i.e. $DR_d = \{r \in \mathcal{R} \mid r$ is dead$\}$.

Definition 5. *A reaction $r \in \mathcal{R}$ is non-reversible if $(0 \leq lb(r) \wedge 0 < ub(r)) \vee (lb(r) < 0 \wedge ub(r) \leq 0)$.*

The set of non-reversible reactions is denoted NR_d, i.e. $NR_d = \{r \in \mathcal{R} \mid r$ is non-reversible $\}$.

Clearly, the sets RR_d, DR_d and NR_d partition the set of reactions $\mathcal{R}$, i.e. $RR_d \cup DR_d \cup NR_d = \mathcal{R}$, $RR_d \cap DR_d = \emptyset$, $DR_d \cap NR_d = \emptyset$, and $RR_d \cap NR_d = \emptyset$.

Non-reversible, reversible and dead reactions will be represented graphically as rectangles, double rectangles, and rectangles with a cross inside respectively.

Example 3. In Fig. 1, the above defined sets are: $RR_d = \{r_8, r_9\, r_{10}\}$, $DR_d = \{r_3\}$, and $NR_d = \{r_1, r_2, r_4, r_5, r_6, r_7, r_{11}, r_{12}, r_{13}\}$.

Given that, in constraint-based models, reactions can be reversible or can proceed only backwards, i.e. $lb(r) \leq ub(r) < 0$, the concepts related to the consumption and production of metabolites must be revisited. Thus, new sets of reactants, products, consumers, and producers which take into account the flux bounds are defined as follows:

- Set of products of r:
$$r_d^{\bullet} = \{m \in \mathcal{M} | (S(m,r) > 0 \land ub(r) > 0) \lor (S(m,r) < 0 \land lb(r) < 0)\}$$
- Set of reactants of r:
$$^{\bullet}r_d = \{m \in \mathcal{M} | (S(m,r) < 0 \land ub(r) > 0) \lor (S(m,r) > 0 \land lb(r) < 0)\}$$
- Set of consumers of m:
$$m_d^{\bullet} = \{r \in \mathcal{R} | (S(m,r) < 0 \land ub(r) > 0) \lor (S(m,r) > 0 \land lb(r) < 0)\}$$
- Set of producers of m:
$$^{\bullet}m_d = \{r \in \mathcal{R} | (S(m,r) > 0 \land ub(r) > 0) \lor (S(m,r) < 0 \land lb(r) < 0)\}$$

Flux dependent definitions of chokepoints and dead-end metabolites can be written as:

Definition 6. *A reaction $r \in \mathcal{R}$ is a flux dependent chokepoint if there exists $m \in M$ such that $m_d^{\bullet} = \{r\}$ or $^{\bullet}m_d = \{r\}$.*

Definition 7. *A metabolite $m \in M$ is a flux dependent dead-end metabolite if $m_d^{\bullet} = \{\}$ or $^{\bullet}m_d = \{\}$.*

The sets of flux dependent chokepoint reactions and dead-end metabolites will be denoted as CP_d and DEM_d respectively.

Example 4. In Fig. 1, r_{12} is a flux dependent chokepoint, i.e. $r_{12} \in CP_d$; and m_h is a flux dependent dead-end metabolite, i.e. $m_h \in DEM_d$.

3 Growth Dependent Chokepoints

In GEMs, unknown flux bounds are given default values, e.g. $lb(r) = -1000$ mmol g^{-1} h^{-1} and $ub(r) = 1000$ mmol g^{-1} h^{-1} (recall that flux bounds establish the direction in which the reaction can proceed). Thus, all the reactions that are given default values are considered as reversible. However, not all the fluxes in the ranges given by the flux bounds of GEMs models are compatible with a positive growth rate. By using Flux Balance Analysis (FBA) [14] and Flux Variability Analysis (FVA) [4] it is possible to obtain tighter flux bounds for a given growth rate. Such tighter bounds could imply that, reactions which were initially considered as reversible, are in fact non-reversible for the given growth rate. This might alter the original set of chokepoints, i.e. the set of chokepoints depend on the growth rate. This section describes how growth dependent chokepoints can be computed.

Flux Balance Analysis (FBA) is a mathematical procedure for the estimation of steady state fluxes in constraint-based models. FBA can be used, for instance, to predict the growth rate of an organism or the rate of production of a given metabolite. Mathematically, FBA is expressed as a linear programming problem that maximises an objective function subject to steady state constraints. In the case of estimating the growth rate, the objective function is biomass production, a reaction that defines the ratios at which metabolites are converted into basic constituents of the cell as nucleic acids or proteins [14].

Let $v \in \mathbb{R}^{|\mathcal{R}|}$ be the vector of fluxes of reactions and $v[r]$ denote the flux of reaction r. At steady state, it holds that $S \cdot v = 0$, where S is the stoichiometric matrix. The steady state fluxes of reactions are also lower and upper bounded by lb and ub. Thus, the FBA linear programming problem is:

$$\max z \cdot v$$
$$st. \quad S \cdot v = 0 \tag{1}$$
$$lb(r) \leq v[r] \leq ub(r) \quad \forall r \in \mathcal{R}$$

where $z \in \mathbb{R}^{|\mathcal{R}|}$ expresses the objective function.

It is a common assumption that the metabolism of prokaryotes has evolved to maximize the growth of the cells. Hence, the growth rate given by the biomass production is an empirically reasonable choice for the objective function of FBA applied to bacteria [17].

A given growth, i.e. a given flux through the reaction modelling biomass production, can be achieved by different fluxes of the reactions. This means that each reaction can have a range of fluxes that is compatible with a given growth. Flux Variability Analysis (FVA) can be used to compute such range of fluxes for each reaction [2].

More precisely, FVA [4] is a mathematical procedure to compute the minimum and maximum fluxes of reactions that are compatible with some state, e.g. supporting 90% of the maximum growth yielded by FBA. Among other applications, FVA can be used to study the network flexibility, and studying the network response under suboptimal conditions.

Let μ_{max} be the maximum growth calculated by FBA. FVA is computed by solving two independent linear programming problems per reaction $r \in \mathcal{R}$. One programming problem maximizes the flux of r, $v[r]$, and the other minimizes $v[r]$. The constraints of both problems are the same: the steady state condition $S \cdot v = 0$, the flux bounds $lb(r) \leq v[r] \leq ub(r)$, and the maintenance of the optimum value given by FBA to a certain degree. This last constraint is expressed as $\gamma \cdot \mu_{max} \leq z \cdot v$ where z is the same vector as in (1) and $\gamma \in [0, 1]$ represents the fraction of optimal value that must be satisfied. Thus, the two programming problems for a given reaction $r \in \mathcal{R}$ can be expressed as:

$$\max / \min v[r]$$
$$st. \quad S \cdot v = 0$$
$$lb(r) \leq v[r] \leq ub(r) \quad \forall r \in \mathcal{R} \tag{2}$$
$$\gamma \cdot \mu_{max} \leq z \cdot v$$

Let $lb_\gamma, ub_\gamma : \mathcal{R} \rightarrow \mathbb{R}$ be the result of running FVA on a constraint-based model $\{\mathcal{R}, \mathcal{M}, \mathcal{S}, lb, ub\}$ for a given γ. If the flux bounds lb, ub of the constrained-based model are replaced by lb_γ, ub_γ, a new constraint-based model, $\{\mathcal{R}, \mathcal{M}, \mathcal{S}, lb_\gamma, ub_\gamma\}$, with more constrained flux bounds is obtained.

Given γ, the sets of flux dependent products, reactants, consumers, and producers of the model $\{\mathcal{R}, \mathcal{M}, \mathcal{S}, lb_\gamma, ub_\gamma\}$ are denoted as $r_\gamma^\bullet, {}^\bullet r_\gamma, m_\gamma^\bullet, {}^\bullet m_\gamma$ respectively. Similarly, the sets of flux dependent reversible reactions, dead reactions, and non-reversible reactions are denoted as RR_γ, DR_γ, NR_γ respectively. The sets of flux dependent chokepoint reactions and dead-end metabolites are denoted as CP_γ and DEM_γ.

In Algorithm 1, an iterative procedure is proposed that, given an input constrained-based model and γ, it produces a list of pairs $(reactant, reaction)$ $((reaction, product))$ where $reaction$ is a growth dependent chokepoint and $reactant(product)$ is the metabolite whose only consumer(producer) is $reaction$.

Algorithm 1. Growth dependent chokepoint reactions computation

---

INPUT: $\{\mathcal{R}, \mathcal{M}, \mathcal{S}, lb, ub\}$, γ.
OUTPUT: List of tuples $(reactant, reaction)$ and $(reaction, product)$ such that $reaction \in CP_\gamma$ and $reaction$ is the only consumer of $reactant$ or the only producer of $product$.

```
 1: procedure FINDCHOKEPOINTREACTIONS
 2:     lb_γ, ub_γ ← FVA({R, M, S, lb, ub}, γ)
 3:     {R, M, S, lb, ub} ← {R, M, S, lb_γ, ub_γ}
 4:
 5:     result ← empty list
 6:     for reaction in R do
 7:         for reactant in •reaction_γ do
 8:             if reactant_γ• = {reaction} then
 9:                 result ← result + (reactant, reaction)
10:             end if
11:         end for
12:         for product in reaction_γ• do
13:             if •product_γ = {reaction} then
14:                 result ← result + (reaction, product)
15:             end if
16:         end for
17:     end for
18:     return result
19: end procedure
```

Prior to the computation of chokepoints (lines 5–18), the algorithm refines the input model as explained by replacing the initial flux bounds by the flux values computed with FVA (lines 2–3). To compute chokepoints, the algorithm iterates over the reactions of the model. For each reaction, the reactants and products involved are iterated. For each pair $(reactant, reaction)$ and $(reaction, product)$, if Definition 6 is satisfied, the reaction is considered a chokepoint reaction with the given metabolite.

Example 5. Let us assume that r_{13} in Fig. 1 represents growth, i.e. the component of z in (1) that corresponds to r_{13} is equal to 1 and the rest of components of z are 0. Let us also assume that it is desired to assess the directionality of the reactions and compute the set of chokepoints when the growth is maximum. This can be achieved by applying Algorithm 1 with $\gamma = 1$. The new flux bounds yielded by the algorithm are shown in Fig. 2.

As a result of the new flux bounds given to the model, reactions r_5, r_7 and r_{11} become dead reactions, i.e. $r_5, r_7, r_{11} \in DR_1$, and reactions r_8, r_9, r_{10}, which where reversible reactions in Fig. 1, become non-reversible reactions, i.e. $r_8, r_9, r_{10} \in NR_1$. This change in the directionality of the reactions involves changes in the set of flux-dependent chokepoints, in particular r_6 becomes a chokepoint, i.e. $r_6 \in CP_1$, and r_{11}, which was a chokepoint in Fig. 1, becomes a dead-reaction, i.e. $r_{11} \in DR_1$.

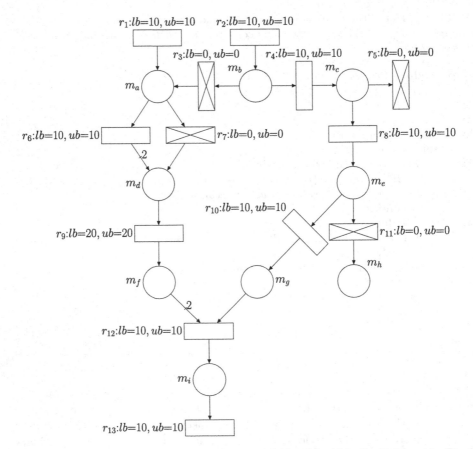

Fig. 2. Petri net resulting from the application of Algorithm 1 to the Petri net in Fig. 1 with r_{13} as the objective function and $\gamma = 1$.

4   Chokepoint Analysis

The computation of flux bounds by means of FVA can be carried out with an optimal state, i.e. $\gamma = 1$ in (2), or with suboptimal states, i.e. $0 \leq \gamma < 1$. While in an optimal state all the fluxes must be optimally directed towards growth, in suboptimal states fluxes are allowed to be diverted towards other functionalities. This section analyses the impact of γ in the sets of reversible, non-reversible, dead and chokepoint reactions of a constraint-based model. To achieve this goal the flux bounds of the model will be refined according to different values of γ and the mentioned sets of reactions will be computed.

All the results presented in this section have been obtained by a software tool which implements Algorithm 1 and computes the sets RR_γ, DR_γ, NR_γ and CP_γ. A description of the tool can be found in Appendix A. The tool has been executed on a number of constraint-based models yielding the results presented in Appendix B.

4.1   Case Study: Mycobacterium Leprae

This subsection presents the results obtained for the constraint-based model of the *in vivo* GEM of *M. leprae* [1,6]. This model is composed of 998 metabolites and 1228 reactions. The sizes of the flux-dependent sets of reactions are $|RR_d| = 288$, $|NR_d| = 938$, $|DR_d| = 2$, and the number of flux-dependent CP is $|CP_d| = 667$. In order to assess the dependence of these sets on γ, the flux bounds of the model have been refined for different values of γ in the interval $[0, 1]$.

Figure 3 shows the sizes of the sets DR_γ, NR_γ and CP_γ, in plot (a), (b) and (c) respectively, for different values of γ. Notice that if $\gamma = 0$ then the constraint $\gamma \cdot \mu_{max} \leq z \cdot v$ in (2) does not impose a minimum growth on the model, and only the steady state condition $S \cdot v = 0$ must be satisfied. In addition to the γ dependent sets, the leftmost value of each plot (depicted in green) in Fig. 3 represents the sizes of the flux-dependent sets prior to FVA, i.e. DR_d, NR_d and CP_d.

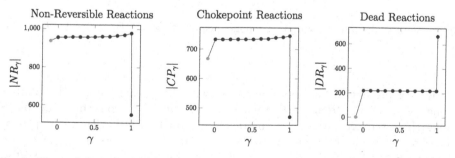

Fig. 3. Sizes of the sets of reactions DR_γ, NR_γ and CP_γ of *M. leprae* for $\gamma \in [0, 1]$. The leftmost value of each plot corresponds to DR_d, NR_d and CP_d respectively.

It can be seen that the set of dead reactions exhibits two major increases, the first one from $|DR_d| = 2$ to $|DR_0| = 219$, and the second one at $\gamma = 1$ ($|DR_1| = 667$), see Fig. 3(c). Given that only the steady state condition is active for $\gamma = 0$, the first increase implies that the fluxes of all the reactions in DR_0 must necessarily be 0 in the long run regardless of the growth rate. Thus, this increase can be due to a shortcoming or incompleteness in the model.

The second increase in the set of dead reactions takes place at $\gamma = 1$ and can be caused by the existence of alternative pathways that consume nutrients, one of them being more efficient than the others in terms of biomass production. The next subsection proposes simplified models that illustrate how these sudden increases take place.

With respect to chokepoints, see Fig. 3(b), the number of flux-dependent chokepoints is $|CP_d| = 668$. This number increases to $|CP_0| = 733$ when the steady state constraint is forced, and increases slowly with γ, at $\gamma = 0.9$ it holds $|CP_{0.9}| = 741$. That is, Algorithm 1 identifies more chokepoints than the ones that are present originally in the model. The sudden drop of chokepoints at $\gamma = 1$, $|CP_1| = 469$, is due to the fact that at the optimal state many chokepoints become dead reactions as discussed previously.

In a similar way to the set of chokepoints, the number of non-reversible reactions increases slowly with γ and it falls abruptly at $\gamma = 1$. This means that the sets NR_γ and CP_γ decrease at the optimal state as many of these reactions become dead reactions.

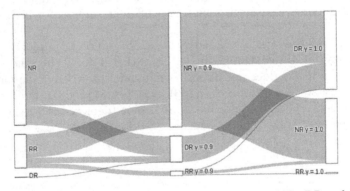

Fig. 4. Sankey diagram showing the dependence of the sets NR, RR and DR on the growth rate.

Figure 4 presents a Sankey diagram showing how reactions are distributed among the sets NR, RR and DR and the flow of transformations that takes place among these sets from the initial model to a model refined with a suboptimal state of $\gamma = 0.9$, and from this suboptimal state model to an optimal state model refined with $\gamma = 1.0$. Notice that at $\gamma = 1.0$ the set of dead reactions becomes the largest set of reactions, and that the set of reversible reactions is vastly reduced at both suboptimal and optimal growth.

As it is reported in Appendix B, similar trends to the ones discussed here for the sets NR, DR and CP are exhibited by other models.

4.2   Dead Reactions and Growth Rate

It has been shown that refining a model with a suboptimal growth can cause the set of dead reactions, DR, to increase, and also that this set increases further with an optimal growth refinement. These changes in DR are caused by particular network structures that can appear in a metabolic network. This subsection illustrates through an abstract model the types of structures that can produce such changes in DR.

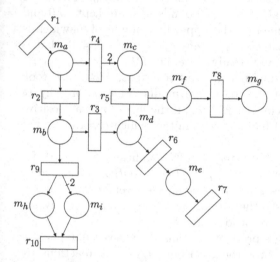

Fig. 5. Petri net illustrating the evolution of dead reactions at $\gamma = 0$ and $\gamma = 1$.

Table 1. Initial flux bounds

r	$lb(r)$	$ub(r)$	type
r_1	0.0	100.0	NR
r_2	0.0	1000.0	NR
r_3	0.0	1000.0	NR
r_4	0.0	1000.0	NR
r_5	0.0	1000.0	NR
r_6	0.0	1000.0	NR
r_7	0.0	1000.0	NR
r_8	0.0	1000.0	NR
r_9	0.0	1000.0	NR
r_{10}	0.0	1000.0	NR

Table 2. Refined model flux bounds

γ	r	$lb_\gamma(r)$	$ub_\gamma(r)$	type
0.0	r_1	0.0	100.0	NR
	r_2	0.0	100.0	NR
	r_3	0.0	100.0	NR
	r_4	0.0	100.0	NR
	r_5	0.0	200.0	NR
	r_6	0.0	200.0	NR
	r_7	0.0	200.0	NR
	r_8	0.0	0.0	DR
	r_9	0.0	0.0	DR
	r_{10}	0.0	0.0	DR
0.9	r_1	90.0	100.0	NR
	r_2	0.0	20.0	NR
	r_3	0.0	20.0	NR
	r_4	80.0	100.0	NR
	r_5	160.0	200.0	NR
	r_6	180.0	200.0	NR
	r_7	180.0	200.0	NR
	r_8	0.0	0.0	DR
	r_9	0.0	0.0	DR
	r_{10}	0.0	0.0	DR
1.0	r_1	100.0	100.0	NR
	r_2	0.0	0.0	DR
	r_3	0.0	0.0	DR
	r_4	100.0	100.0	NR
	r_5	200.0	200.0	NR
	r_6	200.0	200.0	NR
	r_7	200.0	200.0	NR
	r_8	0.0	0.0	DR
	r_9	0.0	0.0	DR
	r_{10}	0.0	0.0	DR

Let us consider the constraint-based model depicted as a Petri net in Fig. 5. The lower and upper flux bounds of the model are reported in Table 1 of the figure. According to such flux bounds all the reactions are non-reversible, i.e. $DR = \{\}$, see column *type*. The exchange reactions are r_1 and r_{10}, which could model nutrient uptake and secretion of a metabolite respectively. The reaction modelling biomass production is r_7, i.e. the flux of r_7 represents the growth rate of system being modelled. Hence, the objective funcion of FBA, see (1), is the maximization of $v[r_7]$. For the given net structure and flux bounds, the maximum growth rate yielded by FBA (1) is $\mu_{max} = 200$.

Table 2 reports the refined flux bounds and the type of each reaction for $\gamma = 0$, $\gamma = 0.9$ and $\gamma = 1$. If $\gamma = 0$ then only the steady state constraint, $S \cdot v = 0$, of FVA (2) is taken into account to compute the refined flux bounds. Thus, any possible steady state must satisfy the bounds, lb_0 and ub_0, of the rows associated with $\gamma = 0$. For such γ, all the lower bounds, lb_0, are kept to 0 and the upper bounds, ub_0, are constrained by the upper flux bound of r_1 which is $ub(r_1) = 100$. It should be noted that $ub_0(r_8) = ub_0(r_9) = ub_0(r_{10}) = 0$, that is reactions r_8, r_9 and r_{10} are dead at any steady state, i.e. $DR_0 = \{r_8, r_9, r_{10}\}$.

Recall that this increase in the number of dead reactions, DR_0, also took place in the previous subsection for the *M. leprae* model. The reason why r_8 becomes dead is because it is producing a DEM m_g, if the flux of r_8 was positive then the concentration of m_g would increase indefinitely which contradicts the existence of a steady state.

Let us now focus on r_9 and r_{10}. The steady state condition, $S \cdot v = 0$, for metabolite m_h imposes $v[r_9] = v[r_{10}]$ (i.e. *input flux = output flux*), while for m_i the steady state condition is $2 \cdot v[r_9] = v[r_{10}]$ (see weight 2 in the arc (r_9, m_i)). The only fluxes that satisfy these two conditions simultaneously are $v[r_9] = v[r_{10}] = 0$, i.e. the reactions are dead at any steady state.

At $\gamma = 0.9$, the flux of r_7 (biomass production) must be at least $0.9 \cdot \mu_{max} = 180$. Notice that, although $DR_0 = DR_{0.9}$, the lower bounds of some reactions are higher at $\gamma = 0.9$ than at $\gamma = 1$, and the upper bounds of some other reactions are lower at $\gamma = 0.9$ than at $\gamma = 1$. In general it holds:

$$ub_{0.9}[r] - lb_{0.9}[r] \geq ub_1[r] - lb_1[r] \quad \forall r \in \mathcal{R}$$

In other words, the range of steady state fluxes allowed for each reaction decreases with γ. This is due to the existence of alternative paths in the metabolic network. In the present example, there are two alternative pathways to biomass production, namely $p_1 = (r_1, r_2, r_3, r_6, r_7)$ and $p_2 = (r_1, r_4, r_5, r_6, r_7)$. Notice that p_2 is more advantageous than p_1 for biomass production as 2 metabolites m_c are produced per each metabolite m_a. Thus, near optimal solutions will tend to exploit p_2 instead of p_1. This implies strictly positive lower bounds, $lb_{0.9}$, for all the reactions in p_2, and decreased upper bounds, $ub_{0.9}$, for the reactions that belong exclusively to p_1, i.e. r_2 and r_3.

The interval $[lb_\gamma[r], ub_\gamma[r]]$ shrinks for every reaction $r \in \mathcal{R}$ as γ increases, and at $\gamma = 1$ the intervals become points. Moreover, at $\gamma = 1$ only the most favourable path p_2 can be used, and hence reactions r_2 and r_3 become dead.

This type of phenomenon causes the sudden increase of DR_1 in the *M. leprae* model of the previous subsection.

5   Conclusions

This work has introduced a computational method to incorporate dynamic information, namely flux bounds, in the computation of chokepoints in metabolic networks expressed as constraint-based models. The goal behind this approach is to obtain more realistic chokepoints, which are known to be potential drug targets, than those based only on the net topology. Given that flux bounds depend on the growth rate of the organism, the concept of growth dependent chokepoints has been defined and an algorithm to compute such chokepoints has been designed.

It was found that the number of chokepoints was seriously affected by the number of dead reactions, i.e. by reactions with null lower and upper flux bounds. Although the number of dead reactions is not relevant in most of the original models, this number increases significantly when dynamic information is accounted for. A major increase takes place when the steady state constraint is enforced on the metabolic network, i.e. at $\gamma = 0$. As discussed in Subsect. 4.2, we hypothesize that such an increase is a sign of some incompleteness in the model such as the existence of dead-end metabolites, or to missing reactions that lead to the incompatibility of positive fluxes with the network stoichiometry. Another major increase of dead reactions takes place when the growth rate is maximum, i.e. at $\gamma = 1$. At such a rate, all the flux is diverted towards the optimal paths for biomass production, and hence, the fluxes of non-optimal alternative paths leading to biomass production and other non-essential paths become 0, i.e. such paths contain dead reactions at $\gamma = 1$. Thus, in this case dead reactions might indicate the existence of alternative, or redundant, paths leading to biomass production. Notice that such redundant paths have the potential to make the metabolism more robust to attacks.

The protocol for chokepoint computation presented on this paper can reduce the time spent identifying drug targets in the process of drug discovery. Drug discovery is a time-consuming process, which involves the identification and validation of drug targets, optimisation, lead discovery and testing before the production of drug candidates. Our protocol can contribute reducing the drug target identification time by prioritising the selection of potential top targets of the pathogenic organism for subsequent validation and optimisation protocols of the drug discovery process.

A   Appendix

The software tool *findCPcli* developed in this work consists of a command line application that, given an input model provided by the user, computes the sizes of the sets of non-reversible reactions, reversible reactions, dead reactions and chokepoint reactions for different values of γ. The results are saved in a spreadsheet file with a format similar to the one presented in Table 3.

The tool *findCPcli* is distributed as a *Python* package and requires *Python 3.5* or a higher version. The source can be found at `github.com/findCP/findCPcli`. *findCPcli* can be installed with the *pip* package management tool:

```
pip install findCPcli
```

Once installed, the results for a given SBML model can be computed running:

```
findCPcli -i <input_file> -cp <output_file>
```

where:

- `<input_file>` is the path of the input SBML model file to be used. The supported file formats are *.xml*, *.json* and *.yml*.
- `<output_file>` is the path of the spreadsheet file that will be saved with the results computed on the model. The available file formats for the spreadsheet file are *.xls*, *.xlsx* and *.ods*.

When the above command is executed, the command line application will inform about the task that will be computed. If the task finishes successfully and the spreadsheet file has been saved, the application will inform about it and will end the execution.

Further information about the operations provided by the application can be found by executing: `findCPcli -h` .

B   Appendix

Table 3 reports the sizes of the sets of reversible, non-reversible, dead and chokepoint reactions for several constraint-based models of the *Biomodels* repository [3]. All the results were computed by the tool *findCPcli*. The maximum CPU time was 82.776 s to compute the results of model MODEL1507180017 in an Intel Core i5-9300H CPU @ 2.40 GHz × 8.

Table 3. Sizes of the sets of reversible, non-reversible, dead and chokepoint reactions for several constraint-based models.

Model	Set	Initial	Set	γ 0.0	0.1	0.2	0.3	0.4	0.5	0.6	0.7	0.8	0.9	1.0
mlep_inVivo_media	RR_d	288	RR_γ	53	52	51	51	51	50	48	48	43	39	10
reactions: 1228	NR_d	938	NR_γ	956	957	958	958	958	959	961	961	966	970	551
metabolites: 998	DR_d	2	DR_γ	219	219	219	219	219	219	219	219	219	219	667
	CP_d	668	CP_γ	733	733	733	733	733	733	735	735	739	741	469
MODEL1507180021	RR_d	208	RR_γ	72	72	72	72	72	72	65	64	63	60	20
reactions: 900	NR_d	692	NR_γ	710	710	710	710	710	710	717	718	719	722	417
metabolites: 688	DR_d	0	DR_γ	118	118	118	118	118	118	118	118	118	118	463
(M. tuberculosis)	CP_d	506	CP_γ	504	504	504	504	504	504	510	510	510	512	350
MODEL1507180007	RR_d	183	RR_γ	27	27	27	26	26	24	24	24	22	20	6
reactions: 554	NR_d	371	NR_γ	377	377	377	378	378	380	380	380	382	384	304
metabolites: 485	DR_d	0	DR_γ	150	150	150	150	150	150	150	150	150	150	244
	CP_d	277	CP_γ	315	315	315	315	315	317	317	317	319	319	271
MODEL1507180063	RR_d	184	RR_γ	32	32	31	31	31	31	28	26	26	13	6
reactions: 556	NR_d	370	NR_γ	399	399	400	400	400	400	403	405	405	418	316
metabolites: 549	DR_d	2	DR_γ	125	125	125	125	125	125	125	125	125	125	234
	CP_d	274	CP_γ	316	316	316	316	316	316	318	321	321	330	291
MODEL1507180030	RR_d	193	RR_γ	28	28	28	28	28	27	27	27	27	27	12
reactions: 560	NR_d	305	NR_γ	317	317	317	317	317	318	318	318	318	318	278
metabolites: 479	DR_d	62	DR_γ	215	215	215	215	215	215	215	215	215	215	270
	CP_d	310	CP_γ	257	257	257	257	257	258	258	258	258	258	258
MODEL1507180070	RR_d	216	RR_γ	25	21	20	18	18	18	17	17	17	17	13
reactions: 743	NR_d	513	NR_γ	425	429	430	432	432	432	433	433	433	433	420
metabolites: 655	DR_d	14	DR_γ	293	293	293	293	293	293	293	293	293	293	310
	CP_d	324	CP_γ	330	333	333	335	335	335	335	335	335	335	336
MODEL1507180045	RR_d	354	RR_γ	171	171	171	171	171	171	171	171	171	169	149
reactions: 778	NR_d	424	NR_γ	487	487	487	487	487	487	487	487	487	489	295
metabolites: 662	DR_d	0	DR_γ	120	120	120	120	120	120	120	120	120	120	334
	CP_d	336	CP_γ	387	387	387	387	387	387	387	387	387	389	229
MODEL1507180024	RR_d	318	RR_γ	75	75	75	75	75	75	75	75	74	62	21
reactions: 832	NR_d	514	NR_γ	510	510	510	510	510	510	510	510	511	523	438
metabolites: 790	DR_d	0	DR_γ	247	247	247	247	247	247	247	247	247	247	373
	CP_d	397	CP_γ	406	406	406	406	406	406	406	406	406	414	380
MODEL1507180036	RR_d	277	RR_γ	103	103	103	103	103	103	99	99	93	89	51
reactions: 870	NR_d	593	NR_γ	593	593	593	593	593	593	597	597	603	607	460
metabolites: 713	DR_d	0	DR_γ	174	174	174	174	174	174	174	174	174	174	359
	CP_d	433	CP_γ	402	402	402	402	402	402	406	406	409	411	346
MODEL1507180049	RR_d	271	RR_γ	189	189	189	189	189	189	189	188	188	184	140
reactions: 971	NR_d	700	NR_γ	717	717	717	717	717	717	717	718	718	722	388
metabolites: 496	DR_d	0	DR_γ	65	65	65	65	65	65	65	65	65	65	443
	CP_d	265	CP_γ	273	273	273	273	273	273	273	273	273	273	156
MODEL1507180068	RR_d	361	RR_γ	84	84	84	84	84	80	80	77	77	77	71
reactions: 1056	NR_d	695	NR_γ	568	568	568	568	568	572	572	575	575	575	463
metabolites: 911	DR_d	0	DR_γ	404	404	404	404	404	404	404	404	404	404	522
	CP_d	549	CP_γ	469	469	469	469	469	471	471	474	474	474	395
MODEL1507180060	RR_d	254	RR_γ	57	57	57	57	54	54	52	50	50	48	8
reactions: 1075	NR_d	821	NR_γ	610	610	610	610	613	613	615	617	617	619	356
metabolites: 761	DR_d	0	DR_γ	408	408	408	408	408	408	408	408	408	408	711
	CP_d	441	CP_γ	363	363	363	363	364	364	366	368	368	370	304

(continued)

Table 3. (*continued*)

Model	Set	Initial	Set	γ 0.0	0.1	0.2	0.3	0.4	0.5	0.6	0.7	0.8	0.9	1.0
MODEL1507180020	RR_d	256	RR_γ	50	48	48	48	47	47	45	45	45	45	12
reactions: 1110	NR_d	751	NR_γ	556	558	558	558	559	559	561	561	561	561	367
metabolites: 879	DR_d	103	DR_γ	504	504	504	504	504	504	504	504	504	504	731
	CP_d	455	CP_γ	396	398	398	398	398	398	400	400	400	400	319
MODEL1507180059	RR_d	630	RR_γ	203	203	203	203	203	203	202	202	200	196	142
reactions: 1112	NR_d	482	NR_γ	511	511	511	511	511	511	512	512	514	518	358
metabolites: 1101	DR_d	0	DR_γ	398	398	398	398	398	398	398	398	398	398	612
	CP_d	470	CP_γ	436	436	436	436	436	436	436	436	436	438	306
MODEL1507180013	RR_d	551	RR_γ	249	249	249	249	249	249	249	249	247	242	49
reactions: 1245	NR_d	694	NR_γ	708	708	708	708	708	708	708	708	710	715	390
metabolites: 987	DR_d	0	DR_γ	288	288	288	288	288	288	288	288	288	288	806
	CP_d	484	CP_γ	533	533	533	533	533	533	533	533	534	536	346
MODEL1507180058	RR_d	452	RR_γ	155	155	155	155	155	155	155	155	150	147	97
reactions: 1285	NR_d	833	NR_γ	904	904	904	904	904	904	904	904	909	912	547
metabolites: 943	DR_d	0	DR_γ	226	226	226	226	226	226	226	226	226	226	641
	CP_d	584	CP_γ	611	611	611	611	611	611	611	611	614	616	409
MODEL1507180015	RR_d	1093	RR_γ	510	510	510	510	510	510	510	510	510	510	334
reactions: 1681	NR_d	588	NR_γ	822	822	822	822	822	822	822	822	822	822	667
metabolites: 1381	DR_d	0	DR_γ	349	349	349	349	349	349	349	349	349	349	680
	CP_d	473	CP_γ	612	612	612	612	612	612	612	612	612	612	551
MODEL1507180054	RR_d	546	RR_γ	85	85	85	82	80	80	77	77	75	71	10
reactions: 2262	NR_d	1716	NR_γ	1138	1138	1138	1141	1143	1143	1146	1146	1148	1152	397
metabolites: 1658	DR_d	0	DR_γ	1039	1039	1039	1039	1039	1039	1039	1039	1039	1039	1855
	CP_d	1039	CP_γ	748	748	748	750	750	750	752	752	754	759	319
MODEL1507180017	RR_d	606	RR_γ	85	85	85	81	80	78	78	77	77	75	13
reactions: 2546	NR_d	1923	NR_γ	1504	1504	1504	1508	1509	1511	1511	1512	1512	1514	525
metabolites: 1802	DR_d	17	DR_γ	957	957	957	957	957	957	957	957	957	957	2008
	CP_d	1112	CP_γ	984	984	984	986	986	988	988	988	988	990	478

References

1. Bannerman, B.P., et al.: Analysis of metabolic pathways in mycobacteria to aid drug-target identification. bioRxiv (2019). https://doi.org/10.1101/535856
2. Burgard, A.P., Vaidyaraman, S., Maranas, C.D.: Minimal reaction sets for escherichia coli metabolism under different growth requirements and uptake environments. Biotechnol. Prog. **17**(5), 791–797 (2001). https://doi.org/10.1021/bp0100880
3. Glont, M., et al.: Biomodels: expanding horizons to include more modelling approaches and formats. Nucleic Acid Res. **46**(D1), D1248–D1253 (2018). https://doi.org/10.1093/nar/gkx1023
4. Gudmundsson, S., Thiele, I.: Computationally efficient flux variability analysis. BMC Bioinf. **11**(1), 489 (2010). https://doi.org/10.1186/1471-2105-11-489
5. Heiner, M., Gilbert, D., Donaldson, R.: Petri nets for systems and synthetic biology. In: Bernardo, M., Degano, P., Zavattaro, G. (eds.) SFM 2008. LNCS, vol. 5016, pp. 215–264. Springer, Heidelberg (2008). https://doi.org/10.1007/978-3-540-68894-5_7

6. Karp, P.D., et al.: The BioCyc collection of microbial genomes and metabolic pathways. Briefings Bioinf. **20**(4), 1085–1093 (2017). https://doi.org/10.1093/bib/bbx085
7. Lamberti, L.M., et al.: Breastfeeding for reducing the risk of pneumonia morbidity and mortality in children under two: a systematic literature review and meta-analysis (2013). https://doi.org/10.1186/1471-2458-13-S3-S18
8. Mackie, A., Keseler, I.M., Nolan, L., Karp, P.D., Paulsen, I.T.: Dead end metabolites - defining the known unknowns of the e. coli metabolic network. PLoS ONE **8**(9), e75210 (2013). https://doi.org/10.1371/journal.pone.0075210
9. Mazurek, S.: Pyruvate kinase type M2: a key regulator of the metabolic budget system in tumor cells. Int. J. Biochem. Cell Biol. **43**(7), 969–980 (2011). https://doi.org/10.1016/j.biocel.2010.02.005
10. Munger, J., et al.: Systems-level metabolic flux profiling identifies fatty acid synthesis as a target for antiviral therapy. Nat. Biotechnol. **26**(10), 1179–1186 (2008). https://doi.org/10.1038/nbt.1500
11. Murata, T.: Petri nets: properties, analysis and applications. Proc. IEEE **77**(4), 541–580 (1989). https://doi.org/10.1109/5.24143
12. Murima, P., McKinney, J.D., Pethe, K.: Targeting bacterial central metabolism for drug development, November 2014. https://doi.org/10.1016/j.chembiol.2014.08.020
13. Orth, J.D., et al.: A comprehensive genome-scale reconstruction of Escherichia coli metabolism-2011. Mol. Syst. Biol. **7**(1), 535 (2011). https://doi.org/10.1038/msb.2011.65
14. Orth, J.D., Thiele, I., Palsson, B.O.: What is flux balance analysis?, March 2010. https://doi.org/10.1038/nbt.1614
15. Rahman, S.A., Schomburg, D.: Observing local and global properties of metabolic pathways: 'load points' and 'choke points' in the metabolic networks. Bioinformatics (Oxford, Engl.) **22**(14), 1767–1774 (2006). https://doi.org/10.1093/bioinformatics/btl181
16. Raman, K., Vashisht, R., Chandra, N.: Strategies for efficient disruption of metabolism in Mycobacterium tuberculosis from network analysis. Mol. BioSyst. **5**(12), 1740–1751 (2009). https://doi.org/10.1039/B905817F
17. Segre, D., Vitkup, D., Church, G.M.: Analysis of optimality in natural and perturbed metabolic networks. Proc. Natl. Acad. Sci. **99**(23), 15112–15117 (2002). https://doi.org/10.1073/pnas.232349399
18. Singh, S., Malik, B.K., Sharma, D.K.: Choke point analysis of metabolic pathways in E. histolytica: a computational approach for drug target identification. Bioinformation **2**(2), 68–72 (2007). https://doi.org/10.6026/97320630002068
19. Varma, A., Palsson, B.Ø.: Metabolic flux balancing: basic concepts, scientific and practical use. Nat. Biotechnol. **12**(10), 994–998 (1994). https://doi.org/10.1038/nbt1094-994
20. WHO: WHO — Causes of death. WHO (2018)
21. Zhang, R., Lin, Y.: DEG 5.0, a database of essential genes in both prokaryotes and eukaryotes. Nucleic Acids Res. **37**(suppl_1), D455–D458 (2008). https://doi.org/10.1093/nar/gkn858

On the Complexity of Quadratization
for Polynomial Differential Equations

Mathieu Hemery, François Fages[(⊠)], and Sylvain Soliman

Inria Saclay Ile de France, EP Lifeware, Palaiseau, France
Francois.Fages@inria.fr

Abstract. Chemical reaction networks (CRNs) are a standard formalism used in chemistry and biology to reason about the dynamics of molecular interaction networks. In their interpretation by ordinary differential equations, CRNs provide a Turing-complete model of analog computation, in the sense that any computable function over the reals can be computed by a finite number of molecular species with a continuous CRN which approximates the result of that function in one of its components in arbitrary precision. The proof of that result is based on a previous result of Bournez et al. on the Turing-completeness of polynomial ordinary differential equations with polynomial initial conditions (PIVP). It uses an encoding of real variables by two non-negative variables for concentrations, and a transformation to an equivalent quadratic PIVP (i.e. with degrees at most 2) for restricting ourselves to at most bimolecular reactions. In this paper, we study the theoretical and practical complexities of the quadratic transformation. We show that both problems of minimizing either the number of variables (i.e., molecular species) or the number of monomials (i.e. elementary reactions) in a quadratic transformation of a PIVP are NP-hard. We present an encoding of those problems in MAX-SAT and show the practical complexity of this algorithm on a benchmark of quadratization problems inspired from CRN design problems.

1 Introduction

Chemical reaction networks (CRNs) are a standard formalism used in chemistry and biology to reason about the dynamics of molecular interaction networks. A CRN over a vector x of molecular species is a finite set of formal chemical reactions of the form

$$r(x) \xrightarrow{f(x)} p(x)$$

composed of a multiset $r(x)$ of reactants (with multiplicity given by stoichiometric coefficients in r), a multiset $p(x)$ of products, and a rate function $f(x)$ on the quantities of reactants. The structure of a CRN is the same as the structure of a Petri net, but the rate functions allow for the definition of continuous-time dynamics in addition to their discrete dynamics: in particular the stochastic semantics which interprets a CRN by a continuous-time Markov chain, and the

© Springer Nature Switzerland AG 2020
A. Abate et al. (Eds.): CMSB 2020, LNBI 12314, pp. 120–140, 2020.
https://doi.org/10.1007/978-3-030-60327-4_7

differential semantics which interprets a CRN by a system of ordinary differential equations (ODEs) [4, 6].

In the differential semantics of a CRN $R = \{r_j(\boldsymbol{x}) \xrightarrow{f_j(\boldsymbol{x})} p_j(\boldsymbol{x})\}$, one associates to each molecular species x_i a concentration also noted x_i by abuse of notation, with the differential function

$$\frac{dx_i}{dt} = \sum_{j \in R} (p_j(x_i) - r_j(x_i)).f_j(\boldsymbol{x}).$$

Mass action law kinetics are monomial rate functions that lead to polynomial ODEs. The other standard rate functions used such as Michaelis-Menten kinetics and Hill kinetics are traditionally obtained by approximations of mass action law systems [11], and can thus be disregarded without loss of generality.

Collision theory shows however that the probabilities of reactions involving three or more reactants are negligible. Hence from a mechanistic point of view, the restriction to reactions involving at most two reactant molecules is of practical importance. We call an elementary CRN (ECRN) a CRN with at most bimolecular reactions and mass action law kinetics. The restriction to at most bimolecular reactions leads to polynomial ODEs of degree at most 2.

With these restrictions, ECRNs have been shown to provide a Turing-complete model of analog computation, in the sense that any computable function over the reals can be computed by an ECRN which approximates the result of that function on one of its components in arbitrary precision [5]. More precisely, we say that a CRN with a distinguished output species x_1 generates a function of time $f : \mathbb{R}_+ \to \mathbb{R}_+$ from initial state $\boldsymbol{x}(0)$ if $\forall t\ x_1(t) = f(t)$. A CRN with distinguished input and output species x_0 and x_1 computes a positive real function $f : \mathbb{R}_+ \to \mathbb{R}_+$ from initial state $\boldsymbol{x}_i(0) = q(x_0(0))$ for some polynomial q and $i \in \{1, \ldots, n\}$, if for any initial concentration $x_0(0)$ of the input species, the concentration of the output molecular species stabilizes at concentration $x_1 = f(x_0(0))$. The proof of Turing-completeness of ECRNs in [5] is based on a previous result of Bournez et al. in [1] on the Turing-completeness of polynomial ordinary differential equations with polynomial initial values (PIVPs) for computing real functions [8]. The proof for ECRNs uses on the one hand, on an encoding of real variables x by the difference of two non-negative variables x^+ and x^- for concentrations, and on the other hand, on a transformation of the PIVP to a quadratic PIVP [2] computing the same function but with degree at most 2.

In this paper, we study the quadratic transformation problem and its computational complexity.

Example 1. The hill function of order 5:

$$H_5(x) = \frac{x^5}{1 + x^5}.$$

is an interesting example because it has been shown to provide a good approximation of the input/output function of the MAPK signalling network which is an ubiquitous CRN structure present in all eukaryote cells and in several copies [9]. That function is a stiff sigmoid function which provides the MAPK network with a switch-like response to the input, ultrasensitivity and an analog/digital converter function. It is thus interesting to compare the MAPK network to the CRN design method above based on the mathematical definition of the H_5 function by ODEs. Following [5], one can easily check that the function $H_5(x)$ is computed by the following PIVP noted in vectorial form for the differential equations[1] and the initial conditions:

$$\frac{d}{dt}\begin{bmatrix} H \\ I \\ T \\ X \end{bmatrix} = \begin{bmatrix} 5.I^2.T^4.X \\ -5.I^2.T^4.X \\ X \\ -X \end{bmatrix}, \quad \begin{bmatrix} 0 \\ 1 \\ 0 \\ x \end{bmatrix}_{t=0}.$$

For any positive value $X(0) = x$ in the initial condition, we have $\lim_{t\to\infty} H(t) = H_5(x)$. However, this PIVP is of order 7 and its direct implementation by CRN would involve non-elementary reactions with 7 reactants. In this example, the proof of existence of a quadratic transformation for PIVPs given in [2] introduces 29 variables, while the MAPK network involves 12 molecular species. In this paper, we consider the quadratic transformation problem as an optimization problem which consists in minimizing the dimension of the quadratic PIVP. The optimization algorithm we propose generates the following optimal ECRN for implementing $h_5(x)$ with only 7 variables (named below by the monomial they represent, e.g. it4x for $I.T^4.X$) and 11 reactions with mass action law kinetics (MA):

```
MA(5.0) for i+it4x=>h+it4x   MA(1.0) for ix=>_
MA(1.0) for x=>_             MA(5.0) for it4x+ix=>it4x
MA(1.0) for 2*x=>tx+2*x      MA(4.0) for ix+t3x=>it4x+ix+t3x
MA(1.0) for tx=>_            MA(1.0) for it4x=>_
MA(3.0) for 2*tx=>t3x+2*tx   MA(5.0) for 2*it4x=>it4x
MA(1.0) for t3x=>_
```

In the following, we show that both problems of minimizing either the dimension (i.e. number of molecular species) or the number of monomials (i.e. number of reactions[2]) in a quadratic transformation of a PIVP are NP-hard. The proof is by reduction of the vertex set covering problem (VSCP). We present an algorithm based on an encoding in a MAX-SAT Boolean satisfiability problem, and

---

[1] More precisely, the first two equations have for solution the Hill function of order 5 as a function of time T, and the last two equations has for effect to stop time T at initial value $X(0)$.

[2] While the correspondance between variables and species is exact, the one between monomials and reactions is in fact more complicated if stoechiometric coefficients and rate constants are exchanged when gathering the monomials appearing in the different differential functions. In the following, we will nonetheless minimize monomials as a proxy for the number of reactions.

show its practicality on a benchmark of quadratization problems inspired from CRN design problems.

The rest of the paper is organized as follows. In the next section, we define the quadratic transformation decision problem (QTDP) as the problem of deciding whether there exists a PIVP quadratization of some given dimension k, and the associated optimization problem (QTP) to determine the minimum number k of variables. We also consider the minimization of the number of monomials. The difficulty of those problems are illustrated with some motivating examples. We distinguish the succinct representation of the input PIVP by a list of monomials, under which QTP is shown to be in NEXP, from the non succinct representation by the full matrix of possible monomials of the input PIVP under which QTP is shown to be in NP. In Sect. 3, we present an encoding of the QTP as a MAX-SAT Boolean satisfiability problem, and derive from that encoding an algorithm to solve QTP and its variant for minimizing the number of monomials.

Then in Sect. 4, we show that the different QTP problems are NP-hard. More precisely, we show that the decision problem in the non-succinct representation of the input PIVP is NP-complete by reduction of the Vertex Set Covering Problem (VSCP), and we conjecture that the decision problem in the succinct representation is NEXP-complete with the argument that some hard instances of QTP require an exponential number of variables in the size of the input PIVP.

Then in Sect. 5, we study the practical complexity of QTP. We propose a benchmark of PIVP quadratization problems inspired from CRN design problems, and show the performance of the MAX-SAT algorithm on this benchmark[3]

2 Quadratic Transformation of PIVPs

2.1 Quadratic Projection Theorem

A PIVP is a system of polynomial differential equations given with initial values. Following the notations of [2], from $\mathcal{A}$ the set of real analytic functions, we say that $f \in \mathcal{A}$ is *projectively polynomial* if f is a component of the solution of a PIVP. We note $\mathcal{P}$ the set of such functions, and $\mathcal{P}_k(n)$ the subset of functions defined by a PIVP of dimension n and degree at most k. $\mathcal{P}_k$ will denote $\bigcup_{n \in \mathbb{N}} \mathcal{P}_k(n)$.

Example 2. The cosine function belongs to the class $\mathcal{P}_1(2)$ since it may be defined over $\mathbb{R}$ through the PIVP:

$$\frac{d}{dt}\begin{bmatrix} x \\ y \end{bmatrix} = \begin{bmatrix} -y \\ x \end{bmatrix}, \quad \begin{bmatrix} 1 \\ 0 \end{bmatrix}_{t=0}.$$

That notation will be kept throughout the article with the last element denoting the initial condition of the PIVP (at $t = 0$ by convention).

---

[3] The benchmark and the implementation in BIOCHAM are available online in a Jupyter notebook at https://lifeware.inria.fr/wiki/Main/Software#CMSB20a.

A folklore theorem of polynomial differential equation systems is that they can be restricted to degree at most 2 without loss of generality on the generated functions:

Theorem 1. $\mathcal{P} = \mathcal{P}_2$: *any function generated by a PIVP can be generated by a PIVP of degree at most two.*

The proof given in [2] is based on Algorithm 1 which consists in introducing as many new variables as the number of possible monomials.

Algorithm 1. Quadratization algorithm of Carothers et al. [2].

Input: PIVP with n variables $\{x_1, \ldots, x_n\}$, and maximum power d_j per variable.
Output: quadratic PIVP with same output function on variable $v_{1,0,\ldots,0}(t)$.

1. Introduce the variables $v_{i_1,\ldots,i_n} = x_1^{i_1} x_2^{i_2}, \ldots, x_n^{i_n}$ for all i_j, $0 \le i_j \le d_j$, $1 \le j \le n$ satisfying $i_k > 0$ for some variable indice k;
2. If the output variable x_1 has a maximum power 0, add the variable $v_{1,0,\ldots} = x_1$.
3. Compute the derivatives of the v variables as functions of the x variables;
4. Replace the monomials in the derivatives of the v variables by monomials of the v variables with degree at most 2.

While it is obvious that the derivatives of the original variables can be rewritten by a sum of the new variables, one must check that the derivatives of the new variables can be written in quadratic form. Let $x_1, x_2 \ldots$ be the variables of the input PIVP, and d_n be the highest degree of x_n among all the monomials of the input PIVP. One new variable is introduced for each monomial $v_{\{i_1,\ldots,i_n\}} = \prod x_n^{i_n}$ that is possible to construct with $i_n \in \{0, \ldots, d_n\}$ and at least one i_n strictly positive[4]. It is then clear that the original function is still computed by the output PIVP of Algorithm 1 since we explicitly introduce it. Furthermore, we can compute the derivative of the new variables:

$$\frac{d}{dt} \prod x_n^{i_n} = \sum_k \left(i_k \frac{dx_k}{dt} x_k^{i_k - 1} \prod_{n \ne k} x_n^{i_n} \right), \tag{1}$$

and it is enough to note that $v_{\{i_1,\ldots,i_k-1,\ldots,i_n\}}$ is one of the new variables and that $\frac{dx_k}{dt}$ has only monomial of degree one in the new set of variables by construction. This derivative is thus quadratic with respect to the new variables.

Proposition 1. *Algorithm 1 introduces $O(d^n)$ variables where n is the number of variables and d the maximum power in the original PIVP.*

---

[4] One can remark that step 2 in Algorithm 1 was omitted in the original proof of [2] but is necessary, as shown for instance for the Hill function given in Sect. 5.

Proof. For a PIVP of n variables x_i with highest degree d_i and with a distinguished output variable x_1, Algorithm 1 introduces $\prod_i (d_i + 1) - 1 + \delta(d_1, 0) = O(d^n)$ variables where δ is the Kronecker delta which is 1 iff $d_i = 0$ and 0 otherwise. The first term in the expression comes from the fact that each old variable x_i may appear in the new set of variables with a power ranging from 0 to d_i. The second term comes from the exclusion of the null variable, and the last one prevents us to delete the distinguished output variable if it does not appear in the derivatives.

However, Algorithm 1 may introduce much more variables than is actually needed as already shown by Example 1 and more precisely by the examples below.

2.2 Examples

Example 3. Applying Algorithm 1 to the PIVP $d_t x = x^k$ with the initial condition $x(t = 0) = x_0$ would introduce k variables for $x, x^2, \ldots, x^k$. But as it can be easily checked, that this PIVP can also be quadratized with only two variables: $x, y = x^{k-1}$ with

$$\frac{d}{dt} \begin{bmatrix} x \\ y \end{bmatrix} = \begin{bmatrix} xy \\ (k-1)y^2 \end{bmatrix}, \quad \begin{bmatrix} x_0 \\ x_0^{k-1} \end{bmatrix}_{t=0}.$$

In the example above, the number of variables needed does not depend on the degree of the input PIVP. More generally it is not always the case that when the degree of a monomial increases, the minimum number of variables in a quadratized form of the PIVP increases:

Example 4. The system:

$$\frac{d}{dt} \begin{bmatrix} x \\ y \end{bmatrix} = \begin{bmatrix} y^3 \\ x^3 + x^2 y^2 \end{bmatrix}$$

needs 7 variables $(x, y, xy, y^2, x^3, y^3, xy^2)$. When increasing the highest degree by one:

$$\frac{d}{dt} \begin{bmatrix} x \\ y \end{bmatrix} = \begin{bmatrix} y^4 \\ x^4 + x^2 y^2 \end{bmatrix}$$

we need only 6 variables $(x, y, x^3, y^3, x^2 y, xy^2)$. But pursuing to increase:

$$\frac{d}{dt} \begin{bmatrix} x \\ y \end{bmatrix} = \begin{bmatrix} y^5 \\ x^5 + x^2 y^2 \end{bmatrix}$$

needs now 9 variables for example: $x, y, x^3, y^3, xy^2, x^4, y^4, x^3 y, xy^3$. This is still far less than the solution given by the mathematical proof with the 35 variables of the monomials smaller than $x^5 y^5$.

Example 5. On the system $\frac{d}{dt}\begin{bmatrix} x \\ y \end{bmatrix} = \begin{bmatrix} y^3 \\ x^3 \end{bmatrix}$, our algorithm presented in the sequel returns the following solution with 5 variables $a = x, b = y, c = x^2$, $d = y^2, e = xy$:

$$d_t a = y^3 = bd, \tag{2}$$
$$d_t b = x^3 = ac, \tag{3}$$
$$d_t c = 2xy^3 = 2de, \tag{4}$$
$$d_t d = 2x^3 y = 2ce, \tag{5}$$
$$d_t e = x^4 + y^4 = c^2 + d^2. \tag{6}$$

A critical aspect of the optimal solution is that it may contain monomials, like xy here, that do not appear in the derivatives of the initial variables and could appear unnecessary at first glance.

Example 6. Interestingly, the PODE

$$\frac{d}{dt}\begin{bmatrix} a \\ b \\ c \end{bmatrix} = \begin{bmatrix} b^2 + a^2 b^2 c^2 \\ c^2 + a^2 b^2 c^2 \\ a^2 + a^2 b^2 c^2 \end{bmatrix}$$

where each derivative is composed of the square of the next variable in addition to a long monomial formed with the square of all possible variables is among the ones needing the most variables. For this example, the optimal set found by our algorithm described in the sequel is:

$$\{a, b, c, a^2, b^2, c^2, abc, ab^2, ac^2, a^2b, a^2c, bc^2, b^2c, ab^2c^2, a^2bc^2, a^2b^2c\},$$

that is 16 variables.

Although we have not been able to prove it, the previous example suggests that a quadratic transformation may effectively need an exponential number of variables. We thus formulate the following conjecture:

Conjecture 1. The quadratization of PIVPs of the form:

$$\frac{dx_i}{dt} = x_{i+1}^2 + \prod x_j^2, \quad x_i(t = 0) = 1,$$

with $i \in (1, \ldots, n)$ and where x_{n+1} denotes x_1, requires an exponential number of variables in n.

2.3   Quadratic Transformation Problems

The quadratic transformation problem (QTP) is the optimization problem of determining the minimum number of variables necessary to define an equivalent quadratic PIVP:

Instance: A PIVP on n variables $X = \{x_i\}_{0 \leq i \leq n-1}$ with a distinguished output variable x_0.
Output: the minimum number k of functions $f_j(X)$ such that $\{x_0, f_j(X)\}$ defines an algebraically equivalent quadratic PIVP.

The associated decision problem (QTDP) is:

Instance: A PIVP on variables $X = \{x_i\}$, a distinguished variable x_0 and an integer k
Output: existence or not of k functions $f_j(X)$ such that $\{x_0, f_j(X)\}_{1 \leq j \leq k}$ defines an algebraically equivalent quadratic PIVP.

It is worth noting that the computational complexity of a decision problem may change drastically, for instance from NP to NEXP, according to the succinct or not representation of the input [10]. The representation of the input PIVP given above by a list of symbolic functions is a succinct representation. A non-succinct representation of the input PIVP is given by the matrix of monomial coefficients $K : \mathbb{R}^n \times \mathbb{R}^m$ where n is the dimension of the PIVP and $m \leq (d+1)^n$ the number of possible monomes to consider (Proposition 1).

Let us denote by nsQTP and nsQTDP the non-succinct variants of the QTP and QTDP problems.

Proposition 2. *nsQTDP $\in$ NP. QTDP $\in$ NEXP.*

Proof. By Proposition 1, the size of a witness for a quadratic PIVP is less than the non-succinct representation of the input PIVP by the full matrix of possible monomials. Given such a witness quadratic PIVP one can check in polynomial time that it defines a quadratic PIVP algebraically equivalent to the original PIVP. For that, we just have to compute the derivatives of all the new variables expressed as functions of the old ones; then to express still with the old variables, all the monomials of degree 2 that may be formed with the new variables (an operation that is clearly quadratic in the number of variables); and finally to rewrite all the new derivatives with monomials or quadratics of the new variables. As each derivative contains only a linear number of monomials, we have a quadratic algorithm to check the validity of a witness, hence we have nsQTDP $\in$ NP.

Now, in the succinct representation of the input PIVP by lists of monomials, the size of the witness is bound by an exponential in the size of the input PIVP, hence we simply get QTDP $\in$ NEXP.

In the following (Theorem 2), we show that nsQTDP is actually NP-complete, and thus nsQTP NP-hard, and we conjecture that QTDP is NEXP-complete by extending our conjecture 1 above to hard instances.

3   MAX-SAT Encoding

The maximum satisfiability problem (MAX-SAT) is a generalization of the Boolean satisfiability problem SAT, where some *soft* clauses, that can be either true or false in a solution, are added to a traditional (*hard*) SAT problem, and where the optimization problem of maximizing the number of soft clauses satisfied is considered.

Algorithm 1 can be reformulated in MAX-SAT form, by expressing the constraints of QTP with Boolean clauses which lead to Algorithm 2.

Algorithm 2. Encoding of QTP in MAX-SAT.

1. For each monomial m in the set M considered in Sec. 2.1 (i.e., all those corresponding to variables v of step 1 of Alg. 1), introduce a Boolean variable x_m representing its presence in the reduced system.
2. For each of those monomials, compute its derivative m' (same as step 2of Alg. 1)
3. For each monomial appearing in any m', compute all the ways to represent it as the product of 0 (constant case), 1 or 2 of the monomials of M.
4. Now add to the MAX-SAT model one hard clause imposing that the output variable is present (i.e., true).
5. Add to the MAX-SAT model one soft clause with the negation of each other variable. The maximization will therefore try to make as few variables present as possible.
6. Add a hard clause for each variable imposing that if it is present, its derivative can be represented (with degree at most 2) in the system. This is done with an implication: if the variable is true, then take (the CNF representation of) the conjunction of all the monomials in its derivative, and for each the disjunction of one of its possible representation computed in step 3 should be true (i.e., present in the system).

An example of what happens in step 3 is as follows: assume you get the monomial ab^2 in the derivative of the monomial m. There are three different ways to represent it: as a single variable x_{ab^2}, or as a product $x_a x_{b^2}$ or $x_{ab}x_b$. Hence in step 6 we will get the CNF representation of $x_m \Rightarrow (x_{ab^2} \vee (x_a \wedge x_{b^2}) \vee (x_{ab} \wedge x_b)) \vee \ldots$ More generally, we have

Proposition 3. *The number of variables in our MAX-SAT model is $|M|$, and the number of clauses, because of the DNF-to-CNF conversion is bounded by $O(|M| + 2^d)$, where d is the highest product of the degrees of any monomial of m'.*

Proof. Indeed there are less than $d = \frac{1}{2} \prod_{1 \leq i \leq n}(d_i + 1)$ ways to represent, as a product (independent of the order) of one or two variables, the monomial $\prod x_i^{d_i}$. This leads, in step 3, to a Boolean representation as a big disjunction of d conjunctions of two variables, which once converted to CNF amounts to at worst 2^d clauses.

4 NP-Hardness

In this section and Appendix 8 we prove the NP-completeness of nsQTDP, through a reduction of the Vertex Set Covering Problem (VSCP) [7], i.e. the problem of determining the minimum number of vertices that touch every edges of a graph.

We give in Appendix 7 a similar, yet simpler, reduction to show the NP-hardness of the Max-Horn-SAT problem (while Horn-SAT and Min-Horn-SAT are in P). It may be useful to the reader to read this proof to help understand the logic of the reduction before getting into the more complicated details of the differential equation setting. In essence both reductions work by translating the choice between the two ends of an edge in a graph in a choice in the other problem. Let us take an edge and its two vertices that we will call V_i and V_j. For the Horn-SAT problem, we introduce a clause of the form $\neg v_i \vee \neg v_j$ that ensures that one of the two variables is set to false in a satisfied instance; setting a variable to false thus indicates that the corresponding vertex is in the covering. For the quadratic reduction problem, we introduce a monomial of degree 3 $(V_i V_j Z)$ in the derivative of an auxiliary variable. To perform a quadratic transformation, we then have to "split" this monomial as the product of two variables: $\overline{V_i Z} \times \overline{V_j}$ or $\overline{V_i} \times \overline{V_j Z}$. The variables of the form $\overline{V_i Z}$ appearing in the reduction will correspond to the vertex V_i in the covering of the graph.

Another way to see the connection between these reductions lies in the parallel between Horn-SAT as a model of theorem prover (if B and C are true then A is true) and the Quadratic Transformation as a model of computation (if variable B and C are computed monomials then A can be).

4.1   Encoding of the Vertex Set Covering Problem

Given a graph $G = (V, E)$, a vertex cover is a subset of vertices, $S \subset V$, so that every edge has at least one endpoint in S:

$$\forall e = (i, j) \in E, (i \in S) \vee (j \in S). \tag{7}$$

The VSCP is the optimization problem of finding the smallest vertex cover in a given graph:

Instance: Graph $G = (V, E)$
Output: Smallest number k such that G has a vertex cover of size k.

The associated decision problem is to determine the existence of a vertex cover of size at most k.

It is well-known that the vertex set covering decision problem is NP-complete. Here, we prove the same for the non-succinct quadratic transformation problem for PIVPs (nsQTDP). The general idea is, starting from a graph G, to construct a PIVP where only the first derivatives contains monomials of degree higher than 2, in such a way that the set of variables of the output is simply linked with the elements of the optimal cover S of G.

Starting from a graph $G = (\{V_1, \ldots, V_n\}, E)$, we construct $\text{PIVP}_3(G)$ with $n + 2$ variables, defined by:

$$\frac{dV_0}{dt} = \sum_{(V_i, V_j) \in E} V_i V_j V_{n+1} + V_1, \tag{8}$$

$$\frac{dV_i}{dt} = \sum_{j=1}^{n+1} a_{i,j} V_i V_j + V_{i+1} \quad \forall i \in [1, n], \tag{9}$$

$$\frac{dV_{n+1}}{dt} = \sum_{j=1}^{n+1} a_{n+1,j} V_{n+1} V_j, \tag{10}$$

$$a_{i,j} = i(n+2) + j \tag{11}$$

and an initial condition of the form: $V_i(t = 0) = \frac{i}{i+1}$.

It is worth noting that the $a_{i,j}$'s (and the initial conditions) are chosen here just to be different in each derivative (and variables), this ensures that no polynomial may be used to quadraticly transformed this PIVP. It is interesting to note that the initial condition are not essential for the proof and that the quadratic transformation is as hard for PODE as it is for PIVP.

This encoding shows with a proof given in Appendix 8 that

Theorem 2. *The nsQTDP (resp. nsQTP) is NP-complete (resp. NP-hard).*

In the succinct representation of the input PIVP by a list of symbolic functions, if Conjecture 1 is true, we get that the witness may have an exponential size in the size of the succinct representation of the input PIVP, which leads us to:

Conjecture 2. The QTDP is NEXP-complete. QTP is NEXP-hard.

4.2 Minimizing the Number of Monomials

It is legitimate to ask if minimizing the number of monomials (i.e. reactions in the ECRN framework) is as hard as minimizing the number of variables (i.e. species). Actually, the proof given above still works for this variant of QTP:

Theorem 3. *Given a PIVP P with variables v_i, determining a set of variables v'_j defines through functions f_j of the v_i: $v'_j = f_j(\{v_i\})$ such that the PIVP P' thus defined is quadratic, encodes the same function as P and has less than k monomials is an NP-complete problem.*

The proof is given in Appendix 8.4.

Now, as shown in the following section, though of same theoretical complexity as minimizing the number of species, minimizing the number of reactions seems a bit easier in practice with the MAX-SAT algorithm.

5   Practical Complexity

5.1   Benchmark of CRN Design Problems

The quadratization problem naturally arises in the synthetic biology perspective for the problem of designing an ECRN to implement a given high-level function presented by a PIVP. We propose here such a benchmark of synthesis problems for sigmoid functions and particularly Hill functions of various order, and other functions of interest to understand the practical complexity of QTP.

For this article, we were particularly interested in the time taken to find the optimal solution of the quadratic transformation and as such report the performance for the resolution of this precise problem. We therefore provide in Table 1 both the total execution time going from the PIVP to the ECRN (Total time) and the time taken by the MAX-SAT solver that solves the quadratic

Table 1. Benchmark of quadratization problems given with computation times in ms for the tranformation to MAX-SAT and for MAX-SAT solving (Algorithm 2, the minimum number of variables compared to the number of variables found by Algorithm 1, and the minimum number of monomials (i.e. elementary reactions).

CRN name	Algorithm 2 total time ms	MAX-SAT time ms	Minimum/Algorithm 1 nb. variables	Min reactions MAX-SAT time ms	Min reactions nb. reactions
circular 2,3	80.35	0.2	5/14	0.2	6
circular 2,4	120.4	0.6	6/23	0.6	8
circular 2,5	869.5	7.2	6/34	6.6	8
circular 2,6	54450	754.5	7/47	945.1	10
hard3	1576	7.3	14/34	7.3	28
hard4	28730	369.3	16/43	297.5	31
hill2	77.74	0.1	3/5	0.1	3
hill2x	90.86	0.1	5/11	0.2	7
hill3	78.06	0.1	4/8	0.1	4
hill3x	103.5	0.2	6/17	0.3	9
hill4	85.18	0.1	5/11	0.1	5
hill4x	152.2	0.7	7/23	0.7	11
hill5	84.7	0.2	5/14	0.2	5
hill5x	543.8	5.2	7/29	3.8	11
hill6	103.4	0.3	6/17	0.3	6
hill6x	3934	60.2	8/35	37.3	13
hill7	112.1	0.5	6/20	0.4	6
hill7x	35130	1016	8/41	338.7	13
hill8	151.1	1.3	7/23	1.0	7
hill10	580.7	10.2	7/29	6.8	7
hill15	92850	6486	8/44	2908	8
monom 2	102.5	0.2	6/7	0.3	14
monom 3	567.0	1.0	16/25	1.9	73
selkov	87.68	0.1	4/4	0.2	12

transformation problem while minimizing the number of species (SAT-Sp time). We also give in the table the number of variables introduced in Algorithm 1. along with the optimal number of variables found by our algorithm (Optimal var.). We finally mention the time taken to minimize the number of reactions (SAT-Reac time) and the resulting number of reactions (Optimal reac.). All computation times are given in milliseconds and were obtained on a personal laptop (Lenovo W530, Intel Core i7-3720QM CPU, 2.60 GHz x 8).

Our protocol to gather these results is as follow. We first time the whole process of compiling the PIVP through the "compile_from_PIVP" command of Biocham, thus giving the Total time. During the process we keep the temporary file that were given to the SAT solver and does a second execution of the SAT solver alone with a verbose output, gathering the information given by the output to determine the SAT time (doing this twice for both SAT-Sp and SAT-Reac). Hence, the total time contains the time it takes to construct and write the cnf file while the MAX-SAT time only measure the resolution of the formulae by the max sat solver. The time taken to convert the resulting PIVP to the ECRN language is essentially negligible.

In Table 1, we use the following nomenclature:

"circular(n, k)" denotes a circular PODE with n variables of degree k:

$$\frac{dX_i}{dt} = X_{i+1}^k, \quad \frac{dX_n}{dt} = X_1^k. \tag{12}$$

it can be check that introducing all monomials of a single variable $(x, x^2, \ldots)$ is sufficient.

"hard$k$" models are designed to be especially demanding in terms of monomials, the input is:

$$\frac{dA}{dt} = C^k + A^2 B^2 C^{k-1}, \quad \frac{dB}{dt} = A^2, \quad \frac{dC}{dt} = B^2, \tag{13}$$

so that while they ask for relatively few variables and are described with a handful of monomials they actually need most of the variables of the proof making them interesting to understand the effective structure of the QTP. The construction is based on the one of circular(n, k) adding a second monomial to the first derivative in order to make mandatory the usage of variables using several of the old variables.

"monom$n$" is one of the most promising model regarding the NEXP complexity as it rely on n variables and a long monomial of size n so that the input is of size n^2. But we suspect it to ask of the order of 2^n variables, the input is:

$$\frac{dX_i}{dt} = X_{(i+1)}^2 + \prod_{j=1}^{n} X_j^2. \tag{14}$$

(for clarity we do not add the modulo in the equation but X_{n+1} is the same as X_1.) We were not able to reduce "long monom 4" despite the reduction being very quick on the $n = 3$ case.

"hill$n$" is the Hill function of order n through the 3 variables PIVP:

$$\frac{dH}{dt} = nI^2T^{n-1}, \quad \frac{dI}{dt} = -nI^2T^{n-1}, \quad \frac{dT}{dt} = 1. \tag{15}$$

so that H is the desired hill function, I is complementary to the hill function ($I + H = 1$) and T is an explicit time variable $T = t$. The "x" after the model indicate that the PODE has been modified to take the desired point of computation as an input, hence the initial concenctration of the X species is now the input of the computation:

$$\frac{dH}{dt} = nI^2T^{n-1}X, \quad \frac{dI}{dt} = -nI^2T^{n-1}X, \quad \frac{dT}{dt} = X, \frac{dX}{dt} = -X. \tag{16}$$

Table 2. Minimal number of variables and optimal solutions found by Algorithm 2 on our benchmark of QTP instances (Table 1 and 2).

Model name	Optimal solution with a minimum number of variables
circular 2,3	$\{x, y, xy, x^2, y^2\}$
circular 2,4	$\{x, y, x^2y, xy^2, x^3, y^3\}$
circular 2,5	$\{x, y, x^3y, xy^3, x^4, y^4\}$
circular 2,6	$\{x, y, x^4y, x^3y^2, xy^4, x^5, y^5\}$
hard3	$\{a, b, c, ac, a^2, b^2, a^2b, ab^2, ab^2c, b^2c, c^3, ac^3, b^2c^3, ab^2c^3\}$
hard4	$\{a, b, c, a^2, b^2, ab^2, a^2b, b^2c, c^3, ac^3, bc^3, a^2c^2, b^2c^2, c^4, bc^4, ab^2c^3\}$
hill2	$\{i, it, h\}$
hill2x	$\{i, h, x, ix, itx\}$
hill3	$\{i, h, it, it^2\}$
hill3x	$\{i, h, x, ix, itx, it^2x\}$
hill4	$\{i, h, t, it^2, it^3\}$
hill4x	$\{i, h, x, ix, tx, itx, it^3x\}$
hill5	$\{i, h, t, t^3, it^4\}$
hill5x	$\{i, h, x, ix, tx, t^3x, it^4x\}$
hill6	$\{i, h, t, it^2, it^4, it^5\}$
hill6x	$\{h, x, ix, tx, it^2x, it^3, it^3x, it^5x\}$
hill7	$\{i, h, t, t^3, it^4, it^6\}$
hill7x	$\{i, h, x, ix, tx, t^3x, it^6x, it^2x\}$
hill8	$\{i, h, t, t^2, t^5, it^6, it^7\}$
hill10	$\{i, h, t, t^3, t^7, it^8, it^9\}$
hill15	$\{i, h, t, t^2, t^5, it^{11}, it^{13}, it^{14}\}$
monom 2	$\{a, b, a^2, b^2, a^2b, ab^2\}$
monom 3	$\{a, b, c, a^2, b^2, c^2, abc, ab^2, ac^2, a^2b, a^2c, bc^2, b^2c, ab^2c^2, a^2bc^2, a^2b^2c\}$
selkov	$\{x, y, xy, x^2\}$

Selkov is a common model of Hopf bifurcation:

$$\frac{dX}{dt} = -X + aY + X^2Y, \quad \frac{dY}{dt} = b - aY - X^2Y, \tag{17}$$

where a and b are tunable parameters

5.2   BioModels Repository

The BioModels database [3] is a repository of models of natural biological processes. Among the 653 models from the curated branch of BioModels, only 232 are reaction models with mass action law kinetics thus leading to polynomial ODEs, among which only 12 are of degrees strictly higher than 2. This is not surprising because the reaction models in BioModels are mechanistic models naturally described by elementary CRNs.

The non elementary CRN with mass actions law kinetics of BioModels are models number: 123, 152, 153, 163, 281, 407, 483, 530, 580, 630, 635, 636. Currently, our MAX-SAT algorithm fails to solve the QTP optimization problem on those instances in less than one hour computation time. The encoding in MAX-SAT is itself very long because of exponential size complexity in those cases.

To take an example, the model label 123 contains 13 species but only 4 of them participate in monomials of degree greater than 2, namely degree 4. Manually restricting the quadratic transformation to this set still gives us a search space of 65 possible variables, a bit larger than what is currently handled by our algorithm. Pruning further to select a smaller subset of the ODE that contains two variables and only one of the two problematic monomials, gives us a model that is easily solved in a few seconds. However, that solution does not solve optimally the complete model.

6   Conclusion

The problem of CRN design for implementing a given computable real function presented as the solution of a PIVP has been solved on the theoretical side by the proof of Turing-completeness for finite continuous CRNs [5]. Nevertheless to make that approach practical, good algorithms are needed to eliminate degrees greater than 2 in the PIVP. Though it is well known in dynamical system theory that there is no loss of generality to consider polynomial ordinary differential equations with degrees at most 2, that quadratization problem has apparently not been studied from a computational point of view.

We have shown the NP-hardness of the quadratization optimization problem in the non succinct representation of the input PIVP by a matrix of monomials, when we want to minimize either the number of species, or the number of reactions. In the succinct symbolic representation of the input PIVP by list of monomials, we conjecture that the problem becomes NEXP-hard. A proof would need to show that the hard instances coming of the vertex set covering problem used in the proof of NP-completeness, may have optimal solutions of exponential size in the succinct representation.

Nevertheless, we have shown that an algorithm based an encoding in MAX-SAT is able to solve interesting CRN design problems in this approach. A particularly interesting example is the automated synthesis of an abstract CRN of 11 reactions over 7 molecular species to implement the Hill function of order 5 which can be compared to the 10 reactions over 12 species of the concrete MAPK signalling CRN implementing a similar input/output function [9].

Acknowledgements. This work was jointly supported by ANR-MOST *BIOPSY Biochemical Programming System* grant ANR-16-CE18-0029 and ANR-DFG *SYMBIONT Symbolic Methods for Biological Networks* grant ANR-17-CE40-0036.

7   Appendix: NP-hardness of MAX-Horn-SAT

A Horn clause is a disjunction of literals with at most one positive literal. Horn-SAT is the problem of deciding the satiafiability of a conjunction of Horn clauses. Such a problem can be easily solved by unit-clause propagation, as follows

1. Ignore the clauses that contain both a variable and its negation
2. Set all variables to false
3. Initialize the score of each clause to its number of negative literals
4. For each unsatisfied clause with 0 score
   (a) If it has no positive literal return *Unsatisfiable*
   (b) Otherwise set the positive literal x to true
   (c) Decrement the score of the other clauses having x as negative literal
5. Return *Satisfiable*

This algorithm clearly shows that Horn-SAT is in P. In addition, this algorithm obviously minimizes the number of variables set to true. Perhaps surprisingly however:

Proposition 4. *Deciding the satisfiability of a Horn-SAT instance while asking that at least k variables are set to true is NP-complete and MAX-Horn-SAT is NP-hard.*

Proof. This can be easily shown by reduction of the *Vertex Set Covering Problem*. Given a graph G with n vertices, we introduce one variable v_i for each vertex, and one clause $\neg v_i \vee \neg v_j$ for each edge (v_i, v_j). A variable set to false indicates that the corresponding vertex is in the covering.

Now, there is a vertex set covering with k vertices if and only if there is a valuation with $n - k$ variables set to true satisfying the Horn-SAT instance.

This concludes the proof of NP-completeness and MAX-Horn-SAT is thus NP-hard.

In essence, the proof of NP-hardness of the non-succinct quadratic transformation follows the same vein but is quite obfuscated by the details of this problem.

8   Appendix: Proof of NP-completeness of nsQTDP

In this appendix we prove the NP-completeness of nsQTDP (Theorem 2). We will construct this proof step by step. In a first time we will describe and study the encoding of the VSCP as a quadratic reduction, then we will prove that choosing an optimal set of variables among the ones introduced by the Algorithm Eq. 1 is an NP-hard problem. Then we will explain why allowing other types of new variables in the output (polynomial or algebraic function) still preserved the NP-hardness of the problem.

By abuse of notation, we use the same names for the vertices of G and the variables of the PIVP. (Except, of course, for V_0 and V_{n+1} that do not exist in the initial graph.) However, to distinguish between the monomials of the various PIVP and the variable of the output of the algorithm, these variables will be indicated with an upper bar like: $\overline{V_i V_j}$ while monomials will not. We will moreover say that a variable is computed while monomial will be designated as reachable given a certain set of variables.

Let us now investigate the structure of the constructed PIVP.

Lemma 1. *Supposing that $\{\overline{V_0}, \ldots, \overline{V_{n+1}}\}$ are already computed, then the derivative of $\overline{V_i V_j}$ is quadratic for the set of variables $\{\overline{V_0}, \ldots, \overline{V_{n+1}}, \overline{V_i V_j}\}$.*

Proof. Denoting $X = \overline{V_i V_j}$, we have:

$$\frac{dX}{dt} = \frac{dV_i}{dt} V_j + V_i \frac{dV_j}{dt} \tag{18}$$

$$= \sum_m (a_{i,m} + a_{j,m}) V_i V_j V_m + V_{i+1} V_j + V_i V_{j+1} \tag{19}$$

$$= \sum_m (a_{i,m} + a_{j,m}) X V_m + V_{i+1} V_j + V_i V_{j+1} \tag{20}$$

where one of the two last term may be missing if i or j is $n+1$. This is quadratic with respect to the aforementioned set.

Hence, if all the initial variables are present, we can add or remove variables of degree two, knowing that there derivatives will always be quadratic. In effect, this allows us to focus on the monomials in the derivative of V_0 as the only monomials that will need new variables to be reachable.

This property does not hold for variable of degree 3 (and all higher degree). Indeed, in that case, the last monomials are of degree 3 (and higher) and so need a way to be computed either by introducing them entirely as new variables or relying on breaking them between variables of lesser degree that may not be already computed. The derivative of the variable of degree 3 $\overline{V_i V_j V_k}$ present for example the non-quadratic monomials: $V_{i+1} V_j V_k$, $V_i V_{j+1} V_k$, $V_i V_j V_{k+1}$.

This is also false for polynomial variables because the derivative all have different rates $a_{i,j}$ that ensure that a polynomial do not appear in its own derivative as a monomial does. Thus computing a polynomial variable may ask us to compute still other variables.

This property is essential for our proof as it allows us to make a direct connexion between vertex covering and quadratic reduction, namely that given a cover $S = S_1, \ldots, S_k$ of G, the set of functions:

$$\{\overline{V_0}, \ldots, \overline{V_{n+1}}, \overline{S_1 V_{n+1}}, \ldots, \overline{S_k V_{n+1}}\} \tag{21}$$

have $n + 2 + k$ elements and defines a quadratic transformation of $PIVP_3(G)$.

It is indeed obvious that the derivative of $\overline{V_0}$ is quadratic using the fact that every edge in E has at least one endpoint in S and so each triplet may be rewritten with two of the new variables. Checking that the other variables also have quadratic derivatives is easy given Lemma 1.

To prove that our transformation is valid however, we need the opposite! We want to check that an optimal transformation of $\text{PIVP}_3(G)$ effectively allows us to find an optimal vertex covering of G. And essentialy, we will do this by showing that optimal reduction are of the form of the set Eq. 21 thus making a direct connexion between optimal covering and quadratic reduction.

Essentialy, the remainder of the proof will be to demonstrate the following lemma:

Lemma 2. *For a given graph G, optimal reductions of $PIVP_3(G)$ may be rewritten in the form of Eq. 21 and thus define an optimal vertex cover of G.*

8.1   Restriction of Variables to Monomials Functions

As explained above, we first prove that finding an optimal set of variables among the monomials described in the paper of Carothers (see Algorithm 1) is an NP-hard problem. This give us the soften version of Lemma 2:

Lemma 3. *Given a graph G, the smallest subset of variables considered in Algorithm 1 that forms a quadratic transformation of $PIVP_3(G)$ gives an optimal vertex cover of G.*

Proof. As expressed above, we want to show that optimal reductions are of the form of Eq. 21, or at least may be easily reshape to be so.

By definition, we need to introduce the first variable $\overline{V_0}$, the derivative of which present the term $\overline{V_1}$, thus asking us to compute it too. Then in turn, it asks us to compute $\overline{V_2}$ and so on until all the variables of degree one are present.

Let us take an optimal quadratic transformation, then the different monomials in the derivative of V_0 are reachable. This means that if we have a monomial like $V_i V_j V_{n+1}$, at least one of the four following variables is present: $\overline{V_i V_{n+1}}$, $\overline{V_j V_{n+1}}$, $\overline{V_i V_j}$ or $\overline{V_i V_j V_{n+1}}$. If we are in the third or fourth case, we can remove this variable and replace it by $\overline{V_i V_{n+1}}$. (As all variables of degree one are present, we know that this new variable preserves the quadraticity of the solution and is thus still optimal.) Moreover, by the structure of $PIVP_3(G)$ variables like these appear only once in the derivative of V_0, thus this transformation still allows us to compute the desired function.

As we then have all variables of degree one and that all the other variables are of degree two, we know that the PIVP is quadratic by Lemma 1, moreover we cannot have increase the number of variables and are thus still optimal. Finally, we have construct a set like Eq. 21 and have thus defined an optimal covering S of G with the optimal transformation of $\text{PIVP}_3(G)$.

To generalize the previous proof to any set of monomials, we note that by construction, the set of Sect. 2.1 overspan the set of variables that may be used to define a quadratic transformation. In particular due to the construction of $PIVP_3$, it contains all monomials of degree one and two that can be formed with the variables of the input PIVP, and all the monomials that appears in the derivatives of $PIVP_3$.

Hence, a monomial that is not present in the mathematical proof can only increase the number of monomials that need to be reached as it appears nowhere and will need to be computed itself. It thus cannot be present in an optimal set.

8.2   Restriction of Variables to Polynomials Functions

Lemma 4. *No polynomial variable is present in an optimal quadratic transformation of $PIVP_3$.*

Proof. The idea is still the same. We want to prove that we need to introduce all the variables of degree one and once this is done, that using only the variables that correspond to the vertex cover is preferable. But it is now more tricky as the ending singulets of the derivative may be added to a polynomial to "save a variable".

For the same reason as before, $\overline{V_0}$ needs to be computed. To investigate why the other variables $\overline{V_i}$ are also needed is more complex.

Suppose we wish to avoid computing the variable V_k so that we add it in an existing polynomial (eventually composed of a single monomial) hence forming the variable $M = P + V_k$, where P is some polynomial. Let us look at its derivative: $\frac{dM}{dt} = \frac{dP}{dt} + \frac{dV_k}{dt}$

As noted above, and due to the presence of the parasitic terms $a_{i,j}$, M do not appear in its own derivative. Thus, a transformation like the one of Lemma 1 is out of hope. Moreover, the derivative of V_k present a term in V_k^2. To compute it we can either add $\overline{V_k}$ to our set of variable which is what we try to avoid, either add $\overline{V_k^2}$ (or a polynomial incorporating it). But you can check that the derivatives of such a variable present a term in V_k^3. So either we abdicate and include the variable of degree one, either you add a polynomial of degree 3, but this polynomial will ask us a new one of degree 4, etc. To avoid an infinite set of variables we have to compute $\overline{V_k}$, and this is true for all k. Thus all variables of the initial PIVP need to be present.

Now, for each monomials in the derivatives of the first variable, we have 2 choices on the way it is computed. Either a single variable is introduced to deal with it and this has already been treated in the previous case. Either all or part of it is computed as part of a polynomial. To prove that this cannot be done in an optimal transformation we need to show that doing so imply to compute additional undesired variables. And once again we can convince ourself by inspecting such derivatives, for example:

$$\frac{d}{dt}(P + V_i V_j) - \frac{dP}{dt} + \frac{dV_i V_j}{dt} \tag{22}$$

$$= \frac{dP}{dt} + \sum_m (a_{i,m} + a_{j,m}) V_i V_j V_m + \ldots \tag{23}$$

cannot be quadratic if the variable $\overline{V_i V_j}$ is not computed, which we try to avoid or another more complex polynomial specificaly tailored for this purpose. Thus, trying to hide a part of a monomial in a polynomial to save a variable always ask at least two variables and cannot be part of an optimal transformation.

8.3   Quadratic Transformation Without Restriction

Finally, we notice that for a function to be in the output, it have to be polynomial as it will actually be used to rewrite polynomial functions. Hence, putting the previous results together we get:

Proposition 5. *A graph G with n vertices has a vertex set cover of size k if and only if $PIVP_3(G)$ has a quadratic transformation to a PIVP of dimension $n + k + 2$.*

which with Proposition 2 concludes the proof of Theorem 2.

8.4   Proof of NP-Hardness for Reactions minimization

Theorem 3

Proof. The core of the proof is similar, using the same reduction from VSCP. Starting from a graph G with n vertices and ℓ edges, we construct $PIVP_3(G)$.

As in the previous case, we still have to introduce all the variables of the form $\overline{V_i}$ giving us a fixed number of monomials upon which no optimization is possible. Let us note $F(n, \ell) = n^2 + 3n + \ell + 2$ this number.

Then, introducing a variable like $\overline{V_i V_j}$ imposes $n + 3$ monomials if $i, j \neq n+1$ and $n + 2$ if $i, j = n + 1$. As we have seen in the previous proof, the optimal cover set may be expressed using only variables like $\overline{V_i}$ and $\overline{V_i V_{n+1}}$, and will thus ask for $k = F(n, \ell) + k_s(n + 2)$ where k_s is the number of vertices in the optimal covering of G. The main difference with the proof for variables is that we do not have to check variables of the form $\overline{V_i V_j}$ as they ask one more monomial than the one with $j = n + 1$ and are thus never optimal.

References

1. Bournez, O., Campagnolo, M.L., Graça, D.S., Hainry, E.: Polynomial differential equations compute all real computable functions on computable compact intervals. J. Complex. **23**(3), 317–335 (2007)

2. Carothers, D.C., Parker, G.E., Sochacki, J.S., Warne, P.G.: Some properties of solutions to polynomial systems of differential equations. Electron. J. Diff. Eqn.**2005**(40), 1–17 (2005)
3. Chelliah, V., Laibe, C., Novère, N.: Biomodels database: a repository of mathematical models of biological processes. In: Schneider, M.V., (ed.) In Silico Systems Biology. Methods in Molecular Biology, vol. 1021, pp. 189–199. Humana Press (2013)
4. Cook, M., Soloveichik, D., Winfree, E., Bruck, J.: Programmability of chemical reaction networks. In: Condon, A., Harel, D., Kok, J., Salomaa, A., Winfree, E. (eds.) Algorithmic Bioprocesses. NCS, pp. 543–584. Springer, Berlin, Heidelberg (2009). https://doi.org/10.1007/978-3-540-88869-7_27
5. Fages, F., Le Guludec, G., Bournez, O., Pouly, A.: Strong turing completeness of continuous chemical reaction networks and compilation of mixed analog-digital programs. In: Feret, J., Koeppl, H. (eds.) CMSB 2017. LNCS, vol. 10545, pp. 108–127. Springer, Cham (2017). https://doi.org/10.1007/978-3-319-67471-1_7
6. Fages, F., Soliman, S.: Abstract interpretation and types for systems biology. Theor. Comput. Sci. **403**(1), 52–70 (2008)
7. Garey, M.R., Johnson, D.S.: Computers and Intractability: A Guide to the Theory of NP-Completeness. Freeman, New York (1979)
8. Graça, D.S., Costa, J.F.: Analog computers and recursive functions over the reals. J. Complex. **19**(5), 644–664 (2003)
9. Huang, C.-Y., Ferrell, J.E.: Ultrasensitivity in the mitogen-activated protein kinase cascade. PNAS **93**(19), 10078–10083 (1996)
10. Papadimitriou, C.H., Yannakakis, M.: A note on succinct representations of graphs. Inf. Control **71**(3), 181–185 (1986)
11. Segel, L.A.: Modeling Dynamic Phenomena in Molecular and Cellular Biology. Cambridge University Press, New York (1984)

Comparing Probabilistic and Logic Programming Approaches to Predict the Effects of Enzymes in a Neurodegenerative Disease Model

Sophie Le Bars[1,2], Jérémie Bourdon[1,2]($\boxtimes$), and Carito Guziolowski[1,3]($\boxtimes$)

[1] LS2N, UMR 6004, Nantes, France
[2] Université de Nantes, Nantes, France
Jeremie.Bourdon@univ-nantes.fr
[3] Ecole Centrale de Nantes, Nantes, France
Carito.Guziolowski@ec-nantes.fr

Abstract. The impact of a given treatment over a disease can be modeled by measuring the action of genes on enzymes, and the effect of perturbing these last over the optimal biomass production of an associated metabolic network. Following this idea, the relationship between genes and enzymes can be established using signaling and regulatory networks. These networks can be modeled using several mathematical paradigms, such as Boolean or Bayesian networks, among others.

In this study we focus on two approaches related to the cited paradigms: a logical (discrete) Iggy, and a probabilistic (quantitative) one Probregnet.

Our objective was to compare the computational predictions of the enzymes in these models upon a model perturbation. We used data from two previously published works that focused on the HIF-signaling pathway, known to regulate cellular processes in hypoxia and angiogenesis, and to play a role in neurodegenerative diseases, in particular on Alzheimer Disease (AD). The first study used Microarray gene expression datasets from the Hippocampus of 10 AD patients and 13 healthy ones, the perturbation and thus the prediction was done *in silico*. The second one, used RNA-seq data from human umbilical vein endothelial cells over-expressing adenovirally HIF1A proteins, here the enzyme was experimentally perturbed and the prediction was done *in silico* too. Our results on the Microarray dataset were that Iggy and Probregnet showed very similar (73.3% of agreement) computational enzymes predictions upon the same perturbation. On the second dataset, we obtained different enzyme predictions (66.6% of agreement) using both modeling approaches; however Iggy's predictions followed experimentally measured results on enzyme expression.

Keywords: Probabilistic modeling · Logical modeling · Neurodegenerative disease · Signaling and regulatory networks

© Springer Nature Switzerland AG 2020
A. Abate et al. (Eds.): CMSB 2020, LNBI 12314, pp. 141–156, 2020.
https://doi.org/10.1007/978-3-030-60327-4_8

1   Introduction

A disease or a treatment has a huge impact on the studied organism that can be observed on the signaling and regulatory network and then on the metabolic network. These networks are typically studied separately to understand the effect of a perturbation and the mechanisms behind. The obvious link between the regulatory and metabolic networks are enzymes [12]. The regulatory network drives enzyme production which will control the biochemical reactions inside the metabolic network. Regulatory networks have been studied for a long time by using different modeling paradigms including Boolean, Neural or Bayesian networks (see [4,13] for a review). At the same time, Constraint based methods, such as Flux Balance Analysis (FBA), have been developed to study metabolic networks.

An integration of these two types of networks may allow a more realistic modeling of the mechanisms triggered during a perturbation. A first approach proposed in this context was rFBA [20]. At each time interval, a consistent regulatory state with metabolic equilibrium state was calculated. Then, FBA was used to find a steady state flow distribution for the current time interval. A new metabolic state lead to a new state of regulation and the process was repeated until it did not evolve anymore. This detailed approach needs an organism easily cultivable (*E. coli* for example). Another approach, SR-FBA [21], expresses the regulatory network in Boolean equations and then translates it into linear equations, added as constraints in the FBA. This approach needs a huge amount of preliminary work to translate all the equations and a well known organism such as *E. coli*. More recent approaches, adaptable to less known organisms, exist such as PROM [15] and FlexFlux [14]. PROM uses probabilities to represent the state of genes that will be used as constraints in FBA but requires hundreds of Microarray data experiments. FlexFlux obtains, for each component of the regulatory network, an equilibrium state constraint defined by an upper and lower bound. These constraints are added to the FBA model. Flexflux needs an SBML-qual model to extract these equations, such a precise representation is not always available for some species.

In this paper we aim to compare a recent tool Probregnet [2] and Iggy [1]. Probregnet uses a Bayesian network model, on which belief propagation techniques are applied to reason over it. Iggy uses a sign-consistency approach, expressed as a logic program in Answer Set Programming [18]. Both tools use prior regulatory knowledge and are able to make computational predictions upon system perturbation using few gene expression datasets. The nature of both approaches, one quantitative, the other discrete, makes it interesting for us to compare them in the context of enzyme prediction.

The results of this comparison were obtained on the HIF signaling pathway, known to be of major importance in neurodegenerative diseases [6]. We applied both tools on a Microarray dataset on Alzheimer's disease and an RNA-Seq dataset on human umbilical vein endothelial cells. We built models upon two regulatory networks of around 80 nodes and 250 edges.

We conclude that Iggy is better suited than Probregnet to compute enzyme predictions $(0.038s$ vs $25s)$[1]. Besides, Iggy and Probregnet showed very similar (73.3% of agreement) computational enzymes predictions upon the same perturbation for Microarray data. On the second dataset, we obtained different enzyme predictions (66.6% of agreement) using both modeling approaches; however Iggy's predictions followed experimentally measured results on enzyme expression.

2   Methods

2.1   Data Sets Used to Conduct Our Comparison

Our study is based on two different datasets. The first one is a Microarray gene expression dataset published in [5] and the second one is an RNA-seq dataset published in [8].

The Microarray data were measured in the Hippocampus brain region of 10 Alzeihmers's patients and 13 healthy patients. The Hippocampus is known to be differentially vulnerable to the histopathological and metabolic features of Alzheimer's disease (AD). An Affymetrix Human Genome Array was used and allowed to collect the expression for 20545 genes.

The RNA-seq data were measured on human umbilical vein endothelial cells (HUVECs) exposed to constitutively active HIF1A over-expression. This data was collected for 3 control cells (with normal expression of HIF1A) and 3 cells with induced over-expression of HIF1A in the form of two types of RNA-seq datasets, one absolute and the other differential. The absolute RNA-seq datasets, consisting of 25691 RNA, were normalized using the edgeR R package [16,17]. These normalized RNA-seq data were used to generate the *in silico* predictions with Probregnet (see Sect. 2.2). The differential RNA-seq datasets were composed of 1854 genes significantly differentially expressed upon HIF1A induction. The genes having a significant differential expression were selected using a cutoff of 1.5, applied on their logarithmic expression. A cutoff of 0.01 was used on the false discovery rate (FDR). This differential RNA-seq dataset was used to generate the *in silico* predictions with Iggy. All the RNA-Seq datasets were extracted from the GEO database[2].

2.2   Regulatory Networks of the HIF-signaling Pathway

In [6] it has been shown that the HIF-signaling pathway is of major importance in neurodegenerative disease, with a key role of the HIF1A protein. In [2] the authors built a gene regulatory network for Alzheimer Disease (AD), focused on the HIF-signaling pathway. We used a signaling and gene regulatory network

---

[1] All computations were performed on a standard laptop machine. Ubuntu 18.04, 64 bits, intel core i7-9850H CPU 2.60 GHz, 32 GB.

[2] https://www.ncbi.nlm.nih.gov/geo/query/acc.cgi?acc=GSE98060.

built upon the same pathway; for this purpose we use the same methods as proposed in the Probregnet [2] pipeline. These steps are explained in the following paragraphs. The retrieved networks were afterward modeled and analyzed with Probregnet and Iggy using two different datasets (see Sect. 2.1).

At first, the HIF-pathway was extracted from the KEGG database thanks to the *graphite* R package[3]. This R package allows to provide networks derived from different databases based on the pathway topology. In this network, all the metabolites nodes have been removed and the edges are propagated through them, and are labeled as indirect processes. The nodes represent either protein or genes. The edges represent multiple biological processes: ubiquitination, phosphorylation, binding, inhibition, activation, expression.

Afterward, we reduced this graph by keeping only the nodes of the network associated with genes present in the gene expression datasets. Since we had two datasets we will retrieve in these step two reduced networks. The first one was based on the Microarray data in [5] extracted from the Hippocampus brain region of healthy and Alzheimer's disease (AD) patients. The genes which were kept were those present in either healthy or AD datasets. In the second network, obtained using the RNA-seq dataset of HUVECS [8], the genes kept were those that were present in either control or HIF1A induced cells.

Finally, both regulatory networks were converted into a directed acyclic graph (DAG) by using the *pcalg* R package[4] that allows to extend a partially directed acyclic graph into a DAG using the algorithm by Dor and Tarsi (1992) [10]. In this algorithm, the DAG will have the same set of vee-structures as the partially directed acyclic graph; where a vee-structure is formed by two edges, directed towards a common head, while their tails are non-adjacent. After this process, the edges in the DAG are not labeled (or signed) anymore. Since Iggy, contrary to Probregnet, needs a signed graph, we took into account the edges that were previously labeled as inhibition in the KEGG database, and the other edges were all labeled as activation. The final regulatory network consists of 86 nodes and 253 edges for the Microarray data and 81 nodes and 233 edges for the RNA-seq data. Both are a reduction of the HIF-signaling pathway adapted to the data.

2.3 The Probregnet Pipeline

The Probregnet[5] pipeline [2] is a complex and global framework that allows to integrate a gene regulatory model (based on graph interactions) into a metabolic network (based on biochemical reactions) using a constraint-based model.

For this paper, we focus on the regulatory network analysis proposed by Probregnet. This analysis is based on Bayesian networks (BN) also called probabilistic directed acyclic models [3]; which allows a representation of conditional dependencies between random variables. Probregnet uses a regulatory network of

---

[3] https://www.bioconductor.org/packages/release/bioc/vignettes/graphite/inst/doc/graphite.pdf.

[4] https://cran.r-project.org/web/packages/pcalg/pcalg.pdf.

[5] https://github.com/hyu-ub/prob_reg_net.

the HIF-signaling pathway converted into a DAG (see Sect. 2.2) in which nodes are genes and edges, interactions between these genes (not signed or labeled). The BayesnetBP R package [11] is used to parametrize the graph with the gene expression data [5] by associating a node with its expression value. In a BN, the value of a child node depends on its parent nodes in the graph. Then, belief propagation is used to establish the repercussion of the perturbation of a given node in the graph over the other nodes. In [2] the perturbed node was HIF1A. The repercussion of the perturbation was monitored thanks to a ratio of the node expression in the perturbed model compared to the node expression in the model without perturbation. They focused on 15 enzymes, present in the network, known to regulate biochemical reactions in the brain.

In our study, we still focus on these 15 enzymes and compute the ratio (or fold-change) of the enzymes' expression in a perturbed model compared to the enzymes' expression in a model without perturbation. For this, we used different BNs parametrized for the two different datasets presented in Sect. 2.1.

Bayesian Networks for Our Case-Studies. Recall that the BN is built using the DAG extracted in Sect. 2.2 parametrized to a specific dataset. Using these DAGs, we obtained from our two datasets (AD and HUVECs) the following BNs:

1. Microarray dataset of the Hippocampus brain region
   (a) BN parametrized using the Microarray data of the 10 AD patients
   (b) Control BN parametrized using the Microarray data of the 13 healthy individuals
2. Rna-seq dataset of HUVECs
   (a) BN parametrized using the RNA-seq data of the 3 adenovirally over-expressed HIF1A cells
   (b) BN parametrized using the RNA-seq data of the 3 HUVECs with normal HIF1A expression

For BN (2a) and BN (2b), the number of cells was not enough for Probregnet in order to parametrize the BN. Therefore, for each of the two conditions (normal and adenovirally over-expressed HIF1A), we completed the 3 HUVECs datasets with 10 artificially generated datasets (by adding an artificial noise in the data of 1%).

Enzymes *in silico* Predictions. For the Microarray dataset of the Hippocampus brain region we computed the fold-change of the 15 enzymes for different types of *in silico* perturbations of the model. Equation 1 describes the expression ratio measured for each enzyme e.

$$y_e = \frac{x_e^{AD_p}}{x_e^C} \tag{1}$$

where y_e refers to the fold-change (FC) expression of enzyme e; $x_e^{AD_p}$, to the expression of enzyme e obtained after simulation of the AD BN (1a) upon perturbation p. This perturbation p was done in three ways: HIF1A over-expressed (set to 13 expression level), under-expressed (set to 8 expression level), and HIF1A unaltered (9.65 expression level). For HIF1A unaltered, enzyme expression is the average expression of the enzyme in the dataset for AD patients. x_e^C refers to the expression of enzyme e in the control BN (1b) without perturbation, that is, the average expression of the enzyme in the dataset for healthy patients.

For the RNA-Seq dataset on HUVECs we still focus on the enzyme and only one *in silico* perturbation. Equation 2 describes the fold-change computed for each enzyme e.

$$z_e = \frac{x_e^{O_p}}{x_e^H} \tag{2}$$

where z_e refers to the FC expression of enzyme e, $x_e^{O_p}$ corresponds to the expression of enzyme e obtained after simulation of the HUVECs over-expressed BN (2a) upon perturbation p. This perturbation p represents an over-expression of HIF1A (set to 17 expression level, when HIF1A average expression across over-expressed samples is 14.5). x_e^H refers to the expression of enzyme e in the BN (2b) without perturbation, that is, the average expression of the enzyme in unaltered cells.

2.4   Iggy

Iggy [1] is a framework, based on Answer Set Programming [18], to test the consistency between a directed and signed graph G and a set of experimental observations μ. G represents a regulatory network, with edges labeled as *activation* ("+") or *inhibition* ("-"). μ defines an initial partial labeling of the nodes in G. It is composed of discrete ("+", over-expressed; "-", under-expressed; 0, no-change) changes associated to some nodes; which represent, for example, the differential expression of a gene between two system conditions. G is consistent with respect to μ if the logic given by its structure *agrees* with the signs in μ. This agreement follows specific constraints over the signs of the edges and the nodes, which need to be verified in all the graphs. The purpose of this verification is to find at least one consistent *global labeling*, which is obtained by assigning "+", "-", or "0" values to all the initially non-observed nodes in the graph, or to decide that this global labeling assignment is not possible. The constraints used by Iggy are:

1. The observations in μ must keep their initial labeling.
2. Each node, labeled respectively as + or −, must be justified by at least one predecessor activating it labeled as + or −, or by one predecessor inhibiting it labeled as − or + respectively.
3. Each node, labeled as 0 must have only one predecessor labeled as 0 or a couple of + and − labeled predecessors.

Iggy automatically detects inconsistencies between G and μ, applies minimal repairs to restore consistency, and predicts the sign of non-observed nodes in G.

Generating Discrete Observations from Datasets to Use Iggy. For the Microarray dataset, we denote as $\bar{y}_g$ the ratio (or fold-change) of the average expression of gene g of AD patients against the average expression of gene g in healthy individuals. To obtain the associated sign for each gene g in the dataset, we discretized $\bar{y}_g$, using the thresholds over the distribution of the expression of the 20545 genes in the dataset as shown in Eq. 3.

$$sign(\bar{y}_g) = \begin{cases} + & \text{if } \bar{y}_g > Q_3 \\ - & \text{if } \bar{y}_g < Q_1 \\ 0 & \text{if } 0.99 \leq \bar{y}_g \leq 1.01 \end{cases} \tag{3}$$

where Q_3 and Q_1 refer to the third and first quartiles of the fold-change gene expression data distribution. From this discretization analysis, the input observations data for Iggy was composed of 16 "+", 24 "-", and 16 "0-changed" nodes. This set, denoted as μ_1, did not include any of the 15 enzymes that will be computationally predicted. Besides, the sign of 3 other nodes (EP300, CREBBP, ARNT in Fig. 2), which are direct predecessors of the enzymes, is set to "0-change" in μ_1 so that we can see only the impact of HIF1A on the enzymes.

To simulate a perturbation in Iggy, there is no change in the regulatory network structure, however the set of observations μ_1 changed slightly:

$$\begin{aligned} S^+ &= \mu_1 + (sign(\bar{y}_H) = +) \\ S^- &= \mu_1 + (sign(\bar{y}_H) = -) \\ S^0 &= \mu_1 + (sign(\bar{y}_H) = 0) \end{aligned} \tag{4}$$

where $\bar{y}_H$ refers to the expression level of HIF1A. We built then three sets of observations, denoted as S^p, where p refers to the type of sign imposed to the HIF1A node to simulate an over-, or under-expression of HIF1A, as well as a *non-change* effect of this protein.

For RNA-Seq data, we used the logFC between HIF1A over-expressed and normally expressed already present in the gene differentially expressed data from RNA-Seq analysis (see Sect. 2.1). We denote this logFC for each gene g as $\bar{z}_g$. To transform the quantitative value of $\bar{z}_g$ in signs we used the same logic as before but with thresholds better adapted to this dataset (Eq. 5).

$$sign(\bar{z}_g) = \begin{cases} + & \text{if } \bar{z}_g > 1.5 \\ - & \text{if } \bar{z}_g < -1.5 \\ 0 & \text{if } -0.15 \leq \bar{z}_g \leq 0.15 \end{cases} \tag{5}$$

From this new discretization analysis, the input observations data for Iggy was composed of 5 "+", 2 "-", and 19 "0-changed" nodes. This set, denoted as μ_2, did not include any of the 12 enzymes that will be computationally predicted and the 3 nodes (EP300, CREBBP, ARNT in Fig. 2) that are direct predecessors of the enzymes are still set to "0-change". Only 12 enzymes were kept and not the 15 initially as three of them (HK3, ENO3 and PDHA2) were not considered to be expressed in the study (count-per-million,counts scaled by total number of reads was under 10) [19]. From μ_2 we built two sets of observations, R^+ and R^0, where the sign imposed to HIF1A, $\bar{z}_H$, was either "+" (over-expressed) or 0 (unaltered), as described by Eq. 6.

$$R^+ = \mu_2 + (sign(\bar{z}_H) = +)$$
$$R^0 = \mu_2 + (sign(\bar{z}_H) = 0)$$

(6)

where $\bar{z}_H$ refers to the expression level of HIF1A. As for the case of Probregnet, we focused on the Iggy's *in silico* prediction of the 12 enzymes upon HIF1A perturbations in the system. We recall in Fig. 1 the different steps described in this Section for Iggy and Probregnet. All scripts and data used in this article are available in: https://gitlab.univ-nantes.fr/E19D080G/comparing_iggy_prob.git

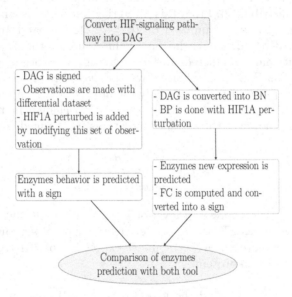

Fig. 1. Diagram representing the different steps in order to compare the two approaches. The steps for both are in blue, those specific to Iggy are on the left in red and those specific to Probregnet are on the right in green. DAG stand for directed acyclic graph, BN for Bayesian network and BP for belief propagation (Color figure online).

3   Results

We focused on the *in silico* computational predictions from both approaches on enzymes involved in biochemical reactions of brain metabolism upon HIF1A stimulation. We illustrate our results in two case studies. The first, uses Microarray gene expression data from the Hippocampus brain region of Alzheimer's Disease (AD) patients and healthy individuals. The second, uses RNA-Seq data of 6 Human umbilical vein endothelial cells (HUVECs) over-expressing adenovirally HIF1A protein or expressing normally HIF1A.

3.1   HIF1A Impact on HIF-signaling Pathway for Alzheimer's Disease Patients

The Microarray data used for this case-study is presented in Sect. 2.1. The network, corresponds to the HIF signaling pathway (see Sect. 2.2). Both data, gene expression datasets and network, were transformed (see Sects. 2.2, 2.3 and 2.4) in order to be used for the comparison of Iggy and Probregnet.

HIF Signaling Pathway. We chose the HIF signaling pathway and focused on the HIF1A protein, which is a potential therapeutic target for neurodegenerative disease [7]. The HIF network, obtained in [2], was extracted from the KEGG database and then reduced (see Sect. 2.2). The resulting graph from this network (86 nodes, 253 edges) was built from the experimental data. The nodes represented genes and proteins, while the edges represented signaling and gene regulatory interactions. In Fig. 2 we show a subgraph of this HIF graph, focusing on the genes of the network that are directly connected to the enzymes.

Evolution of Enzyme Production According to HIF1A Fluctuation with the Bayesian Approach. We present here the results obtained with the Probregnet pipeline (see Sect. 2.3 and Table 1). We compared three particular (perturbed) states with respect to an *unaltered state* of the system, and computed the predictions of the fold-change of the enzymes level for each comparison (see Eq. 1).

Table 1. The three compared model states. The name of this comparison, used in the rest of this Section, appears in the first column.

Name	Description
HIF1A -	AD model with HIF1A under-expressed (HIF1A expression set to 8) against healthy model without perturbation (HIF1A normal expression)
HIF1A 0	AD model without perturbation (HIF1A normal expression) against healthy model without perturbation (HIF1A normal expression)
HIF1A +	AD model with HIF1A over-expressed (HIF1A expression set to 13) against healthy model without perturbation (HIF1A normal expression)

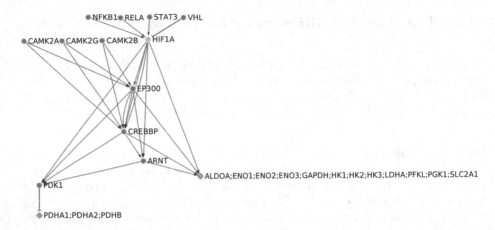

Fig. 2. Subgraph regulatory network of HIF-pathway. Only the enzymes and their predecessors are represented in this schema. The enzymes are represented as orange diamonds, the predecessors genes as blue circles, and the perturbed node, HIF1A, as a yellow circle. The edges represent either activation in green or inhibition in red (Color figure online).

As we can see in Fig. 3, the predicted fold-change of 9 out of 15 enzymes increases across the three comparative states of the system ordered as: *HIF1A -*, *HIF1A 0*, *HIF1A +*.

Evolution of Enzyme Production According to HIF1A Fluctuation with the Logical Approach. Here we used Iggy with the same regulatory network as Probregnet with 3 sets of observations (see Eq. 4) that correspond to the genes variations in each of the three perturbed states (see Table 1). Our results (see Table 2), focus on the *sign prediction* of the 15 enzymes. The sign represents the over-expression ("+", green), under-expression ("-", red), and the no-variation (0, blue) of the level of the enzymes upon each comparative case detailed in Table 1. All but three of the enzymes are over-expressed when HIF1A is over-expressed. The three enzymes that are evolving with a contradictory sign are the ones inhibited by PDK1 (see Fig. 2), this goes in agreement with the sub-graph topology.

Comparison of the Enzymes Computational Predictions Using Iggy and Probregnet. Recall that Iggy predicted discrete signs of the nodes in the graph whereas Probregnet, quantitative values. Thus, for each enzyme, we compared Iggy's predicted sign against the derivative sign of the mathematical curve represented in the plots of Fig. 4. If the sign of the derivative is the same as the tendencies observed for Iggy in the 3 comparisons, then the name of the enzyme will appear in green, else, in red. 11 enzymes will evolve in the same way with the two approaches except for HK1, PFKL, ENO2 and PDHA2. Probregnet fold-change expression of 3 out of 4 of these enzymes will remain

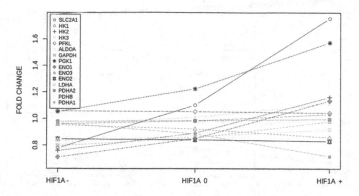

Fig. 3. Probregnet computational predictions using three perturbed states of the HIF model. Evolution of the fold-change of the 15 enzymes across the perturbed system states detailed in Table 1. In the X-axis we show the 3 perturbed states of the system, in the Y-axis, the value of the predicted fold-change.

Table 2. Iggy's sign prediction of the 15 enzymes after perturbing HIF1A.

HIF1A -	HIF1A 0	HIF1A +
PDHA1 = +	PDHA1 = 0	PDHA1 = -
PDHA2 = +	PDHA2 = 0	PDHA2 = -
PDHB = +	PDHB = 0	PDHB = -
LDHA = -	LDHA = 0	LDHA = +
GAPDH = -	GAPDH = 0	GAPDH = +
HK1 = -	HK1 = 0	HK1 = +
HK2 = -	HK2 = 0	HK2 = +
HK3 = -	HK3 = 0	HK3 = +
ENO1 = -	ENO1 = 0	ENO1 = +
ENO2 = -	ENO2 = 0	ENO2 = +
ENO3 = -	ENO3 = 0	ENO3 = +
PGK1 = -	PGK1 = 0	PGK1 = +
SLC2A1 = -	SLC2A1 = 0	SLC2A1 = +
PFKL = -	PFKL = 0	PFKL = +
ALDOA = -	ALDOA = 0	ALDOA = +

unaltered (difference in fold-change expression of less than 0.1) across the three comparative cases. Besides, the probabilistic approach does not take the inhibiting effect of PDK1 on the three PDH enzymes into account as it adds a manual correction by multiplying the fold-change of these three enzymes by the inverse of the fold-change predicted for PDK1 in [2]. The only one that is significantly decreasing and opposite to Iggy's prediction is HK1.

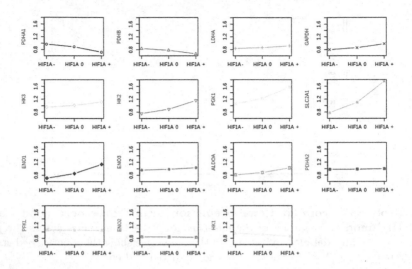

Fig. 4. Probregnet fold-change evolution (Y-axis) for each of the 15 enzymes compared to Iggy's predicted sign. Three comparative cases are studied corresponding to *HIFA1 -*, *HIF1A 0*, and *HIF1A +*, as detailed in Table 1 (X-axis). The 11 enzymes in green are evolving in the same way as the predicted sign of Iggy, while the 4 red ones are evolving in a different way (Color figure online).

3.2 *in vitro* Over-Expression of HIF1A in HUVECS (Human Umbilical Vein Endothelial Cells)

The induced over-expression of HIF1A adenovirally allows us to do a comparison between Iggy and Probregnet with another dataset, for which experimental perturbation results are available.

Regulatory Network from HIF Signaling Pathway. As explained in Sect. 2.2, we converted the HIF signaling pathway into a regulatory network adapted to the RNA-Seq data. We obtained a new regulatory network of 81 nodes and 233 edges. Its structure is strongly similar to the precedent one and the enzymes neighbourhood is the same as Fig. 2. The main difference is that there are new regulators of HIF1A in this regulatory network (7 nodes are predecessors of HIF1A and not only 4 as shown in Fig. 2).

Comparison Between Real Experimental Data and Iggy's and Probregnet Computational Predictions. We used the absolute normalized RNA-Seq dataset for Probregnet; while the differentially one for Iggy (see Sect. 2.1). The studied condition was the comparison between the enzymes expression in a model with HIF1A protein induction with respect to a model without HIF1A induction. Once the graph was made and data transformed we were able to apply Probregnet and Iggy on these data. Our results are shown in Fig. 5. We exclude for this study the enzymes HK3, ENO3 and PDHA2 because their expression level was too low in HUVECs cells. Therefore, we will study the expression of only 12 enzymes.

For Probregnet predictions (blue bars in Fig. 5), we used the normalized dataset (Sect. 2.1) and computed the fold-change for each enzyme (see Eq. 2). For Iggy predictions, we generated a new set of observations (see Eq. 6) and computed the predictions for each enzyme. Recall that the 12 enzymes sign was not contained in the observation dataset. In Fig. 5, we present only the Iggy predictions using the observation dataset R^+ (see Sect. 2.4). We obtained 10 "+" predictions and 2 "-" predictions in the PDHA1 and PDHB enzymes. The observation dataset R^0, generates "0" predictions (unchanged behaviors) for all of the 12 enzymes.

For the experimental observations (pink bars in Fig. 5), we used the normalized dataset and computed the fold-change of each enzyme as the average enzyme expression across HIF1A induced cells against the average enzyme expression across normal cells.

In Fig. 5 we can see that the enzyme levels evolve in the same way for Iggy and the real experimental data but slightly differently with Probregnet (8 of 12 have the same tendency). In addition, we compared all the signs predictions for all the nodes present in the graph (81) and Iggy predicted 65% in the same way as the real data, while Probregnet only 43.75%.

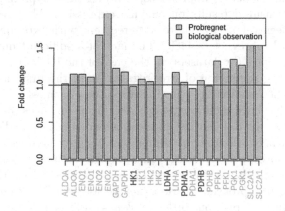

Fig. 5. Comparison between Probregnet, biological observation and Iggy. The FC of each enzyme computed with Probregnet is represented by blue bars, while the FC of biological observation by pink bars. If the Probregnet prediction or the biological observation of the enzyme agrees with Iggy's prediction sign, that is $FC > 1$ agrees with "+" and $FC < 1$ agrees with "-", the enzyme name is colored in green, else, in red (Color figure online).

4   Discussion

In this study, we are comparing two different modeling approaches, Iggy and Probregnet, on two datasets. These approaches perform enzyme *in silico* predictions, upon network stimulation. Both require a prior regulatory network: directed acyclic graph for Probregnet, and directed and signed graph for Iggy; and few experimental samples: 2 samples in two different conditions for Iggy and at least 10 samples in one condition for Probregnet to parametrize the BN. These methods are intrinsically different in the way their predictions are obtained. Iggy, models network structure and experimental dataset as facts in a logic program that when executed decides if these information is consistent, performs repairs to the data, and when consistent, deduces *coloring models* (solutions) that explain the qualitative signs (or shifts-of-expression) in some nodes of the graph, given a graph topology and an initial set of observations describing a shift of equilibrium (two conditions comparison). Whereas, Probregnet is a two step process : (i) it learns the Bayesian network parameters from a graph topology and multiple experimental datasets, (ii) it computes a belief propagation to predict the quantitative outcome of a system perturbation. We chose these methods since we want to investigate the benefit of a discrete and logical approach, such as Iggy, on the context of gene regulatory and metabolic network integration.

We used as case-study the HIF-signaling pathway. Our results on the Microarray dataset were that Iggy and Probregnet showed very similar (73.3% of agreement) computational enzymes predictions upon the same perturbation. On the second dataset, we obtained different enzyme predictions (only 66.6% of agreement) using both modeling approaches; however Iggy's predictions followed experimentally measured results on enzyme expression. Moreover, concerning other network species, Iggy was more in agreement with experimental observations (65%) than Probregnet ($\approx$44%). The lack of a sufficient number (>10) of gene expression profiles (or datasets) in the case of the HUVECs data may have impacted the wrong prediction of Probregnet. As in the first case study, some of the wrong predictions were concerning inhibited enzymes.

Both approaches have their advantages and inconveniences. Probregnet, does not need a relative (or differential) dataset under another condition. It needs, however, a small network (tens of components). Iggy handles large-scale networks and it has proven its efficiency on networks with more than a thousand of nodes [9]. Interestingly, the integration process between network and datasets is different for both approaches. Probregnet performs a linear regression of the datasets, and requires a previous order (acyclic condition) of the network edges; while Iggy does not impose this acyclic condition, but will raise places (datapoints) in the dataset where the observation does not agree with the network structure and proposes automatic repairs. In this context, Iggy performs less pre-treatment on the network structure. Furthermore, the nature of the computational predictions of both approaches is different. Iggy predicts a discrete tendency (sign) for the unobserved nodes and not a precise quantitative measure as given by Probregnet. However, Iggy is able to take into account different natures of biological interactions such as complex-formations (modeled with a Boolean *and* gate), activations, or inhibitions. Regarding the computation-time, a test was made for this case-study with a network of more than 4000 edges and 1000 nodes where Iggy's analysis finished in 0.47 s, while Probregnet, after 2 h.

This lower computation-time allows Iggy to run several benchmarks of *in silico* perturbations.

As a continuation of this work we plan to propose a way to integrate Iggy's predictions as constraints in the metabolic reaction equations. The small computational time of Iggy to propose enzyme predictions upon system perturbations encourages us to evolve our system. This new version will study the impact, on the metabolic behavior, of several treatments proposed and in this way will establish a feed-back of metabolic network biomass prediction on regulatory networks. This feedback is currently absent in Probregnet.

References

1. Thiele, S., Cerone, L., Saez-Rodriguez, J., Siegel, A., Guziołowski, C., Klamt, S.: Extended notions of sign consistency to relate experimental data to signaling and regulatory network topologies. BMC Bioinform. **16**, 345 (2015). https://doi.org/10.1186/s12859-015-0733-7
2. Yu, H., Blair, R.H.: Integration of probabilistic regulatory networks into constraint-based models of metabolism with applications to Alzheimer's disease. BMC Bioinform. **20**, 386 (2019)
3. Cowell, R.G.: Local propagation in conditional Gaussian Bayesian networks. J. Mach. Learn. Res. **6**, 1517–1550 (2005)
4. Yaghoobi, H., Haghipour, S., Hamzeiy, H., Asadi-Khiavi, M.: A review of modeling techniques for genetic regulatory networks. J. Med. Signals Sens. **2**(1), 61–70 (2012)
5. Liang, W.S., Dunckley, T., Beach, T.G., et al.: Gene expression profiles in anatomically and functionally distinct regions of the normal aged human brain. Physiol. Genomics **28**(3), 311–322 (2007)
6. Zhang, Z., Yan, J., Chang, Y., ShiDu Yan, S., Shi, H.: Hypoxia Inducible Factor-1 as a Target for Neurodegenerative Diseases. Curr. Med. Chem. **18**(28), 4335–4343 (2011)
7. Ogunshola, O., Antoniou, X.: Contribution of hypoxia to Alzheimer's disease: is HIF-1 α a mediator of neurodegeneration? Cell Mol. Life Sci. **66**(22), 3555–63 (2009)
8. Downes, N., Laham-Karam, N., Kaikkonen, M., Ylä-Herttuala, S.: Differential but complementary HIF1α and HIF2α transcriptional regulation. Mol. Ther. J. Am. Soc. Gene Ther. **26**(7), 1735–1745 (2018)
9. Folschette, M., Legagneux, V., Poret, A., Chebouba, L., Guziolowski, C., Théret, N.: A pipeline to create predictive functional networks: application to the tumor progression of hepatocellular carcinoma. BMC Bioinform. **21**, 18 (2020)
10. Dor, D., Tarsi, M.: A simple algorithm to construct a consistent extension of a partially orientedgraph. Technical report R-185, Cognitive Systems Laboratory, UCLA (1992)
11. Yu, H., Moharil, J., Blair, R.H.: BayesNetBP: an R package for probabilistic reasoning in Bayesian networks. In editing
12. Hao, T., Wu, D., Zhao, L., Wang, Q., Wang, E., Sun, J.: The genome-scale integrated networks in microorganisms. Front. Microbiol. **9**, 296 (2018). https://doi.org/10.3389/fmicb.2018.00296
13. Angione, C.: Human systems biology and metabolic modelling: a review-from disease metabolism to precision medicine. BioMed. Res. Int. **2019**, Article ID 8304260 (2019). https://doi.org/10.1155/2019/8304260

14. Marmiesse, L., Peyraud, R., Cottret, L.: FlexFlux: combining metabolic flux and regulatory network analyses. BMC Syst. Biol. **9**, 93 (2015). https://doi.org/10.1186/s12918-015-0238-z

15. Chandrasekaran, S., Price, N.D., : Probabilistic integrative modeling of genome-scale metabolic and regulatory networks in Escherichia coli and Mycobacterium tuberculosis. Proc. Natl. Acad. Sci. USA **107**(41), 17845–1750 (2010). https://doi.org/10.1073/pnas.1005139107

16. Robinson, M.D., McCarthy, D.J., Smyth, G.K.: edgeR: a bioconductor package for differential expression analysis of digital gene expression data. Bioinformatics **26**(1), 139–140 (2010). https://doi.org/10.1093/bioinformatics/btp616

17. McCarthy, D.J., Chen, Y., Smyth, G.K.: Differential expression analysis of multifactor RNA-Seq experiments with respect to biological variation. Nucleic Acids Res. **40**(10), 4288–4297 (2012). https://doi.org/10.1093/nar/gks042

18. Lifschitz, V.: What is answer set programming? In: Third AAAI Conference on Artificial Intelligence (2008)

19. Chen, Y, Lun, A.T.L., Smyth, G.K.: From reads to genes to pathways: differential expression analysis of RNA-Seq experiments using Rsubread and the edgeR quasi-likelihood pipeline. F1000Research **5**, 1438 (2016). http://f1000research.com/articles/5-1438

20. Covert, M.W., Schilling, C.H., Palsson, B.: Regulation of gene expression in flux balance models of metabolism. J. Theor. Biol. **213**(1), 73–88 (2001)

21. Shlomi, T., Eisenberg, Y., Sharan, R., Ruppin, E.: A genome-scale computational study of the interplay between transcriptional regulation and metabolism. Mol. Syst. Biol. **3**, 101 (2007)

Boolean Networks

Control Strategy Identification via Trap Spaces in Boolean Networks

Laura Cifuentes Fontanals[1,2(✉)], Elisa Tonello[1], and Heike Siebert[1]

[1] Freie Universität Berlin, Berlin, Germany
l.cifuentes@fu-berlin.de
[2] Max Planck Institute for Molecular Genetics, Berlin, Germany

Abstract. The control of biological systems presents interesting applications such as cell reprogramming or drug target identification. A common type of control strategy consists in a set of interventions that, by fixing the values of some variables, force the system to evolve to a desired state. This work presents a new approach for finding control strategies in biological systems modeled by Boolean networks. In this context, we explore the properties of trap spaces, subspaces of the state space which the dynamics cannot leave. Trap spaces for biological networks can often be efficiently computed, and provide useful approximations of attraction basins. Our approach provides control strategies for a target phenotype that are based on interventions that allow the control to be eventually released. Moreover, our method can incorporate information about the attractors to find new control strategies that would escape usual percolation-based methods. We show the applicability of our approach to two cell fate decision models.

Keywords: Boolean network · Control strategy · Trap space · Phenotype

1 Introduction

The control of biological systems presents interesting applications such as cell fate reprogramming, drug target identification for disease treatments or stem cells programming [5,17]. Controlling a cell fate decision network could for instance allow, in the case of cancer cells, to lead the system to an apoptotic state and, therefore, evolve towards the elimination of pathological cells [1]. Finding adequate candidates for control is a complex problem, in particular since the experimental testing of all the possibilities is not feasible. Mathematical modeling can help address this problem by enabling *in silico* identification of possible effective candidates.

Electronic supplementary material The online version of this chapter (https://doi.org/10.1007/978-3-030-60327-4_9) contains supplementary material, which is available to authorized users.

A. Abate et al. (Eds.): CMSB 2020, LNBI 12314, pp. 159–175, 2020.
https://doi.org/10.1007/978-3-030-60327-4_9

Modeling of biological processes is often challenged by the lack of information about kinetic parameters or specific reaction mechanisms. The Boolean formalism aims at capturing the qualitative behavior of systems via a coarse representation of the relationship between the species of interest. Mechanisms underlying activation and inhibition processes are summarized in logical functions, allowing for two activity levels for each variable. The two values can represent for example if a gene is expressed or not, or if the concentration of a protein is above or below a certain threshold. Boolean modeling has in many instances been shown to capture the fundamental behaviors and dynamics of biological systems and has been widely used to make predictions or design strategies for therapeutic interventions [3, 6, 7].

Control of biological systems is a broad field that encompasses a variety of approaches and goals. Attractor control aims at leading the system to a desired attractor, starting from a particular initial state (source-target control) [14] or from all possible initial states (full-network control) [20]. However, it is often useful to induce a desired phenotype rather than a specific attractor. Phenotypes are usually defined in terms of some biomarkers i.e., observable and measurable components that represent the main characteristics of biological processes. The approach that focuses control on a set of relevant variables is also known as target control [16,18]. In this work, we are interested in full-network control for a target phenotype.

There are different approaches for system interventions, that is, the way the control is applied to biological systems. In the context of Boolean modeling, we consider as interventions the perturbations or modifications that fix the value of some components (node control) [14,20]. In the example of a gene regulatory network, fixing a variable to a certain value can be understood as the knockout or permanent activation of a gene. Among other approaches to Boolean network control is edge control, which targets the interactions between variables [2,15]. For a gene regulatory network, edge control can be interpreted for instance as the modification of a protein to alter its interaction with a certain gene.

Control of dynamical systems has been a popular research field in systems biology in the last years, also in the Boolean setting. Many approaches focus on the structure and topology of the network, for example by looking at feedback loops [19] or stable motifs [20], and several studies discuss the complexity and characteristics of such problems [9,13]. Other approaches include techniques based on topological information to reduce the size of the search space [16] or computational algebra methods [15]. Recent works have explored attractor control through the characterization of basins of attraction, that is, sets of states from which only a certain attractor can be reached [14]. However, the identification of basins of attraction might require the exploration of the complete state space. Attractor reachability can be investigated using trap spaces, which are subspaces that trajectories cannot leave. By definition, every trap space contains at least one attractor and, therefore, in some cases minimal trap spaces can be good approximations for the attractors [11]. The identification of trap spaces in biological systems can often be performed efficiently by exploiting properties of the prime implicants [10].

Our approach aims to identify strategies for phenotype control by exploiting properties of trap spaces. We introduce the concept of space of attraction, a subspace that approximates the basin of attraction, to find control strategies without the need of computing the whole basin. We extend this idea to define spaces of attraction for trap spaces and relate them to control strategies, which are defined as sets of constraints that fix the value of some variables and induce a certain target phenotype. We exploit properties of trap spaces and computation techniques for target control to define a new method to compute control strategies that do not require a permanent intervention and allow the control to be eventually released. Our approach can incorporate information about the attractors to obtain new control strategies that might escape percolation-based target control techniques. The method presented here is widely applicable to Boolean models of biological systems and can provide, under certain conditions, control strategies that are independent of the type of update used in the model.

We start by giving a general overview about Boolean modeling (Sect. 2). Then we introduce the concepts of control strategy and space of attraction in this setting (Sect. 3), providing the theoretical bases for the computation of some types of control strategies. In Sect. 4, we present a method to compute control strategies based on the theoretical principles explained in Sect. 3 and implemented using the prime implicants of the function. Lastly, in Sect. 5 we show the applicability of our method to two cell fate decision networks [7,21].

2   Background: Boolean Networks and Dynamics

A *Boolean network* on n variables is defined as a function $f\colon \mathbb{B}^n \to \mathbb{B}^n$, where $\mathbb{B} = \{0, 1\}$. $V = \{1, ..., n\}$ is the set of variables of f, $\mathbb{B}^n$ is the *state space* of the Boolean network and every $x \in \mathbb{B}^n$ is a *state* of the state space. For any $x \in \mathbb{B}^n$ and $I \subseteq V$, $\bar{x}^I$ is defined as $\bar{x}_i^I = x_i$ for $i \in V \backslash I$ and $\bar{x}_i^I = 1 - x_i$ for $i \in I$. If $I = \{i\}$, $\bar{x}^I$ is written as $\bar{x}^i$.

A *dynamics* on $\mathbb{B}^n$ or *state transition graph* is a directed graph with vertex set $\mathbb{B}^n$. There are several ways of associating a dynamics to a Boolean network f. In the *general asynchronous dynamics* or *general asynchronous state transition graph* $GD(f)$ there exists an edge from a vertex x to a vertex y if and only if there exists $\emptyset \neq I \subseteq V$ such that $\bar{x}^I = y$ and $f_i(x) = y_i$ for every $i \in I$. Note that the general asynchronous dynamics considers transitions which update subsets of components simultaneously in a non-deterministic way. By choosing different types of updates, other state transition graphs can be defined. The *asynchronous dynamics* $AD(f)$ is defined by considering the transitions updating only one component at a time and the *synchronous dynamics* $SD(f)$ considers only the transitions where all the components that can be updated are updated at once. Note that $AD(f)$ and $SD(f)$ are subgraphs of $GD(f)$. To simplify the notation, $D(f)$ will denote any of these dynamics associated to f. The choice of asynchronous and general asynchronous updates is motivated by the attempt to capture different, and sometimes unknown, time scales that might coexist in the modeled system. An example of asynchronous dynamics of a Boolean network is shown in Fig. 1.

A *trap set* $T \subseteq \mathbb{B}^n$ is a set such that for all $x \in T$, if y is a successor of x in the dynamics, then $y \in T$. A minimal trap set under inclusion is an *attractor*. An attractor can be a *stable state* (or *fixed point*), when it consists only of one state, or a *cyclic* (or *complex*) *attractor* when it is larger. In biological systems, stable states can be identified with different cell fates or cell types, and cyclic attractors with cell cycles or specific cell processes. Given a Boolean function f and an attractor A, the *weak basin of attraction* of A is defined as the set of states x such that there exists a path from x to an element of A in $D(f)$. The *strong basin of attraction* of A is the set of states in the weak basin of A that do not belong to the weak basin of attraction of any other attractor different from A. Figure 1 shows the weak and strong basins for an attractor in an asynchronous state transition graph.

The control interventions considered in this work consist in fixing the values of some components. Formally, given a state $c \in \mathbb{B}^n$ and a subset of variables $I \subseteq V$, we define the *subspace* induced by c and I as the set $\Sigma(I, c) = \{x \in \mathbb{B}^n | \forall i \in I, x_i = c_i\}$. The variables in I are called *fixed variables*, while the other variables are called *free*. We denote subspaces as states, using the symbol $*$ for the free variables. For example, the subspace $\{x \in \mathbb{B}^4 | x_1 = 1 \text{ and } x_3 = 0\}$ is denoted as $1 * 0*$.

The identification of control variables requires examining the effect that fixing certain variables has on the dynamics. Given a Boolean function f and a subspace $\Theta = \Sigma(I, c)$, the restriction of the function f to the subspace Θ is defined as:

$$f_{\restriction_\Theta} : \Theta \to \Theta, \text{ where for all } i \in V, (f_{\restriction_\Theta})_i(x) = \begin{cases} f_i(x), & i \notin I, \\ c_i, & i \in I. \end{cases}$$

Note that $f_{\restriction_\Theta} : \Theta \to \Theta$ can be identified with a Boolean network $g : \mathbb{B}^m \to \mathbb{B}^m$, where $m = n - |I|$. Via this identification, we extend all the definitions that apply to a Boolean network to such restrictions. For example, the state transition graph corresponding to $f_{\restriction_\Theta} : \Theta \to \Theta$ is defined as usual, only with vertex set Θ instead of $\mathbb{B}^n$ (see Fig. 2). Moreover, if T is a trap set in $D(f)$, then $T \cap \Theta$ is a trap set in $D(f_{\restriction_\Theta})$.

A subspace that is also a trap set is called a *trap space*. While trap sets and attractors might vary when considering different types of dynamics, trap spaces are independent of the type of update. The Boolean function represented in Fig. 1 has four trap spaces: 000, 111, $0 * 0$, $* * *$.

In this work we aim at using trap spaces to find control strategies for phenotypes. Phenotypes are usually defined in terms of the state of some measurable components called biomarkers, which are observable components that can be used as indicators of different cell types or cell fates or to distinguish between healthy and pathological conditions. Although the notion of phenotype is usually related to stability, we extend this concept to consider any possible state in order to allow non-attractive states satisfying the phenotype characteristics to become attractors in the controlled system. Thus, in this work, we define a *phenotype* as a subspace.

3   Spaces of Attraction and Control Strategies

The strong basin of attraction of an attractor A can be naturally related to control since, by definition, it contains all the states that have paths to A but not to any other attractor. In contrast to methods requiring basin exploration, we use subspace approximation of the basins combined with trap spaces computation. To do so, we extend the notion of basin of attraction to trap sets. We then exploit useful properties of trap spaces, e.g. independence of the update, efficient identification and potential approximation of attractors, to develop a new approach for the identification of control strategies.

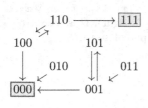

Basins of attraction of A_1:

- $Strong(A_1) = \{000, 001, 010, 011, 101\}$
- $Weak(A_1) = \{000, 001, 010, 011, 101, 100, 110\}$

Spaces of attraction of A_1:

- $\Omega_1 = 0**$, $\Omega_2 = 00*$, $\Omega_3 = 01*$, $\Omega_4 = 0*0$, $\Omega_5 = 0*1$, $\Omega_6 = *01$, $\Omega_7 = 000$, $\Omega_8 = 001$, $\Omega_9 = 010$, $\Omega_{10} = 011$, $\Omega_{11} = 101$, with $\Omega_i \subsetneq Strong(A_1)$ for all $1 \leq i \leq 11$.

Fig. 1. Asynchronous dynamics of the Boolean function $f(x) = (\bar{x}_1\bar{x}_2x_3 \vee x_1x_2, x_1\bar{x}_2\bar{x}_3 \vee x_1x_2x_3, x_1x_2 \vee x_1x_3 \vee x_2x_3)$, with attractors $A_1 = 000$ and $A_2 = 111$ and trap spaces 000, 111, $0*0$, $***$. All the spaces of attraction of A_1 are included in its strong basin (in red) while the basin itself is not a space of attraction. (Color figure online)

3.1   Control Strategies

We now formalise the notion of control strategy. A control strategy is a subspace defined by a set of interventions that fix the value of some variables and thus force all attractors to be contained in the subspace defining the phenotype.

Definition 1. *Given a Boolean function f and a subspace $P \subseteq \mathbb{B}^n$, a control strategy (CS) for the phenotype P in $D(f)$ is a subspace $\Theta \subseteq \mathbb{B}^n$ such that, for any attractor A of $D(f_{\restriction\Theta})$, $A \subseteq P$.*

If the desired phenotype is a stable state in the original dynamics ($P = \{y\}$, $y \in \mathbb{B}^n$), a control strategy for P is a subspace Θ such that y is the only attractor of $f_{\restriction\Theta}$. Figure 2 shows an example of a control strategy for a stable state. The size of the subspace defining a control strategy represents the number of interventions in the system. Therefore, the most interesting control strategies are the subspaces that are maximal with respect to inclusion.

A common approach in the context of control is the use of value percolation [16,18]. Different combinations of variables to be fixed are considered, and their values propagated iteratively until an invariant subspace is reached. A combination of variables and values is an intervention strategy if the subspace obtained at the end of the iterative percolation process is contained in the target phenotype. Strategies obtained with this approach satisfy the conditions of Definition 1. However, the class of control strategies identified by the definition is larger, as we will discuss in the following.

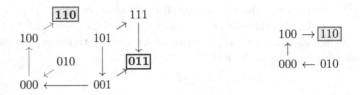

Fig. 2. Asynchronous dynamics of the Boolean function $f(x) = (x_1\bar{x}_3 \vee \bar{x}_2\bar{x}_3, x_1 \vee x_3, x_1x_3 \vee x_2x_3)$ (left) and $f_{\restriction_\Omega}(x) = (x_1 \vee \bar{x}_2, x_1, 0)$ with $\Omega = {*}{*}0$ (right). Ω is a control strategy for the phenotype $P = \{110\}$ in $AD(f)$. Ω does not percolate to P.

3.2   Spaces of Attraction

Trap sets are sets of states that the dynamics cannot leave. Each trap set contains, as a consequence, at least one attractor. The concept of basin of attraction defined for an attractor can be naturally extended to trap sets. As mentioned before, we wish to approximate basins of attraction by subspaces. Combining these two ideas, we introduce the concept of space of attraction of a trap set T as a subspace Ω such that from any state in Ω there exists a path to T and no trap set disjoint from T is reachable from Ω.

Definition 2. *Let f be a Boolean function and T be a trap set of f. A space of attraction of the trap set T in $D(f)$ is a subspace Ω such that for all $x \in \Omega$ and for any trap set S, if there exists a path in $D(f)$ from x to an element of S, then $S \cap T \neq \emptyset$.*

Definition 2 implies the existence of a path from the space of attraction Ω to the trap set T since, for any state in Ω, the trap set consisting of all the states reachable from it cannot be disjoint from T. A trap set can have many spaces of attraction. In fact, any subspace contained in a space of attraction is also a space of attraction. Moreover, if there is only a unique trap set T_m minimal with respect to inclusion contained in a trap set T, any space of attraction of T is also a space of attraction of T_m. Both trap spaces and spaces of attraction are subspaces that characterize the long term behavior of the system. However, in contrast to trap spaces, spaces of attraction can depend on the update.

If a trap set is an attractor, its spaces of attraction can be related to its basins of attraction. The spaces of attraction of an attractor A are clearly contained in the strong basin of A since, by Definition 2, none of the other attractors can be reached from any state inside the space of attraction. However, the strong basin of attraction of A might not be a space of attraction (see Fig. 1).

Spaces of attraction, as well as basins, might include paths crossing non-attractive cycles in the state transition graph. As a consequence, some paths starting in the space of attraction (or basin) might not reach the trap set (or attractor), staying indefinitely in non-attractive cycles. While in very specific circumstances such behavior might be relevant, generally it constitutes an artifact arising from the non-deterministic update. Here, we extend the standard view on basins of attraction to spaces of attraction, assuming the trajectories of interest will eventually leave non-attractive strongly connected components in the state transition graph.

The condition that a subspace needs to satisfy to be a space of attraction of a trap set T gets simplified when the subspace considered is the entire state space. In this case, it is only required that the trap set T can be reached from every state in the state space (see Lemma 1). This condition already implies that there cannot be a trap set disjoint from T.

Lemma 1. *Let f be a Boolean function and T a trap set of f. Then $\mathbb{B}^n$ is a space of attraction of the trap set T in $D(f)$ if and only if for all $x \in \mathbb{B}^n$ there exists a path in $D(f)$ from x to some $y \in T$.*

The application of Lemma 1 to the restriction on a subspace immediately yields the following corollary.

Corollary 1. *Let f be a Boolean function, T a trap set of f and Ω a subspace such that $T \subseteq \Omega$. Then Ω is a space of attraction of T in $D(f_{\restriction\Omega})$ if and only if for all $x \in \Omega$ there exists a path in $D(f_{\restriction\Omega})$ from x to some $y \in T$.*

In other words, a space of attraction of a trap set T for the Boolean function restricted to that subspace defines the restrictions that we can impose on the function f to lead the dynamics to T. If T is a trap space, there is always a trivial space of attraction for the restricted function which is T itself.

Note that a subspace Ω that is a space of attraction of T for the Boolean function f is not necessarily a space of attraction for $f_{\restriction\Omega}$ (see Fig. 3).

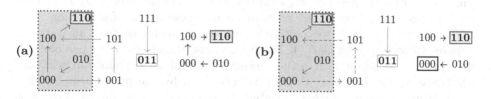

Fig. 3. (a) $\Omega = **0$ is a space of attraction for $AD(f)$ and $AD(f_{\restriction\Omega})$, with $f(x) = (\bar{x}_2 \vee x_1\bar{x}_3, x_1\bar{x}_3 \vee x_2x_3, \bar{x}_1\bar{x}_2 \vee x_2x_3)$ and $f_{\restriction\Omega}(x) = (\bar{x}_2 \vee x_1, x_1, 0)$. (b) $\Omega = **0$ is a space of attraction for $AD(g)$ but not for $AD(g_{\restriction\Omega})$, with $g(x) = (x_1\bar{x}_3 \vee \bar{x}_2x_3 \vee x_1\bar{x}_2, x_1\bar{x}_3 \vee x_2x_3, \bar{x}_1\bar{x}_2 \vee x_2x_3)$ and $g_{\restriction\Omega}(x) = (x_1, x_1, 0)$.

Given a trap space T that only contains attractors belonging to a certain phenotype P, any space of attraction that leads the system to T would also lead it to an attractor belonging to P. In other words, any space of attraction for a trap space T is also a control strategy for a phenotype P if T only contains attractors belonging to P. The following proposition formalizes this idea.

Proposition 1. *Let $P \subseteq \mathbb{B}^n$ be a subspace and f a Boolean function. Let T be a trap space such that if $A \subseteq T$ is an attractor of $D(f)$, then $A \subseteq P$. Let Ω be a space of attraction of T in $D(f_{\restriction \Omega})$ such that $T \subseteq \Omega$. Then Ω defines a control strategy in $D(f)$ for P.*

Proof. Let A be an attractor for $D(f_{\restriction \Omega})$. Then $A \subseteq \Omega$. Since Ω is a space of attraction of T in $D(f_{\restriction \Omega})$ and A is a trap set in $D(f_{\restriction \Omega})$, $T \cap A \neq \emptyset$. As T and A are trap sets, $T \cap A$ is also a trap set in $D(f_{\restriction \Omega})$. Since A is minimal, $A = T \cap A \subseteq T$. Then, since T is a trap space and for all $x \in T, f_{\restriction \Omega}(x) = f(x)$, A is also an attractor of $D(f)$ and, therefore, $A \subseteq P$. □

Since a trap space is always a space of attraction of itself, given a subspace $P \subseteq \mathbb{B}^n$, any trap space containing only attractors in P is a control strategy for P. Note, however, that Proposition 1 does not characterize all the control strategies satisfying Definition 1. The example in Fig. 3 (a) shows a control strategy Ω that does not percolate to any trap space containing only the attractor 110.

From a theoretical standpoint, the type of control strategies identified by Proposition 1 allow the interventions to be released after a certain number of steps. That is because these control strategies induce the target phenotype by leading the system to a trap space. Once the trap space is reached, the control can be released and the system will remain in the trap space, eventually evolving to the phenotype of interest. This additional theoretical property could widen the range of possible choices for system control. Provided that the time scales of the processes involved are sufficiently understood, it could allow for instance to apply interventions relying on agents that decay over time.

3.3 Identification of Spaces of Attraction

As explained in the previous section, control strategies for a phenotype P can be found by identifying spaces of attraction of trap spaces containing only attractors in P. In this section, we explore ways of finding these spaces of attraction.

Given a trap space T, we look for a subspace Ω such that from all states in Ω there is a path to T in $D(f_{\restriction \Omega})$. To do so, we use the idea of value percolation, which is a common approach in the context of control. As explained in Sect. 3.1, it is based on the fact that the constraints given by the fixed variables of a subspace might induce further variables to get fixed. Thus, in our setting, a subspace $\Omega = \Sigma(W, c)$ that percolates to the trap space $T = \Sigma(U, c)$ is a space of attraction of T in $f_{\restriction \Omega}$. The following lemma formalizes this idea.

Lemma 2. *Let $f \colon \mathbb{B}^n \to \mathbb{B}^n$ be a Boolean function, $c \in \mathbb{B}^n$ and $S = \Sigma(U, c)$, $\Omega = \Sigma(W, c)$ subspaces of $\mathbb{B}^n$ such that $S \subseteq \Omega$ and $W \subseteq U \subseteq V$. If for all*

$s \in U \backslash W$, $f_s(x) = c_s$ for all $x \in \Omega$, then for all $x \in \Omega$ there exists a path in $D(f_{\restriction \Omega})$ from x to some $y \in S$.

Proof. Since the proof depends on the update, we treat each case separately.

$D = AD$: For each $x \in \Omega$ and for each $s \in U \backslash W$ such that $x_s \neq c_s$, $f_s(x) = c_s$. Therefore, x admits a successor y in $AD(f_{\restriction \Omega})$ with $y_s = c_s$. This implies the existence of a path in $AD(f_{\restriction \Omega})$ from any state in Ω to S.

$D = SD$: For each $x \in \Omega$ and for each $s \in U$, $f_s(x) = c_s$. Therefore, x admits a successor $y \in S \subseteq \Omega$ in $SD(f_{\restriction \Omega})$.

$D = GD$: Since all the paths in $AD(f)$ and $SD(f)$ are also paths in $GD(f)$, the conclusion follows from the previous cases.   □

Lemma 2 can be extended with Corollary 1 to provide conditions that allow the identification of spaces of attraction.

Lemma 3. *Let $f \colon \mathbb{B}^n \to \mathbb{B}^n$ be a Boolean function and $T = \Sigma(U, c)$ a trap space of f with $U \subseteq V$ and $c \in \mathbb{B}^n$. Let $\Omega = \Sigma(W, c)$ be a subspace of $\mathbb{B}^n$ such that $T \subseteq \Omega$ and $W \subseteq U \subseteq V$. If $f_s(x) = c_s$ for all $x \in \Omega$ and $s \in U \backslash W$, then Ω is a space of attraction of T for $D(f_{\restriction \Omega})$.*

To improve the spaces of attraction obtained with Proposition 1, we can extend Lemma 3 applying the idea used in Lemma 2 several times, building a path of percolated subspaces ending in the trap space T.

Proposition 2. *Let $f \colon \mathbb{B}^n \to \mathbb{B}^n$ be a Boolean function and let $c \in \mathbb{B}^n$. Let $T = \Sigma(U, c)$ be a trap space and $\Omega = \Sigma(W, c)$ be a subspace containing T with $W \subseteq U \subseteq V$. Let $I_0 = W$ and $I_{k+1} = \{s \in U | s \in I_k$ or $f_s(x) = c_s$ for all $x \in S_k\}$, where $S_k = \Sigma(I_k, c)$. If there exists a k_T such that $I_{k_T} = U$, then Ω is a space of attraction of T for $D(f_{\restriction \Omega})$.*

Proposition 2 gives sufficient conditions for a subspace to be a space of attraction of a trap space in the restriction and, together with Proposition 1, provides a way to identify control strategies for a given phenotype. However, not all spaces of attraction fall under the conditions given by Proposition 2. The example in Fig. 3 (a) shows a space of attraction $\Omega = **0$ for a trap space $T = 110$, which is a control strategy for $P = \{110\}$, where Ω does not percolate to T.

Sometimes the attractors of a system of interest are known. In other cases they are not known but can be approximated by minimal trap spaces [11], that is, each minimal trap space contains only one attractor and every attractor is included in a minimal trap space. This information is not usually exploited by target control methods, which often rely solely on percolation-like techniques. The approach described in this work can use this knowledge to find additional control strategies. If the attractors are known or they can be approximated by minimal trap spaces, we can easily find trap spaces satisfying the conditions of Proposition 1 by simply checking whether these attractors or minimal trap spaces are included in a trap space. Therefore, larger trap spaces containing only attractors of the target phenotype can be identified. By Proposition 1, spaces of attraction for these trap spaces are also control strategies for the phenotype. These control strategies do not necessarily percolate to the phenotype and, therefore, might not be identified by usual percolation techniques. Figure 2 shows an example of such a control strategy, where $\Omega = T = **0$ is a space of attraction for the trap space T, which contains only the attractor $A = 110$, and so, is a control strategy for the phenotype $P = A$. Note that Ω does not percolate to A.

The attractors of a Boolean network might vary in different dynamics. Therefore, the trap spaces satisfying Proposition 1 and the control strategies characterized by them might also be dependent on the dynamics. Conversely, the spaces of attraction obtained by Proposition 2 are independent of the update. Thus, if the trap spaces considered satisfy the conditions of Proposition 1 in all the dynamics, the control strategies identified are also independent of the update.

4   Computation of Control Strategies

We propose a method to find control strategies for a given phenotype, using the ideas explained in the previous section. The main steps of the method are represented in Fig. 4 and the detailed procedure is shown in Algorithm 1.

In order to implement the computation of the control strategies, we use the prime implicants of the function. Given a Boolean function $f\colon \mathbb{B}^n \to \mathbb{B}^n$, a c-*implicant* of f_i, with $c \in \mathbb{B}$ and $i \in V$, is a subspace Q such that $f_i(x) = c$ for all $x \in Q$. A *prime implicant* is an implicant that is maximal under inclusion. Given $T = \Sigma(U, c)$, finding a subspace satisfying the hypothesis of Lemma 3 is equivalent to finding a subspace that is a c_i-implicant of f_i for all $i \in U$. Moreover, prime implicants can also be used to compute the trap spaces [10]. The computation of the prime implicants of a Boolean function is in general a hard problem. However, networks modeling biological systems are usually relatively sparse, since the number of components regulating a variable is relatively small compared to the size of the network. Therefore, they are rather tractable in terms of prime implicants computation. Several tools are available for the computation of prime implicants and trap spaces. We use *PyBoolNet* [12], a Python package that allows generation and analysis of Boolean networks and provides an efficient computation of prime implicants and trap spaces for quite large networks.

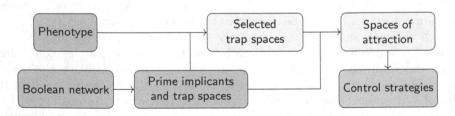

Fig. 4. Main steps of the method for finding control strategies for a phenotype, represented in color boxes according to their role: inputs (blue), precomputation (green), main computation (beige), output (red). (Color figure online)

We describe now the main steps of the method, outlined in Fig. 4.

Inputs. The inputs are the Boolean function describing the system and the subspace of the target phenotype P. The attractors, if known, are also used as input. Prime implicants and trap spaces can be given as input or computed from the Boolean function.

Selection of Trap Spaces. Trap spaces of interest are divided into two types: trap spaces contained in P (Type 1) and trap spaces not contained in P but containing only attractors in P (Type 2). As trap spaces have been identified in the previous step, this selection only requires checking whether a trap space belongs to one of the types (Algorithm 1: 3–5). Trap spaces of Type 2 are only identified when all the attractors are known or can be approximated by minimal trap spaces. In order to avoid unnecessary calculations, we do not consider trap spaces that percolate to smaller ones, since if a trap space T_1 percolates to a trap space T_2, all spaces of attractions of T_1 are also spaces of attraction of T_2.

Algorithm 1. Control strategies for a phenotype P

Input: f Boolean function, P phenotype, $attr$ attractors of f (optional), m limit size of control strategies (optional)
Output: control strategies for P

1: **function** CONTROLSTRATEGIES(f, P, $attr$)
2:     $\mathbf{T} \leftarrow$ trapSpaces(f)
3:     **selTS** $\leftarrow$ selectedTrapSpaces1($\mathbf{T}$, P)
4:     **if** $attr \neq \emptyset$ **then**:
5:         **selTS** $\leftarrow$ selTS + selectedTrapSpaces2($\mathbf{T}$, P, $attr$)
6:     $\mathbf{CA} \leftarrow \emptyset$
7:     **for** i in $\{1, \ldots, \min(m, n)\}$ **do**:   ▷ n total number of variables
8:         $\mathbf{S} \leftarrow \{$S subspace: $|\text{fixed(S)}| = i, \exists T \in \mathbf{selTS}$ with $T \subseteq S\}$
9:         **for** S in $\mathbf{S}$ **do**:
10:            **if** (S $\not\subseteq$ S' for all S' in $\mathbf{CA}$) **and** isSpaceAttraction(f, S, **selTS**) **then**:
11:                add S to $\mathbf{CA}$
12:    **return CA**

Algorithm 2. Subspace is a space of attraction

Input: f Boolean function, S subspace, $\mathbf{TS}$ trap spaces
Output: *True* if S is space of attraction of a trap space in $\mathbf{TS}$. *False* otherwise.

1: **function** ISSPACEATTRACTION(f, S, $\mathbf{TS}$)
2:     f' $\leftarrow$ percolateFunction(f, S)
3:     **return** isNotEmpty($\{$T in $\mathbf{TS}$: T $\subseteq$ S and fixed(T) $\subseteq$ fixed(f')$\}$)

Computation of Spaces of Attraction. Spaces of attraction for the trap spaces from the previous step are computed using the theoretical principles described in Proposition 2. The detailed procedure is shown in Algorithm 1:

6–11. For each subspace S that contains at least one of the selected trap spaces (Algorithm 1: 8), it is checked whether it is a space of attraction for one of the selected trap spaces (Algorithm 1: 10). To do so, the percolated function of f obtained by fixing the variables in S is calculated (Algorithm 2: 2). If T is contained in the subspace generated by S and all the variables fixed in T are also fixed in the percolated function, then the subspace generated by S is a space of attraction of T (Algorithm 2: 3). Since the aim is to find maximal spaces of attraction satisfying this property, the subspaces S are taken randomly fixing an increasing number of variables, so that supersets of sets already defining a space of attraction are not considered (Algorithm 1: 8, 10).

Output The obtained spaces of attraction are control strategies for the phenotype P by Proposition 1 and, therefore, are returned as output.

The method also allows to include some constraints on the control strategies. One example is the exclusion of some components, which can be taken into account when selecting the subspaces S (Algorithm 1: 8). In addition, limiting the number of interventions (Algorithm 1: 7) might allow to reduce the computational cost without losing interesting solutions, since small control strategies are usually the most relevant.

The computation of trap spaces scales well with the size of the network [10]. On the other hand, the identification of control strategies is based on the exploration of all possible candidate subspaces and might pose a difficulty in terms of scalability over large networks. Possible approaches that address this point are suggested in the discussion.

5   Application: Cell Fate Decision Networks

In this section we discuss the application of our method to two Boolean networks describing cell fate decision processes. In the first case study we consider two different control problems, one having a phenotype as target for the control, the second targeting single attractors. The second case study focuses on phenotype control. We compare the control strategies identified by our approach to the ones obtained using exclusively value percolation, as described in Sect. 3.1. We show that, for both examples, new control strategies can be identified with the procedure introduced in this work.

All computations in this section were done on an 8-processor computer, Intel®Core™ i7-2600 CPU at 3.40 GHz, 16 GB memory, without any use of parallelization. The running times indicated are for the full procedure described in Section 4, with the time for computing and selecting trap spaces being a negligible fraction of the total time in all cases. The source code is available at http://github.com/lauracf/trap-space-control.

5.1   MAPK Network

The network considered in this case study was introduced by Grieco et al. (2013) [7] to model the effect of the Mitogen-Activated Protein Kinase (MAPK) pathway on cell fate decisions taken in pathological cells (see Fig. 5). It uses 53

Boolean variables, four being inputs (DNA-damage, EGFR-stimulus, FGFR3-stimulus and TGFBR-stimulus) and three outputs (Apoptosis, Proliferation and Growth-Arrest). The asynchronous dynamics has 18 attractors, 12 being stable states and 6 cyclic attractors. All of them can be approximated by minimal trap spaces, since each minimal trap space only contains one attractor and there is no attractor that is not contained in a minimal trap space [11]. Therefore, we can use trap spaces of both Type 1 and Type 2 to compute control strategies.

The phenotype chosen as target for the control is the apoptosis phenotype, which is defined in [7] as the states fixing Apoptosis and Growth Arrest to 1 and Proliferation to 0. There are 103 non-percolating trap spaces, which are trap spaces that do not percolate to smaller ones, containing only attractors in the apoptosis phenotype. Of these, 64 are of Type 1 and 39 of Type 2. We set an upper bound of four components to the size of the control strategies, since generally only small control strategies are of interest and this limit already allows to find relevant ones. In addition, we exclude interventions that fix any of the output nodes of the network. In this setting, we identify two control strategies of size 1 ({TGFBR-stimulus = 1} and {DNA-damage = 1}) and no control strategies of size 2, 3 and 4. The running time is around 6 min.

Using exclusively the percolation of the fixed values we identify two control strategies of size 1 ({TGFBR-stimulus = 1} and {TGFBR = 1}), 121 control strategies of size 2, 164 of size 3 and 139 of size 4. Looking at the Boolean function, we observe that TGFBR is uniquely regulated by TGFBR-stimulus, so fixing TGFBR-stimulus to 1 implies that TGFBR is also fixed to 1 and, therefore, these interventions are equivalent in terms of their effect on the apoptosis phenotype. However, it is obvious that if the control fixing TGFBR to 1 is released, TGFBR could be updated to zero again by TGFBR-stimulus, and this change would induce the system to leave the apoptosis phenotype. Therefore, the control of TGFBR requires a permanent intervention.

Our method uncovers the control strategy {DNA-damage = 1}, which is not obtained by using solely value percolation. In fact, the percolation of the subspace defined by this strategy does not reach the phenotype, but stops at the subspace $T = \{$DNA-damage = 1, ATM = 1, TAOK = 1$\}$. This implies that the components defining the phenotype can still oscillate in the restricted system. However, since T is one of the trap spaces selected by our method, all the attractors inside T belong to the apoptosis phenotype and, therefore, the constraint {DNA-damage = 1} is identified as a control strategy.

Of the control strategies of size 2, 3 and 4 that can be identified by percolation, 18, 13 and 7 respectively are supersets of the control strategy {DNA-damage = 1} identified by our method. For this reason, the subspaces obtained by percolating these interventions are contained in the trap space T mentioned above and therefore the associated control can be eventually released, without affecting the reachability of the target. The remaining control strategies are not guaranteed to lead to a trap space. As a consequence, in these cases, an early release of the control could lead to the loss of the control goal. This illustrates how our method can complement previous approaches, by identifying new control strategies of low complexity, while at the same time providing information about the effects of a possible release of the control.

The components appearing in the minimal control strategies identified (DNA-damage and TGFBR-stimulus) correspond to two inputs of the model. These inputs represent anti-proliferative stimuli from the MAPK network [7] and, therefore, can be expected to play an important role in the phenotype decision. It is, however, certainly interesting that they are capable of fully inducing the apoptosis phenotype without further conditions on internal processes.

In addition to the control problem for the apoptosis phenotype, we also searched for control strategies for the 10 apoptotic stable states. We set the maximum size of control strategies to five. For eight stable states (A_1 to A_8 in Table 1) exactly one control strategy of size 4 is obtained. For stable state A_9, two control strategies of size 5 are found, and for A_{10} no control strategies up to size 5 are identified. The list of stable states and their control strategies can be found in the supplementary material. The running time for one stable state is around 21 min.

Since the chosen stable states belong to the apoptosis phenotype, all the selected trap spaces are also considered when computing the control strategies for the apoptosis phenotype. Therefore, the control strategies of the stable states are subspaces of the ones obtained for the apoptosis phenotype. One of the main differences is that the four inputs are present in all the control strategies of the stable states. The input variables are, by definition, not regulated by any component, and therefore must be directly controlled if the value in a given steady state is to be achieved. The analysis of the control problem for the phenotype revealed that fixing DNA-damage to 1 is enough to lead the system to the apoptosis subspace, but fixing the additional inputs is necessary to obtain a specific steady state. Fixing the four inputs is already enough to induce the stable states A_1 to A_8 solely by percolation. However, the stable states A_9 and A_{10} require additional internal processes to be controlled. For A_9, the two control strategies identified do not percolate directly to the attractor, but lead the dynamics to one of the selected trap spaces. For A_{10}, no control strategies up to size 5 are found neither by our method nor percolation techniques, suggesting that a higher number of interventions might be necessary. These observations show that control for a phenotype can be more achievable than for a specific attractor, and thus in some cases more interesting for application.

5.2   T-LGL Network

We now consider a control problem for the network introduced by Zhang et al. (2008) [21] to model the T cell large granular lymphocyte (T-LGL) survival signaling network (see Fig. 6). It consists of 60 Boolean variables, six being inputs (CD45, IL15, PDGF, Stimuli, Stimuli2 and TAX) and three readouts (Apoptosis, Proliferation and Cytoskeleton-signaling). The asynchronous dynamics has 156 attractors, 86 being stable states and 70 cyclic attractors. As in the previous network, all of them can be approximated by minimal trap spaces [11]. Thus, we can use trap spaces of both Type 1 and Type 2 to compute control strategies.

We consider the apoptosis phenotype defined by fixing Apoptosis to 1 and Proliferation to 0. Note that the third readout, Cytoskeleton signaling, is forced to 0 by its regulator Apoptosis having value 1. There are 883 non-percolating trap spaces containing only attractors in the apoptosis phenotype. 729 trap spaces are of Type 1 and 154 of Type 2. As in the previous case study, we set an upper bound of four components to the size of the control strategies and we exclude interventions that fix any of the readout nodes of the network. In this setting, six control strategies are identified: three of size 3 ({CD45 = 0, IL15 = 0, PDGF = 1}, {CD45 = 0, IL15 = 0, Stimuli = 1}, {CD45 = 0, IL15 = 0, TAX = 1}) and three of size 4 ({CD45 = 1, PDGF = 0, PDGFR = 0, Stimuli2 = 1}, {CD45 = 1, PDGF = 0, S1P = 0, Stimuli2 = 1}, {CD45 = 1, PDGF = 0, SPHK1 = 0, Stimuli2 = 1}). The running time is around 12 min.

The three control strategies of size 3 consist only of input components. All the control strategies of size 4 have three components in common while the fourth varies within PDGFR, S1P and SPHK1, suggesting that these three interventions might be equivalent in terms of their effect on the apoptosis phenotype. In fact, by looking at the Boolean function, we observe that fixing PDGFR = 0, implies SPHK1 = 0, which also implies S1P = 0. Identifying equivalent interventions a priori might allow to reduce the computational cost of the method.

Using only percolation we find exactly one control strategy of size 1 ({Caspase = 1}) and none of size 2, 3 or 4. However, this control strategy is relatively trivial since the Caspase component is directly regulating Apoptosis. The control strategies identified by our method do not percolate directly to the phenotype. At the end of the percolation process, the dynamics reaches one of the trap spaces selected as containing only attractors in the apoptosis phenotype. This case study highlights the added value of our approach which can uncover relevant system interventions not identified by usual percolation approaches.

6   Discussion

In this work, we considered properties of trap spaces and principles of target control to introduce a new approach to compute control strategies. The procedure proposed is applicable to both phenotype and attractor control and allows the interventions to be released after a certain amount of time, in contrast to usual target control methods that require permanent interventions.

The approach presented here is widely applicable to Boolean models of biological systems and can provide intervention strategies that are independent of the type of update considered in the modeling. Moreover, restrictions on the control strategies, in the form of variables to be excluded, can be added. Our approach also allows to incorporate information about the attractors, with the possibility to obtain control strategies that escape regular percolation-based techniques. As demonstrated with the two case studies, our method can identify new control strategies that require a small number of control variables, and thus revealing potentially valuable intervention approaches.

Our approach efficiently identifies control strategies for relatively large biological networks. A naturally important further step is a rigorous comparison with existing methods, for instance approaches based on stable motifs [18,20]. Furthermore, the performance of the method could benefit from the adoption of fine-tuning strategies developed to speed up some procedures involved in candidate screening. For instance, we could consider the reduction of the size of the search space by identifying a priori equivalent interventions, adapting existing approaches [16]. Moreover, approaches based on answer set programming have been used to efficiently compute minimal intervention strategies [8]. One could investigate extending such approaches to the detection of the control strategies characterized in our work. Further steps also include the extension of the method to other types of control, such as edge interventions or sequential control.

Acknowledgements. E.T. was funded by the Volkswagen Stiftung (Volkswagen Foundation), project ID 93063.

References

1. Baig, S., Seevasant, I., Mohamad, J., Mukheem, A., Huri, H.Z., Kamarul, T.: Potential of apoptotic pathway-targeted cancer therapeutic research: where do we stand? Cell Death Dis. **7**(1), e2850 (2016). https://doi.org/10.1038/cddis.2015.275
2. Biane, C., Delaplace, F.: Causal reasoning on boolean control networks based on abduction: theory and application to cancer drug discovery. IEEE/ACM Trans. Comput. Biol. Bioinform. **16**(5), 1574–1585 (2019). https://doi.org/10.1109/TCBB.2018.2889102
3. Calzone, L., et al.: Mathematical modelling of cell-fate decision in response to death receptor engagement. PLOS Comput. Biol. **6**(3), 1–15 (2010). https://doi.org/10.1371/journal.pcbi.1000702
4. Chaouiya, C., Naldi, A., Thieffry, D.: Logical modelling of gene regulatory networks with GINsim. Bacterial Mol. Netw. **804**, 463–479 (2012)
5. Csermely, P., Korcsmáros, T., Kiss, H.J., London, G., Nussinov, R.: Structure and dynamics of molecular networks: a novel paradigm of drug discovery: a comprehensive review. Pharmacol. Therapeutics **138**(3), 333–408 (2013). https://doi.org/10.1016/j.pharmthera.2013.01.016
6. Flobak, Å., et al.: Discovery of drug synergies in gastric cancer cells predicted by logical modeling. PLOS Comput. Biol. **11**(8), 1–20 (2015). https://doi.org/10.1371/journal.pcbi.1004426
7. Grieco, L., Calzone, L., Bernard-Pierrot, I., Radvanyi, F., Kahn-Perlès, B., Thieffry, D.: Integrative modelling of the influence of MAPK network on cancer cell fate decision. PLOS Comput. Biol. **9**(10), 1–15 (2013). https://doi.org/10.1371/journal.pcbi.1003286
8. Kaminski, R., Schaub, T., Siegel, A., Videla, S.: Minimal intervention strategies in logical signaling networks with asp. Theor. Pract. Logic Program. **13**, 675–690 (2013). https://doi.org/10.1017/S1471068413000422
9. Kim, J., Park, S.M., Cho, K.H.: Discovery of a kernel for controlling biomolecular regulatory networks. Sci. Rep. **3**, 2223 (2013). https://doi.org/10.1038/srep02223
10. Klarner, H., Bockmayr, A., Siebert, H.: Computing maximal and minimal trap spaces of Boolean networks. Natural Comput. **14**(4), 535–544 (2015). https://doi.org/10.1007/s11047-015-9520-7

11. Klarner, H., Siebert, H.: Approximating attractors of Boolean networks by iterative ctl model checking. Front. Bioeng. Biotechnol. **3**, 130 (2015). https://doi.org/10. 3389/fbioe.2015.00130
12. Klarner, H., Streck, A., Siebert, H.: PyBoolNet: a python package for the generation, analysis and visualization of Boolean networks. Bioinformatics **33**(5), 770–772 (2016). https://doi.org/10.1093/bioinformatics/btw682
13. Liu, Y.Y., Slotine, J.J., Barabási, A.L.: Controllability of complex networks. Nature **473**, 167–173 (2011). https://doi.org/10.1038/nature10011
14. Mandon, H., Su, C., Haar, S., Pang, J., Paulevé, L.: Sequential reprogramming of boolean networks made practical. In: Bortolussi, L., Sanguinetti, G. (eds.) Computational Methods in Systems Biology. vol. 11773, pp. 3–19. Springer International Publishing, Cham (2019). https://doi.org/10.1007/978-3-030-31304-3_1
15. Murrugarra, D., Veliz-Cuba, A., Aguilar, B., Laubenbacher, R.: Identification of control targets in boolean molecular network models via computational algebra. BMC Syst. Biol. **10**(1), 94 (2016). https://doi.org/10.1186/s12918-016-0332-x
16. Samaga, R., Kamp, A.V., Klamt, S.: Computing combinatorial intervention strategies and failure modes in signaling networks. J. Comput. Biol. **17**(1), 39–53 (2010). https://doi.org/10.1089/cmb.2009.0121
17. Takahashi, K., Yamanaka, S.: A decade of transcription factor-mediated reprogramming to pluripotency. Nat. Rev. Mol. Cell Biol. **17**(3), 183–193 (2016). https://doi.org/10.1038/nrm.2016.8
18. Yang, G., Gómez Tejeda Zañudo, J., Albert, R.: Target control in logical models using the domain of influence of nodes. Front. Physiol. **9**, 454 (2018). DOI: https:// doi.org/10.3389/fphys.2018.00454
19. Zañudo, J.G.T., Yang, G., Albert, R.: Structure-based control of complex networks with nonlinear dynamics. Proc. Natl. Acad. Sci. **114**(28), 7234–7239 (2017). https://doi.org/10.1073/pnas.1617387114
20. Zañudo, J.G.T., Albert, R.: Cell fate reprogramming by control of intracellular network dynamics. PLOS Comput. Biol. **11**(4), 1–24 (2015). https://doi.org/10. 1371/journal.pcbi.1004193
21. Zhang, R., et al.: Network model of survival signaling in large granular lymphocyte leukemia. Proc. Natl. Acad. Sci. **105**(42), 16308–16313 (2008). https://doi.org/10. 1073/pnas.0806447105

Qualitative Analysis of Mammalian Circadian Oscillations: Cycle Dynamics and Robustness

Ousmane Diop[1(✉)], Madalena Chaves[2], and Laurent Tournier[1]

[1] MaIAGE, INRAE, Université Paris-Saclay, 78350 Jouy-en-Josas, France
`ousmane.diop@inrae.fr`
[2] Université Côte d'Azur, Inria, INRAE, CNRS, Sorbonne Université, Biocore Team, Sophia Antipolis, France

Abstract. In asynchronous Boolean models, periodic solutions are represented by terminal strongly connected graphs, which are typically composed of hundreds of states and transitions. For biological systems, it becomes a challenging task to compare such mathematical objects with biological knowledge, or interpret the transitions inside an attractor in terms of the sequence of events in a biological cycle. A recent methodology generates summary graphs to help visualizing complex asynchronous attractors and order the dynamic progression based on known biological data. In this article we apply this method to a Boolean model of the mammalian circadian clock, for which the summary graph recovers the main phases of the cycle, in the expected order. It also provides a detailed view of the attractor, suggesting improvements in the design of the model's logical rules and highlighting groups of transitions that are essential for the attractor's robustness.

Keywords: Mammalian circadian clock · Asynchronous Boolean network · Complex attractor · Summary graph

1 Introduction

The analysis of periodic orbits and their properties remains a most challenging problem in dynamical systems theory. Many living systems exhibit periodical dynamics and the current literature covers a large diversity of mathematical models used to represent, explore, and study the mechanisms leading to physical or biological rhythms [7]. A thorough analysis of such cyclic attractors opens the door to a whole family of meaningful questions related to the robustness of the oscillatory behavior, the estimation and control of the period or amplitude of oscillations in terms of the parameters of the system, the location of the orbit in the state space, *etc.*

Supported by the ANR (French agency for research) through project ICycle ANR-16-CE33-0016-01.

Very little is known on how to express the properties of a periodic solution in terms of the system's parameters but qualitative models, such as piecewise linear or Boolean models, suggest some ideas. Piecewise linear systems partition the state space into regions where solutions of the system can be explicitly computed leading, in some examples, to the estimation of the period and other quantities in terms of the parameters [13,14]. Boolean models provide an ideal framework to analyze qualitative dynamical properties by enabling algorithmic approaches to characterize, for instance, the location of a periodic orbit in the state space, or the influence of the interaction graph on the dynamical behavior [17,21].

In an asynchronous Boolean network, a periodic orbit corresponds to a terminal strongly connected component of the state transition graph. If the computation of such an object is not an issue form a theoretical point of view, its size can grow very large, strongly limiting the biological interpretation of the states and transitions inside the attractor. To tackle this issue, general approaches can be used such as model checking techniques [23]. Recently, a dedicated method developed by Diop et al. [5] proposes to generate a *summary graph* of an asynchronous attractor, based on a classification of its states according to experimentally observed phases of the biological system. The summary graph provides a qualitative view of the general progression along the periodic orbit, capturing the underlying dynamics within the cyclic attractor. In [5] it was successfully applied to a Boolean model of the mammalian cell cycle [6].

In this paper, we propose to apply the summary graph method to the mammalian circadian clock, a biological rhythm which is based on the interactions among five main proteins CLOCK:BMAL1, REV-ERBα, ROR, PER, and CRY. The core of the clock mechanism is formed by three feedback loops. First, the BMAL1 complex promotes the transcription of *Per* and *Cry* genes. The corresponding proteins bind to form a complex PER:CRY, which then translocates to the nucleus where it will block the transcriptional activity of BMAL1, thus forming a first negative feedback loop. In addition, BMAL1 also promotes the transcription of the two genes *Rev-erb* and *Ror*. Eventually, *Bmal1* transcription will be inhibited by REV-ERB and activated by ROR, leading to a second negative loop and a new positive loop, respectively. The periodic behavior of the clock system is determined in large part by the phase opposition between CLOCK:BMAL1 and PER:CRY, which corresponds to the day/night succession (BMAL1 peaks during the day).

Examples of Boolean models for circadian clocks include one for the plant *Arabidopsis thaliana* [1] and a compact mammalian clock model [4] which reproduces the interplay between the negative feedback loops induced by BMAL1 activity. However, our present objective is to have a deeper understanding of the circadian cycle recently developed by Almeida et al. [2]. This is a continuous model of the core clock mechanism that faithfully reproduces the circadian rhythm, by including not only the five main proteins but also their transcription regulated by clock controlled elements (CCE). To use the graph method [5], we first construct a Boolean version of the continuous model in [2] that exhibits one cyclic attractor and, in addition, correctly reproduces the effect of some well known gene knock-outs. Next, Sect. 3 analyzes the cyclic attractor with the summary graph method, by classifying groups of Boolean states according to their corresponding circadian time zones. Finally, in Sect. 4 the summary graph is further used to identify key groups of transitions within the attractor and relate them back to parts of the network's logical rules. This analysis has two outcomes, first suggesting an improved Boolean rule for one of the variables where the model lacked clear information. As a second outcome, our analysis predicts that some transitions, while allowing some short-circuits between phases, seem to contribute to the overall robustness of the attractor and globally ensure the good progression of the cell clock.

2   Proposing a New Boolean Model of the Circadian Clock

As evoked previously, the following Boolean model of the mammalian circadian clock is highly inspired by the continuous model developed in [2]. In addition to the three feedback loops already described, this model takes into account the transcription of the five clock proteins, each regulated through a particular combination of transcription factors. The latter bind to specific sites on the promoters, called clock controlled elements (CCEs): Ebox (enhancer box), Dbox (DBP/E4BP4 response element), and Rbox (REV-ERB/ROR response element). With the introduction of Dbox (which activates both REV-ERB and PER), two new proteins are added to the model, DBP and E4BP4, each also regulated by one of the CCE.

A complete and detailed justification of the continuous model assumptions and construction can be found in [2], but we provide a brief summary of the eight variables and corresponding differential equations in Table 1.

2.1   Construction of the Boolean Model

The design of a Boolean or of a continuous model for the same biological system differs in some fundamental aspects and, in general, there is no direct equivalence between terms in the two frameworks. For instance, activation and inhibition links typically have clear logical representations, but the effects of detailed mass-action kinetics or mass conservation laws are harder to represent in a Boolean model, and may require the definition of new variables. The purpose of our Boolean model is to transcribe as closely as possible the interactions in the continuous model in [2], as described below. The continuous equations and corresponding logical rules are shown in Table 1, for a clear comparison between the two models.

In [2], there are three CCE named *Dbox*, *Ebox*, and *Rbox*, each responding to the conjugation of two components, an activator and an inhibitor. More precisely, *Dbox* is activated by DBP and inhibited by E4, *Ebox* is activated by Bmal and inhibited by CRY and *Rbox* is activated by ROR and inhibited by REV. The continuous equations are formed by a synthesis term depending on one or two

Table 1. Differential equations of the circadian clock model in [2] and corresponding logical rules in the Boolean model. The term $M_{PC} = \gamma_{pc} PER \cdot CRY - \gamma_{cp} PC$ represents the kinetics of the complex binding: $PER + CRY \rightleftharpoons PC$ (mass action law). We use classical Boolean operators: $X \vee Y$ (X or Y), $X \wedge Y$ (X and Y) and $\overline{X}$ (not X).

Continuous equation	Logical rule
$dBMAL1/dt = Rbox - \gamma_{bp} BMAL1 \cdot PC$	$Bmal' = Rbox \wedge \overline{PC}$
$dROR/dt = Ebox + Rbox - \gamma_{ror} ROR$	$ROR' = Rbox2$
	$ROR2' = Ebox \wedge Rbox \wedge ROR$
$dREV/dt = 2Ebox + Dbox - \gamma_{rev} REV$	$REV' = Ebox \vee REV2$
	$REV2' = Ebox \wedge Dbox \wedge REV$
$dDBP/dt = Ebox - \gamma_{dbp} DBP$	$DBP' = Ebox$
$dE4/dt = 2Rbox - \gamma_{e4} E4$	$E4' = Rbox$
$dCRY/dt = Ebox + 2Rbox - M_{PC} - \gamma_c CRY$	$CRY' = Ebox \vee CRY2$
	$CRY2' = Ebox \wedge Rbox \wedge CRY$
$dPER/dt = Ebox + Dbox - M_{PC} - \gamma_p PER$	$PER' = Ebox \vee Dbox$
$dPC/dt = M_{PC} - \gamma_{bp} BMAL1 \cdot PC$	$PC' = PER \wedge CRY$

CCE and a degradation term. The CRY, PER and PC equations also contain the binding and dissociation terms denoted M_{PC} in Table 1.

In general, Boolean variables are assumed to degrade when not updated in the next step and, therefore, linear degradation terms do not appear explicitly. Thus, a naive approach to construct a corresponding Boolean model is to combine the synthesis terms as logical conjugations or disjunctions of the given variables and directly obtain the rule, for instance: $DBP' = Ebox$ or $E4' = Rbox$. However, not all continuous equations follow this simple construction and other properties that strongly contribute to the dynamics must be taken into account.

Indeed, (i) some variables have nonlinear degradation (cf. $BMAL1$), (ii) others contain mass-action terms (cf. CRY, PER, PC) and (iii) three of the variables are regulated by two CCEs and are themselves regulators of CCEs (ROR, REV, CRY). This last property implies the existence of different thresholds for the different regulatory activities. For this reason, ROR, REV, and CRY are assumed to have an extra discrete level, here represented by extra Boolean variables $ROR2$, $REV2$, and $CRY2$ (following [24]), under the assumption

$$x = \begin{cases} 0 & \text{when } X = X2 = 0, \\ 1 & \text{when } X = 1, X2 = 0, \\ 2 & \text{when } X = X2 = 1. \end{cases}$$

The states corresponding to $X = 0$, $X2 = 1$ have no biological meaning (also called "forbidden") and trajectories towards these states from the other "biological" states must be excluded. This can be achieved by using the method described in [3] which complements the rules for $X' = (\cdots)$ and $X2' = (\cdots)$ as follows: $X' = (\cdots) \vee X2$ and $X2' = (\cdots) \wedge X$. Since the CCEs appear in additive form in the continuous equations, we assume that only one CCE is sufficient to trigger the activity of X, while the two CCEs are needed to trigger the activity of $X2$. The order in which the CCEs are activated and in turn activate each variable, was decided by comparing to the continuous solutions. The expressions for the CCEs are written as logical conjunctions:

$$Ebox = Bmal \wedge \overline{CRY2}, \quad Dbox = DBP \wedge \overline{E4},$$
$$Rbox = ROR \wedge \overline{REV}, \quad Rbox_2 = ROR2 \wedge \overline{REV}.$$

In the continuous model, there is only one nonlinear degradation term, corresponding to the inhibition of BMAL1 transcription by PC ($-\gamma_{bp}BMAL1 \cdot PC$). Accordingly, we assume $Bmal$ is explicitly repressed by PC but, conversely, PC is not strongly affected (indeed, our analysis showed that a rule of the form $PC = PER \wedge CRY \wedge \overline{Bmal}$ prevents oscillatory behavior).

The binding of PER and CRY to form the complex PC is described by mass-action kinetics in the continuous model (see term M_{PC} in Table 1). In the Boolean model, for simplicity, we assumed that PC is produced when both PER and CRY are available, leading to the rule $PC = PER \wedge CRY$, but no explicit effect from PC on PER or CRY. Although there is no systematic way to translate mass-action kinetics or other mass conservation laws into Boolean factors, and these are usually treated on a case-by-case basis, we will see in Sect. 4 that the analysis by the summary graph method suggests a refinement of the PER rule. This refinement can be interpreted as a more suitable way to include the mass-action terms into the circadian clock Boolean model.

In this way, we obtained a qualitative multi-valued model which closely translates the differential model of [2]. The multi-valued model is equivalent to the 11-dimensional Boolean network depicted in Table 2, to which the methodology in [5] can now be applied.

Table 2. Logical rules of the Boolean clock model, with 11 variables.

$Bmal' = ROR \wedge \overline{REV} \wedge \overline{PC}$
$ROR' = ROR2 \vee \overline{REV}$
$REV' = (Bmal \wedge \overline{CRY2}) \vee REV2$
$DBP' = Bmal \wedge \overline{CRY2}$
$E4' = ROR \wedge \overline{REV}$
$CRY' = Bmal \vee CRY2$
$PER' = (Bmal \wedge \overline{CRY2}) \vee (DBP \wedge \overline{E4})$
$PC' = PER \wedge CRY$
$REV2' = Bmal \wedge \overline{CRY2} \wedge DBP \wedge \overline{E4} \wedge REV$
$CRY2' = Bmal \wedge \overline{CRY2} \wedge ROR \wedge \overline{REV} \wedge CRY$
$ROR2' = Bmal \wedge \overline{CRY2} \wedge ROR \wedge \overline{REV}$

2.2   First Dynamical Analysis of the Boolean Model

To compute and analyze the dynamical behavior of a Boolean model $X_i' = F_i(X_1, \ldots, X_n)$, $i = 1, \ldots, n$, we need to specify an updating order for the variables. Applying the logical rules in the order defined by the updating schedule leads to a *state transition graph* (STG) with 2^n states, where a sequence of transitions represents a trajectory of the system. To analyze the STG, we first compute its strongly connected components (SCCs) which are defined as sets of states C, such that for every pair of states $x, y \in C$, there exist two paths (or sequences of transitions) in C leading from x to y and from y to x. SCCs may consist of single or multiple states and may have incoming and outgoing transitions, but two distinct SCCs can not be mutually connected, otherwise they would form a single SCC. The asymptotic behavior of the system is thus characterized by the SCCs without outgoing transitions, also called terminal SCCs or *attractors*. An attractor with multiple states represents a periodic orbit of the system.

As a preliminary analysis, we considered the basic synchronous updating schedule, where all variables are simultaneously updated: $X_i[t + 1] = F_i(X[t])$. The synchronous STG of the model contains a single attractor, composed of only five states. This simple cyclic attractor captures the Bmal/PER:CRY phase opposition, a central feature in the circadian clock (see Fig. 1).

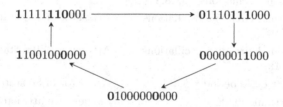

Fig. 1. Synchronous attractor of the Boolean model (variables are ordered as in Table 2). Bold digits indicate the succession of states corresponding to the *Bmal/PC* phase opposition: in the left column, *Bmal* is expressed while *PC* is turned off, the former leading to expression of *PER* and *CRY* (top left); next *PC* becomes expressed and *Bmal* is turned off (right column), and *CRY* and *PER* eventually turn off; at the bottom both *PC* and *Bmal* are off, before a new cycle begins.

In this paper, we will prefer an asynchronous updating schedule as it is much more realistic from a biological perspective. In this scheme, at most one variable is updated at each instant. To construct the STG for this asynchronous schedule, for each state $X = (X_1, \ldots, X_n)$ define the subset of variables $I_X = \{i : X_i \neq F_i(X)\}$. Then, for each $i \in I_X$, add a transition $X \to Y$ where $Y_i = F_i(X)$ and $Y_j = X_j$ for all $j \neq i$. More details on the asynchronous strategy can be found for instance in [21]. From a biological point of view, asynchronous updating is preferable as it allows for variability and different timescales in the network interactions. From a graph theoretical point of view, synchronous and asynchronous STGs are very different as the asynchronous STG is non determin-istic, *i.e.* a state no longer has one and only one successor. Usually, this generates lots of transitions, making biological interpretation more difficult.

When applied to our circadian model, the asynchronous strategy still gives a single attractor, but the terminal SCC is now composed of 442 states and 1737 transitions. Interestingly, the five states of the synchronous attractor are all present in the asynchronous attractor, but the simple cyclic trajectory has now been replaced by a complex graph with a much larger amount of details. To further analyze this attractor and validate our model, we consider six different versions of the model representing six known mutations (see Table 3). For some of them, circadian oscillations may be completely lost, which in our model trans-lates to the attractor shrinking to a single fixed state. For others, oscillations endure but are somewhat degraded (*eg.* with a shorter period); in those cases our model conserves a complex attractor, but with less states than the original. These results seem to confirm that the model in Table 2 reproduces the essential core of the circadian clock dynamics.

Table 3. Effects of some mutations on the dynamics of the Boolean model.

Mutation	Biological phenotype	Effect on the attractor
Bmal=0	Arrhythmic (complete loss of circadian oscillations) [9,18,19]	Single state attractor 0100100000
PER=0	Abnormal circadian oscillations [9,18,19]	Attractor of 114 states
CRY=0	Abnormal circadian oscillations [9,18,19]	Attractor of 106 states
REV=0	Shorter cycle period [15,18]	Attractor of 80 states
ROR=0	Arrhythmic [9,18]	Single state attractor 00000000000
REV=1	Arrhythmic [10]	Single state attractor 00100000000

Mutant analysis constitutes an interesting way to validate a discrete dynam-ical model. Indeed, the comparison between wild type and mutant phenotypes usually provides qualitative differences, such as the disappearance of oscillations for instance, that can be well captured by a Boolean model. Nevertheless, when dealing with such a complex attractor (hundreds of states, thousands of transi-tions), a more direct comparison of the attractor with biological data is rapidly limited, hindering model validation. For example, in Table 3 the degradation of circadian oscillations is paralleled with the number of states in the attractor, which is questionable. In the following we use the methodology proposed in [5], constructing a reduced version of the attractor based on biological knowledge. This summary graph leads to refine the analysis of the attractor, providing fur-ther validation of the model.

3   Comparing the Attractor with Circadian Oscillations

3.1   Dividing the Circadian Cycle into Qualitative Phases

In order to further analyze the model's attractor, a necessary first step is to classify its states into groups that will correspond to different stages of the circadian clock. This is essentially a modeling step, therefore there is not a unique way to make this classification, as it is based on a compromise between available biological data on the one hand, and the different variables and interactions included in the model on the other. In the following, we give a brief description of the main regulatory events during the circadian cycle, together with the modeling choices we made to deduce the corresponding partition of the attractor's states. For a comprehensive biological review we mainly referred to [20]. Note that this article is based on data at the transcriptional level; therefore, we sometimes used other sources to complete our classification (see [2,16] and references therein).

As already mentioned, a hallmark of circadian rhythm progression is the phase opposition between the CLOCK:BMAL1 complex on the one hand and the PER:CRY complex on the other. This opposition divides the cycle into two major steps, approximately correlated with the day/night separation (see Fig. 2). More precisely, [20] introduces two biological phases, respectively called *Activation* and *Repression*. The first one takes place during the day and corresponds to the activation of Bmal, while PER:CRY is absent. The second one sees the repression of Bmal and takes place during the night. Projecting this on the variables of our model, this leads to consider two groups of states: one where $Bmal = 1$, $PC = 0$ and one where $Bmal = 0$, $PC = 1$.

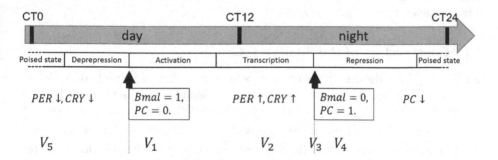

Fig. 2. Description of the main qualitative stages of the circadian clock. CT stands for Circadian Time and is an standard marker of time arbitrarily starting (CT0) at the beginning of activity for a diurnal organism. Top: temporal succession of biological phases as described in [20]; middle: projection of the main regulatory events on the model's variables; bottom: corresponding qualitative phases V_i defined in (1).

These two main phases are separated by intermediate phases called *Transcription* and *Poised state-Derepression* in [20]. In the *Transcription* phase, *Per* and *Cry* genes are transcribed, followed by their complexation and the translocation of the complex into the nucleus. With the lack of precise timing of these events, we decided to subdivide this into a first step where PER or CRY are present but not at the same time, followed by a second step where they are both present. The latter corresponds to the pre-formation step of the PER:CRY complex, *i.e.* PER and CRY have sufficiently accumulated and the complex is about to form in the cytosol [19]. Finally, after the repression phase, the PER:CRY complex disappears due to auto-repression of PER and CRY [8]. Again, without knowing the precise timing of the disappearance of PER, CRY and the complex, we simply consider intermediary states where $Bmal = PC = PER = CRY = 0$, just before the activation of Bmal and the subsequent beginning of a new cycle.

This description leads to the definition of five groups of states in the attractor, or qualitative phases, denoted by $(V_i)_{1 \le i \le 5}$ and defined as follows.

$$
\begin{cases}
V_1 = (Bmal \wedge \overline{PC}) \wedge (\overline{PER} \vee \overline{CRY}) & \text{(afternoon)}, \\
V_2 = (Bmal \wedge \overline{PC}) \wedge PER \wedge CRY & \text{(late afternoon)}, \\
V_3 = Bmal \wedge PC \wedge PER \wedge CRY & \text{(transition day-night)}, \\
V_4 = \overline{Bmal} \wedge PC & \text{(night)}, \\
V_5 = \overline{Bmal} \wedge \overline{PC} \wedge \overline{PER} \wedge \overline{CRY} & \text{(late night to next morning)}.
\end{cases}
\tag{1}
$$

These phases are defined by taking into account the main variables of our Boolean model. They are based on qualitative and not temporal considerations; however, thanks to the description in [20] we were able to approximately place them along the circadian time scale (see Fig. 2).

Remark 1. For the sake of convenience, we use the same symbol V_i to designate both the subset of states in the attractor and the Boolean formula describing those states. For instance, $V_4 = \overline{Bmal} \wedge PC$ denotes the set of states in the attractor such that $Bmal = 0$ and $PC = 1$.

3.2   Construction of the Summary Graph

Let $A = (V, E)$ denote the attractor, which is a directed graph over $|V| = 442$ states. The sets V_i defined by (1) are, by definition, subsets of V that are mutually exclusive. To complete them into a partition of V, introduce the set $U = V \backslash \left(\bigcup_{i=1}^{5} V_i \right)$, containing "unclassified" states. The first thing to note with this partition is that every V_i is actually not empty, confirming the attractor accurately captures all important phases of the circadian clock. To be more precise, $|V_1| = 77$, $|V_2| = 31$, $|V_3| = 31$, $|V_4| = 136$ and $|V_5| = 30$ and there are $|U| = 137$ unclassified states. We now briefly recall the definition of a summary graph (interested reader may refer to [5] for more details).

Definition 1. *The* summary graph *of the graph $A = (V, E)$ on a partition $\mathcal{P} = \{V_1, V_2, \ldots, V_k\}$ of V is the directed graph $\mathcal{G} = (\mathcal{P}, \mathcal{E})$, whose vertices are the V_i, $i = 1, \ldots, k$ and where there is an edge from V_i to V_j iff $i \ne j$ and there exist $x \in V_i$ and $y \in V_j$ such that $(x, y) \in E$.*

From a graph theoretical point of view, $\mathcal{G}$ is simply the quotient graph of A on the partition $\mathcal{P}$. For the sake of simplicity, we use the same symbol V_i to design the set of states V_i and the vertex V_i of the summary graph. This enables an easy description of sets of trajectories in the attractor A: actually, each edge of the summary graph is unequivocally associated with a subset of asynchronous transitions in A. The summary graph of the attractor is depicted in Fig. 3 (top left).

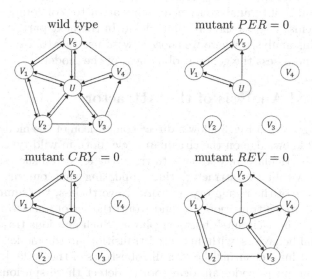

Fig. 3. Summary graph of the attractor on the partition $\{U, V_i, i = 1 \ldots 5\}$ defined in (1), in wild type condition as well as for mutants $PER = 0$, $CRY = 0$ and $REV = 0$.

The summary graph in Fig. 3(top left) shows that the expected succession of phases $V_1 \to V_2 \to V_3 \to V_4 \to V_5 \to V_1$ is actually present, confirming the existence of accurate circadian oscillations in the attractor, with respect to regulators Bmal, PER , CRY and PER:CRY. Moreover, three one-directional transitions $V_2 \to V_3$, $V_3 \to V_4$, and $V_4 \to V_5$ even indicate irreversible progression through the cycle. The edge $V_2 \to V_3$ corresponds to the formation and translocation of the PC complex, following the accumulation of PER and CRY. This irreversibility is in adequacy with [19]: as the main role of PC is to inhibit $Bmal$, it will switch off only after $Bmal = 0$. Similarly, the edge from V_3 to V_4 corresponds to transitions where $Bmal$ is repressed, which is known to be irreversible. As for the edge $V_4 \to V_5$, it is associated to a set of transitions where PC switches off and can be viewed as a consequence of PER and CRY auto-repression [8]. From a graph point of view, the presence of these three one-directional edges imposes a general orientation for the cycle, consistent with what is known on the opposition between the *Activation* and *Repression* phases of [20].

In addition to the summary graph in wild type condition, Fig. 3 also shows the summary graphs for the three mutants that exhibited oscillatory behaviors in Table 3: $PER = 0$, $CRY = 0$ and $REV = 0$. Interestingly, the mutation of REV has a limited impact on the succession of the phases, showing this perturbation does not impair the cycle progression, as it was observed in [15, 18]. For PER and CRY mutants, the perturbation is stronger and the cycle is impaired. However, we still observe a $Bmal$ oscillation between V_1 and V_5 (notably through U), confirming circadian oscillations may persist, although in a degraded form [9,18,19].

Whether in wild type or in mutant conditions, the summary graph reveals to be a powerful tool to analyze asynchronous attractors of complex oscillatory biological systems, such as the circadian clock. In the next part, we propose an extension of this analysis, where we show how to use this tool to help redesign the network, and assess the general robustness of the model.

4   Advanced Analysis of the Attractor

The summary graphs in Fig. 3 allow a direct comparison of the model's attractor with biological knowledge on the circadian cycle, both in wild type and mutant conditions. In particular, we were able to track the expected succession of qualitative phases within the attractor, thus validating that our model captures the essential core of the regulatory network. Nevertheless, the summary graphs also point to spurious transitions to (and from) the set of unclassified states U, responsible for "short-circuits" between phases. Such spurious trajectories may reveal abnormal behaviors, with unwanted transitions in the model, or they may catch important links ensuring the overall robustness of the cycle. In both cases, the summary graph provides an ideal tool to detect those spurious transitions and analyze them in a biological context.

4.1   Adjustment of the Attractor and Refinement of the Model

We start with a more in-depth examination of the set U. In Sect. 3 it is defined as $U := V \setminus (\bigcup_i V_i) \subseteq \{0,1\}^{11}$ and is a subset of 137 states in the attractor that we were not able to classify according to the general description of the circadian cycle. By assigning the value 1 to the states in U and the value 0 to the states in $V \setminus U$, we obtain a partial Boolean function (PBF) that we can identify using Boolean inference techniques (see [21,22]). The inference consists in identifying minimal supports[1], i.e. minimal subsets of variables that are sufficient to reproduce the PBF. This function provides an easy way to characterize the unclassified states, showing they can be decomposed into two components:

$$U = \underbrace{\overline{Bmal} \wedge \overline{PC} \wedge (PER \vee CRY)}_{U_1} \vee \underbrace{Bmal \wedge PC \wedge \overline{PER}}_{U_2}. \tag{2}$$

While the inference of a PBF does not lead to a unique solution in general, here (2) is actually unique, as the set $\{Bmal, PC, PER, CRY\}$ is the only minimal

---

[1] In [21], this step is performed by the algorithm REVEAL [11], in [22] it is performed by the algorithm presented in [12]. Overall, all the inferences performed in the present paper are fast, due to the small dimension (11) of the network. A general discussion about the complexity of the inference problem can be found for instance in [22].

support of the PBF U (it will be the case for all PBF identified in this article). As before, we conveniently use the same symbol U to designate the set of unclassified states (within the attractor) and the Boolean function that characterizes them (see Remark 1 above).

The Boolean formula of U is decomposed as in (2) to highlight two separate subsets of unclassified states: the first one U_1 is composed of states where variables $Bmal$ and PC are both off whereas the second one U_2 is composed of states where variables $Bmal$ and PC are both on. Specifically, in this second subset the PER:CRY complex is forming (since it has not repressed Bmal yet) while PER is switched off. Clearly, such states should not exist in the attractor since PER is essential for complex formation. In order to observe the interplay between U_1, U_2 with the different phases V_i, we reconstruct the summary graph of the attractor, using Definition 1 on the new partition $\{V_1 \ldots, V_5, U_1, U_2\}$ (Fig. 4, left).

By looking at this graph, one can see that the set U_2 is connected to the rest of the attractor mainly through phase V_3. More precisely, only six transitions in the attractor are responsible for the entry into U_2, all coming from V_3. Since an asynchronous strategy is used, all these transitions can be traced back to situations where PER has disappeared before the complex PER:CRY has completed the repression of Bmal, leading to a contradiction. Therefore, we decide to suppress these transitions in the attractor, leading to the disconnection of the set U_2 (Fig. 4, center).

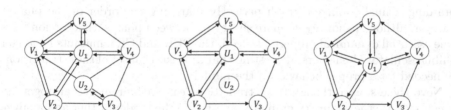

Fig. 4. Refined summary graphs. Left: initial attractor's summary graph with U decomposed in (U_1, U_2); center: summary graph once the six transitions $V_3 \rightarrow U_2$ have been removed; right: summary graph of the new amended model's attractor.

After the removal of the six transitions in the graph of the attractor, we obtain a new amended model by applying a few steps. First, we recompute the strongly connected component decomposition of the truncated attractor. We find a unique terminal SCC of 393 states, in which the states in U_2 have disappeared. From this SCC we reconstruct the partial truth table of the network on these 393 states, and then use an inference technique to identify the logical rules corresponding to this partial truth table. The last two steps (construction of partial truth table, followed by inference) are described in more details in [5] and, in a different context, in [21]. Finally, the only affected rule of the network is the rule of variable PER, with the addition of a new (unique) clause:

$$PER' = (Bmal \wedge \overline{CRY2}) \vee (DBP \wedge \overline{E4}) \vee (\mathbf{Bmal} \wedge \mathbf{PC}). \qquad (3)$$

To verify that this modification did not alter other parts of the dynamics, we made the same analyses as in Sects. 2 and 3, confirming that the slight modification in (3) was sufficient to get rid of unwanted states U_2 while conserving good dynamical properties, in wild type and mutant conditions. The summary graph of this new amended model is depicted in Fig. 4, on the right.

The modified rule (3) adds a single new interaction in the network, which is a positive effect of PC onto PER (the positive effect of $Bmal$ to PER was already present in the original network). Interestingly, this effect was present in the ODE system of [2], as an unbinding term of the PER:CRY complex. As already said, such mass action law kinetic terms are often made implicit by default in the Boolean framework. Here, our method points to the importance of this particular one to avoid unwanted transitions within the attractor. From a modeling point of view, the summary graph thus provides a valuable help in the design of Boolean models of complex, oscillating biological systems.

4.2 A Tool to Assess the General Robustness of the Attractor

The remaining set of unclassified states U_1 is composed of states where $Bmal$ and PC are off while PER or CRY can be on. Contrary to U_2, this set is highly connected to almost every phases of the attractor (see Fig. 4). In total, the four edges entering in U_1 amount to 117 transitions. As before, we tried to remove those transitions in order to disconnect U_1 from the rest of the attractor, thus obtaining a final summary graph perfectly matching the order of the phases. However, although showing no more short-cut between phases, this version loses some essential dynamical property. Indeed, the PER and CRY mutants no longer exhibit oscillatory behaviors, suggesting that at least a subset of these transitions are needed for a proper behavior of the model.

Nevertheless, even though those transitions cannot be removed all together, we can still use the summary graph to investigate the model further and analyze each edge entering U_1 separately, in the context of the circadian cycle.

1. The edge $V_4 \rightarrow U_1$ is associated with 63 transitions in which the variable PC switches off while PER and CRY are not both deactivated. From a biological point of view, it means that the PER:CRY complex disappears before PER and CRY have disappeared, indicating a short half-life of the complex. These transitions are to be compared with the $V_4 \rightarrow V_5$ transitions where PC switches off after PER and CRY, suggesting a longer half-life. In our model the two types of transitions coexist since the relative times of disappearance of the three components, PER, CRY and PER:CRY are not taken explicitly into account. Interestingly, note that the removal of the 63 transitions does not affect the global dynamical properties of the attractor, as the mutants are not affected.

2. The edge $V_5 \rightarrow U_1$ corresponds to 9 transitions in the attractor, in which the variable PER switches on while $Bmal$ is not yet active. Since PER transcription is activated through the CLOCK:BMAL1 complex, these transitions

seem rather unrealistic. To investigate further, we removed the transitions and applied the same technique as before, to obtain the modified PER rule:

$$PER' = (Bmal \wedge \overline{CRY2}) \vee (Bmal \wedge PC)$$
$$\vee \left[(DBP \wedge \overline{E4}) \wedge (\mathbf{PER} \vee \mathbf{CRY} \vee PC) \right].$$

Dynamically, the removal does not affect the main properties of the attractor (the summary graphs of wild type and mutant conditions are similar) however, the inferred logical rule exhibits new and undocumented interactions, namely an auto-activation of PER and a positive effect of CRY on PER. This points to a specific part of the model that will need a closer look in the future.

3. Finally, the edges $V_1 \rightarrow U_1$ and $V_2 \rightarrow U_1$ correspond to a total of 45 transitions in which the variable $Bmal$ switches to 0 in the absence of PC. With respect to circadian clock events, these transitions describe an early deactivation of Bmal, leading the cycle to bypass the important steps of PER:CRY formation and translocation (phase V_3). However, it is the removal of these transitions that directly alter the behaviors of the mutants, suggesting they are necessary to ensure the robustness of the attractor. Their removal leads to a new rule for $Bmal$:

$$Bmal' = (ROR \wedge \overline{REV} \wedge \overline{PC}) \vee \left[(\mathbf{Bmal} \wedge (\mathbf{PER} \vee \mathbf{CRY}) \wedge \overline{PC} \right],$$

that highlights, besides an auto-activation term, a direct positive effect of PER and CRY on $Bmal$. Those new interactions generate two positive feedback loops involving $Bmal$, PER and CRY that directly interfere with the negative feedback loop between $Bmal$ and PC:

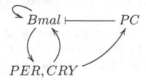

This alteration of the negative loop is directly responsible for the loss of robustness of the model.

The objective of this section was to analyse the "short-cut" transitions in the circadian attractor (see Fig. 4 left). As shown by the three examples above, the summary graph combined with Boolean inference indicates that some of these unwarranted transitions should not be removed and, in fact, appear to play a significant role during the progression of the cycle, namely in the system's response to gene knock-outs.

5   Conclusions and Perspectives

In this article we used the summary graph, a novel tool introduced in [5], to analyze a Boolean model of the mammalian circadian cycle regulation. This tool is well adapted to study biological oscillations, as it provides a rational way to compare a complex Boolean attractor (with hundreds of states) with an oscillating phenomenon. Combined with Boolean inference techniques, it becomes particularly useful in model design as it allows to relate a local dynamical property of the attractor, such as the transition from one phase to the next, to the part of the network's topology directly responsible for it.

After verifying the model's attractor correctly captures the correct succession of phases, the summary graph was further analyzed to provide more insight on the model.

In a first step, our analysis indicates that the attractor contains many "short-cuts", that is transitions between states (essentially towards sets U_1 and U_2) which may lead to distorted cycles with very short days or very short nights. In a second step, the analysis shows that some of these short-cuts (those passing through set U_2) can be removed by a refinement of the logical rules. Namely, the rule of PER should be modified to include the effect of PC dissociation which, for simplicity, was not taken into account in the initial Boolean model (compare Table 2 and Eq. (3)). Finally, in a third step, our analysis suggests that some short-cuts, specifically $V_1 \rightarrow U_1$ and $V_2 \rightarrow U_1$, are necessary for a correct performance. These transitions are characterized by an early BMAL1 deactivation, and they are responsible also for generating the short cycle mutants observed in the PER and CRY knock-outs. The transitions through U_1 may thus be necessary to generate circadian cycle robustness in response to perturbations in gene expression.

More generally, the summary graph provides an efficient way to tackle complex qualitative attractors, by testing the effect of specific perturbations on the dynamics. In future works we plan to further investigate the role of the states U_1 in maintaining circadian oscillations, by studying the links between state transitions and the topology of the circadian network.

Acknowledgements. The authors would like to thank Franck Delaunay for many useful discussions on the circadian clock events, as well as for pointers to relevant references.

References

1. Akman, O., Watterson, S., Parton, A., Binns, N., Millar, A., Ghazal, P.: Digital clocks: Simple boolean models can quantitatively describe circadian systems. J. R. Soc. Interface **9**, 2365–2382 (2012)
2. Almeida, S., Chaves, M., Delaunay, F.: Transcription-based circadian mechanism controls the duration of molecular clock states in response to signaling inputs. J. Theor. Biol. **484**, 110015 (2020)

3. Chaves, M., Tournier, L., Gouzé, J.L.: Comparing Boolean and piecewise affine differential models for genetic networks. Acta Biotheor. **58**(2–3), 217–232 (2010)
4. Comet, J.P., Bernot, G., Das, A., Diener, F., Massot, C., Cessieux, A.: Simplified models for the mammalian circadian clock. Procedia Comput. Sci. **11**, 127–138 (2012)
5. Diop, O., Tournier, L., Fromion, V.: Summarizing complex asynchronous Boolean attractors, application to the analysis of a mammalian cell cycle model. In: 18th European Control Conference (ECC), Naples, Italy, pp. 1677–1682 (2019)
6. Fauré, A., Naldi, A., Chaouiya, C., Thieffry, D.: Dynamical analysis of a generic boolean model for the control of the mammalian cell cycle. Bioinformatics **22**(14), e124–e131 (2006)
7. Forger, D.B.: Biological Clocks, Rhythms, and Oscillations: The Theory of Biological Timekeeping. MIT Press (2017)
8. Gallego, M., Virshup, D.M.: Post-translational modifications regulate the ticking of the circadian clock. Nat. Rev. Mol. Cell Biol. **8**(2), 139–148 (2007)
9. Ko, C.H., Takahashi, J.S.: Molecular components of the mammalian circadian clock. Hum. Mol. Genet. **15**(suppl_2), R271–R277 (2006)
10. Kornmann, B., Schaad, O., Bujard, H., Takahashi, J.S., Schibler, U.: System-driven and oscillator-dependent circadian transcription in mice with a conditionally active liver clock. PLoS Biol. **5**(2), e34 (2007)
11. Liang, S., Fuhrman, S., Somogyi, R.: Reveal, a general reverse engineering algorithm for inference of genetic network architectures. In: Pacific Symposium on Biocomputing, vol. 3, pp. 18–29 (1998)
12. Murakami, K., Uno, T.: Efficient algorithms for dualizing large-scale hypergraphs. Discrete Appl. Math. **170**, 83–94 (2014)
13. Ndiaye, I., Chaves, M., Gouzé, J.L.: Oscillations induced by different timescales in signal transduction modules regulated by slowly evolving protein-protein interactions. IET Syst. Biol. **4**(4), 263–276 (2010)
14. Poignard, C., Chaves, M., Gouzé, J.L.: A stability result for periodic solutions of nonmonotonic smooth negative feedback systems. SIAM J. Appl. Dyn. Syst. **17**(2), 1091–1116 (2018)
15. Preitner, N., et al.: The orphan nuclear receptor rev-erbα controls circadian transcription within the positive limb of the mammalian circadian oscillator. Cell **110**(2), 251–260 (2002)
16. Relógio, A., Westermark, P.O., Wallach, T., Schellenberg, K., Kramer, A., Herzel, H.: Tuning the mammalian circadian clock: robust synergy of two loops. PLoS Comput. Biol. **7**(12), e1002309 (2011)
17. Remy, E.: Mossé B., Thieffry D.: Boolean dynamics of compound regulatory circuits. In: Rogato, A., Zazzu, V., Guarracino, M. (eds.) Dynamics of Mathematical Models in Biology. Springer, Cham (2016)
18. Ripperger, J.A., Jud, C., Albrecht, U.: The daily rhythm of mice. FEBS Lett. **585**(10), 1384–1392 (2011)
19. Rosensweig, C., Green, C.B.: Periodicity, repression, and the molecular architecture of the mammalian circadian clock. Eur. J. Neurosci. **51**(1), 139–165 (2018)
20. Takahashi, J.S.: Transcriptional architecture of the mammalian circadian clock. Nat. Rev. Genet. **18**(3), 164 (2017)
21. Tournier, L., Chaves, M.: Uncovering operational interactions in genetic networks using asynchronous Boolean dynamics. J. Theor. Biol. **260**(2), 196–209 (2009)
22. Tournier, L., Goelzer, A., Fromion, V.: Optimal resource allocation enables mathematical exploration of microbial metabolic configurations. J. Math. Biol. **75**, 1349–1380 (2017). https://doi.org/10.1007/s00285-017-1118-5

23. Traynard, P., Feillet, C., Soliman, S., Delaunay, F., Fages, F.: Model-based investigation of the circadian clock and cell cycle coupling in mouse embryonic fibroblasts: Prediction of reverb-α up-regulation during mitosis. BioSyst. **149**, 59–69 (2016)
24. Van Ham, P.: How to deal with variables with more than two levels. In: Kinetic Logic a Boolean Approach to the Analysis of Complex Regulatory Systems, pp. 326–343. Springer, Berlin, Heidelberg (1979)

Synthesis and Simulation of Ensembles of Boolean Networks for Cell Fate Decision

Stéphanie Chevalier[1], Vincent Noël[2], Laurence Calzone[2], Andrei Zinovyev[2,3], and Loïc Paulevé[4(✉)]

[1] LRI, CNRS, UMR8623, Univ. Paris-Saclay, Orsay, France
[2] Institut Curie, INSERM, U. PSL, Mines ParisTech, Paris, France
[3] Lobachevsky University, 603000 Nizhny Novgorod, Russia
[4] Univ. Bordeaux, Bordeaux INP, CNRS, LaBRI, UMR5800,
33400 Talence, France
loic.pauleve@labri.fr

Abstract. The construction of models of biological networks from prior knowledge and experimental data often leads to a multitude of candidate models. Devising a single model from them can require arbitrary choices, which may lead to strong biases in subsequent predictions.

We introduce here a methodology for a) synthesizing Boolean model ensembles satisfying a set of biologically relevant constraints and b) reasoning on the dynamics of the ensembles of models. The synthesis is performed using Answer-Set Programming, extending prior work to account for solution diversity and universal constraints on reachable fixed points, enabling an accurate specification of desired dynamics. The sampled models are then simulated and the results are aggregated through averaging or can be analyzed as a multi-dimensional distribution.

We illustrate our approach on a previously published Boolean model of a molecular network regulating the cell fate decisions in cancer progression. It appears that the ensemble-based approach to Boolean modelling brings new insights on the variability of synergistic interacting mutations effect concerning propensity of a cancer cell to metastasize.

1 Introduction

The ability to derive one single model from observations of a biological system usually faces arbitrary choices, sometimes referred to as *art*.

Computational models of molecular interaction networks are usually built from data related to the architecture of the network from known interactions; and data related to its dynamics, such as measurements of gene expressions or proteins activity at different times and/or conditions. However, despite huge advances in experimental technologies, observations of the biological processes stay very scarce, either in terms of temporal resolution, number of observed

---

S. Chevalier and V. Noël—Co-first authors.

A. Abate et al. (Eds.): CMSB 2020, LNBI 12314, pp. 193–209, 2020.
https://doi.org/10.1007/978-3-030-60327-4_11

entities, synchronisation between measure points, or a variety of experimental conditions. Combined with complex structures for molecular interactions, the model engineering problem, in this case, appears to be largely under-specified, leading to (too) many potential candidate models.

Boolean Networks (BNs), and logical models in general, are widely adopted for the modelling of signalling pathways and gene and transcription factors networks [3,5,28]. With BNs, the activity of components is caricatured to "off" and "on" , and their evolution is computed according to logical rules (e.g., gene 1 can be active only whenever its activators 2 and 3 are active). However, in practice, biological data still let open a multitude of candidate BNs. Thus, arbitrary modelling choices have to be made, e.g., by prioritizing certain logics between regulators or by preferring smallest/largest models, which may introduce biases in subsequent model predictions.

In this paper, we present an approach aiming at reducing modelling biases by constructing and reasoning on dynamics of *ensembles* of BNs. The idea of ensemble modelling has recently gained momentum with machine learning, notably with random forests. By analogy, we constitute ensembles of BNs sampled from the whole multitude of models compatible with network architecture and dynamical properties. They are then simulated asynchronously and the simulations are aggregated through averaging. The obtained results allow an interpretation at the level of cell population and take into account its potential heterogeneity.

In the literature, ensembles of *random* BNs have been employed to show emerging properties of families of BNs sharing properties related to their architecture or logic rules [14,16]. In [21], ensembles of BNs sharing a network architecture are used to assess dynamical properties of qualitative differential equations. In contrast, our approach is focused on ensembles of models which satisfy a set of constraints both on their architecture and on their dynamics. Synthesis of BNs from such constraints received a lot of interest in the literature [2,6,7,13,24,27], and methods like [2,6,13,27] allows reasoning implicitly on ensembles of models, notably by enabling checking for their emptiness.

Here, we extend prior work on BNs synthesis from reachability and attractor properties with Most Permissive semantics [2] to support universal properties on (reachable) fixed points and the specification of network perturbations. The synthesis is performed with the logic framework of Answer-Set Programming [1] (ASP). Then, we use heuristics to drive the ASP solver in different regions of the solution space to sample ensembles of BNs capturing the diversity of the comprehensive solution set. Dynamics of ensembles are then explored through stochastic simulations for quantifying the propensities of reachable attractors, subject to different network perturbations. To that aim, we extended the simulator MaBoSS [22] to support ensembles of BNs as input.

We illustrate our approach on a model of molecular pathways regulating tumour invasion and migration [4]. We sampled ensembles of BNs sharing the same network architecture as the original model and constrained by the dynamical properties related to attractor reachability. Then, as in the original study, we evaluated the shift of reachable phenotypes caused by an epistatic interaction between mutations in model genes (gain of Notch function and loss of p53 function). It appears that, contrary to the initial single model analysis, the ensemble approach reveals a potential variability in the effectiveness of the double mutant to enhance the metastasis potential.

2 Background

2.1 Boolean Networks

A *Boolean network* (BN) of dimension n is a function

$$f : \mathbb{B}^n \to \mathbb{B}^n \tag{1}$$

where $\mathbb{B} := \{0, 1\}$. For all $i \in [n]$, $f_i : \mathbb{B}^n \to \mathbb{B}$ denotes the *local function* of the i-th component. A vector $x \in \mathbb{B}^n$ is called a *configuration* of the BN f. The set of components which value differs between two configurations $x, y \in \mathbb{B}^n$ is denoted by $\Delta(x, y) := \{i \in [n] \mid x_i \neq y_i\}$.

A BN f is said *locally monotonic* whenever each of its local functions is monotonic (this does not imply f monotonicity). Intuitively, local monotonicity imposes that a variable always appears with the same sign in a minimal disjunctive/conjunctive normal form of the local functions.

Figure 1 is an example of locally-monotonic BN with $n = 3$.

Mutations. In the following, we will consider the analysis of a BN f subject to some *permanent* perturbations of its components, that we refer to as mutations, being either a *gain of function* (GoF; locked to 1) and *loss of function* (LoF; locked to 0). A mutation is specified by a couple (i, v), where $i \in [n]$ is a component and $v \in \mathbb{B}$ is its forced value. Given a BN f and a set of mutations $M \subseteq [n] \times \mathbb{B}$, we denote by f/M the mutated BN, where, for each $i \in [n]$, $(f/M)_i(x) := v$ if $(i, v) \in M$, and $(f/M)_i(x) := f_i(x)$ otherwise.

Influence Graph. For each $i \in [n]$, f_i typically depends on a small subset of components of the BN. The *influence graph* summarizes these dependencies with a positive (resp. negative) edge from node j to i if there are configurations in which the sole increase of j would strictly increase (resp. decrease) the value of f_i. A node can have both positive and negative influences on i, indicating that f_i is non-monotonic. Remark that different BNs can have the same influence graph. Fig. 1 (right) shows the influence graph of the BN example.

$$f_1(x) := \neg x_2$$
$$f_2(x) := \neg x_1$$
$$f_3(x) := \neg x_1 \wedge x_2$$

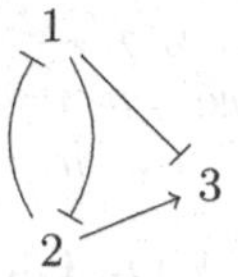

Fig. 1. Example of Boolean network f and its influence graph $G(f)$ where positive edges are with normal tip and negative edges are with bar tip.

Definition 1. *Given a BN f of dimension n, its* influence graph $G(f)$ *is a directed graph* $([n], E_+, E_-)$ *with* positive *and* negative *edges such that* $(j, i) \in E_+$ *(resp.* $(j, i) \in E_-$*) iff* $\exists x, y \in \mathbb{B}^n$ *s.t.* $\Delta(x, y) = \{j\}$, $x_j < y_j$, *and* $f_i(x) < f_i(y)$ *(resp.* $f_i(x) > f_i(y)$*). The influence graph* $\mathcal{G} = ([n], E_+, E_-)$ *is a subgraph of* $\mathcal{G}' = ([n], E'_+, E'_-)$, *denoted by* $\mathcal{G} \subseteq \mathcal{G}'$, *iff* $E_+ \subseteq E'_+$ *and* $E_- \subseteq E'_-$.

2.2   BN Semantics

From a configuration $x \in \mathbb{B}^n$, semantics of BNs specify how to compute the next possible configurations. One of the most classical semantics is the fully-asynchronous (often simply called asynchronous), where only one component i is updated at a time (to the value $f_i(x)$). It can be defined as a binary relation $\xrightarrow[\text{a1}]{f}$ between configurations:

Definition 2 (Fully-Asynchronous Semantics).

$$\forall x, y \in \mathbb{B}^n, \quad x \xrightarrow[\text{a1}]{f} y \ \text{iff} \ \exists i \in [n] : \Delta(x, y) = \{i\} \wedge y_i = f_i(x).$$

We write $\rho^f_{\text{a1}}(x) := \{y \in \mathbb{B}^n \mid x \xrightarrow[\text{a1}]{f}{}^* y\}$ the set of configurations in transitive relation with x, with $\xrightarrow[\text{a1}]{f}{}^*$ the reflexive and transitive closure of $\xrightarrow[\text{a1}]{f}$.

However, as demonstrated in [19], the fully-asynchronous semantics of BNs, as the synchronous and (general) asynchronous, are not faithful abstractions of quantitative systems: they can both introduce spurious behaviours (as expected with qualitative models) and miss others.

The *Most Permissive* (MP) semantics of BNs [19] offers the guarantees to not preclude any behaviour realisable in any quantitative refinement of the model, thus providing a formal over-approximation of dynamics. Moreover, the abstraction is minimal: any behaviour it predicts is realisable by a quantitative refinement of the BN using the asynchronous semantics. Importantly, the complexity for deciding main dynamical properties is considerably lower than with (a)synchronous semantics, as we will mention in the next subsection.

MPBNs can be defined by the means of hypercubes (partially) closed by f, a hypercube being specified by a vector associating each component to either a fixed Boolean value or free ($*$).

Definition 3 (Hypercube). *A hypercube h of dimension n is a vector in* $(\mathbb{B} \cup \{*\})^n$. *The set of its associated configurations is denoted by* $c(h) := \{x \in \mathbb{B}^n \mid \forall i \in [n], h_i \neq * \Rightarrow x_i = h_i\}$.
Given two hypercubes $h, h' \in (\mathbb{B} \cup \{*\})^n$, h is smaller *than h' if and only if* $\forall i \in [n], h'_i \neq * \Rightarrow h_i = h'_i$.

Definition 4 (K-Closed Hypercube). *Given a subset of components $K \subseteq [n]$, a hypercube $h \in (\mathbb{B} \cup \{*\})^n$ is K-closed by f whenever for each configuration $x \in c(h)$, for each component $i \in K$, $h_i \in \{*, f_i(x)\}$.*
It is minimal *whenever no different K-closed hypercube is smaller than it.*

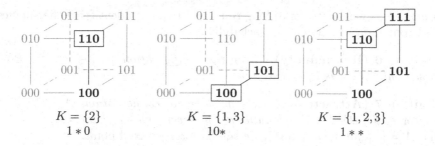

Fig. 2. Examples of smallest K-closed hypercubes containing the configuration 100 for the BN f of dimension 3 defined by $f_1(x) := 1$, $f_2(x) := x_1$, $f_3(x) := x_1 \wedge \neg x_3$. Configurations belonging to the hypercube are in bold; these verifying the MP reachability property are boxed. The hypercube 11* is the only one which is closed by f and minimal.

A hypercube $[n]$-closed by f is also known as a *trap space* [15].

Example 1. Let us consider the BN $f : \mathbb{B}^3 \to \mathbb{B}^3$ with $f_1(x) := 1$, $f_2(x) := x_1$, and $f_3(x) := x_1 \wedge \neg x_3$. The hypercube $1 * *$ is closed by f, with $c(1 * *) = \{100, 101, 110, 111\}$. The hypercube $1 * 0$ is the smallest hypercube $\{2\}$-closed by f containing 100; it is not closed by f, nor the smallest hypercube $\{2\}$-closed by f containing 110.

Starting from a configuration $x \in \mathbb{B}^n$, the MP semantics allows transitions towards any configuration y which is present in at least one smallest K-closed hypercube h containing x, for some $K \subseteq n$, and so that the state of each component $i \in K$ of y can be computed by f_i from a configuration of h.

Definition 5 (Most-Permissive Semantics). *Given a BN f of dimension n and two configurations $x, y \in \mathbb{B}^n$, $y \in \rho_{\mathrm{mp}}^f(x)$ if and only if there exists $K \subseteq [n]$ such that the smallest K-closed hypercube h containing x verifies (1) $y \in c(h)$, and (2) $\forall i \in K$, there exists a configuration $z \in c(h)$ such that $f_i(z) = y_i$.*

Figure 2 gives examples of computations of ρ_{mp}^f.

A way to interpret the MP semantics is to see the components free in a hypercube as being in the course of changing of state, while other components can independently consider them either as 0 or 1. This abstracts the missing information on the ordering of thresholds of activation/inhibition between components: while the quantitative value of component u progressively increases, at a given time it can be high enough to activate a component (i.e., 1) but not yet high enough to activate another one (i.e., 0). These *dynamic* states are overlooked by asynchronous semantics, making it an incorrect over-approximation of quantitative systems, contrary to the MP semantics [19].

2.3   Dynamical Properties

In the following, we will focus on two main dynamical properties of BNs: *reachability* which relates to the existence of trajectories between two configurations, and *attractors* which relates to long-run behaviours by identifying the smallest sets of configurations closed by reachability.

Definition 6 (Reachability). *Given two configurations* $x, y \in \mathbb{B}^n$ *of a BN* f *with semantics* σ, y *is reachable from* x *whenever* $y \in \rho_\sigma^f(x)$.

Definition 7 (Attractor). *A non-empty set of configurations* $A \subseteq \mathbb{B}^n$ *is an* attractor *of the BN* f *with semantics* σ *whenever* $\forall x \in A, \rho_\sigma^f(x) = A$.
When $A = \{x\}$ *for some* $x \in \mathbb{B}^n$, *we say that* x *is a* fixed point.

With MP semantics, attractors match with minimal trap spaces. With fully-asynchronous semantics, deciding if $y \in \rho_{a1}^f(x)$ or if x belongs to an attractor are both PSPACE-complete problems. With MP semantics, deciding if $y \in \rho_{mp}^f(x)$ is PTIME if f is locally-monotonic and $\mathrm{P^{NP}}$ otherwise; deciding if x belongs to an attractor is coNP if f is locally-monotonic and $\mathrm{coNP^{coNP}}$ otherwise. Deciding if there exists a fixed point is NP-complete with both semantics [19].

Notice the following relations between MP and fully-asynchronous semantics:

- $x \in \mathbb{B}^n$ is a fixed point with MP semantics if and only if it is a fixed point with full-asynchronous semantics (iff $f(x) = x$);
- $y \in \mathbb{B}^n$ is reachable from $x \in \mathbb{B}^n$ with the fully-asynchronous semantics *only if* it is reachable with MP semantics ($\rho_{a1}^f(x) \subseteq \rho_{mp}^f(x)$);
- the number of attractors with MP semantics is less than or equal to the number of attractors with fully-asynchronous semantics.

2.4   Answer-Set Programming

Answer Set Programming (ASP; [1,10]) is a declarative approach to solving combinatorial satisfaction problems. It is close to SAT (propositional satisfiability) [17] and known to be efficient for enumerating solutions of NP problems comprising up to tens of millions of variables while providing a convenient language for specifying the problem. We give a very brief overview of ASP syntax and semantics that we use in the next sections; see [10] for more details.

An ASP program is a Logic Program (LP) being a set of logical rules with first order logic predicates of the form:

1 $a_0 \leftarrow a_1, \ldots, a_n, \text{ not } a_{n+1}, \ldots, \text{ not } a_{n+k}.$

where a_i are (variable-free) atoms, i.e., elements of the Herbrand base, which is built from all the possible predicates of the LP. The Herbrand base is built by instantiating the LP predicates with the LP terms (constants or elements of the Herbrand universe).

Essentially, such a logical rule states that when all $a_1, \ldots, a_n$ are true and none of $a_{n+1}, \ldots, a_{n+k}$ can be proven to be true, then a_0 has to be true as well. Whenever a_0 is $\perp$ (false), the rule, also called integrity constraint, becomes:

2 $\leftarrow a_1, \ldots, a_n, \text{ not } a_{n+1}, \ldots, \text{ not } a_{n+k}.$

Such a rule is satisfied only if the right-hand side of the rule is false (at least one of $a_1, \ldots, a_n$ is false or at least one of $a_{n+1}, \ldots, a_{n+k}$ is true). On the other hand, a_0:- $\top$ (a_0 is always true) is abbreviated as a_0. A solution (answer set) is a *stable* Herbrand model, that is, a minimal set of true atoms where all the logical rules are satisfied. For instance, consider the following program:

```
3 a.
4 b ← a.
5 d ← a, c.
```

It has for unique solution {a, b}: indeed, whereas {a,b,d} does not contradict the rules, d is not a fact and cannot be derived from a rule; so it is not stable.

ASP allows using variables (starting with an upper-case) instead of terms/predicates: these *template* declarations will be expanded before the solving. We also use the notation a(X): b(X) which is satisfied when for each b(X) true, a(X) is true. If any term follows such a condition, it is separated with ;.

ASP can express *disjunctive* logic programs [18], by the means of disjunctions in the rule head (";"-separated atoms):

```
6 a; b ← body.
```

Such a disjunctive rule implies that solutions are subset minimal: an answer set is a solution only if none of its subsets is itself a solution [9]. For instance, let's consider the disjunction:

```
7 a; b; c.
```

The interpretation $I = \{a, b\}$ is a model but not minimal: both interpretations $\{a\}$ and $\{b\}$ are smaller than I and satisfy the rule. Hence I is not a solution. As showed in [8], the complexity of problems addressed with ASP can be extended thanks to disjunctive rules up to 2QBF, i.e. a two quantification levels Boolean formula ($\forall x \exists y.\phi$ or $\exists y \forall x.\phi$ where ϕ is a quantifier-free propositional formula). Indeed, 2QBF can be reduced to the problem of verifying the existence of an answer set of a disjunctive ASP program.

3   BN Synthesis from Architecture and Dynamical Properties

We formulate the problem of BN synthesis as a Boolean satisfiability problem encoded in ASP. With this approach, we leverage a priori knowledge and experimental data as constraints on the network architecture and the dynamical properties of the models under the MP semantics. Our method is based on [2], which implements constraints on existence and absence of trajectories between partially-specified configurations, existence of (reachable) fixed points and trap spaces. In biological applications, these constraints match well the observed properties of cell populations evolving towards mutually exclusive phenotypes.

In this paper, we extend [2] to support universal properties on (reachable) attractors. This enables specifying tight dynamical constraints. For instance, given a set of experimentally observed phenotypes, existential constraints guarantee that at least one attractor of the model dynamics match with each phenotype, whereas a universal constraint ensures that every attractor matches with at least one of the phenotype.

A universal property involves by nature universal quantifiers. ASP can address formulas implying one level of universal quantifier (i.e., of the form $\exists x \forall y : P(x, y)$) thanks to a technique presented in [8]. To explore a set of values and check the respect of a property for each, it uses a disjunctive rule and a saturation on the same term. A disjunctive rule implies the subset minimality semantics. This minimality ensures an answer set is a solution only if none of its subsets is itself a solution [9]. Hence, saturating the answer set with the predicates of the disjunction cleverly exploits this minimality: the solver is forced to ensure that no strict subset of these predicates form a solution.

3.1 Universal Constraints on Fixed Points

We exploit this *saturation technique* [8,9] for ensuring universal constraints on the fixed points or fixed points reachable from a given configuration. We describe here the ASP rules for the universal fixed point constraint, which ensures that all the fixed points of the BN are compatible with a given set of markers (observations). To that purpose, we let the solver deduce a configuration z by the disjunctive rule:

```
8 cfg(z,N,-1) ; cfg(z,N,1) ← node(N).
```

The predicate template cfg(X,N,V) assigns the value V to the literal N in the configuration X. Through the above rule, a set of node values is thus constituted to define a configuration z, with the predicate cfg(z,N,_) subject to the subset minimality semantics. To respect the desired property, each configuration z is either not a fixed point ($f(z) \neq z$) or has the same component states than the ones expressed in a dedicated predicate. A configuration is not a fixed point whenever at least one of its component can change of state:

```
9 mcfg(z,N,V) ← cfg(z,N,V).
10 valid ← cfg(z,N,V) ; eval(z,N,-V).
```

mcfg(X,N,V) predicate template leads to the evaluation of the configuration X given the Boolean rules of the network [2]. The reachable values are then stored in the predicate eval(X,N,V). Whenever it is possible to evaluate a component N to the opposite value than in z, then z is not a fixed point, making valid true.

Otherwise, z has to have the same component states than those specified by an observation X marked by the predicate is_universal_fp(X), which is expressed by the following ASP rule:

```
11 valid ← cfg(z,N,V):obs(X,N,V); is_universal_fp(X).
```

Observe in l.10 and l.11 that each time an assignment is in agreement with the desired property, a predicate valid is deduced, which triggers the saturation of the configuration z:

₁₂ `cfg(z,N,-V) ← cfg(z,N,V), valid.`

Thus, when `valid` is deduced, the answer set contains all possible component values for z. According to the subset minimality semantics, the solver is then forced to ensure that there is no sub-answer set. And the only way to find such a smaller answer-set is to find a z from which `valid` cannot be deduced, i.e., which is a counter-example to the universal property: in that case, l.13 eliminates the answer set:

₁₃ `← not valid.`

A variant of this constraint enables to restrict the universal property to fixed points that are reachable from a given initial configuration. This is specified by `is_universal_fp(X,S)` predicates, where `S` points the initial configuration, and `X` to an observation, as used above. By combining such predicates, one can then specify sets of phenotypes reachable from a given configuration. The encoding of this variant contains a third way to deduce `valid`: the non-reachability of the configuration z from `S`.

Our implementation also offers to specify mutations, which can be combined with reachability and with universal constraints on reachable fixed points to leverage observations about cell fates in different mutation conditions.

3.2   Synthesis Problem

Synthesis requires (i) an influence graph to delimit the interactions that can be used by the BNs and (ii) the dynamical properties of the behaviours that have to be reproduced. For modelling the tumour invasion and migration as in [4], the dynamical properties refer to cell fate observations in different mutation conditions. These fates are described by sets of markers (i.e. a set of values for some nodes of the network) which constitutes partial observations of genes activity. In term of dynamics, these observations are related to reachable attractors in the corresponding mutated BNs.

A (partial) observation o of a configuration of dimension n is specified by a set of couples associating a component to a Boolean value: $o \subseteq [1]n \times \mathbb{B}$, assuming there is no $i \in [n]$ such that $\{(i, 0), (i, 1)\} \subseteq o$.

Formally, the synthesis problem we tackle is the following.
Given

- an influence graph $\mathcal{G} = \{[n], E_+, E_-)$
- p partial observations $o^1, \ldots, o^p$
- sets FP, UFP and UA of indices of observations
- sets PR, URFP and URA of couples of indices of observations: URFP $\subseteq [p]^2$

find a locally-monotonic BN f of dimension n such that

- $G(f) \subseteq \mathcal{G}$,
- there exist p configurations $x^1, \ldots, x^p$ such that:
  - (observations) $\forall m \in [p], \forall (i, v) \in o^m, x_i^m = v$,

- (positive reachability) $\forall (m, m') \in \mathsf{PR}$, $x^{m'} \in \rho_{\mathrm{mp}}^f(x^m)$,
- (fixed points) $\forall m \in \mathsf{FP}$, $f(x^m) = x^m$,
- (universal fixed point) $\forall z \in \mathbb{B}^n$, $f(z) = z \Rightarrow \exists m \in \mathsf{UFP} : \forall (i, v) \in o^m$, $z_i = v$;
- (universal reachable fixed point) $\forall z \in \mathbb{B}^n$, $f(z) = z \Rightarrow \exists (x, s) \in \mathsf{URFP} :$ $z \notin \rho_{\mathrm{mp}}^f(s) \vee \forall (i, v) \in x$, $z_i = v$;

Each of these constraints can be parametrized by mutations, in which case, the properties have to be verified on the mutated f.

Remark that such a problem can be non-satisfiable.

Our encoding also offers constraints related to the absence of paths between configurations (*negative reachability*) and to *trap space* where a set of components have a fixed state matching with a given observation [2]. Moreover, one can optionally impose that the influence graph of f is equal to the input $\mathcal{G}$.

Our implementation avoids redundancy in the models by enumerating only among non-equivalent BNs (i.e., their values differ for at least one configuration). This is achieved by using a canonical representation of Boolean functions in disjunctive normal form with a total ordering between clauses.

In total, our encoding generates $O(ndk^2)$ atoms and $O(nd^2k^2)$ rules, where d is the in-degree of nodes in the influence graph, and k is a fixed bound on the number of clauses of Boolean functions. Whenever k is set to $\binom{d}{\lfloor d/2 \rfloor}$, the complete set of solutions can be enumerated.

3.3 Sampling the Diversity of All Solutions

The whole set of constraints, comprising the domain of admissible BNs and the dynamical properties they should satisfy, is represented by a single logic program expressed in ASP, such that each solution corresponds to a distinct BN.

Whereas the enumeration of ASP solutions is known to be efficient, typical solvers will enumerate solutions by slightly varying parts of a firstly identified one. Thus, a partial enumeration will very likely give a set of solutions which are all look alike, e.g., where the Boolean function of only one component varies.

Inspired by [20], we tweak heuristics of the solver `clingo` [11] to stir it towards distant solutions: at each solution, we randomly select a subset of variables assignments and ask the solver to avoid them in the next iterations. At the cost of enumeration speed, this allows sampling ensembles of *diverse* BNs.

4 Stochastic Simulations of Ensembles of BNs

4.1 Continuous-Time Boolean Modelling

We first recall the continuous-time Markov chain interpretation of BNs introduced in [23]. Considering a BN f of dimension of n, we represent the state evolution by a Markov process $s : t \rightarrow s(t)$ defined on $t \in I \subset \mathbb{R}$ applied on the network state space, with I the simulation interval. This process is defined by:

1. Its initial condition:
$$P[s(0) = x], \quad \forall x \in \mathbb{B}^n$$

2. Its conditional probabilities (of a single condition):

$$P[s(t) = y | s(t') = x], \quad \forall x, y \in \mathbb{B}^n, \forall t', t \in I, t' < t$$

In continuous-time, these conditional probabilities are defined as transition rates [25]: $\rho(x \rightarrow y)(t) \in [0, \infty]$. Because we want a generalization of the fully-asynchronous Boolean dynamics, transition rates $\rho(x \rightarrow y)$ are non-zero only if $x \xrightarrow[a1]{f} y$, i.e., a single component $i \in [n]$ is changing of value. In that case, each local function $f_i(x)$ is replaced by two functions $R_i^{up/down}(x) \in [0, \infty]$. The transition rates are defined as follows:

$$\rho(x \rightarrow y) = \begin{cases} R_i^{up}(x) & \text{if } x_i = 0 \\ R_i^{down}(x) & \text{if } x_i = 1 \end{cases} \text{ with } \Delta(x,y) = \{i\}$$

where R_i^{up} corresponds to the activation rate of node i, and R_i^{down} corresponds to the inactivation rate of node i. Therefore, the continuous Markov process is completely defined by all these $R^{up/down}$ and an initial condition. By default, the value of these rates is set to 1, but they can be modified to represent the time scales of different processes.

To explore the probability space of this Markov process, we use the Gillespie algorithm [12]. This algorithm produces a set of realizations or stochastic trajectories of the Markov process. From this finite set, probabilities can be estimated.

To relate continuous-time probabilities to real processes, an observable time window δt is defined. A discrete-time $\tau \in \mathbb{N}$ stochastic process $s(\tau)$ can be extracted from the continuous-time Markov process:

$$P[s(\tau) = x] \equiv \frac{1}{\delta t} \int_{\tau \delta t}^{(\tau+1)\delta t} P[s(t) = x] dt$$

For each trajectory j, we compute the time for which the system is in state x in the window $[\tau \delta t, (\tau + 1)\delta t]$, and divide it by δt. We obtain an estimate of $P[s(\tau) = x]$ for trajectory j, i.e. $\hat{P}_j[s(\tau) = x]$. Then to compute the estimate of a set of trajectories, we compute the average over j of all $\hat{P}_j[s(\tau) = x]$.

4.2   Lifting to Ensembles of BNs

To simulate an ensemble of BNs, we first choose a total number of stochastic trajectories M. We generate M/k stochastic trajectories for each model and compute the average $\hat{P}_k[s(\tau) = x]$ for all models k. We then compute the average over k of all $\hat{P}_k[s(\tau) = x]$, to obtain the $P[s(\tau) = x]$ for the ensemble of boolean networks. We also keep the option to export the individual probability distributions $\hat{P}_k[s(\tau) = x]$ to allow us analyzing the composition of the ensemble. The approach results in time-series of the probability for each observed state. The

case study hereafter focuses on steady-state analysis. This imposes to simulate
the ensemble long enough to reach stationarity, requiring a preliminary analysis.
We can then study the proportion of each attractor for our ensemble.

We implemented this new feature in the MaBOSS simulation software [22][1].

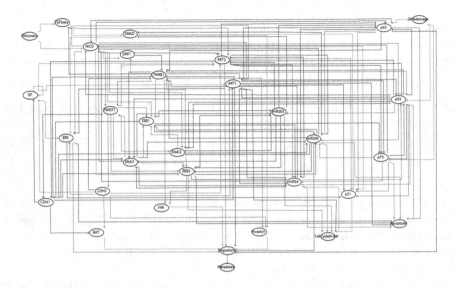

Fig. 3. Influence graph of Cohen's model relating 32 nodes with 159 edges, where
positive edges are in green and negative in red. (Color figure online)

5   Case Study on Cell Fate Decision Modelling

5.1   Background Model

We illustrate our ensemble modelling approach on a published model of cell fate
decision leading to the early events of the metastasis or cell death through apoptosis [4]. Initial triggers, such as DNA damage or micro-environmental cues, and
the activity of some genes or proteins participating in the process affect the final
decision. The signalling pathway involves TGFbeta, WNT, beta-catenin, p53
and its homologs, selected miRNA, and transcription factors of the epithelial to
mesenchymal (EMT) transitions. Figure 3 shows the influence graph of the BN.
The functions of the BN, we refer to as "Cohen's model", have been designed
manually so the simulations fit with experimental data related to stable phenotypes under different single mutations. Then, the initial publication explored the
synergy between mutants that led to metastatic phenotypes.

---

[1] https://maboss.curie.fr, https://github.com/colomoto/pyMaBoSS.

5.2   Single Model Analysis

We first reproduced part of the analysis of [4] on the original Cohen's model by computing the propensities of attractors reachable from 4 possible initial conditions, where all nodes are inactive, except miRNAs that are active, and the 2 nodes modelling DNA damage and micro-environment cues that are free. We considered the wild-type condition (Fig. 4(a)) with no mutation, and the double-mutant of p53 LoF and Notch GoF (Fig. 4(b)).

The wild-type model has 9 fixed points that each correspond to one of the 4 identified physiological phenotypes: Apoptosis, EMT, Metastasis (or equivalent to Migration) and Homeostatic State (HS). The double-mutant shows exclusively the Metastasis phenotype.

5.3   Ensemble Analysis

Synthesis. To test the impact of alternative Boolean functions, we synthesised ensembles of BNs that share the same influence graph as Cohen's model and reproduce the desired dynamics. We synthesized two ensembles of 1,000 diverse BNs each, where we disallowed having cyclic attractors. The first ensemble ensures only the wild-type (WT) behaviour, meaning that all the fixed points match with one of the 4 physiological phenotypes, and each physiological phenotype is reachable from at least one of the initial condition. The second ensemble adds further constraints related to the single mutations of p53 LoF, which should show the same behaviour as WT, and Notch GoF, where only 2 of the WT phenotypes and a third different one should be observed[2].

Ensemble Simulations. With the same settings as with Cohen's model, we performed stochastic simulations of the two synthesized ensembles, with uniform activation and de-activation rates. The WT behaviours look similar, with some differences in propensities of phenotypes (Fig. 4(c,e)). The double-mutant on the ensembles shows a much less contrasted picture than on the Cohen's model. While Migration becomes the most likely outcome, several other phenotypes are observed, suggesting a potential variability of the effect of the double-mutation. Interestingly, even the single mutant constraints of the second ensemble are not sufficiently restrictive to guarantee the behaviour observed in Cohen's model.

Variability of Propensities of Phenotypes. To study the ensemble composition, we want to analyze the steady-state probabilities for each model. Depending on the results we might have a lot of visited states, which bring a dimensionality issue. We choose to represent these results using Principal Component Analysis (PCA) [26], which allows us to visualize the distribution of attractor's proportions in a reduced number of dimensions.

---

[2] Code, data, and notebooks at https://doi.org/10.5281/zenodo.3938904; Synthesis has been performed on 36-cores CPUs @ 2.6 Ghz with 192 Go of RAM; first ensemble was generated at a rate of 5 s/model/CPU; second ensemble was generated at a rate of 3 min/model/CPU.

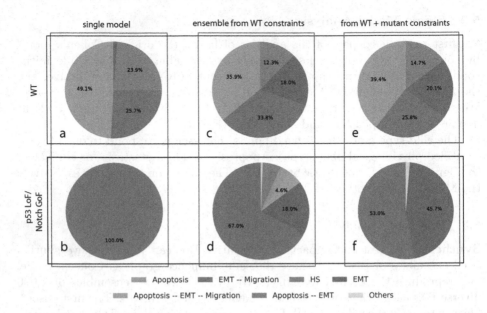

Fig. 4. Simulations results for phenotypes propensities in Cohen's model (a, b), ensemble from WT constraints (c, d), and ensemble from WT and single mutants constraints (e, f), in wild-type condition (a, c, e) and double-mutant p53 LoF/Notch GoF (b, d, f).

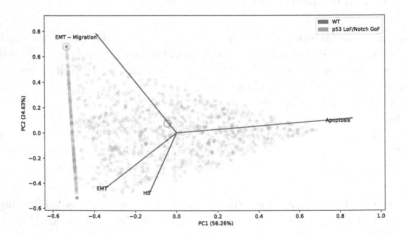

Fig. 5. PCA representation of the steady-state distribution of each model of the ensemble from WT and single mutants constraints. Each point represents the result of one model simulation (blue one are from WT simulations, orange one from p53 LoF/Notch GoF). Large blue and orange circles highlight the position of the original single Cohen's model simulation. The triangular pattern of the distribution comes from the fact that the phenotype probabilities are located in the n-dimensional simplex. (Color figure online)

We apply PCA to the probability distribution of each model within the ensemble from WT and single mutants constraints, allowing us to represent their respective probability distributions (Fig. 5). The first component, representing 56% of the observed variance, shows a negative correlation between apoptotic and EMT phenotypes. The second component, representing 24% of observed variance, shows a negative correlation between EMT without Migration and EMT with Migration. The distribution of the ensemble's probability distributions is diverse, illustrating the performance of the all possible model diversity sampling. The p53 LoF/Notch GoF double mutant shows a shift towards EMT and/or Migration phenotypes, away from Apoptotic phenotypes. The alignment of models on the top-left corresponds to models which don't show any apoptotic phenotypes ($\sim$96% of the models).

6 Conclusion

The synthesis of BNs from network architecture and dynamical constraints can lead to a multitude of admissible solutions.

In this work, we employed Answer-Set Programming to sample ensembles of diverse BNs, all possessing the same network architecture and satisfying the same set of dynamical constraints. We significantly extend the previously described methodology with the new type of biologically relevant universal constraints.

Our synthesis framework enables specifying existence and absence of reachability properties between (partial) configurations of the BN, existence of fixed points and cyclic attractors matching with observations, and universal properties on the fixed points and reachable fixed points; all these properties can be parametrized by mutation settings.

The dynamics of ensembles is explored by stochastic simulations using the new Ensemble MaBoSS simulator, which is introduced here for the first time. The ensemble-based simulations are used for computing and comparing propensities of reachable attractors under different mutations or their combinations. The result of an ensemble-based simulation represents a multidimensional distribution of the vectors of attractor probabilities, which can be aggregated by computing its mean point. Moreover, the multi-variate variance of the distribution can be explored, e.g. by applying the standard machine learning methods such as Principal Component Analysis, which can lead to the insights about the diversity of possible modelling scenarios compatible with available biological knowledge and the experimental data.

As illustrated on a biological case study, BN ensemble modelling brings an insight into the potential variability of predictions subject to model uncertainty.

In future work, we plan to address the evaluation of the diversity of sampled ensembles, with metrics helping estimate required sample size and compare sampling heuristics.

Acknowledgements. This work has been partially supported by Agence Nationale de la Recherche in the program Investissements d'Avenir (project No. ANR-19-P3IA-0001; PRAIRIE 3IA Institute), by ANR-FNR project "AlgoReCell" (ANR-16-CE12-0034), by ITMO Cancer, and by the Ministry of Science and Higher Education of the Russian Federation (project No. 14.Y26.31.0022). Experiments were carried out using the PlaFRIM experimental testbed,supported by Inria, CNRS (LABRI and IMB), Université de Bordeaux, Bordeaux INP and Conseil Régionald'Aquitaine (see https://www.plafrim.fr).

References

1. Baral, C.: Knowledge Representation, Reasoning and Declarative Problem Solving. Cambridge University Press, Cambridge (2003)
2. Chevalier, S., Froidevaux, C., Paulevé, L., Zinovyev, A.: Synthesis of Boolean networks from biological dynamical constraints using answer-set programming. In: 2019 IEEE 31st International Conference on Tools with Artificial Intelligence (ICTAI), pp. 34–41 (2019). https://doi.org/10.1109/ICTAI.2019.00014
3. Clarke, M.A., Fisher, J.: Executable cancer models: successes and challenges. Nat. Rev. Cancer **20**, 343–354 (2020). https://doi.org/10.1038/s41568-020-0258-x
4. Cohen, D.P.A., Martignetti, L., Robine, S., Barillot, E., Zinovyev, A., Calzone, L.: Mathematical modelling of molecular pathways enabling tumour cell invasion and migration. PLoS Comput. Biol. **11**(11), e1004571 (2015). https://doi.org/10.1371/journal.pcbi.1004571
5. Collombet, S., et al.: Logical modeling of lymphoid and myeloid cell specification and transdifferentiation. Proc. Nat. Acad. Sci. **114**(23), 5792–5799 (2017). https://doi.org/10.1073/pnas.1610622114
6. Corblin, F., Tripodi, S., Fanchon, E., Ropers, D., Trilling, L.: A declarative constraint-based method for analyzing discrete genetic regulatory networks. Biosystems **98**(2), 91–104 (2009). https://doi.org/10.1016/j.biosystems.2009.07.007
7. Dorier, J., Crespo, I., Niknejad, A., Liechti, R., Ebeling, M., Xenarios, I.: Boolean regulatory network reconstruction using literature based knowledge with a genetic algorithm optimization method. BMC Bioinform. **17**(1), 410 (2016). https://doi.org/10.1186/s12859-016-1287-z
8. Eiter, T., Gottlob, G.: On the computational cost of disjunctive logic programming: propositional case. Ann. Math. Artif. Intell. **15**(3), 289–323 (1995). https://doi.org/10.1007/BF01536399
9. Eiter, T., Ianni, G., Krennwallner, T.: Answer set programming: a primer. In: Tessaris, S., et al. (eds.) Reasoning Web 2009. LNCS, vol. 5689, pp. 40–110. Springer, Heidelberg (2009). https://doi.org/10.1007/978-3-642-03754-2_2
10. Gebser, M., Kaminski, R., Kaufmann, B., Schaub, T.: Answer set solving in practice. Synth. Lect. Artif. Intell. Mach. Learn. **6**, 1–23 (2012)
11. Gebser, M., Kaminski, R., Kaufmann, B., Schaub, T.: Clingo = ASP + control: preliminary report. CoRR abs/1405.3694 (2014)
12. Gillespie, D.T.: A general method for numerically simulating the stochastic time evolution of coupled chemical reactions. J. Comput. Phys. **22**(4), 403–434 (1976). https://doi.org/10.1016/0021-9991(76)90041-3

13. Goldfeder, J., Kugler, H.: BRE: IN - a backend for reasoning about interaction networks with temporal logic. In: Bortolussi, L., Sanguinetti, G. (eds.) CMSB 2019. LNCS, vol. 11773, pp. 289–295. Springer, Cham (2019). https://doi.org/10.1007/978-3-030-31304-3_15

14. Kauffman, S.: A proposal for using the ensemble approach to understand genetic regulatory networks. J. Theor. Biol. **230**(4), 581–590 (2004). https://doi.org/10.1016/j.jtbi.2003.12.017

15. Klarner, H., Bockmayr, A., Siebert, H.: Computing maximal and minimal trap spaces of Boolean networks. Nat. Comput. **14**(4), 535–544 (2015). https://doi.org/10.1007/s11047-015-9520-7

16. Krawitz, P., Shmulevich, I.: Basin entropy in Boolean network ensembles. Phys. Rev. Lett. **98**(15), 158701 (2007). https://doi.org/10.1103/physrevlett.98.158701

17. Lin, F., Zhao, Y.: ASSAT: computing answer sets of a logic program by SAT solvers. Artif. Intell. **157**(1), 115–137 (2004). https://doi.org/10.1016/j.artint.2004.04.004

18. Lobo, J., Minker, J., Rajasekar, A.: Foundations of Disjunctive Logic Programming. MIT Press, Cambridge (1992)

19. Paulevé, L., Kolčák, J., Chatain, T., Haar, S.: Reconciling qualitative, abstract, and scalable modeling of biological networks. bioRxiv (2020). https://doi.org/10.1101/2020.03.22.998377

20. Razzaq, M., Kaminski, R., Romero, J., Schaub, T., Bourdon, J., Guziolowski, C.: Computing diverse Boolean networks from phosphoproteomic time series data. In: Češka, M., Šafránek, D. (eds.) CMSB 2018. LNCS, vol. 11095, pp. 59–74. Springer, Cham (2018). https://doi.org/10.1007/978-3-319-99429-1_4

21. Schwieger, R., Siebert, H.: Graph representations of monotonic Boolean model pools. In: Feret, J., Koeppl, H. (eds.) CMSB 2017. LNCS, vol. 10545, pp. 233–248. Springer, Cham (2017). https://doi.org/10.1007/978-3-319-67471-1_14

22. Stoll, G., et al.: MaBoSS 2.0: an environment for stochastic Boolean modeling. Bioinformatics **33**(14), 2226–2228 (2017). https://doi.org/10.1093/bioinformatics/btx123

23. Stoll, G., Viara, E., Barillot, E., Calzone, L.: Continuous time Boolean modeling for biological signaling: application of Gillespie algorithm. BMC Syst. Biol. **6**(1), 116 (2012). https://doi.org/10.1186/1752-0509-6-116

24. Terfve, C., et al.: CellNOptR: a flexible toolkit to train protein signaling networks to data using multiple logic formalisms. BMC Syst. Biol. **6**(1), 133 (2012). https://doi.org/10.1186/1752-0509-6-133

25. Van Kampen, N.G.: Stochastic Processes in Physics and Chemistry, vol. 1. Elsevier, Amsterdam (1992)

26. Wold, S., Esbensen, K., Geladi, P.: Principal component analysis. Chemom. Intell. Lab. Syst. **2**(1–3), 37–52 (1987). https://doi.org/10.1016/0169-7439(87)80084-9

27. Yordanov, B., Dunn, S.J., Kugler, H., Smith, A., Martello, G., Emmott, S.: A method to identify and analyze biological programs through automated reasoning. Syst. Biol. Appl. **2**(1), 1–16 (2016). https://doi.org/10.1038/npjsba.2016.10

28. Zañudo, J.G., Steinway, S.N., Albert, R.: Discrete dynamic network modeling of oncogenic signaling: mechanistic insights for personalized treatment of cancer. Curr. Opin. Syst. Biol. **9**, 1–10 (2018). https://doi.org/10.1016/j.coisb.2018.02.002

Classifier Construction in Boolean Networks Using Algebraic Methods

Robert Schwieger[1](✉)(iD), Matías R. Bender[2](iD), Heike Siebert[1],
and Christian Haase[1]

[1] Freie Universität Berlin, Berlin, Germany
rschwieger@zedat.fu-berlin.de
[2] Technische Universität Berlin, Berlin, Germany
mbender@math.tu-berlin.de

Abstract. We investigate how classifiers for Boolean networks (BNs) can be constructed and modified under constraints. A typical constraint is to observe only states in attractors or even more specifically steady states of BNs. Steady states of BNs are one of the most interesting features for application. Large models can possess many steady states. In the typical scenario motivating this paper we start from a Boolean model with a given classification of the state space into phenotypes defined by high-level readout components. In order to link molecular biomarkers with experimental design, we search for alternative components suitable for the given classification task. This is useful for modelers of regulatory networks for suggesting experiments and measurements based on their models. It can also help to explain causal relations between components and phenotypes. To tackle this problem we need to use the structure of the BN and the constraints. This calls for an algebraic approach. Indeed we demonstrate that this problem can be reformulated into the language of algebraic geometry. While already interesting in itself, this allows us to use Gröbner bases to construct an algorithm for finding such classifiers. We demonstrate the usefulness of this algorithm as a proof of concept on a model with 25 components.

Keywords: Boolean networks · Algebraic geometry · Gröbner bases · Classifiers

1 Motivation

For the analysis of large regulatory networks so called *Boolean networks (BNs)* are used among other modeling frameworks [1,26,32]. They have been applied

Supported by the DFG-funded Cluster of Excellence MATH+: Berlin Mathematics Research Center, Project AA1-4. Matías R. Bender was supported by the ERC under the European's Horizon 2020 research and innovation programme (grant agreement No. 787840).

A. Abate et al. (Eds.): CMSB 2020, LNBI 12314, pp. 210–233, 2020.
https://doi.org/10.1007/978-3-030-60327-4_12

frequently in the past [3, 18, 22, 34]. In this approach interactions between different components of the regulatory networks are modeled by logical expressions. Formally, a Boolean network is simply a Boolean function $f : \{0,1\}^n \to \{0,1\}^n$, $n \in \mathbb{N}$. This Boolean function contains the information about the interactions of the components in the network. It is then translated into a so called *state transition graph (STG)*. There are several slightly different formalisms for the construction of the STG of a BN. In all cases, the resulting state transition graph is a directed graph over the set of vertices $\{0,1\}^n$. The vertices of the STG are also called *states* in the literature about Boolean networks.

Modelers of regulatory networks are frequently – if not to say almost always – confronted with uncertainties about the exact nature of the interactions among the components of the network. Consequently, in many modeling approaches models may exhibit alternative behaviors. In so called *asynchronous* Boolean networks for example each state in the state transition graph can have many potential successor states (see e.g. [9]). More fundamentally, alternative models are constructed and then compared with each other (see e.g. [33, 37]).

To validate or to refine such models we need to measure the real world system and compare the results with the model(s). However, in reality for networks with many components it is not realistic to be able to measure all the components. In this scenario there is an additional step in the above procedure in which the modeler first needs to select a set of components to be measured which are relevant for the posed question. This scenario motivates our problem here. How can a modeler decide which components should be measured? It is clear that the answer depends on the question posed to the model and on the prior knowledge or assumptions assumed to be true.

When formalizing this question we are confronted with the task to find different representations of partially defined Boolean functions. In the field of *logical analysis of data (LAD)* a very similar problem is tackled [2, 4, 11, 23]. Here a list of binary vectorized samples needs to be extended to a Boolean function – a so called *theory* (see e.g. [11, p. 160]). In the literature of LAD this problem is also referred to as the Extension-Problem [11, p. 161 and p. 170]. Here reformulations into linear integer programs are used frequently [11]. However, they are more tailored to the case where the partially defined Boolean functions are defined explicitly by truth tables. In contrast to the scenario in LAD in our case the sets are typically assumed to be given implicitly (e.g. by so called *readout components*).

A common assumption in the field of Boolean modeling is that *attractors* play an important role. Attractors of BNs are thought to capture the long term behavior of the modeled regulatory network. Of special interest among these attractors are *steady states* (defined by $f(x) = x$ for a BN f). Consequently, a typical scenario is that the modeler assumes to observe only states of the modeled network which correspond to states belonging to attractors or even only steady states of the STG. The state space is then often partitioned by so-called *readout components* into *phenotypes*.

Our first contribution will be a reformulation of the above problem into the language of algebraic geometry. For this purpose we focus on the case of classification into two phenotypes. This is an important special case. Solutions to the more general case can be obtained by performing the algorithm iteratively. The two sets of states A_1 and A_2 in $\{0, 1\}^n$ describing the phenotypes will be defined by some polynomial equations in the components of the network. This algebraic reformulation is possible since we can express the Boolean function $f : \{0, 1\}^n \to \{0, 1\}^n$ with polynomials over $\mathbb{F}_2[x_1, \ldots, x_n]$ – the polynomial ring over the finite field of cardinality two (see Sect. 2). In this way we relate the problem to a large well-developed theoretical framework. Algebraic approaches for the construction and analysis of BNs and chemical reaction systems have been used in the past already successfully (see e.g. [24,27,40]). Among other applications they have been applied to the control of BNs [30] and to the inference of BNs from data [38,41].

Our second contribution will be to use this algebraic machinery to construct a new algorithm to find alternative classifiers. To our knowledge this is the first algorithm that is able to make use of the implicit description of the sets that should be classified. For this algorithm we use Gröbner bases. Gröbner bases are one of the most important tools in computational algebraic geometry and they have been applied in innumerous applications, e.g. cryptography [17], statistics [15], robotics [7], biological dynamical systems [14,25,30,39]. Specialized algorithms for the computations of Gröbner bases have been developed for the Boolean case and can be freely accessed [5]. They are able to deal with with systems of Boolean polynomials with up to several hundreds variables [5] using a specialized data structure (so called zero-suppressed binary decision diagram (ZDD) [29]). Such approaches are in many instances competitive with conventional solvers for the Boolean satisfiability problem (SAT-solvers) [5].

Our paper is structured in the following way. We start by giving the mathematical background used in the subsequent sections in Sect. 2. In Sect. 3 we formalize our problem. We then continue in Sect. 4 to give a high-level description of the algorithm we developed for this problem. More details about the used data structures and performance can be found in Sect. 5. As a proof of concept we investigate in Sect. 6 a BN of 25 components modeling cell-fate decision [8]. We conclude the paper with discussing potential ways to improve the algorithm.

2   Mathematical Background

In the course of this paper we need some concepts and notation used in computational algebraic geometry. For our purposes, we will give all definitions for the field of cardinality two denoted by $\mathbb{F}_2$ even though they apply to a much more general setting. For a more extensive and general introduction to algebraic geometry and Gröbner bases we refer to [12].

We denote the ring of polynomials in $x_1, \ldots, x_n$ over $\mathbb{F}_2$ with $\mathbb{F}_2[x_1, \ldots, x_n]$. For $n \in \mathbb{N}$, let $[n] := \{1, \ldots, n\}$. Given $\alpha = (\alpha_1, \ldots, \alpha_n) \in \mathbb{Z}_{\geq 0}^n$, we denote by x^α the monomial $\prod_i x_i^{\alpha_i}$ in $\mathbb{F}_2[x_1, \ldots, x_n]$. For $f_1, \ldots, f_k$ in $\mathbb{F}_2[x_1, \ldots, x_n]$

we denote with $\langle f_1, \ldots, f_k \rangle$ a so-called *ideal* in $\mathbb{F}_2[x_1, \ldots, x_n]$ – a subset of polynomials which is closed under addition and multiplication with elements in $\mathbb{F}_2[x_1, \ldots, x_n]$ – generated by these polynomials. The set of Boolean functions – that is the set of functions from $\mathbb{F}_2^n$ to $\mathbb{F}_2$ – will be denoted by $\mathbb{B}(n)$. When speaking about Boolean functions and polynomials in $\mathbb{F}_2[x_1, \ldots, x_n]$ we need to take into account that the set of polynomials $\mathbb{F}_2[x_1, \ldots, x_n]$ does not coincide with the set of Boolean functions. This is the case since the so-called *field polynomials* $x_1^2 - x_1, \ldots, x_n^2 - x_n$ evaluate to zero over $\mathbb{F}_2^n$ [20]. Consequently, there is not a one-to-one correspondence between polynomials and Boolean functions. However, we can say that any two polynomials whose difference is a sum of field polynomials corresponds to the same Boolean function (see e.g. [10]). In other words we can identify the ring of Boolean functions $\mathbb{B}(n)$ with the quotient ring $\mathbb{F}_2[x_1, \ldots, x_n]/\langle x_1^2 - x_1, \ldots, x_n^2 - x_n \rangle$. We will denote both objects with $\mathbb{B}(n)$. A canonical system of representatives of $\mathbb{B}(n)$ is linearly spanned by the the square-free monomials in $\mathbb{F}_2[x_1, \ldots, x_n]$. Hence, in what follows when we talk about a Boolean function $f \in \mathbb{B}(n)$ as a polynomial in the variables $x_1, \ldots, x_n$ we refer to the unique polynomial in $\mathbb{F}_2[x_1, \ldots, x_n]$ which involves only monomials that are square-free and agrees with f as a Boolean function.

Since we are interested in our application in subsets of $\mathbb{F}_2^n$, we need to explain their relationship to the polynomial ring $\mathbb{F}_2[x_1, \ldots, x_n]$. This relationship is established using the notion of the vanishing ideal. Instead of considering a set $B \subseteq \mathbb{F}_2^n$ we will look at its *vanishing ideal* $\mathcal{I}(B)$ in $\mathbb{B}(n)$. The vanishing ideal of B consists of all Boolean functions which evaluate to zero on B. Conversely, for an ideal $\mathcal{I}$ in $\mathbb{B}(n)$ we denote with $\mathcal{V}(\mathcal{I})$ the set of points in $\mathbb{F}_2^n$ for which every Boolean function in $\mathcal{I}$ evaluates to zero. Due to the Boolean Nullstellensatz (see [19,35]) there is an easy relation between a set $B \subseteq \mathbb{F}_2^n$ and its vanishing ideal $\mathcal{I}(B)$: For an ideal $\mathcal{I}$ in $\mathbb{B}(n)$ such that $\mathcal{V}(\mathcal{I}) \neq \emptyset$ and for any polynomial $h \in \mathbb{B}(n)$ it holds

$$h \in \mathcal{I} \Leftrightarrow \forall v \in \mathcal{V}(\mathcal{I}) : h(v) = 0.$$

In this paper, we will consider Boolean functions whose domain is restricted to certain states (e.g. attractors or steady states). Hence, there are different Boolean functions that behave in the same way when we restrict their domain.

Example 1. Consider the set $B := \{000, 110, 101, 011\}$. Consider the Boolean function $f := x_1$ and $g := x_2 + x_3$. Both Boolean functions are different, i.e., $f(1, 1, 1) = 1$ and $g(1, 1, 1) = 0$, but they agree over B.

	000	110	101	011
f	0	1	1	0
g	0	1	1	0

Note that $\mathcal{I}(B) = \langle x_1 + x_2 + x_3 \rangle$, that is, the ideal $\mathcal{I}(B)$ is generated by the Boolean function $x_1 + x_2 + x_3$ since it is the unique Boolean function vanishing only on B.

Given a set B, we write $\mathbb{B}(n)/\mathcal{I}(B)$ to refer to the set of all the different Boolean functions on B. As we saw in the previous example, different Boolean functions agree on B. Hence, we will be interested in how to obtain certain representatives of the Boolean function in $\mathbb{B}(n)/\mathcal{I}(B)$ algorithmically. In our application, the set $\mathbb{B}(n)/\mathcal{I}(B)$ will become the set of all possible classifiers we can construct that differ on B. To obtain specific representatives of a Boolean function in $\mathbb{B}(n)/\mathcal{I}(B)$ we will use Gröbner bases. A Gröbner basis of an ideal is a set of generators of the ideal with some extra properties related to *monomial orderings*. A monomial ordering is a total ordering on the set of monomials in $\mathbb{F}_2[x_1, \ldots, x_n]$ satisfying some additional properties to ensure the compatibility with the algebraic operations in $\mathbb{F}_2[x_1, \ldots, x_n]$ (see [12, p. 69] for details).

For any polynomial in $p \in \mathbb{F}_2[x_1, \ldots, x_n]$ and monomial ordering $\prec$, we denote the *initial monomial* of p by $in_\prec(p)$, that is the largest monomial appearing in p with respect to $\prec$. We are interested in specific orderings – the lexicographical orderings – on these monomials. As we will see, the usage of lexicographical orderings in the context of our application will allow us to look for classifiers which are optimal in a certain sense.

Definition 1 ([12, p. 70]). *Let $\alpha = (\alpha_1 \ldots \alpha_n)$ and $\beta = (\beta_1 \ldots \beta_n)$ be two elements in $\mathbb{Z}_{\geq 0}^n$. Given a permutation σ of $\{1, \ldots, n\}$, we say $x^\alpha \succ_{lex(\sigma)} x^\beta$ if there is $k \in [n]$ such that*

$$(\forall i < k : \alpha_{\sigma(i)} = \beta_{\sigma(i)}) \text{ and } \alpha_{\sigma(k)} > \beta_{\sigma(k)}.$$

Definition 2 ([36, p. 1]). *Let $\prec$ be any monomial ordering. For an ideal $\mathcal{I} \subseteq \mathbb{F}_2[x_1, \ldots, x_n]$ we define its* initial ideal *as the ideal*

$$in_\prec(\mathcal{I}) := \langle in_\prec(f) | f \in \mathcal{I} \rangle.$$

A finite subset $G \subseteq \mathcal{I}$ is a Gröbner basis *for $\mathcal{I}$ with respect to $\prec$ if $in_\prec(\mathcal{I})$ is generated by $\{in_\prec(g) | g \in G\}$. If no element of the Gröbner basis G is redundant, then G is* minimal. *It is called* reduced *if for any two distinct elements $g, g' \in G$ no monomial in g' is divisible by $in_\prec(g)$. Given an ideal and a monomial ordering, there is a unique minimal reduced Gröbner basis involving only monic polynomials; we denote it by $G_\prec(\mathcal{I})$. Every monomial not lying in $in_\prec(\mathcal{I})$ is called* standard monomial.

We can extend the monomial orderings to partial orderings of polynomials on $\mathbb{F}_2[x_1, \ldots, x_n]$. Consider polynomials $f, g \in \mathbb{F}_2[x_1, \ldots, x_n]$ and a monomial ordering $\succ$. We say that $f \succ g$ if $in_\prec(f) \succ in_\prec(g)$ or $f - in_\prec(f) \succ g - in_\prec(g)$. The division algorithm rewrites every polynomial $f \in \mathbb{F}_2[x_1, \ldots, x_n]$ modulo $\mathcal{I}$ uniquely as a linear combination of these standard monomials [12, Ch. 2]. The result of this algorithm is called Normal form. For the convenience of the reader, we state this in the following lemma and definition.

Lemma 1 (Normal Form). *Given a monomial ordering $\succ$, $f \in \mathbb{B}(n)$ and an ideal $\mathcal{I}$ there is a unique $g \in \mathbb{B}(n)$ such that f and g represent the same Boolean function in $\mathbb{B}(n)/\mathcal{I}$ and g is minimal with respect to the ordering $\succ$ among all the Boolean functions equivalent to f in $\mathbb{B}(n)/\mathcal{I}$. We call g the normal form of f modulo I denoted by $\mathrm{NF}_\mathcal{I}(\succ, f)$.*

Example 2 (Cont.). Consider the permutation σ of $\{1, 2, 3\}$ such that $\sigma(i) := 4 - i$. Then, $\mathrm{NF}_{\mathcal{I}}(\succ_\sigma, f) = \mathrm{NF}_{\mathcal{I}}(\succ_\sigma, g) = x_1$. Hence, if we choose a "good" monomial ordering, we can get simpler Boolean functions involving less variables.

3 Algebraic Formalization

As discussed in Sect. 1, we start with the assumption that we are given a set of Boolean vectors B in $\mathbb{F}_2^n$ representing the observable states. In our applications these states are typically attractors or steady states of a BN. We also assume that our set B is partitioned into a set of phenotypes, i.e. $B = A_1 \cup \cdots \cup A_k$. Our goal is then to find the components that allow us to decide for a vector in B to which set A_i, $i \in [k]$, it belongs. For a set of indices $I \subseteq [n]$ and $x \in \mathbb{F}_2^n$, let us denote with $\mathrm{proj}^I(x)$ the projection of x onto the components I. Our problem could be formalized in the following way.

Problem 1 (State-Discrimination-Problem). For a given partition of non-empty sets $A_1, \ldots, A_k$ $(k \geq 2)$ of states $B \subseteq \{0, 1\}^n$, find the sets of components $\emptyset \neq I \subseteq [n]$ such that $\mathrm{proj}^I(A_1), \ldots, \mathrm{proj}^I(A_k)$ forms a partition of $\mathrm{proj}^I(B)$.

Clearly, since the sets $A_1, \ldots, A_k$ form a partition of B, we can decide for each state x in B to which set A_i it belongs. If $I \subseteq [n]$ is a solution to Problem 1, this decision can be solely based on $\mathrm{proj}^I(x)$ since $\mathrm{proj}^I(A_1), \ldots, \mathrm{proj}^I(A_k)$ form a partition of $\mathrm{proj}^I(B)$. As we discussed in Sect. 1, Problem 1 is equivalent to Extension-Problem [11, p. 161 and p. 170]. However, in our case the sets in Problem 1 are typically given implicitly. That is, we are given already some information about the structure of the sets in the above problem. This calls for an algebraic approach. We consider the case in Problem 1 where k equals two. This is an important special case since many classification problems consist of two sets (e.g. healthy and sick). Furthermore, solutions to the more general case can be obtained by considering iteratively the binary case (see also the case study in Sect. 6).

Let $\mathcal{I}(B) \subseteq \mathbb{F}_2[x_1, \ldots, x_n]$ be the vanishing ideal of a set $B \subseteq \mathbb{F}_2^n$. Let $f : \mathbb{F}_2^n \to \mathbb{F}_2$ be a Boolean function which can be identified with an element in $\mathbb{B}(n) := \mathbb{F}_2[x_1, \ldots, x_n] / \langle x_1^2 - x_1, \ldots, x_n^2 - x_n \rangle$. We want to find representatives of f in $\mathbb{B}(n)/\mathcal{I}(B)$ which depend on a minimal set of variables with respect to set inclusion or cardinality. We express this in the following form:

Problem 2. For $f \in \mathbb{B}(n)$ and $B \subseteq \mathbb{F}_2^n$, find the representatives of f in $\mathbb{B}(n)/\mathcal{I}(B)$ which depend on a set of variables satisfying some minimality criterion.

It is clear that Problem 2 is equivalent to Problem 1 for the case $k = 2$ since due to the Boolean Strong Nullstellensatz (see [35]) a Boolean function f is zero on B if and only if f is in $\mathcal{I}(B)$. Therefore, all Boolean functions which are in the same residue class as f agree with it as Boolean functions on B and vice versa. The sets of variables each representative depends on are the solutions to Problem 2. The representatives are then the classifiers.

Here we will focus on solutions of Problem 2 which are minimal with respect to set inclusion or cardinality. However, also other optimality criteria are imaginable. For example one could introduce some weights for the components. Let us illustrate Problem 2 with a small example.

Example 3. Consider the set $B = \{000, 111, 011, 101\} \subset \mathbb{F}_2^3$. Then $\mathcal{I}(B) \subseteq \mathbb{B}(3)$ $\underbrace{\qquad\qquad}_{=:f}$ is given by $\langle x_1x_2 + x_1 + x_2 + x_3 \rangle$ since f is the unique Boolean function that is zero on B and one on it complement. Let $\varphi(x) = x_1x_2x_3$. It is easy to check that for example $x_1x_3 + x_2x_3 + x_3$ and x_1x_2 are different representatives of φ in $\mathbb{B}(3)/\mathcal{I}(B)$. The representative x_1x_2 depends only on two variables while the other two representatives depend on three.

We can obtain a minimal representative of f in Problem 2 by computing $\mathrm{NF}_{\mathcal{I}}(\prec, f)$ for a suitable lexicographical ordering $\prec$.

Proposition 1. *Given a set of points $B \subset \mathbb{F}_2^n$ and a Boolean function $f \in \mathbb{B}(n)$ assume, with no loss of generality, that there is an equivalent Boolean function $g \in \mathbb{F}_2[x_1, \ldots, x_n]$ modulo $\mathcal{I}(B)$ involving only $x_k, \ldots, x_n$. Consider a permutation σ of $\{1, \ldots, n\}$ such that $\sigma(\{k, \ldots, n\}) = \{k, \ldots, n\}$. Then, the only variables appearing in $\mathrm{NF}_I(\prec_{lex(\sigma)}, f) = \mathrm{NF}_I(\prec_{lex(\sigma)}, g)$ are the ones in $\{x_k, \ldots, x_n\}$. In particular, if there is no Boolean function equivalent to f modulo $\mathcal{I}(B)$ involving a proper subset of $\{x_k, \ldots, x_n\}$, then $\mathrm{NF}_I(\prec_{lex(\sigma)}, f) = \mathrm{NF}_I(\prec_{lex(\sigma)}, g)$ involves all the variables in $\{x_k, \ldots, x_n\}$.*

Proof. The proof follows from the minimality of $\mathrm{NF}_I(\prec_{lex(\sigma)}, f)$ with respect to $\prec_{lex(\sigma)}$. Note that, because of the lexicographical ordering $\prec_{lex(\sigma)}$, any Boolean function equivalent to f modulo $\mathcal{I}(B)$ involving variables in $\{x_1, \ldots, x_{k-1}\}$ will be bigger than g, so it cannot be minimal.

4   Description of the Algorithm

Clearly we could use Proposition 1 to obtain an algorithm that finds the minimal representatives of φ in $\mathbb{B}(n)/\mathcal{I}(B)$ by iterating over all lexicographical orderings in $\mathbb{F}_2[x_1, \ldots, x_n]$. However, this naive approach has several drawbacks:

1. The number of orderings over $\mathbb{B}(n)$ is growing rapidly with n since there are $n!$ many lexicographical orderings over $\mathbb{B}(n)$ to check.
2. We do not obtain for every lexicographical ordering a minimal representative. Excluding some of these orderings "simultaneously" could be very beneficial.
3. Different monomial orderings can induce the same Gröbner bases. Consequently, the normal form leads to the same representative.
4. Normal forms with different monomial orderings can result in the same representative. If we detect such cases we avoid unnecessary computations.

We describe now an algorithm addressing the first two points. Recall that for a monomial ordering $\succ$ and $f \in \mathbb{B}(n)$, we use the notation $\mathrm{NF}_{\mathcal{I}}(\succ, f)$ to denote the normal form of f in $\mathbb{B}(n)/\mathcal{I}$ with respect to $\succ$. When $\mathcal{I}$ is clear from the context, we write $\mathrm{NF}(\succ, f)$. We denote with φ any representative of the indicator function of A in $\mathbb{B}(n)/\mathcal{I}(B)$. Let $\mathrm{Var}(\varphi)$ be the variables occurring in φ and $\mathrm{Comp}(\varphi)$ be its complement in $\{x_1, \ldots, x_n\}$. Instead of iterating through the orderings on $\mathbb{B}(n)$, we consider candidate sets in the power set of $\{x_1, \ldots, x_n\}$, denoted by $\mathscr{P}(x_1, \ldots, x_n)$ (i.e. we initialize the family of candidate sets P with $P \leftarrow \mathscr{P}(x_1, \ldots, x_n)$). We want to find the sets A in the family of candidate sets P for which the equality $A = \mathrm{Var}(\varphi)$ holds for some minimal solution φ to Problem 2, that is, involving the minimal amount of variables. For each candidate set A involving k variables we pick a lexicographical ordering $\succ$ for which it holds $A^c \succ A$, i.e. for every variable $x_i \in A^c$ and $x_j \in A$ it holds $x_i \succ x_j$, where $A^c := \{x_1, \ldots, x_n\} \setminus A$. This approach is sufficient to find the minimal solutions as we will argue below. This addresses the first point above since there are 2^n candidate sets to consider while there are $n!$ many orderings.[1]

4.1 Excluding Candidate Sets

To address the second point we will exclude after each reduction step a family of candidate sets. If, for an ordering $\succ$, we computed a representative $\mathrm{NF}(\succ, \varphi)$ we can, independently of the minimality of $\mathrm{NF}(\succ, \varphi)$, exclude some sets in P. To do so, we define for any set $A \subseteq \{x_1, \ldots, x_n\}$ the following families of sets:

$$\mathrm{FORWARD}(A) := \{B \subseteq \{x_1, \ldots, x_n\} | A \subset B\},$$
$$\mathrm{FORWARDEQ}(A) := \{B \subseteq \{x_1, \ldots, x_n\} | A \subseteq B\},$$
$$\mathrm{BACKWARD}(A) := \{B \subseteq \{x_1, \ldots, x_n\} | B \subseteq A\},$$
$$\mathrm{SMALLER}(A, \succ) := \{x \in \{x_1, \ldots, x_n\} | \exists y \in A : y \succ x\},$$
$$\mathrm{SMALLEREQ}(A, \succ) := \{x \in \{x_1, \ldots, x_n\} | \exists y \in A : y \succeq x\}.$$

It is clear that, if we obtain in a reduction step a representative $\phi = \mathrm{NF}(\succ, \varphi)$, we can exclude the sets in $\mathrm{FORWARD}(\mathrm{Var}(\phi))$ from the candidate sets P. But, as we see in the following lemma, we can exclude even more candidate sets.

Lemma 2. *Let $\succ$ be a lexicographical ordering and let $\phi = \mathrm{NF}(\succ, \varphi)$ be the corresponding normal form of φ. Then none of the sets $A \subseteq \mathrm{SMALLER}(\mathrm{Var}(\phi), \succ)$ can belong to a minimal solution to Problem 2.*

Proof. Assume the contrary, that is there is a minimal solution ψ with $\mathrm{Var}(\psi) \subseteq \mathrm{SMALLER}(\mathrm{Var}(\phi), \succ)$. It follows that, by the definition of lexicographical orderings, there is at least one $y \in \mathrm{Var}(\phi)$ with $y \succ \mathrm{Var}(\psi)$. Consequently, ψ is smaller than ϕ with respect to $\succ$ which cannot happen by the definition of the normal form. $\qquad \square$

---

[1] Note that $\lim_{n \to \infty} \frac{n!}{2^n} = \infty$, so it is more efficient to iterate through 2^n candidate sets than through $n!$ orderings.

If we also take the structure of the polynomials into account, we can improve Lemma 2 further. For this purpose, we look at the initial monomial of $\phi = \mathrm{NF}(\succ, \varphi)$ with respect to $\succ$, $M := in_\succ(\phi)$. We consider the sets $\mathrm{Var}(M)$ and $\mathrm{Comp}(M)$. Given a variable $x_i \in \{x_1, \ldots, x_n\}$ and a subset $S \subseteq \{x_1, \ldots, x_i\}$, let $S_{\succ x_i}$ be the set of variables in S bigger than x_i, i.e.

$$S_{\succ x_i} := \{x_j \in S | x_j \succ x_i\}.$$

Lemma 3. *Consider a lexicographical ordering $\succ$. Let $\phi = \mathrm{NF}(\succ, \varphi)$ and $M = in_\succ(\phi)$. If $x_i \in \mathrm{Var}(M)$, then any set $S \subseteq SMALLEREQ(\mathrm{Var}(\phi), \succ)$ with $x_i \notin S$ and $S \cap Comp_{\succ x_i}(M) = \emptyset$ cannot belong to a minimal solution to Problem 2.*

Proof. Note that $\prod_{j \in S} x_j \prec M$ as M involves x_i but $\prod_{j \in S} x_j$ only involves variables smaller than x_i. Then, the proof is analogous to Lemma 2 using the fact that any minimal solution involving only variables in S has monomials smaller or equal than $\prod_{j \in S} x_j \prec M$. Hence, ϕ is not minimal with respect to $\prec$.

In particular, Lemma 3 entails the case where the initial monomial of $\mathrm{NF}(\succ, \varphi)$ is a product of all variables occurring in $\mathrm{NF}(\succ, \varphi)$. In this case, for every subset $S \subseteq \mathrm{Var}(\phi) \subseteq SMALLEREQ(\mathrm{Var}(\phi), \succ)$ it holds $S \cap \mathrm{Comp}(M) = S \cap \mathrm{Comp}(\phi) = \emptyset$. Therefore, according to Lemma 3 $\mathrm{NF}(\succ, \varphi)$ is minimal.

For a lexicographical ordering $\succ$ and a normal form $\phi = \mathrm{NF}(\succ, \varphi)$ we can, using Lemma 3, exclude the families of sets in (1) from the set of candidates P.

$$\mathrm{BACKWARD}(S_i) \text{ with } x_i \in \mathrm{Var}(in_\succ(\phi)) \text{ and} \tag{1}$$
$$S_i := SMALLEREQ(\mathrm{Var}(\phi), \succ) \backslash (\{x_i\} \cup Comp_{\succ x_i}(in_\succ(\phi)))$$

We illustrate this fact with a small example:

Example 4. Consider a lexicographical ordering $\succ$ with $x_4 \succ \cdots \succ x_1$ and a normal form $\phi = \mathrm{NF}(\succ, \varphi)$ with initial monomial $x_4 x_2$. Then, we can exclude from P the sets in $\mathrm{BACKWARD}(\{x_3, x_2, x_1\})$ and $\mathrm{BACKWARD}(\{x_4, x_1\})$.

Note that if we consider, instead of lexicographical orderings, graded monomial orderings, then we obtain the following version of Lemma 3. This is useful to lower bound the number of variables in a minimal solution. Also, it could be useful when considering different optimality criteria.

Lemma 4. *Let $\succ$ be a graded monomial ordering [12, Ch. 8.4]. Then, the total degree d of $\mathrm{NF}(\succ, \varphi)$ is smaller or equal to the number of variables involved in any minimal representation of φ.*

Proof. Assume that φ has a representation involving less than d variables. Then, this representation has to have degree less than d (because every monomial is square-free). Hence, we get a contradiction because $\mathrm{NF}(\succ, \varphi)$ is not minimal.

We can now use the results above to construct Algorithm 1. In each step of our algorithm we choose a candidate set A of P and an ordering $\succ$ satisfying $A^c \succ A$. Then we compute the reduction of φ with respect to $\succ$ with the

Algorithm 1. compute_solutions(φ, $\{x_1, \ldots, x_n\}$, $\mathcal{I}$)

1: $P \leftarrow \mathscr{P}(\{x_1, \ldots, x_n\})$
2: $S \leftarrow \mathscr{P}(\{x_1, \ldots, x_n\})$
3: **while** $P \neq \emptyset$ **do**
4:    $A \leftarrow$ any set in P
5:    $\succ \leftarrow$ any lexicographical ordering satisfying $\mathrm{Comp}(A) \succ A$
6:    $\varphi \leftarrow \mathrm{NF}(\succ, \varphi)$
7:    $V \leftarrow \mathrm{Var}(\varphi)$
8:    $P \leftarrow P - \mathrm{FORWARDEQ}(V)$
9:    $S \leftarrow S - \mathrm{FORWARD}(V)$
10:    **for all** x_i in $\mathrm{Var}(in_{\prec}(\varphi))$ **do**
11:      $S_i \leftarrow$ Compute S_i according to Eq.(1)
12:      $P \leftarrow P - \mathrm{BACKWARD}(S_i)$
13:      $S \leftarrow S - \mathrm{BACKWARD}(S_i)$
14:    **end for**
15: **end while**
16: **return** S

---

corresponding Gröbner basis. Let us call the result ϕ. After each reduction in Algorithm 1 we exclude from P the sets that we already checked and the sets we can exclude with the results above. That is, we can exclude from P the candidate sets $\mathrm{FORWARDEQ}(\mathrm{Var}(\phi))$ (Line 9 in Algorithm 1) and according to Lemma 3 the family of sets $\mathrm{BACKWARD}(S_i)$ with $x_i \in \mathrm{Var}(in_{\prec}(\phi))$ where S_i is defined according to (1). The algorithm keeps doing this until the set of candidate sets is empty. To be able to return the solutions we keep simultaneously track of the set of potential solutions denoted by S. Initially this set equals P. But since we subtract from S not the set $\mathrm{FORWARDEQ}(\mathrm{Var}(\phi))$ but $\mathrm{FORWARD}(\mathrm{Var}(\phi))$ we keep some of the sets that we checked already in S. This guarantees that S will contain all solutions when P is empty.

5 Implementation and Benchmarking

When implementing Algorithm 1 the main difficulties we face is an effective handling of the candidate sets. In each step in the loop in Algorithm 1 we need to pick a new set A from the family of candidate sets P. Selecting a candidate set from P is not a trivial task since it structure can become very entangled. The subtraction of the sets $\mathrm{FORWARD}(\cdot)$ and $\mathrm{BACKWARD}(\cdot)$ from P can make the structure of the candidate sets very complicated. In practice this a very time-consuming part of the algorithm. To tackle this problem we use a specialized data structure – so-called Zero-suppressed decision diagram (ZDDs) [28,29] – to represent P. ZDDs are a type of Binary Decision Diagrams (BDDs). A binary decision diagram represents a Boolean function or a family of sets as a rooted directed acyclic graph. Specific reduction rules are used to obtain a compact, memory efficient representation. ZDDs can therefore effectively store families of sets. Furthermore, set operations can be computed directly on ZDDs.

This makes them an ideal tool for many combinatorial problems [28]. We refer to the literature for a more detailed introduction to ZDDs [28,29].

Our implementation in Python can be found at https://git.io/Jfmuc. For the Gröbner basis calculations as well as the ZDDs we used libraries from PolyBoRi (see [5]). The computation time for the network with 25 components considered in the case study in Sect. 6 was around 10 s on a personal computer with an intel core i5 vPro processor. Other networks we created for test purposes resulted in similar results (around 30 s for example1.py in the repository). However, computation time depends highly on the structure of the network and not only on its size. For a network even larger (example2.py in the repository with 38 components) computations took around two seconds while computations for a slightly different network of the same size (see example3.py in the repository) took around a minute. For a similar network (example4.py in the repository) we aborted the computation after one hour. In general, the complexity of computing Gröbner bases is highly influenced by algebraic properties such as the regularity of the vanishing ideal of the set we restrict our classifiers to. If the shapes of the sets in our algorithm are more regular (e.g. some components are fixed to zero or one) the number of candidate sets is reduced much faster by the algorithm. Similar, computations seem to be also often faster for networks with fewer regulatory links.

6   Case Study

Let us consider the Boolean model constructed in [8] modeling cell-fate decision. These models can be used to identify how and under which conditions the cell chooses between different types of cell deaths and survival. The complete model can be found in the BioModels database with the reference MODEL0912180000. It consists of 25 components. The corresponding Boolean function is depicted in Table 1. While in [8] the authors use a reduced model (see also [31]) to make their analysis more tractable, we can and do work with the complete model here.

The Boolean network depicted in Table 1 models the effect of cytokines such as TNF and FASL on cell death. In the Boolean model they correspond to input components. These cytokines can trigger cell death by apoptosis or necrosis (referred to as non-apoptotic cell death abbreviated by NonACD). Under different cellular conditions they lead to the activation of pro-survival signaling pathway(s). Consequently, the model distinguishes three phenotypes: Apoptosis, NonACD and Survival. Three corresponding signaling pathways are unified in their model. Finally, specific read-out components for the three phenotypes were defined. The activation of CASP3 is considered a marker for apoptosis. When MPT occurs and the level of ATP drops the cell enters non-apoptotic cell death. If NfκB is activated cells survive [8, p. 4]. This leads to the three classifiers in the model depicted in Table 2. Each classifier tells us to which cell fate (apoptosis, NonACD, survival) a state belongs.

We are interested in alternative classifiers on the set of attractors of the Boolean network. Let us denote the union of these attractors[2] with B (in agreement with the notation in Problem 2). In this case all attractors are steady states (see [8, p. 4] for details). For illustrating our results we computed the steady states of the network using GINsim [21] (see Fig. 1). But this is not necessary for our calculations here. However, we can see that the classifiers given in [8] indeed result in disjoint sets of phenotypes.

Since the Boolean network in Table 1 possesses only steady states as attractors we can represent the ideal $\mathcal{I}(B)$ in Problem 2 as $\langle f_1(x) + x_1, \ldots, f_n(x) + x_n \rangle$ where f is the Boolean function depicted in Table 1.

Next, we computed for each of the classifiers alternative representations. In Table 3, we present the nine different minimal representations on B of the classifier for NonACD. Among these options there are three ways how to construct a classifier based on one component (that is ATP, MPT or ROS). Also interestingly none of the components in the Boolean network is strictly necessary for the classification of the phenotypes. Consequently, there are potentially very different biological markers in the underlying modeled regulatory network. Despite this, there are some restrictions on the construction of the classifier, e.g., if we want to use the component Cytc, MOMP or SMAC we need to use also the component labeled as apoptosome. In total, the components useful for the classification of NonACD are ATP, CASP3, Cytc, MOMP, apoptosome, MPT, ROS and SMAC. The remaining 17 components are redundant for this purpose.

We obtain similar results for the other two classifiers. For apoptosis we found 17 alternative classifiers depicted in Table 4 involving the nine components (ATP, BAX, CASP8, Cytc, MOMP, SMAC, MPT, ROS, CASP3 and apoptosome). For the classifier for survival of the cell depicted in Table 5 we found much more alternative classifiers (84 alternative classifiers). Most classifiers depend on four components. But we can observe that each of the components IKK, BCL2, NFKB1, RIP1ub, XIAP, cFLIP can be used for classification. Computations for each of the three classifiers took around 10–30 s on a personal computer with an intel core i5 vPro processor in each case.

7 Possible Further Improvements

There is still some room for further improvement of the above algorithm. We address the third point in the beginning of Sect. 4. We can represent the lexicographical orderings on $\mathbb{B}(n)$ using weight vectors $w \in \mathbb{N}^n$. More precisely, let $\succ$ be any monomial ordering in $\mathbb{F}_2[x_1, \ldots, x_n]$ and $w \in \mathbb{N}^n$ any weight vector. Then we define $\succ_w$ as follows: for two monomials x^α and x^β, $\alpha, \beta \in \mathbb{N}^n$ we set

$$x^\alpha \succ_w x^\beta \Leftrightarrow w \cdot \alpha > w \cdot \beta \text{ or } (w \cdot \alpha = w \cdot \beta \text{ and } x^\alpha \succ x^\beta).$$

---

[2] An attractor of a Boolean network is a terminal strongly connected component of the corresponding state transition graph.

According to [36, Prop 1.11] for every monomial ordering $\succ$ and for every ideal in $\mathbb{F}_2[x_1, \ldots, x_n]$ there exists a non-negative integer vector $w \in \mathbb{N}^n$ s.t. $in_w(\mathcal{I}) = in_{\succ}(\mathcal{I})$ where $in_w(\mathcal{I})$ is the ideal generated by the initial forms $in_w(f), f \in \mathcal{I}$ – that is, the sum of monomials x^α in f which are maximal with respect to the inner product $\alpha \cdot w$. We say also in this case w *represents* $\succ$ *for* $\mathcal{I}$. The following lemma shows how we can construct weight vectors representing lexicographical

Table 1. Boolean network with 25 components given in [8].

Component	Update function
ATP	$1 + MPT$
BAX	$CASP8 \cdot (1 + BCL2)$
$BCL2$	$NFKB1$
$CASP3$	$(1 + XIAP) \cdot apoptosome$
$CASP8$	$((1 + DISCTNF) \cdot (1 + DISCFAS) \cdot CASP3 \cdot (1 + cFLIP) + (1 + DISCTNF) \cdot DISCFAS \cdot (1 + cFLIP) + (1 + DISCTNF) \cdot (1 + DISCFAS) \cdot CASP3 \cdot (1 + cFLIP) + (1 + DISCTNF) \cdot DISCFAS \cdot (1 + cFLIP)) \cdot DISCTNF \cdot (1 + cFLIP) + ((1 + DISCTNF) \cdot (1 + DISCFAS) \cdot CASP3 \cdot (1 + cFLIP) + (1 + DISCTNF) \cdot DISCFAS \cdot (1 + cFLIP) + (1 + DISCTNF) \cdot (1 + DISCFAS) \cdot CASP3 \cdot (1 + cFLIP) + (1 + DISCTNF) \cdot DISCFAS \cdot (1 + cFLIP)) + DISCTNF \cdot (1 + cFLIP)$
$Cytc$	$MOMP$
$DISCFAS$	$FASL \cdot FADD$
$DISCTNF$	$TNFR \cdot FADD$
$FADD$	$FADD$
$FASL$	$FASL$
IKK	$RIP1ub$
$MOMP$	$((1 + BAX) \cdot MPT) \cdot BAX + ((1 + BAX) \cdot MPT) + BAX$
MPT	$(1 + BCL2) \cdot ROS$
$NFKB1$	$IKK \cdot (1 + CASP3)$
$NonACD$	$1 + ATP$
$RIP1$	$(1 + TNFR) \cdot DISCFAS \cdot (1 + CASP8) \cdot TNFR \cdot (1 + CASP8) + (1 + TNFR) \cdot DISCFAS \cdot (1 + CASP8) + TNFR \cdot (1 + CASP8)$
$RIP1K$	$RIP1$
$RIP1ub$	$RIP1 \cdot cIAP$
ROS	$(1 + RIP1K) \cdot MPT \cdot NFKB1 \cdot RIP1K \cdot (1 + NFKB1) + RIP1K \cdot (1 + NFKB1) + (1 + RIP1K) \cdot MPT \cdot NFKB1$
$SMAC$	$MOMP$
TNF	TNF
$TNFR$	TNF
$XIAP$	$(1 + SMAC) \cdot NFKB1$
$apoptosome$	$ATP \cdot Cytc \cdot (1 + XIAP)$
$cFLIP$	$NFKB1$
$cIAP$	$(1 + NFKB1) \cdot (1 + SMAC) \cdot cIAP \cdot NFKB1 \cdot (1 + SMAC) + (1 + NFKB1) \cdot (1 + SMAC) \cdot cIAP + NFKB1 \cdot (1 + SMAC)$

Table 2. Classifiers for the Boolean network depicted in Table 1.

Bio. interpretation of classifier	Classifier
Survival	$NFKB1$
Apoptosis	$CASP3$
NonACD	$1 + ATP$

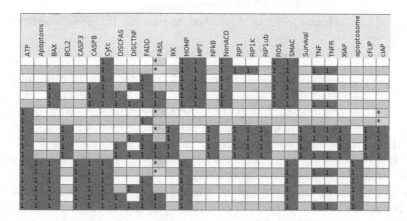

Fig. 1. 27 steady states of the complete BN computed with GINsim [21]. Components marked with * can either be set to zero or one. Steady states are grouped into phenotypes. Six steady states are not corresponding to any phenotype.

orderings. Note that each ideal in $\mathbb{B}(n)$ is a principal ideal[3] and each ideal $\langle f \rangle$ in $\mathbb{B}(n)$ corresponds to an ideal $\langle f, x_1^2 + x_1, \ldots, x_n^2 + x_n \rangle$ in $\mathbb{F}_2[x_1, \ldots, x_n]$. Let us also for simplicity consider the lexicographical ordering defined by $x_n \succ x_{n-1} \succ \cdots \succ x_1$. The general case can be obtained by permutation.

Lemma 5. *Consider an ideal of the form* $\mathcal{I} = \langle f, x_1^2 + x_1, \ldots, x_n^2 + x_n \rangle \subseteq \mathbb{F}_2[x_1, \ldots, x_n]$ *and the lexicographical ordering* $\succ$ *defined by* $x_n \succ x_{n-1} \succ \cdots \succ x_1$. *Then* $\succ$ *is represented by the weight vector* $w \in \mathbb{Q}^n$ *with* $w_k = 1 + \sum_{j=1}^{k-1} w_j$ *or alternatively* $w_k = 2^{k-1}$.

Proof. Let $\succ$ be as above and w the corresponding weight vector defined there. We first show that $in_w(\mathcal{I}) \subseteq in_\succ(\mathcal{I})$. This is true since by definition of $\mathcal{I}$ we know that the monomials $x_1^2, \ldots, x_n^2$ are contained in $in_w(\mathcal{I})$ (and obviously in $in_\succ(\mathcal{I})$). Consequently, we can represent $in_w(\mathcal{I})$ in the form $in_w(\mathcal{I}) = \langle in_w(g_1), \ldots, in_w(g_k), x_1^2, \ldots, x_n^2 \rangle$ with square free polynomials $g_1, \ldots, g_k$ in $\mathcal{I}$. Now by construction of w for square free polynomials $g \in \mathbb{F}_2[x_1, \ldots, x_n]$ the equality $in_w(g) = in_\succ(g)$ holds. It follows $in_w(\mathcal{I}) \subseteq in_\succ(\mathcal{I})$. Analogously it holds $in_\succ(\mathcal{I}) \subseteq in_w(\mathcal{I})$.

Let us call a reduced Gröbner basis with a distinguished initial monomial a *marked Gröbner basis* in accordance with [13, p. 428]. Next, we form equivalence classes of weight vectors which will lead to the same marked Gröbner bases.

Definition 3 ([13, p. 429]). *Let* G *be any marked Gröbner basis for an ideal* $\mathcal{I}$ *consisting of* t *polynomials*

$$g_i = x^{\alpha(i)} + \sum_\beta c_{i,\beta} x^\beta,$$

---

[3] This follows from the identity $f \cdot (f + g + f \cdot g) = f$ for $f, g \in \mathbb{B}(n)$.

where $i \in \{1, \ldots, t\}$ and $x^{\alpha(i)}$ is the initial monomial. We denote with C_G the set

$$C_G = \left\{ w \in (\mathbb{R}^n)^+ : (\alpha(i) - \beta) \cdot w \geq 0 \text{ whenever } c_{i,\beta} \neq 0 \right\}.$$

We can combine Lemma 5 and Definition 3 to potentially improve our algorithm. If we computed for a lexicographical ordering in Algorithm 1 a Gröbner basis G we can compute and save the equivalence class C_G. Now proceeding with the algorithm, for a new lexicographical ordering we need to check, we can create the corresponding weight vector w using Lemma 5 and check if w is in any of the previously computed equivalence classes C_G. If this is the case we can use the result of the previous computation.

Another aspect of the algorithm we can improve is the conversion of Gröbner bases. For an ideal $\mathcal{I}(B) = \langle f, x_1^2 + x_1, \ldots, x_n^2 + x_n \rangle$ the ring $\mathbb{F}_2[x_1, \ldots, x_n]/\mathcal{I}(B)$ is zero-dimensional, and so a finite dimensional vector space. Therefore, it is possible to use linear algebra for the conversion of Gröbner bases. This leads to the Faugère-Gianni-Lazard-Mora algorithm (FLGM algorithm) [13,16, p. 49].

8   Conclusion

We reformulated Problem 1 into the language of algebraic geometry. To do so we described the set of potential classifiers using residue classes modulo the vanishing ideal of the attractors or steady states of the Boolean network. This enabled us to construct an algorithm using normal forms to compute optimal classifiers. Subsequently we demonstrated the usefulness of this approach by creating an algorithm that produces the minimal solutions to Problem 1. We showed that it is possible to apply this algorithm to a model for cell-fate decision with 25 components from [8]. Especially, in combination with reduction algorithms for Boolean networks this allows us to investigate larger networks.

We hope that it will be also possible to exploit the algebraic reformulation further to speed up computations to tackle even larger networks. Some parts in the algorithm can be improved to obtain potentially faster computation times. For example the conversion between different Gröbner bases can be done more efficiently using the FLGM algorithm (see [13,16, p. 49–54]) which uses linear algebra for the conversion of Gröbner bases. Since at the moment of writing this article there was no implementation of this available in PolyBoRi we did not use this potential improvement.

However, the main bottleneck for the speed of the algorithm seems to be the enumeration of possible orderings (or more precisely candidate sets). Therefore, we believe that this will not lead to a significant increase in speed but this remains to be tested. Instead we believe, that for larger networks heuristics should be investigated. Here ideas from the machine learning community could be useful. Potential crosslinks to classifications problems considered there should be explored in the future.

Also different optimality criteria for picking classifiers might be useful. For example one could try to attribute measurement costs to components and pick polynomial orderings which lead to optimal results in such a context as well.

In the introduction we mentioned the relationship of Problem 1 to problems in the LAD community. There one starts typically with a data set of Boolean vectors. Here we focused on the case where our sets to be classifies are implicitly given. However, approaches developed for interpolation of Boolean polynomials from data points such as [6] could be used in the future to tailor our approach to such scenarios as well.

Appendix

Table 3. 9 different representations of classifier for NonACD (see Sect. 6).

Components	Expression
ATP	$ATP + 1$
$CASP3, Cytc$	$Cytc + CASP3$
$CASP3, MOMP$	$MOMP + CASP3$
$CASP3, SMAC$	$SMAC + CASP3$
$Cytc, apoptosome$	$Cytc + apoptosome$
$MOMP, apoptosome$	$apoptosome + MOMP$
MPT	MPT
ROS	ROS
$SMAC, apoptosome$	$apoptosome + SMAC$

Table 4. 17 different representations of the classifier for apoptosis (see Sect. 6)

Components	Expression
ATP, BAX	$BAX \cdot ATP$
$ATP, CASP8$	$ATP \cdot CASP8$
$ATP, Cytc$	$Cytc + ATP + 1$
$ATP, MOMP$	$ATP + MOMP + 1$
$ATP, SMAC$	$ATP + SMAC + 1$
BAX, MPT	$BAX \cdot MPT + BAX$
BAX, ROS	$BAX \cdot ROS + BAX$
$CASP3,$	$CASP3$
$CASP8, MPT$	$MPT \cdot CASP8 + CASP8$
$CASP8, ROS$	$ROS \cdot CASP8 + CASP8$
$Cytc, MPT$	$Cytc + MPT$
$Cytc, ROS$	$Cytc + ROS$
$MOMP, MPT$	$MPT + MOMP$
$MOMP, ROS$	$ROS + MOMP$
$MPT, SMAC$	$MPT + SMAC$
$ROS, SMAC$	$ROS + SMAC$
$apoptosome,$	$apoptosome$

Table 5. 84 different representations of the classifier of survival of the cell. (see Sect. 6)

Components	Expression
$ATP, BAX,$ $DISCFAS, TNF$	$DISCFAS \cdot BAX + DISCFAS \cdot ATP + DISCFAS \cdot$ $TNF + BAX \cdot ATP + BAX \cdot TNF + BAX + ATP \cdot TNF$
$ATP, BAX,$ $DISCFAS,$ $TNFR$	$DISCFAS \cdot BAX + DISCFAS \cdot ATP + DISCFAS \cdot$ $TNFR + BAX \cdot ATP + BAX \cdot TNFR + BAX + ATP \cdot TNFR$
$ATP, BAX,$ $FADD, FASL,$ TNF	$BAX \cdot FASL \cdot FADD + BAX \cdot TNF + ATP \cdot FASL \cdot$ $FADD \cdot TNF + ATP \cdot FADD \cdot TNF + ATP \cdot TNF +$ $FASL \cdot FADD + FADD \cdot TNF$
$ATP, BAX,$ $FADD, FASL,$ $TNFR$	$BAX \cdot TNFR + BAX \cdot FASL \cdot FADD + ATP \cdot TNFR \cdot$ $FASL \cdot FADD + ATP \cdot TNFR \cdot FADD + ATP \cdot TNFR +$ $TNFR \cdot FADD + FASL \cdot FADD$
$ATP, CASP3,$ $DISCFAS, TNF$	$DISCFAS \cdot ATP \cdot TNF + DISCFAS \cdot ATP + DISCFAS \cdot$ $CASP3 + ATP \cdot TNF + TNF \cdot CASP3$
$ATP, CASP3,$ $DISCFAS,$ $TNFR$	$DISCFAS \cdot ATP \cdot TNFR + DISCFAS \cdot ATP +$ $DISCFAS \cdot CASP3 + ATP \cdot TNFR + TNFR \cdot CASP3$
$ATP, CASP3,$ $FADD, FASL,$ TNF	$ATP \cdot FASL \cdot FADD \cdot TNF + ATP \cdot FASL \cdot FADD +$ $ATP \cdot TNF + FASL \cdot FADD \cdot CASP3 + TNF \cdot CASP3$
$ATP, CASP3,$ $FADD, FASL,$ $TNFR$	$ATP \cdot TNFR \cdot FASL \cdot FADD + ATP \cdot TNFR + ATP \cdot$ $FASL \cdot FADD + TNFR \cdot CASP3 + FASL \cdot FADD \cdot CASP3$
$ATP, CASP8,$ $DISCFAS, TNF$	$DISCFAS \cdot TNF \cdot CASP8 + DISCFAS \cdot TNF + DISCFAS \cdot$ $CASP8 + DISCFAS + ATP \cdot TNF \cdot CASP8 + ATP \cdot TNF$
$ATP, CASP8,$ $DISCFAS,$ $TNFR$	$DISCFAS \cdot TNFR \cdot CASP8 + DISCFAS \cdot TNFR +$ $DISCFAS \cdot CASP8 + DISCFAS + ATP \cdot TNFR \cdot$ $CASP8 + ATP \cdot TNFR$
$ATP, CASP8,$ $FADD, FASL,$ TNF	$ATP \cdot TNF \cdot CASP8 + ATP \cdot TNF + FASL \cdot FADD \cdot$ $TNF \cdot CASP8 + FASL \cdot FADD \cdot TNF + FASL \cdot FADD \cdot$ $CASP8 + FASL \cdot FADD$
$ATP, CASP8,$ $FADD, FASL,$ $TNFR$	$ATP \cdot TNFR + ATP \cdot FASL \cdot CASP8 + ATP \cdot CASP8 +$ $TNFR \cdot FASL \cdot FADD + TNFR \cdot CASP8 + FASL \cdot FADD \cdot$ $CASP8 + FASL \cdot FADD + FASL \cdot CASP8 + CASP8$
$ATP, DISCFAS,$ $TNF, apoptosome$	$DISCFAS \cdot ATP \cdot TNF + DISCFAS \cdot ATP + DISCFAS \cdot$ $apoptosome + ATP \cdot TNF + apoptosome \cdot TNF$
$ATP, DISCFAS,$ $TNFR,$ $apoptosome$	$DISCFAS \cdot ATP \cdot TNFR + DISCFAS \cdot ATP + DISCFAS \cdot$ $apoptosome + ATP \cdot TNFR + apoptosome \cdot TNFR$
$ATP, FADD,$ $FASL, TNF,$ $apoptosome$	$ATP \cdot FASL \cdot FADD \cdot TNF + ATP \cdot FASL \cdot FADD + ATP \cdot$ $TNF + apoptosome \cdot FASL \cdot FADD + apoptosome \cdot TNF$

(continued)

Table 5. (*continued*)

Components	Expression
ATP, FADD, FASL, TNFR, apoptosome	$ATP \cdot TNFR \cdot FASL \cdot FADD + ATP \cdot TNFR + ATP \cdot FASL \cdot FADD + apoptosome \cdot TNFR + apoptosome \cdot FASL \cdot FADD$
ATP, RIP1	$ATP \cdot RIP1$
ATP, RIP1K	$ATP \cdot RIP1K$
BAX, DISCFAS, MPT, TNF	$DISCFAS \cdot BAX + DISCFAS \cdot MPT + DISCFAS \cdot TNF + DISCFAS + BAX \cdot MPT + BAX \cdot TNF + MPT \cdot TNF + TNF$
BAX, DISCFAS, MPT, TNFR	$DISCFAS \cdot BAX + DISCFAS \cdot TNFR + DISCFAS \cdot MPT + DISCFAS + BAX \cdot TNFR + BAX \cdot MPT + TNFR \cdot MPT + TNFR$
BAX, DISCFAS, ROS, TNF	$DISCFAS \cdot BAX + DISCFAS \cdot ROS + DISCFAS \cdot TNF + DISCFAS + BAX \cdot ROS + BAX \cdot TNF + ROS \cdot TNF + TNF$
BAX, DISCFAS, ROS, TNFR	$DISCFAS \cdot BAX + DISCFAS \cdot TNFR + DISCFAS \cdot ROS + DISCFAS + BAX \cdot TNFR + BAX \cdot ROS + TNFR \cdot ROS + TNFR$
BAX, FADD, FASL, MPT, TNF	$BAX \cdot FASL \cdot FADD + BAX \cdot TNF + MPT \cdot FASL \cdot FADD \cdot TNF + MPT \cdot FADD \cdot TNF + MPT \cdot TNF + FASL \cdot FADD \cdot TNF + FASL \cdot FADD + TNF$
BAX, FADD, FASL, MPT, TNFR	$BAX \cdot TNFR + BAX \cdot FASL \cdot FADD + TNFR \cdot MPT \cdot FASL \cdot FADD + TNFR \cdot MPT \cdot FADD + TNFR \cdot MPT + TNFR \cdot FASL \cdot FADD + TNFR + FASL \cdot FADD$
BAX, FADD, FASL, ROS, TNF	$BAX \cdot FASL \cdot FADD + BAX \cdot TNF + FASL \cdot FADD \cdot ROS \cdot TNF + FASL \cdot FADD \cdot TNF + FASL \cdot FADD + FADD \cdot ROS \cdot TNF + ROS \cdot TNF + TNF$
BAX, FADD, FASL, ROS, TNFR	$BAX \cdot TNFR + BAX \cdot FASL \cdot FADD + TNFR \cdot FASL \cdot FADD \cdot ROS + TNFR \cdot FASL \cdot FADD + TNFR \cdot FADD \cdot ROS + TNFR \cdot ROS + TNFR + FASL \cdot FADD$
BCL2	$BCL2$
CASP3, DISCFAS, MPT, TNF	$DISCFAS \cdot MPT \cdot TNF + DISCFAS \cdot MPT + DISCFAS \cdot TNF + DISCFAS \cdot CASP3 + DISCFAS + MPT \cdot TNF + TNF \cdot CASP3 + TNF$
CASP3, DISCFAS, MPT, TNFR	$DISCFAS \cdot TNFR \cdot MPT + DISCFAS \cdot TNFR + DISCFAS \cdot MPT + DISCFAS \cdot CASP3 + DISCFAS + TNFR \cdot MPT + TNFR \cdot CASP3 + TNFR$
CASP3, DISCFAS, ROS, TNF	$DISCFAS \cdot ROS \cdot TNF + DISCFAS \cdot ROS + DISCFAS \cdot TNF + DISCFAS \cdot CASP3 + DISCFAS + ROS \cdot TNF + TNF \cdot CASP3 + TNF$
CASP3, DISCFAS, ROS, TNFR	$DISCFAS \cdot TNFR \cdot ROS + DISCFAS \cdot TNFR + DISCFAS \cdot ROS + DISCFAS \cdot CASP3 + DISCFAS + TNFR \cdot ROS + TNFR \cdot CASP3 + TNFR$
CASP3, FADD, FASL, MPT, TNF	$MPT \cdot FASL \cdot FADD \cdot TNF + MPT \cdot FASL \cdot FADD + MPT \cdot TNF + FASL \cdot FADD \cdot TNF + FASL \cdot FADD \cdot CASP3 + FASL \cdot FADD + TNF \cdot CASP3 + TNF$

(*continued*)

Table 5. (*continued*)

Components	Expression
CASP3, FADD, FASL, MPT, TNFR	$TNFR \cdot MPT \cdot FASL \cdot FADD + TNFR \cdot MPT + TNFR \cdot FASL \cdot FADD + TNFR \cdot CASP3 + TNFR + MPT \cdot FASL \cdot FADD + FASL \cdot FADD \cdot CASP3 + FASL \cdot FADD$
CASP3, FADD, FASL, ROS, TNF	$FASL \cdot FADD \cdot ROS \cdot TNF + FASL \cdot FADD \cdot ROS + FASL \cdot FADD \cdot TNF + FASL \cdot FADD \cdot CASP3 + FASL \cdot FADD + ROS \cdot TNF + TNF \cdot CASP3 + TNF$
CASP3, FADD, FASL, ROS, TNFR	$TNFR \cdot FASL \cdot FADD \cdot ROS + TNFR \cdot FASL \cdot FADD + TNFR \cdot ROS + TNFR \cdot CASP3 + TNFR + FASL \cdot FADD \cdot ROS + FASL \cdot FADD \cdot CASP3 + FASL \cdot FADD$
CASP8, DISCFAS, MPT, TNF	$DISCFAS \cdot TNF \cdot CASP8 + DISCFAS \cdot TNF + DISCFAS \cdot CASP8 + DISCFAS + MPT \cdot TNF \cdot CASP8 + MPT \cdot TNF + TNF \cdot CASP8 + TNF$
CASP8, DISCFAS, MPT, TNFR	$DISCFAS \cdot TNFR + DISCFAS \cdot MPT + DISCFAS \cdot CASP8 + DISCFAS + TNFR \cdot MPT + TNFR \cdot CASP8 + TNFR + MPT \cdot CASP8$
CASP8, DISCFAS, ROS, TNF	$DISCFAS \cdot TNF \cdot CASP8 + DISCFAS \cdot TNF + DISCFAS \cdot CASP8 + DISCFAS + ROS \cdot TNF \cdot CASP8 + ROS \cdot TNF + TNF \cdot CASP8 + TNF$
CASP8, DISCFAS, ROS, TNFR	$DISCFAS \cdot TNFR + DISCFAS \cdot ROS + DISCFAS \cdot CASP8 + DISCFAS + TNFR \cdot ROS + TNFR \cdot CASP8 + TNFR + ROS \cdot CASP8$
CASP8, FADD, FASL, MPT, TNF	$MPT \cdot TNF \cdot CASP8 + MPT \cdot TNF + FASL \cdot FADD \cdot TNF \cdot CASP8 + FASL \cdot FADD \cdot TNF + FASL \cdot FADD \cdot CASP8 + FASL \cdot FADD + TNF \cdot CASP8 + TNF$
CASP8, FADD, FASL, MPT, TNFR	$TNFR \cdot MPT + TNFR \cdot FASL \cdot FADD + TNFR \cdot CASP8 + TNFR + MPT \cdot FASL \cdot CASP8 + MPT \cdot CASP8 + FASL \cdot FADD \cdot CASP8 + FASL \cdot FADD$
CASP8, FADD, FASL, ROS, TNF	$FASL \cdot FADD \cdot TNF + FASL \cdot FADD \cdot CASP8 + FASL \cdot FADD + FASL \cdot ROS \cdot CASP8 + ROS \cdot TNF + ROS \cdot CASP8 + TNF \cdot CASP8 + TNF$
CASP8, FADD, FASL, ROS, TNFR	$TNFR \cdot FASL \cdot FADD + TNFR \cdot ROS + TNFR \cdot CASP8 + TNFR + FASL \cdot FADD \cdot CASP8 + FASL \cdot FADD + FASL \cdot ROS \cdot CASP8 + ROS \cdot CASP8$
Cytc, DISCFAS, TNF	$Cytc \cdot DISCFAS \cdot TNF + Cytc \cdot DISCFAS + Cytc \cdot TNF + DISCFAS \cdot TNF + DISCFAS + TNF$
Cytc, DISCFAS, TNFR	$Cytc \cdot DISCFAS \cdot TNFR + Cytc \cdot DISCFAS + Cytc \cdot TNFR + DISCFAS \cdot TNFR + DISCFAS + TNFR$
Cytc, FADD, FASL, TNF	$Cytc \cdot FASL \cdot FADD \cdot TNF + Cytc \cdot FASL \cdot FADD + Cytc \cdot TNF + FASL \cdot FADD \cdot TNF + FASL \cdot FADD + TNF$

(*continued*)

Table 5. (*continued*)

Components	Expression
Cytc, FADD, FASL, TNFR	$Cytc \cdot TNFR \cdot FASL \cdot FADD + Cytc \cdot TNFR + Cytc \cdot FASL \cdot FADD + TNFR \cdot FASL \cdot FADD + TNFR + FASL \cdot FADD$
Cytc, RIP1	$Cytc \cdot RIP1 + RIP1$
Cytc, RIP1K	$Cytc \cdot RIP1K + RIP1K$
DISCFAS, MOMP, TNF	$DISCFAS \cdot MOMP \cdot TNF + DISCFAS \cdot MOMP + DISCFAS \cdot TNF + DISCFAS + MOMP \cdot TNF + TNF$
DISCFAS, MOMP, TNFR	$DISCFAS \cdot TNFR \cdot MOMP + DISCFAS \cdot TNFR + DISCFAS \cdot MOMP + DISCFAS + TNFR \cdot MOMP + TNFR$
DISCFAS, MPT, TNF, apoptosome	$DISCFAS \cdot apoptosome + DISCFAS \cdot MPT \cdot TNF + DISCFAS \cdot MPT + DISCFAS \cdot TNF + DISCFAS + apoptosome \cdot TNF + MPT \cdot TNF + TNF$
DISCFAS, MPT, TNFR, apoptosome	$DISCFAS \cdot apoptosome + DISCFAS \cdot TNFR \cdot MPT + DISCFAS \cdot TNFR + DISCFAS \cdot MPT + DISCFAS + apoptosome \cdot TNFR + TNFR \cdot MPT + TNFR$
DISCFAS, ROS, TNF, apoptosome	$DISCFAS \cdot apoptosome + DISCFAS \cdot ROS \cdot TNF + DISCFAS \cdot ROS + DISCFAS \cdot TNF + DISCFAS + apoptosome \cdot TNF + ROS \cdot TNF + TNF$
DISCFAS, ROS, TNFR, apoptosome	$DISCFAS \cdot apoptosome + DISCFAS \cdot TNFR \cdot ROS + DISCFAS \cdot TNFR + DISCFAS \cdot ROS + DISCFAS + apoptosome \cdot TNFR + TNFR \cdot ROS + TNFR$
DISCFAS, SMAC, TNF	$DISCFAS \cdot SMAC \cdot TNF + DISCFAS \cdot SMAC + DISCFAS \cdot TNF + DISCFAS + SMAC \cdot TNF + TNF$
DISCFAS, SMAC, TNFR	$DISCFAS \cdot TNFR \cdot SMAC + DISCFAS \cdot TNFR + DISCFAS \cdot SMAC + DISCFAS + TNFR \cdot SMAC + TNFR$
DISCFAS, TNF, cIAP	$DISCFAS \cdot TNF \cdot cIAP + DISCFAS \cdot cIAP + TNF \cdot cIAP$
DISCFAS, TNFR, cIAP	$DISCFAS \cdot TNFR \cdot cIAP + DISCFAS \cdot cIAP + TNFR \cdot cIAP$
FADD, FASL, MOMP, TNF	$FASL \cdot FADD \cdot MOMP \cdot TNF + FASL \cdot FADD \cdot MOMP + FASL \cdot FADD \cdot TNF + FASL \cdot FADD + MOMP \cdot TNF + TNF$
FADD, FASL, MOMP, TNFR	$TNFR \cdot FASL \cdot FADD \cdot MOMP + TNFR \cdot FASL \cdot FADD + TNFR \cdot MOMP + TNFR + FASL \cdot FADD \cdot MOMP + FASL \cdot FADD$
FADD, FASL, MPT, TNF, apoptosome	$apoptosome \cdot FASL \cdot FADD + apoptosome \cdot TNF + MPT \cdot FASL \cdot FADD \cdot TNF + MPT \cdot FASL \cdot FADD + MPT \cdot TNF + FASL \cdot FADD \cdot TNF + FASL \cdot FADD + TNF$
FADD, FASL, MPT, TNFR, apoptosome	$apoptosome \cdot TNFR + apoptosome \cdot FASL \cdot FADD + TNFR \cdot MPT \cdot FASL \cdot FADD + TNFR \cdot MPT + TNFR \cdot FASL \cdot FADD + TNFR + MPT \cdot FASL \cdot FADD + FASL \cdot FADD$

(*continued*)

Table 5. (*continued*)

Components	Expression
$FADD, FASL,$ $ROS, TNF,$ $apoptosome$	$apoptosome \cdot FASL \cdot FADD + apoptosome \cdot TNF + FASL \cdot$ $FADD \cdot ROS \cdot TNF + FASL \cdot FADD \cdot ROS + FASL \cdot$ $FADD \cdot TNF + FASL \cdot FADD + ROS \cdot TNF + TNF$
$FADD, FASL,$ $ROS, TNFR,$ $apoptosome$	$apoptosome \cdot TNFR + apoptosome \cdot FASL \cdot FADD + TNFR \cdot$ $FASL \cdot FADD \cdot ROS + TNFR \cdot FASL \cdot FADD + TNFR \cdot$ $ROS + TNFR + FASL \cdot FADD \cdot ROS + FASL \cdot FADD$
$FADD, FASL,$ $SMAC, TNF$	$FASL \cdot FADD \cdot SMAC \cdot TNF + FASL \cdot FADD \cdot SMAC +$ $FASL \cdot FADD \cdot TNF + FASL \cdot FADD + SMAC \cdot TNF + TNF$
$FADD, FASL,$ $SMAC, TNFR$	$TNFR \cdot FASL \cdot FADD \cdot SMAC + TNFR \cdot FASL \cdot$ $FADD + TNFR \cdot SMAC + TNFR + FASL \cdot FADD \cdot$ $SMAC + FASL \cdot FADD$
$FADD, FASL,$ $TNF, cIAP$	$FASL \cdot FADD \cdot TNF \cdot cIAP + FASL \cdot FADD \cdot cIAP +$ $TNF \cdot cIAP$
$FADD, FASL,$ $TNFR, cIAP$	$TNFR \cdot FASL \cdot FADD \cdot cIAP + TNFR \cdot cIAP + FASL \cdot$ $FADD \cdot cIAP$
IKK	IKK
$MOMP, RIP1$	$MOMP \cdot RIP1 + RIP1$
$MOMP, RIP1K$	$RIP1K \cdot MOMP + RIP1K$
$MPT, RIP1$	$MPT \cdot RIP1 + RIP1$
$MPT, RIP1K$	$RIP1K \cdot MPT + RIP1K$
$NFKB1$	$NFKB1$
$RIP1, ROS$	$ROS \cdot RIP1 + RIP1$
$RIP1, SMAC$	$SMAC \cdot RIP1 + RIP1$
$RIP1, cIAP$	$RIP1 \cdot cIAP$
$RIP1K, ROS$	$RIP1K \cdot ROS + RIP1K$
$RIP1K, SMAC$	$RIP1K \cdot SMAC + RIP1K$
$RIP1K, cIAP$	$RIP1K \cdot cIAP$
$RIP1ub$	$RIP1ub$
$XIAP$	$XIAP$
$cFLIP$	$cFLIP$

References

1. Albert, R., Thakar, J.: Boolean modeling: a logic-based dynamic approach for understanding signaling and regulatory networks and for making useful predictions. Wiley Interdiscip. Rev.: Syst. Biol. Med. **6**(5), 353–369 (2014)
2. Alexe, S., Blackstone, E., Hammer, P.L., Ishwaran, H., Lauer, M.S., Snader, C.E.P.: Coronary risk prediction by logical analysis of data. Ann. Oper. Res. **119**(1–4), 15–42 (2003)

3. Bonzanni, N., et al.: Hard-wired heterogeneity in blood stem cells revealed using a dynamic regulatory network model. Bioinformatics **29**(13), i80–i88 (2013)
4. Boros, E., Hammer, P.L., Ibaraki, T., Kogan, A., Mayoraz, E., Muchnik, I.: An implementation of logical analysis of data. IEEE Trans. Knowl. Data Eng. **12**(2), 292–306 (2000)
5. Brickenstein, M., Dreyer, A.: Polybori: a framework for gröbner-basis computations with Boolean polynomials. J. Symb. Comput. **44**(9), 1326–1345 (2009)
6. Brickenstein, M., Dreyer, A.: Gröbner-free normal forms for Boolean polynomials. J. Symb. Comput. **48**, 37–53 (2013)
7. Buchberger, B.: Applications of Gröbner bases in non-linear computational geometry. In: Rice, J.R. (ed.) Mathematical Aspects of Scientific Software. The IMA Volumes in Mathematics and its Applications, pp. 59–87. Springer, New York (1988). https://doi.org/10.1007/978-1-4684-7074-1_3
8. Calzone, L., et al.: Mathematical modelling of cell-fate decision in response to death receptor engagement. PLOS Comput. Biol. **6**(3), 1–15 (2010). https://doi.org/10.1371/journal.pcbi.1000702
9. Chaouiya, C., Remy, E., Mossé, B., Thieffry, D.: Qualitative analysis of regulatory graphs: a computational tool based on a discrete formal framework. In: Benvenuti, L., De Santis, A., Farina, L. (eds.) Positive Systems. Lecture Notes in Control and Information Science, vol. 294. Springer, Heidelberg. https://doi.org/10.1007/978-3-540-44928-7_17
10. Cheng, D., Qi, H.: Controllability and observability of Boolean control networks. Automatica **45**(7), 1659–1667 (2009)
11. Chikalov, I., et al.: Logical analysis of data: theory, methodology and applications. In: Three Approaches to Data Analysis. Intelligent Systems Reference Library, vol. 41. Springer, Heidelberg (2013). https://doi.org/10.1007/978-3-642-28667-4_3
12. Cox, D.A., Little, J., O'Shea, D.: Ideals, Varieties, and Algorithms. UTM. Springer, Cham (2015). https://doi.org/10.1007/978-3-319-16721-3
13. Cox, D.A., Little, J., O'Shea, D.: Using Algebraic Geometry (2004)
14. Dickenstein, A., Millán, M.P., Shiu, A., Tang, X.: Multistationarity in structured reaction networks. Bull. Math. Biol. **81**(5), 1527–1581 (2019). https://doi.org/10.1007/s11538-019-00572-6
15. Drton, M., Sturmfels, B., Sullivant, S.: Lectures on Algebraic Statistics. Oberwolfach Seminars, Birkhäuser Basel (2009). https://www.springer.com/gp/book/9783764389048
16. Faugère, J.C., Gianni, P., Lazard, D., Mora, T.: Efficient computation of zero-dimensional Gröbner bases by change of ordering. J. Symb. Comput. **16**(4), 329–344 (1993)
17. Faugère, J.-C., Joux, A.: Algebraic cryptanalysis of hidden field equation (HFE) cryptosystems using Gröbner bases. In: Boneh, D. (ed.) CRYPTO 2003. LNCS, vol. 2729, pp. 44–60. Springer, Heidelberg (2003). https://doi.org/10.1007/978-3-540-45146-4_3
18. Fauré, A., Vreede, B.M., Sucena, É., Chaouiya, C.: A discrete model of drosophila eggshell patterning reveals cell-autonomous and juxtacrine effects. PLoS Comput. Biol. **10**(3), e1003527 (2014)
19. Gao, S., Platzer, A., Clarke, E.M.: Quantifier elimination over finite fields using Gröbner bases. In: Winkler, F. (ed.) CAI 2011. LNCS, vol. 6742, pp. 140–157. Springer, Heidelberg (2011). https://doi.org/10.1007/978-3-642-21493-6_9
20. Germundsson, R.: Basic results on ideals and varieties in finite fields. Tech. rep. S-581 83 (1991)

21. Gonzalez, A.G., Naldi, A., Sánchez, L., Thieffry, D., Chaouiya, C.: GIN-sim: a software suite for the qualitative modelling, simulation and analysis of regulatory networks. Biosystems **84**(2), 91–100 (2006). https://doi.org/10.1016/j.biosystems.2005.10.003. http://www.sciencedirect.com/science/article/pii/S0303264705001693

22. González, A., Chaouiya, C., Thieffry, D.: Logical modelling of the role of the Hh pathway in the patterning of the drosophila wing disc. Bioinformatics **24**(16), i234–i240 (2008)

23. Hammer, P.L., Bonates, T.O.: Logical analysis of data - an overview: from combinatorial optimization to medical applications. Ann. Oper. Res. **148**(1), 203–225 (2006)

24. Jarrah, A.S., Laubenbacher, R.: Discrete models of biochemical networks: the toric variety of nested canalyzing functions. In: Anai, H., Horimoto, K., Kutsia, T. (eds.) AB 2007. LNCS, vol. 4545, pp. 15–22. Springer, Heidelberg (2007). https://doi.org/10.1007/978-3-540-73433-8_2

25. Laubenbacher, R., Stigler, B.: A computational algebra approach to the reverse engineering of gene regulatory networks. J. Theor. Biol. **229**(4), 523–537 (2004). https://doi.org/10.1016/j.jtbi.2004.04.037. http://www.sciencedirect.com/science/article/pii/S0022519304001754

26. Le Novere, N.: Quantitative and logic modelling of molecular and gene networks. Nat. Rev. Genet. **16**(3), 146–158 (2015)

27. Millán, M.P., Dickenstein, A., Shiu, A., Conradi, C.: Chemical reaction systems with toric steady states. Bull. Math. Biol. **74**(5), 1027–1065 (2012)

28. Minato, S.I.: Zero-suppressed BDDs for set manipulation in combinatorial problems. In: Proceedings of the 30th International Design Automation Conference, pp. 272–277. ACM (1993)

29. Mishchenko, A.: An introduction to zero-suppressed binary decision diagrams. In: Proceedings of the 12th Symposium on the Integration of Symbolic Computation and Mechanized Reasoning, vol. 8, pp. 1–15. Citeseer (2001)

30. Murrugarra, D., Veliz-Cuba, A., Aguilar, B., Laubenbacher, R.: Identification of control targets in Boolean molecular network models via computational algebra. BMC Syst. Biol. **10**(1), 94 (2016). https://doi.org/10.1186/s12918-016-0332-x

31. Naldi, A., Remy, É., Thieffry, D., Chaouiya, C.: Dynamically consistent reduction of logical regulatory graphs. Theor. Comput. Sci. **412**(21), 2207–2218 (2011). https://doi.org/10.1016/j.tcs.2010.10.021. http://www.sciencedirect.com/science/article/pii/S0304397510005839

32. Samaga, R., Klamt, S.: Modeling approaches for qualitative and semi-quantitative analysis of cellular signaling networks. Cell Commun. Signal. **11**(1), 43 (2013)

33. Samaga, R., Saez-Rodriguez, J., Alexopoulos, L.G., Sorger, P.K., Klamt, S.: The logic of EGFR/ERBB signaling: theoretical properties and analysis of high-throughput data. PLoS Comput. Biol. **5**(8), e1000438 (2009)

34. Sánchez, L., Chaouiya, C., Thieffry, D.: Segmenting the fly embryo: logical analysis of the role of the segment polarity cross-regulatory module. Int. J. Dev. Biol. **52**(8), 1059–1075 (2002)

35. Sato, Y., Inoue, S., Suzuki, A., Nabeshima, K., Sakai, K.: Boolean Gröbner bases. J. Symb. Comput. **46**(5), 622–632 (2011)

36. Sturmfels, B.: Gröbner Bases and Convex Polytopes, vol. 8. American Mathematical Society, Providence (1996)

37. Thobe, K., Sers, C., Siebert, H.: Unraveling the regulation of mTORC2 using logical modeling. Cell Commun. Signal. **15**(1), 6 (2017)

38. Veliz-Cuba, A.: An algebraic approach to reverse engineering finite dynamical systems arising from biology. SIAM J. Appl. Dyn. Syst. **11**(1), 31–48 (2012)
39. Veliz-Cuba, A., Aguilar, B., Hinkelmann, F., Laubenbacher, R.: Steady state analysis of Boolean molecular network models via model reduction and computational algebra. BMC Bioinform. **15**(1), 221 (2014). https://doi.org/10.1186/1471-2105-15-221
40. Veliz-Cuba, A., Jarrah, A.S., Laubenbacher, R.: Polynomial algebra of discrete models in systems biology. Bioinformatics **26**(13), 1637–1643 (2010). https://doi.org/10.1093/bioinformatics/btq240
41. Vera-Licona, P., Jarrah, A., Garcia-Puente, L.D., McGee, J., Laubenbacher, R.: An algebra-based method for inferring gene regulatory networks. BMC Syst. Biol. **8**(1), 37 (2014)

Sequential Temporary and Permanent Control of Boolean Networks

Cui Su[1] and Jun Pang[1,2(✉)]

[1] Interdisciplinary Centre for Security, Reliability and Trust,
University of Luxembourg, Esch-sur-Alzette, Luxembourg
{cui.su,jun.pang}@uni.lu
[2] Faculty of Science, Technology and Medicine, University of Luxembourg,
Esch-sur-Alzette, Luxembourg

Abstract. Direct cell reprogramming makes it feasible to reprogram abundant somatic cells into desired cells. It has great potential for regenerative medicine and tissue engineering. In this work, we study the control of biological networks, modelled as Boolean networks, to identify control paths driving the dynamics of the network from a source attractor (undesired cells) to the target attractor (desired cells). Instead of achieving the control in one step, we develop attractor-based sequential temporary and permanent control methods (AST and ASP) to identify a sequence of interventions that can alter the dynamics in a stepwise manner. To improve their feasibility, both AST and ASP only use biologically observable attractors as intermediates. They can find the shortest sequential control paths and guarantee 100% reachability of the target attractor. We apply the two methods to several real-life biological networks and compare their performance with the attractor-based sequential instantaneous control (ASI). The results demonstrate that AST and ASP have the ability to identify a richer set of control paths with fewer perturbations than ASI, which will greatly facilitate practical applications.

Keywords: Boolean networks · Cell reprogramming · Attractors · Node perturbations · Sequential control

1 Introduction

Direct cell reprogramming, also called transdifferentiation, has provided a great opportunity for treating the most devastating diseases that are caused by a deficiency or defect of certain cells. It allows us to harness abundant somatic cells and transform them into desired cells to restore the structure and functions of damaged organs. However, the identification of efficacious intervention targets hinders the practical application of direct cell reprogramming.

Conventional experimental approaches are usually prohibited due to the high complexity of biological systems and the high cost of biological experiments [28]. Mathematical modelling of biological systems paves the way to study

© Springer Nature Switzerland AG 2020
A. Abate et al. (Eds.): CMSB 2020, LNBI 12314, pp. 234–251, 2020.
https://doi.org/10.1007/978-3-030-60327-4_13

mechanisms of biological processes and identify therapeutic targets with formal reasoning and tools. Among various modelling frameworks, Boolean network (BN) has a distinct advantage [6,7]. It provides a qualitative description of biological systems and thus evades the parametrisation problem, which often occurs in quantitative modelling, such as networks of ordinary differential equations (ODEs). In BNs, molecular species (genes, transcription factors, etc.) are assigned binary-valued nodes, being either '0' or '1'. The value of '0' describes the absence or inactive state of a species, whereas '1' represents the presence or active state. Activation/inhibition regulations between species are encoded as Boolean functions, which determine the evolution of the nodes. The dynamics of a BN evolves in discrete time under one of the updating schemes, such as *synchronous* or *asynchronous* updating schemes. The asynchronous updating scheme is considered more realistic than the synchronous one, since it non-deterministically updates one node at each time step and therefore can capture different biological processes at different time scales [19]. The long-run behaviour of the network dynamics is described as *attractors*, to one of which the network eventually settles down. Attractors are used to characterise cellular phenotypes or functional cellular states [5], such as proliferation, differentiation or apoptosis etc. In the context of BNs, direct cell reprogramming is equivalent to a *source-target control* problem: identifying a set of nodes, the perturbation of which can drive the network dynamics from a source attractor to the desired attractor.

The non-determinism of the asynchronous dynamics of BNs contributes to a better depiction of biological systems. As a result, it makes the control problem more challenging and renders the control methods designed for synchronous BNs inapplicable [8,20,31]. Another major obstacle to the control of BNs is the infamous state explosion problem — the state space is exponential in the size of the network. It prohibits the scalability and minimality of the control methods for asynchronous BNs [10,30]. The limitations of the existing methods motivate us to work on efficient and efficacious methods for the minimal source-target control of asynchronous BNs. There are different strategies to solve the control problem. Based on the control steps, we have *one-step control* and *sequential control*. One-step control applies all the perturbations simultaneously for one time, while sequential control identifies a sequence of perturbations that are applied at different time steps. In particular, we are interested in the sequential control that only adopts attractors as intermediates, called *attractor-based sequential control*. Rapid development of gene editing techniques enables us to silence or overexpress the expression of genes for different periods of time, thus, we have *instantaneous, temporary and permanent perturbations*. Instantaneous perturbations are applied instantaneously; temporary perturbations are applied for sufficient time steps and then released; permanent perturbations are applied for all the following steps. So far, we have developed methods for the minimal one-step instantaneous, temporary and permanent control (OI, OT and OP) [21,22,26] and the attractor-based sequential instantaneous control (ASI) [11]. In this work, we focus on the attractor-based sequential temporary and permanent control (AST and ASP).

Due to the intrinsic diversity and complexity of biological systems, no single control method can perfectly suit all cases. Thus, it is of great importance to explore more strategies to provide a number of cautiously selected candidates for later clinical validations. AST and ASP integrate promising factors: attractor-based sequential control and temporary/permanent control. Attractor-based sequential control is more practical than the general sequential control [12], where any state can play the role of intermediate states. Moreover, temporary and permanent controls have proved their potential in reducing the number of perturbations [26]. In this work, we continue to develop efficient methods to solve the AST and ASP control problems. We apply our methods for AST and ASP to several biological networks to show their ability in finding new control paths with fewer perturbations compared to our previous methods [11,21,22,26]. We believe our new methods can provide a better understanding of the mechanism-of-action of interventions and improve the efficiency of translating identified reprogramming paths into practical applications.

2   Preliminaries

In this section, we give preliminary notions of Boolean networks.

2.1   Boolean Networks

A Boolean network (BN) describes elements of a dynamical system with binary-valued nodes and interactions between elements with Boolean functions. It is formally defined as:

Definition 1 (Boolean networks). *A Boolean network is a tuple $G = (X, F)$ where $X = \{x_1, x_2, \ldots, x_n\}$, such that $x_i, i \in \{1, 2, \ldots, n\}$ is a Boolean variable and $F = \{f_1, f_2, \ldots, f_n\}$ is a set of Boolean functions over X.*

For the rest of the exposition, we assume that an arbitrary but fixed network $G = (X, F)$ of n variables is given to us. For all occurrences of x_i and f_i, we assume x_i and f_i are elements of X and F, respectively. A *state* s of G is an element in $\{0, 1\}^n$. Let S be the set of states of G. For any state $s = (s[1], s[2], \ldots, s[n])$, and for every $i \in \{1, 2, \ldots, n\}$, the value of $s[i]$, represents the value that the variable x_i takes when the network is in state s. For some $i \in \{1, 2, \ldots, n\}$, suppose f_i depends on $x_{i_1}, x_{i_2}, \ldots, x_{i_k}$. Then $f_i(s)$ denotes the value $f_i(s[i_1], s[i_2], \ldots, s[i_k])$. For two states $s, s' \in S$, the *Hamming distance* between s and s' is denoted as $\mathsf{hd}(s, s')$.

Definition 2 (Control). *A control C is a tuple $(\mathbb{0}, \mathbb{1})$, where $\mathbb{0}, \mathbb{1} \subseteq \{1, 2, \ldots, n\}$ and $\mathbb{0}$ and $\mathbb{1}$ are mutually disjoint (possibly empty) sets of indices of nodes of a Boolean network G. The size of the control C is defined as $|C| = |\mathbb{0}| + |\mathbb{1}|$. Given a state $s \in S$, the application of C to s is defined as a state $s' = C(s)$ ($s' \in S$), such that $s'[i] = 0 = 1 - s[i]$ if $i \in \mathbb{0}$, and $s'[i] = 1 = 1 - s[i]$ if $i \in \mathbb{1}$.*

Definition 3 (Boolean networks under control). *Let* $C = (0, 1)$ *be a control and* $G = (X, F)$ *be a Boolean network. The Boolean network* G *under control* C, *denoted as* $G|_C$, *is defined as a tuple* $G|_C = (\hat{X}, \hat{F})$, *where* $\hat{X} = \{\hat{x}_1, \hat{x}_2, \ldots, \hat{x}_n\}$ *and* $\hat{F} = \{\hat{f}_1, \hat{f}_2, \ldots, \hat{f}_n\}$, *such that for all* $i \in \{1, 2, \ldots, n\}$:
(1) $\hat{x}_i = 0$ *if* $i \in 0$, $\hat{x}_i = 1$ *if* $i \in 1$, *and* $\hat{x}_i = x_i$ *otherwise;*
(2) $\hat{f}_i = 0$ *if* $i \in 0$, $\hat{f}_i = 1$ *if* $i \in 1$, *and* $\hat{f}_i = f_i$ *otherwise.*

The state space of $G|_C$, denoted $S|_C$, is derived by fixing the values of the variables in the set C to their respective values and is defined as $S|_C = \{s \in S \mid s[i] = 1 \text{ if } i \in 1 \text{ and } s[j] = 0 \text{ if } j \in 0\}$. Note that $S|_C \subseteq S$. For any subset S' of S we let $S'|_C = S' \cap S|_C$.

2.2 Dynamics of Boolean Networks

In this section, we define several notions that can be interpreted on both G and $G|_C$. We use the generic notion $G = (X, F)$ to represent either $G = (X, F)$ or $G|_C = (\hat{X}, \hat{F})$. A Boolean network $G = (X, F)$ evolves in discrete time steps from an initial state s_0. Its state changes in every time step according to the update functions F and the update scheme. Different updating schemes lead to different dynamics of the network [14,32]. In this work, we are interested primarily in the *asynchronous updating scheme* – at each time step, one node is randomly selected to update its value based on its Boolean function. We define asynchronous dynamics formally as follows:

Definition 4 (Asynchronous dynamics of Boolean networks). *Suppose* $s_0 \in S$ *is an initial state of* G. *The asynchronous evolution of* G *is a function* $\xi : \mathbb{N} \to \wp(S)$ *such that* $\xi(0) = \{s_0\}$ *and for every* $j \geq 0$, *if* $s \in \xi(j)$ *then* $s' \in \xi(j+1)$ *is a possible next state of* s *iff either* $\mathsf{hd}(s, s') = 1$ *and there exists an* i *such that* $s'[i] = f_i(s) = 1 - s[i]$ *or* $\mathsf{hd}(s, s') = 0$ *and there exists an* i *such that* $s'[i] = f_i(s) = s[i]$.

It is worth noting that the asynchronous dynamics is non-deterministic and thus it can capture biological processes happening at different classes of time scales. Henceforth, when we talk about the dynamics of G, we shall mean the asynchronous dynamics as defined above. The dynamics of a Boolean network can be described as a *transition system (TS)*.

Definition 5 (Transition system of Boolean networks). *The transition system of a Boolean network* G, *denoted as* TS, *is a tuple* (S, E), *where the vertices are the set of states* S *and for any two states* s *and* s' *there is a directed edge from* s *to* s', *denoted* $s \to s'$ *iff* s' *is a possible next state of* s *according to the asynchronous evolution function* ξ *of* G.

A *path* σ from a state s to a state s' is a (possibly empty) sequence of transitions from s to s'. Thus, $\sigma = s_0 \to s_1 \to \ldots \to s_k$, where $s_0 = s$ and $s_k = s'$. A path from a state s to a subset S' of S is a path from s to any state $s' \in S'$. For a state $s \in S$, $\mathsf{reach}(s)$ denotes the set of states S' such that there is a path from s to any $s' \in S'$ in TS and can be defined as the fixpoint of the successor operation which is often denoted as post*. Thus, reach$(s) = $ post*(s).

The long-run behaviour of the dynamics of a Boolean network is characterised as *attractors*, defined as follows.

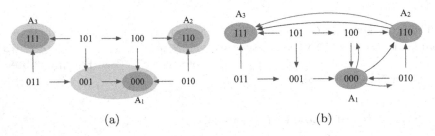

(a)   (b)

Fig. 1. (a) The transition system of the BN of Example 1; and (b) the control paths of Example 2 from attractor A_1 to attractor A_3. Paths indicated with blue (red) arrows are control paths with instantaneous (temporary/permanent) perturbations. (Color figure online)

Definition 6 (Attractor). *An attractor A of TS is a minimal non-empty subset of states of S such that for every $s \in A$, reach$(s) = A$.*

Any state which is not part of an attractor is a transient state. An attractor A of TS is said to be reachable from a state s if reach$(s) \cap A \neq \emptyset$. The network starting at any initial state $s_0 \in S$ will eventually end up in one of the attractors of TS and remain there forever unless perturbed. Thus, attractors are used to hypothesise cellular phenotypes or cell fates. We can easily observe that any attractor of TS is a bottom strongly connected component of TS.

Let $\mathcal{A}$ be the set of attractors of TS. For an attractor $A \in \mathcal{A}$, we define *the weak basin and the strong basin* of A to imply the commitment of states to A in Definition 7. Intuitively, the weak basin of A, $bas_{TS}^{W}(A)$, includes all the states s from which there exists at least one path to A. It is possible that there also exist paths from s to other attractor A' ($A' \neq A$) of TS, while the notion of strong basin does not allow this. The strong basin of A, $bas_{TS}^{S}(A)$, consists of all the states from which there only exist paths to A.

Definition 7 (Weak basin and strong basin). *The weak basin of A is defined as $bas_{TS}^{W}(A) = \{s \in S \mid$ reach$(s) \cap A \neq \emptyset\}$; and the strong basin of A is defined as $bas_{TS}^{S}(A) = \{s \in S \mid$ reach$(s) \cap A \neq \emptyset$ and reach$(s) \cap A' = \emptyset$ for $A' \in \mathcal{A}, A' \neq A\}$.*

Example 1. Consider a network $G = (X, F)$, where $X = \{x_1, x_2, x_3\}$, $F = \{f_1, f_2, f_3\}$, and $f_1 = x_2$, $f_2 = x_1$ and $f_3 = x_2 \wedge x_3$. Its transition system TS is given in Fig. 1a. This network has three attractors that are marked with dark grey nodes, including $A_1 = \{000\}$, $A_2 = \{110\}$, and $A_3 = \{111\}$. The strong basin of each attractor is marked as the light grey region. The weak basin of A_1 includes all the states except for states 110 and 111. The weak basin of A_2 and A_3 are $bas_{TS}^W(A_2) = \{010, 100, 101, 110\}$ and $bas_{TS}^W(A_3) = \{011, 101, 111\}$.

3 Sequential Temporary and Permanent Control

3.1 The Control Problem

As discussed in the introduction, direct cell reprogramming harnesses abundant somatic cells and reprograms them into desired cells. However, a major obstacle to the application of this novel technique lies in the identification of effective targets, the intervention of which can lead to desired changes. We aim to solve this problem by identifying key molecules based on BNs that model gene regulatory networks, such that the control of these molecules can drive the dynamics of a given network from a source attractor to the desired target attractor. We call it *source-target control* of BNs.

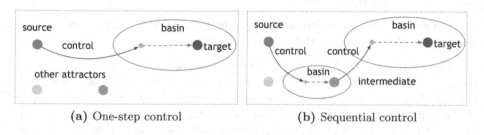

(a) One-step control   (b) Sequential control

Fig. 2. Two control strategies. (Color figure online)

Thanks to the rapid advances in gene editing techniques, the control can be applied for different periods of time. Thus, we have *instantaneous control*, *temporary control* and *permanent control*. Let A_s and A_t denote the source and target attractors, respectively.

Definition 8 (Instantaneous, temporary and permanent controls).
(1) An instantaneous control is a control $C = (0, 1)$, such that by applying C to a state $s \in A_s$ instantaneously, the network always reaches the target attractor A_t.
(2) A temporary control is a control $C = (0, 1)$, such that there exists a $t_0 \geq 0$, for all $t \geq t_0$, the network always reaches the target attractor A_t on the application of C to a state $s \in A_s$ for t steps.
(3) A permanent control is a control $C = (0, 1)$, such that the network always reaches the target attractor A_t on the permanent application of C to a state $s \in A_s$.

Temporary control applies perturbations for sufficient time and then is released, while permanent control maintains the perturbations for all the following time steps. Benefited from the extended intervention effects, temporary and permanent controls can potentially reduce the number of perturbations, which makes experiments easier to carry out and less costly [26].

The source-target control can also be achieved in one step or in multiple steps, called *one-step control* and *sequential control*, respectively. As illustrated in Fig. 2a, one-step control simultaneously applies all the required perturbations for one time (red arrow) to drive the network from a source state (blue node) to a state (yellow node), from which the network will converge to the target attractor in finite time steps (dashed line). In Fig. 2b, sequential control utilises other states as intermediates and identifies a sequence of perturbations, the application of which guides the network towards the target attractor in a stepwise manner. Considering difficulties in conducting clinical experiments, we are interested in *attractor-based sequential control*, where only biologically observable attractors can act as intermediates.

Given a source attractor A_s and a target attractor A_t of TS, the one-step control is formally defined as:

Definition 9 (One-step control). *Compute a control $C_{A_s \to A_t}$, such that the application of $C_{A_s \to A_t}$ to a state $s \in A_s$ can drive the network towards A_t.*

When the control $C_{A_s \to A_t}$ is the instantaneous, temporary or permanent control, we call it one-step instantaneous, temporary or permanent control (OI, OT or OP), respectively. To minimise the experimental costs, we are interested in the minimal solution $C_{A_s \to A_t}^{\min}$, which is the minimal such subset of $\{1, 2, \ldots, n\}$. Let $\mathcal{A}$ be the attractors of TS. The attractor-based sequential control is defined as follows:

Definition 10 (Attractor-based sequential control). *Find a sequence of attractors of TS, i.e. $\{A_1, A_2, \ldots, A_m\}$, where $A_1 = A_s, A_m = A_t, A_i \neq A_j$ for any $i, j \in [1, m]$ and $2 \leq m \leq |\mathcal{A}|$, such that after the application of a sequence of minimal one-step controls $\{C_{A_1 \to A_2}^{\min}, C_{A_2 \to A_3}^{\min}, \ldots, C_{A_{m-1} \to A_m}^{\min}\}$, the network always eventually reaches A_m, i.e. A_t. We call it an attractor-based sequential control path, denoted as*

$$\rho : A_1 \xrightarrow{C_{A_1 \to A_2}^{\min}} A_2 \xrightarrow{C_{A_2 \to A_3}^{\min}} A_3 \xrightarrow{\cdots} \cdots \xrightarrow{C_{A_{m-1} \to A_m}^{\min}} A_m$$

$(|C_{A_1 \to A_2}^{\min}| + |C_{A_2 \to A_3}^{\min}| + \cdots + |C_{A_{m-1} \to A_m}^{\min}|)$ is the total number of perturbations.

Similarly, when the control $C_{A_s \to A_t}$ is the instantaneous, temporary or permanent control, we call it attractor-based sequential instantaneous, temporary or permanent control (ASI, AST or ASP), respectively.

We have developed efficient methods to tackle the minimal OI, OT and OP control [21, 22, 26], as well as ASI control [11, 12]. Considering the advantages of sequential control and temporary and permanent perturbations, in this paper, we shall develop methods to solve the AST and ASP control problems based on the methods for the minimal OT and OP control. Since the problems of the minimal OT and OP control are at least PSPACE-hard [26], the AST and ASP control are also computational difficult. We will demonstrate that based on efficient computation of the minimal OT and OP control, our methods for the AST and ASP control can also achieve a significant level of efficiency.

3.2 Attractor-Based Sequential Temporary Control

Algorithm 1 describes a procedure COMP_SEQ_TEMP to compute AST control paths within k perturbations. This algorithm is based on our previously proposed methods, including the computation of weak basin and strong basin [21,22], denoted COMP_WEAK_BASIN and COMP_STRONG_BASIN, and the computation of the minimal OT control [26], denoted COMP_TEMP_CONTROL. Particularly, the procedure COMP_TEMP_CONTROL is based on Theorem 1.

Theorem 1. A control $C = (0, 1)$ is a minimal temporary control from s to A_t iff (1) $bas^S_{TS}(A_t) \cap S|_C \neq \emptyset$ and $C(s) \in bas^S_{TS|_C}(bas^S_{TS}(A_t) \cap S|_C)$ and (2) C is a minimal such subset of $\{1, 2, \ldots, n\}$.

The procedure for the computation of AST control, COMP_SEQ_TEMP, takes as inputs the Boolean functions F, a threshold k of the number of perturbations, a source attractor A_s, a target attractor A_t, and the set of attractors $\mathcal{A}$ of TS. It contains two parts.

The first part includes lines 2–13. For each attractor A ($A \in \mathcal{A}$ and $A \neq A_t$), we generate a dictionary L_A to save all the valid sequential control paths from A to A_t (line 5). We compute the minimal OT control set from A to A_t, denoted $C_{A \to A_t}$. $C_{A \to A_t}$ is considered valid and saved to L_A (line 8) if (1) A is the source attractor A_s and the number of perturbations $|C_{A \to A_t}|$ is not greater than k; or (2) A is not A_s and $|C_{A \to A_t}|$ is less or equal to $(k - 1)$. If A is an intermediate attractor ($A \neq A_s$), $C_{A_s \to A}$ requires at least one perturbation. Therefore, the size of $C_{A \to A_t}$ should not exceed $(k - 1)$. A is saved to I as an intermediate attractor if $A \neq A_s$ and $|C_{A \to A_t}| \leq k - 1$.

The second part includes lines 14–31. We extend the control paths computed in the previous part by iteratively taking every intermediate attractors $A'_t \in I$ as a new target and computing the minimal temporary control from an attractor A'_s ($A'_s \in (\mathcal{A} \setminus (A'_t \cup A_t))$) to A'_t. Specifically, for each new target attractor A'_t, we compute the minimal temporary control set $C_{A'_s \to A'_t}$ from A'_s to A'_t (line 20). Then, for every sequential path from A'_t to A_t, for instance $(\Delta_{A'_t \to A_t}, \rho_{A'_t \to A_t})$, we verify whether A'_s can be appended to the beginning of $\Delta_{A'_t \to A_t}$ to form a new path from A'_s to A'_t based on the following two conditions: (1) A'_s is not an intermediate attractor in the path $A'_t \to \ldots \to A_t$ (line 22); and (2) the total number of perturbations of the new path $\Delta_{A'_s \to A_t}$ should not exceed k (or $k - 1$) if $A'_s = A_s$ (or $A'_s \neq A_s$) (line 25). If both conditions are satisfied, we save the new path to $L_{A'_s}$ (line 29) and add A'_s to I' as a new candidate intermediate if $A'_s \neq A_s$ (line 30). After going through all the intermediate attractors in I (lines 16–30), we update the set of intermediate attractors I and repeat steps at lines 14–31 until I is an empty set.

Algorithm 1. Attractor-based sequential control of BNs

1: **procedure** COMP_SEQ_TEMP$(F, k, A_s, A_t, \mathcal{A})$
2:     Initialise a list $I := \emptyset$ to store possible intermediate attractors.
3:     $WB_{A_t} :=$ COMP_WEAK_BASIN(F, A_t) // *weak basin of the target*
4:     $SB_{A_t} :=$ COMP_STRONG_BASIN(F, A_t) // *strong basin of the target*
5:     Initialise a dictionary to store paths $\mathcal{L} := \{L_{A_1}, L_{A_2}, \ldots, L_{A_m}\}, A_i \in \mathcal{A}, A_i \neq A_t$.
6:     **for** $A \in (\mathcal{A} \setminus A_t)$ **do**   //*find attractors that have shorter paths to A_t*
7:         $C_{A \to A_t} :=$ COMP_TEMP_CONTROL(A, WB_{A_t}, SB_{A_t})
8:         **if** $(A = A_s$ and $|C_{A \to A_t}| \leq k)$ or $(A \neq A_s$ and $|C_{A \to A_t}| \leq k - 1)$ **then**
9:             // $C_{A_s \to A}$ *needs at least one perturbation*
10:             $\Delta_{A \to A_t}.\text{add}(A_t)$
11:             $\rho_{A \to A_t}.\text{add}(C_{A \to A_t})$
12:             Add the path $(\Delta_{A \to A_t}, \rho_{A \to A_t})$ to L_A
13:             Add A to I as a candidate intermediate if $A \neq A_s$.
14:     **while** $I \neq \emptyset$ **do**
15:         Initialise a new list $I' := \emptyset$
16:         **for** $A'_t \in I$ **do**   // *new target*
17:             $WB_{A'_t} :=$ COMP_WEAK_BASIN(F, A'_t)
18:             $SB_{A'_t} :=$ COMP_STRONG_BASIN(F, A'_t)
19:             **for** $A'_s \in (\mathcal{A} \setminus (A'_t \cup A_t))$ **do**   // *new source*
20:                 $C_{A'_s \to A'_t} :=$ COMP_TEMP_CONTROL$(A'_s, WB_{A'_t}, SB_{A'_t})$
21:                 **for** $(\Delta_{A'_t \to A_t}, \rho_{A'_t \to A_t}) \in L_{A'_t}$ **do**
22:                     **if** $A'_s \notin \Delta_{A'_t \to A_t}$ **then**
23:                         h: the total number of perturbations required by $\rho_{A'_t \to A_t}$.
24:                         $h = h + |C_{A'_s \to A'_t}|$
25:                         **if** $(A'_s = A_s$ and $h \leq k)$ or $(A'_s \neq A_s$ and $h \leq k - 1)$ **then**
26:                             $\rho_{A'_s \to A_t} := \rho_{A'_t \to A_t}; \Delta_{A'_s \to A_t} := \Delta_{A'_t \to A_t}$
27:                             Insert $C_{A'_s \to A'_t}$ to the beginning of $\rho_{A'_s \to A_t}$.
28:                             Insert A'_t to the beginning of $\Delta_{A'_s \to A_t}$.
29:                             Add the extended path $(\Delta_{A'_s \to A_t}, \rho_{A'_s \to A_t})$ to $L_{A'_s}$.
30:                             Add A'_s to I' as a candidate intermediate if $A'_s \neq A_s$.
31:         $I := I'$
32:     Return L_{A_s}
33: **procedure** PERM_CONTROL_VALIDATION$(C_{A'_s \to A'_t}, A'_t, \Delta_{A'_t \to A_t}, \rho_{A'_t \to A_t})$
34:     $A_1 := \Delta_{A'_t \to A_t}[0]$ // *the first intermediate A_1 in $\Delta_{A'_t \to A_t}$*
35:     $C_{A'_t \to A_1} := \rho_{A'_t \to A_t}[0]$ // *the first control set $C_{A'_t \to A_1}$ in $\rho_{A'_t \to A_t}$*
36:     $\Delta' := \Delta_{A'_t \to A_t}.\text{pop}(), \rho' := \rho_{A'_t \to A_t}.\text{pop}()$ //*delete the first element*
37:     $C'' := C_{A'_s \to A'_t} \setminus C_{A'_t \to A_1}$
38:     $isValid := True$
39:     **if** $A'_t|_{C''} = A_1|_{C''}$ and $\Delta' \neq \emptyset$ **then**
40:         $isValid :=$ PERM_CONTROL_VALIDATION$(C'', A'_t, \Delta', \rho')$
41:     **else if** $A'_t|_{C''} \neq A'|_{C''}$ **then**
42:         $isValid := False$
        return $isValid$

3.3   Attractor-Based Sequential Permanent Control

In this section, we develop an algorithm to solve the ASP control problem. We have developed an algorithm to compute the minimal OP control [26], denoted as COMP_PERM_CONTROL, based on Theorem 2.

Theorem 2. *A control $C = (0, 1)$ is a minimal permanent control from s to A_t iff (1) $C(s) \in bas_{TS|_C}^S(A_t)$ and (2) C is a minimal such subset of $\{1, 2, \ldots, n\}$.*

The procedure for ASP control explores in the same way as the procedure for AST control, COMP_SEQ_TEMP in Algorithm 1, to construct sequential paths, but it is more involved. It can be achieved by modifying procedure COMP_SEQ_TEMP as follows. First, at lines 7 and 20, we simply replace the procedure COMP_TEMP_CONTROL with the procedure COMP_PERM_CONTROL. Second, when extending the sequential paths (lines 14–31), besides the conditions at line 26, we add the procedure PERM_CONTROL_VALIDATION in Algorithm 1 to verify whether the control $C_{A'_s \to A'_t}$ can be inserted to the beginning of the path from A'_t to A_t. Because for each control step of AST, the temporary perturbations are released at one time point to retrieve the original transition system and let the network evolve spontaneously to the the intermediate/target attractor. But ASP adopts permanent control that will be maintained for all the following time steps. Therefore, when extending a permanent control C to the beginning of a sequential path, it has to be verified whether the application of C will affect the reachability of the following control steps. To avoid duplication, here we only give the explanations of the procedure PERM_CONTROL_VALIDATION. The purpose of this procedure is to verify whether the control $C_{A'_s \to A'_t}$ can be added to the beginning of $\rho_{A'_t \to A_t}$ to form a new path $\rho_{A'_t \to A_t}$ The verification is carried out recursively. Let us assume $\Delta_{A'_t \to A_t} = \{A_1, A_2, \ldots, A_t\}$. The first intermediate attractor is A_1 and the control from A'_t to A_1 is $C_{A'_t \to A_1}$. Since $C_{A'_s \to A'_t}$ and $C_{A'_t \to A_1}$ may require to perturb the same node in the opposite way, we compute $(C_{A'_s \to A'_t} \setminus C_{A'_t \to A_1})$ and denote it as C''. If the projections of A'_t and A_1 to C'' are the same, A_1 is preserved under the permanent control C'' and we proceed to the remaining control steps (lines 39–40); otherwise, $C_{A'_s \to A'_t}$ is not a valid control (lines 41–42).

Example 2. To continue with Example 1, we compute the control paths from A_1 to A_3. As shown in Fig. 1b, the control path indicated with blue arrows, $A_1 \xrightarrow{\{x_1, x_2\}} A_2 \xrightarrow{\{x_3\}} A_3$, is the shortest ASI control, which requires three perturbations. AST and ASP have the same results indicated with red arrows in Fig. 1b: $A_1 \xrightarrow[\{x_1\}]{\{x_2\}} A_2 \xrightarrow{\{x_3\}} A_3$, which require two perturbations in total.

4   Evaluation

In this section, we evaluate the performance of AST and ASP on several real-life biological networks. To demonstrate their efficacy, we compare their performance with ASI [11]. The minimal number of perturbations required by OI, OT and OP is set as the threshold k of the number of perturbations for ASI, AST and ASP, respectively. In this way, the results will demonstrate whether AST and ASP can find sequential paths with fewer perturbations than ASI. All the methods are implemented as an extension of our software tool ASSA-PBN [14–16]. All the experiments are performed on a high-performance computing (HPC) platform, which contains CPUs of Intel Xeon Gold 6132 @2.6 GHz. We describe and discuss the results of the myeloid differentiation network [9] and the Th cell differentiation network [17] in detail (Sects. 4.1 and 4.2), and we give an overview of the results of the other networks (Sect. 4.3).

4.1   The Myeloid Differentiation Network

The myeloid differentiation network is constructed to model the differentiation process of common myeloid progenitors (CMPs) into four types of mature blood cells [9]. With our attractor detection method [13], we identify six single-state attractors of the network, five of which are non-zero attractors (not all the nodes have a value of '0'). It has been validated that expressions of four attractors correspond to microarray expression profiles of megakaryocytes, erythrocytes, granulocytes and monocytes [9]. The fifth attractor with the activation of PU1, cJun and EgrNab might be caused by pathological alterations [9] and the sixth attractor is an all-zero attractor, where all the nodes have a value of '0'.

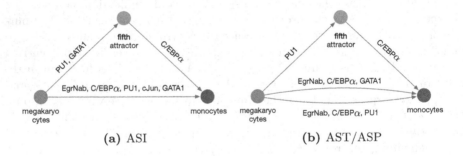

Fig. 3. Sequential control of the myeloid differentiation network.

We take the conversion from megakaryocytes to monocytes as an example to show the performance of the methods. Note that the sixth attractor does not have a biological interpretation and mature erythrocytes in mammals do not have cell nucleus, therefore we do not consider these two attractors as intermediate attractors. Under this condition, the three methods (ASI, AST, ASP) identify both one-step and sequential paths as illustrated in Fig. 3. In particular, the results of AST and ASP are identical. We can see that the minimal OI control requires the activation of EgrNab, C/EBPα, PU1, cJun and the inhibition of GATA1 (Fig. 3a); while OT or OP can achieve the goal by either (1) the activation of EgrNab, C/EBPα and PU1; or (2) the activation of EgrNab and C/EBPα, together with the inhibition of GATA1 (Fig. 3b). All the sequential paths need two steps, where the fifth attractor is adopted as an intermediate attractor. For the first step, ASI activates PU1 and inhibits GATA1, while AST or ASP only needs to activate PU1. When the network converges to the fifth attractor, all the three methods require to activate C/EBPα. After that, the network will evolve spontaneously to the target attractor monocytes. Figure 3 shows that AST and ASP are able to identify a path with only two perturbations, while ASI requires at least three perturbations.

The efficacy of the identified sequential temporary/permanent path is confirmed by the predictions in [9]. According to the expression profiles, both PU1 and C/EBPα are not expressed in MegE lineage (megakaryocytes and erythrocytes), while they are expressed in GM lineage (monocytes and granulocytes). In this network, no regulator can activate C/EBPα and PU1 is primarily activated by C/EBPα. Therefore, C/EBPα has to be altered externally to reprogram MegE lineage to GM lineage. However, more perturbations are necessary to accurately reach the monocytes lineage. Sustained activation of PU1 and the absence of C/EBPα guide the network to the fifth attractor, the expression of which differs with monocytes only in C/EBPα [9].

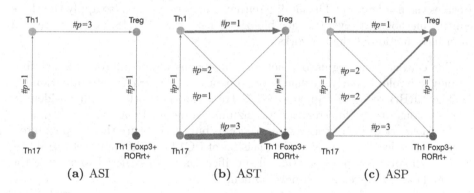

Fig. 4. Sequential control of the Th cell differentiation network.

4.2   The Th Cell Differentiation Network

The T-helper (Th) cell differentiation network is a comprehensive model integrating regulatory network and signalling pathways that regulate Th cell differentiation [17]. This network consists of 12 single-state attractors under one initial condition and the attractors can be classified into different Th subtypes based on the expression of four master regulators [17].

Let Th17 and a Th1 subtype (Th1 Foxp3+ RORrt+) be the source and target attractors, respectively. For the purpose of illustration, we limit the number of control paths by only adopting Th1 and Treg as intermediate attractors. In addition, we set the node 'proliferation' as a non-perturbed node, since it denotes a cell fate and thus cannot be perturbed in reality. Figure 4 describes the control paths identified by the three methods. The thickness of arrows implies the number of control sets and the equations $\#p = m$ above each arrow denotes the number of perturbations required by each step. All the methods identify sequential paths passing through Th1 and/or Treg. Figure 4a only shows the shortest ASI path with five perturbations, while AST and ASP provide multiple paths with only two or three perturbations (Fig. 4b and Fig. 4c), demonstrating the advantages of AST and ASP in reducing the number of perturbations. Among the sequential paths of AST and ASP, only the AST path, Th17 $\xrightarrow{\text{IL27R}}$ Treg $\xrightarrow{\text{TBET}}$ the Th1 subtype, perturbs two nodes, all the other paths using either temporary or permanent perturbations need to perturb at least three nodes. This shows that AST has the potential to further reduce the number of perturbations compared to ASP. Moreover, in terms of the number of solutions, it is obvious that the arrows in Fig. 4b are thicker than those in Fig. 4c, which indicates that AST provides more solutions than ASP.

4.3   Other Biological Networks

Besides the myeloid and Th cell differentiation networks, we also apply the three control methods to several other biological networks [1–4,18,23,24]. Here is a brief introduction of the networks.

- The cardiac gene regulatory network integrates key regulatory factors that play key roles in early cardiac development and FHF/SHF determination [4].
- The ERBB receptor-regulated G1/S transition network is built to identify efficacious targets for treating trastuzumab resistant breast cancer [24].
- The network of PC12 cell differentiation is built to capture the complex interplay of molecular factors in the decision of PC12 cell differentiation [18].
- The network of hematopoietic cell specification is constructed to capture the lymphoid and myeloid cell development [2].
- The network of bladder tumour is constructed to study mutually exclusivity and co-occurrence in genetic alterations [23].
- The pharmacodynamic model of bortezomib responses integrates major survival and apoptotic pathways in U266 cells to connect bortezomib exposure to multiple myeloma cellular proliferation [1].
- The network of a CD4$^+$ immune effector T cell is constructed to capture cellular dynamics and molecular signalling under both immunocompromised and healthy settings [3].

Columns 2–4 of Table 1 summarise the number of nodes, edges and attractors contained in each network. For each network, we choose a pair of source and target attractors and compute control paths with ASI, AST and ASP.

Efficacy. For each pair of source and target attractors, all the control paths with at most k perturbation are computed. For the purpose of comparison, in Table 1, columns 5–7 only summarise the minimal number of perturbations needed by each control method and columns 8–10 summarise the number of corresponding control paths. The results show that by extending the period of control time, AST and ASP have the ability to compute more control paths with fewer perturbations than ASI. This brings significant benefits for practical applications. First, fewer perturbations can reduce the experimental costs and make the experiments easier to conduct. Second, a richer set of control paths provides biologists more options to tackle diverse biological systems.

To further compare AST and ASP, AST is more appealing than ASP. As discussed in the previous subsection, the control of Th cell differentiation network (T-diff in Table 1) shows that AST has the potential to identify smaller control sets than ASP. For the other cases listed in Table 1, although AST requires the same number of perturbations as ASP, AST identifies more solutions than ASP. Apart from that, AST has an intrinsic advantage compared to ASP – temporary control will eventually be released and therefore can eliminate risks of unforeseen consequences, which may be caused by the permanent shift of the dynamics.

Efficiency. The last three columns of Table 1 give the computation time of ASI, AST and ASP. Although AST and ASP take longer time than ASI, they are still quite efficient and are capable of handling large-scale and comprehensive networks. In general, the computational time of the methods depends on the size of the network, the threshold of the number of perturbations k and the number of existing solutions within the threshold. By increasing the threshold k, our methods can identify more candidate solutions at the cost of longer computational time. Currently, due to the lack of large and well-behaved networks, we are not yet able to find out the precise limit of our methods on the size of networks.

Table 1. Sequential control of several biological networks.

Network	$|V|$	$|E|$	$|\mathcal{A}|$	#perturbations			#paths			Time (seconds)		
				ASI	AST	ASP	ASI	AST	ASP	ASI	AST	ASP
Myeloid	11	30	6	3	2	2	1	1	1	0.006	0.034	0.038
Cardiac	15	39	6	3	2	2	1	3	2	0.018	0.658	0.653
ERBB	20	52	3	8	3	3	2	3	3	0.007	0.249	0.319
PC12	33	62	7	8	2	2	3	50	30	0.050	1.188	1.462
HSC	33	88	5	12	2	2	2	12	6	0.406	12.217	8.879
Bladder	35	116	4	5	2	2	2	2	2	0.139	0.709	0.676
Bortezomib	67	135	5	3	2	2	1	4	2	1.900	105.184	119.138
T-diff	68	175	12	5	2	3	4	1	14	9.713	95.211	71.044
CD4$^+$	188	380	6	3	2	2	3	48	6	256.492	539.868	1304.490

5 Discussion

We have demonstrated the potential strengths of AST and ASP, however, they are not warranted to be the best methods for all kinds of biological systems.

Indeed, there is no control method that can perfectly solve all the control problems due to the intrinsic diversity and complexity of biological systems. Given a specific task, it is thus recommended to compute all the control paths with available control methods. Various sets of identified therapeutic targets serve as candidates, such that biologists can choose appropriate targets, the modulation of which will not disrupt physiological functions of biological systems.

Although the dynamics of asynchronous BNs are non-deterministic, our methods guarantee to find the shortest control paths with 100% reachability *in silico*. Experimental validation is necessary to verify their therapeutic efficacy *in vivo*. It is worth noticing that the consistency of the efficacy *in silico* and *in vivo* highly relies on the quality of the constructed BNs. The identified perturbations can effectively modulate the dynamics as expected, provided that the adopted network well captures the structural and dynamical properties of the real-life biological system. However, mathematical modelling of vastly complex biological systems is already a challenging task by itself in systems biology. We have spotted some flaws of the constructed networks in the literature during analysis, summarised as follows.

First, simulation is often used to evaluate the stable behaviour of dynamics in most of the works. However, simulation can hardly cover the entire transition system of a BN, which is exponential in the size of the network. As a consequence, the information on attractors is usually incomplete, especially for networks of medium or large sizes. This problem can be solved by using our attractor detection method [13, 29] to identify all the exact attractors of a network.

Second, we noticed that the attractors of some large networks are purely induced by input nodes. For instance, given a network with 2 input nodes (nodes without upstream regulators), it has 2^2 attractors. Each attractor corresponds to one combination of the input nodes $(00, 01, 10, 11)$. For such networks, the input nodes, that have different values in the source and target attractors, are the key nodes for modulating the dynamics. Such kind of networks may capture some activation or inhibition regulations, but they fail to depict the intrinsic mechanisms of biological processes.

Third, in some networks, cell phenotypes or cell fates, such as apoptosis, proliferation, and differentiation, are represented as marker nodes. Benefited from this, attractors can be classified based on the expressions of those nodes. A problem that often occurs is that there does not exist any control sets without perturbing these marker nodes, however, these nodes can not be perturbed in reality. Again, we hypothesise that these constructed networks do not reflect the intrinsic properties of biological systems.

Our methods [11, 13, 22, 26, 29] can provide accurate information of the networks, such as the number and size of the attractors and potential sets of control nodes. Such information related to the network dynamics should be taken into account when inferring the networks by updating the Boolean functions or adding/deleting regulators.

6   Conclusion and Future Work

In this work, we have developed the AST and ASP control methods to identify sequential control paths for modulating the dynamics of biological systems. To make it practical, only biologically observable attractors are served as intermediates. We compared the performance of the two methods with ASI on a variety of biological networks. The results show that our new methods have apparent advantages in reducing the number of perturbations and enriching the diversity of solutions. Among the three sequential control methods (ASI, AST and ASP), AST is more preferable because it requires the fewest number of perturbations and it adopts temporary perturbations which will eventually be released and thus can evade unforeseen consequences that might be caused by permanent perturbations.

Until now, we have developed source-target control methods to alter the dynamics of BNs in different ways. Currently, we are working on a target control method to identify a subset of nodes, the intervention of which can transform any somatic cells to the desired cell. We also plan to study the control of probabilistic Boolean networks [25, 27] based on our control methods for BNs. We believe our works can provide deep insights into regulatory mechanisms of biological processes and facilitate direct cell reprogramming.

Acknowledgements. This work was partially supported by the project SEC-PBN funded by University of Luxembourg and the ANR-FNR project AlgoReCell (INTER/ANR/15/11191283).

References

1. Chudasama, V., Ovacik, M., Abernethy, D., Mager, D.: Logic-based and cellular pharmacodynamic modeling of bortezomib responses in u266 human myeloma cells. J. Pharmacol. Experimental Therapeutics **354**(3), 448–458 (2015)
2. Collombet, S., et al.: Logical modeling of lymphoid and myeloid cell specification and transdifferentiation. Proc. National Acad. Sci. **114**(23), 5792–5799 (2017)
3. Conroy, B.D., et al.: Design, assessment, and in vivo evaluation of a computational model illustrating the role of CAV1 in CD4+ T-lymphocytes. Front. Immunol. **5**, 599 (2014)
4. Herrmann, F., Groß, A., Zhou, D., Kestler, H.A., Kühl, M.: A Boolean model of the cardiac gene regulatory network determining first and second heart field identity. PLoS ONE **7**, 1–10 (2012)
5. Huang, S.: Genomics, complexity and drug discovery: insights from Boolean network models of cellular regulation. Pharmacogenomics **2**(3), 203–222 (2001)
6. Kauffman, S.A.: Homeostasis and differentiation in random genetic control networks. Nature **224**, 177–178 (1969)
7. Kauffman, S.A.: Metabolic stability and epigenesis in randomly constructed genetic nets. J. Theor. Biol. **22**(3), 437–467 (1969)
8. Kim, J., Park, S., Cho, K.: Discovery of a kernel for controlling biomolecular regulatory networks. Sci. Rep. **3**(2223) 156–216 (2013)

9. Krumsiek, J., Marr, C., Schroeder, T., Theis, F.J.: Hierarchical differentiation of myeloid progenitors is encoded in the transcription factor network. PLoS ONE **6**(8), e22649 (2011)
10. Mandon, H., Haar, S., Paulevé, L.: Relationship between the reprogramming determinants of boolean networks and their interaction graph. In: Cinquemani, E., Donzé, A. (eds.) HSB 2016. LNCS, vol. 9957, pp. 113–127. Springer, Cham (2016). https://doi.org/10.1007/978-3-319-47151-8_8
11. Mandon, H., Su, C., Haar, S., Pang, J., Paulevé, L.: Sequential reprogramming of boolean networks made practical. In: Bortolussi, L., Sanguinetti, G. (eds.) CMSB 2019. LNCS, vol. 11773, pp. 3–19. Springer, Cham (2019). https://doi.org/10.1007/978-3-030-31304-3_1
12. Mandon, H., Su, C., Pang, J., Paul, S., Haar, S., Paulevé, L.: Algorithms for the sequential reprogramming of Boolean networks. IEEE/ACM Trans. Comput. Biol. Bioinform. **16**(5), 1610–1619 (2019)
13. Mizera, A., Pang, J., Qu, H., Yuan, Q.: Taming asynchrony for attractor detection in large Boolean networks. IEEE/ACM Trans. Comput. Biol. Bioinform. **16**(1), 31–42 (2019)
14. Mizera, A., Pang, J., Su, C., Yuan, Q.: ASSA-PBN: a toolbox for probabilistic Boolean networks. IEEE/ACM Trans. Computat. Biol. Bioinform. **15**(4), 1203–1216 (2018)
15. Mizera, A., Pang, J., Yuan, Q.: ASSA-PBN: a tool for approximate steady-state analysis of large probabilistic Boolean networks. In: Proceedings 13th International Symposium on Automated Technology for Verification and Analysis. LNCS, vol. 9364, pp. 214–220. Springer (2015)
16. Mizera, A., Pang, J., Yuan, Q.: ASSA-PBN 2.0: a software tool for probabilistic boolean networks. In: Bartocci, E., Lio, P., Paoletti, N. (eds.) CMSB 2016. LNCS, vol. 9859, pp. 309–315. Springer, Cham (2016). https://doi.org/10.1007/978-3-319-45177-0_19
17. Naldi, A., Carneiro, J., Chaouiya, C., Thieffry, D.: Diversity and plasticity of th cell types predicted from regulatory network modelling. PLoS Computat. Biol. **6**(9) 1256 (2010)
18. Offermann, B., et al.: Boolean modeling reveals the necessity of transcriptional regulation for bistability in PC12 cell differentiation. Front. Genetics **7**, 44 (2016)
19. Papin, J.A., Hunter, T., Palsson, B.O., Subramaniam, S.: Reconstruction of cellular signalling networks and analysis of their properties. Nat. Rev. Molecular Cell Biol. **6**(2), 99 (2005)
20. Pardo, J., Ivanov, S., Delaplace, F.: Sequential reprogramming of biological network fate. In: Bortolussi, L., Sanguinetti, G. (eds.) CMSB 2019. LNCS, vol. 11773, pp. 20–41. Springer, Cham (2019). https://doi.org/10.1007/978-3-030-31304-3_2
21. Paul, S., Su, C., Pang, J., Mizera, A.: A decomposition-based approach towards the control of Boolean networks. In: Proceedings 9th ACM Conference on Bioinformatics, Computational Biology, and Health Informatics, pp. 11–20. ACM Press (2018)
22. Paul, S., Su, C., Pang, J., Mizera, A.: An efficient approach towards the source-target control of Boolean networks. IEEE/ACM Transactions on Computational Biology and Bioinformatics (2020), accepted
23. Remy, E., Rebouissou, S., Chaouiya, C., Zinovyev, A., Radvanyi, F., Calzone, L.: A modeling approach to explain mutually exclusive and co-occurring genetic alterations in bladder tumorigenesis. Cancer Res. **75**(19), 4042–4052 (2015)
24. Sahin, O., et al.: Modeling ERBB receptor-regulated G1/S transition to find novel targets for de novo trastuzumab resistance. BMC Syst. Biol. **3**(1), 1 (2009)

25. Shmulevich, I., Dougherty, E.R.: Probabilistic Boolean Networks: The Modeling and Control of Gene Regulatory Networks. SIAM Press (2010)
26. Su, C., Paul, S., Pang, J.: Controlling large boolean networks with temporary and permanent perturbations. In: ter Beek, M.H., McIver, A., Oliveira, J.N. (eds.) FM 2019. LNCS, vol. 11800, pp. 707–724. Springer, Cham (2019). https://doi.org/10.1007/978-3-030-30942-8_41
27. Trairatphisan, P., Mizera, A., Pang, J., Tantar, A.A., Schneider, J., Sauter, T.: Recent development and biomedical applications of probabilistic Boolean networks. Cell Commun. Signal. **11**, 46 (2013)
28. Wang, L.Z., et al.: A geometrical approach to control and controllability of nonlinear dynamical networks. Nat. Commun. **7**, 11323 (2016)
29. Yuan, Q., Mizera, A., Pang, J., Qu, H.: A new decomposition-based method for detecting attractors in synchronous Boolean networks. Sci. Comput. Program. **180**, 18–35 (2019)
30. Zañudo, J.G.T., Albert, R.: Cell fate reprogramming by control of intracellular network dynamics. PLoS Comput. Biol. **11**(4), e1004193 (2015)
31. Zhao, Y., Kim, J., Filippone, M.: Aggregation algorithm towards large-scale Boolean network analysis. IEEE Trans. Automatic Control **58**(8), 1976–1985 (2013)
32. Zhu, P., Han, J.: Asynchronous stochastic Boolean networks as gene network models. J. Comput. Biol. **21**(10), 771–783 (2014)

Inference and Identification

ABC(SMC)²: Simultaneous Inference and Model Checking of Chemical Reaction Networks

Gareth W. Molyneux[✉] and Alessandro Abate

Department of Computer Science, University of Oxford, Oxford OX1 3QD, UK
{gareth.molyneux,alessandro.abate}@cs.ox.ac.uk

Abstract. We present an approach that simultaneously infers model parameters while statistically verifying properties of interest to chemical reaction networks, which we observe through data and we model as parametrised continuous-time Markov Chains. The new approach simultaneously integrates learning models from data, done by likelihood-free Bayesian inference, specifically Approximate Bayesian Computation, with formal verification over models, done by statistically model checking properties expressed as logical specifications (in CSL). The approach generates a probability (or credibility calculation) on whether a given chemical reaction network satisfies a property of interest.

1 Introduction

Contribution. We introduce a framework that integrates Bayesian inference and formal verification that additionally employs supervised machine learning, which allows for model-based probabilistic verification of data-generating stochastic biological systems. The methodology performs data-driven inference of accurate models, which contributes to the verification of whether or not the underlying stochastic system satisfies a given formal property of interest. Verification entails the estimation of the probability that models of the system satisfy a formal specification. Our framework accommodates partially known systems that might only generate finite, noisy observations. These systems are captured by parametric models, with uncertain rates within a known stoichiometry.

Related Work. Bayesian inference techniques [9,10] have been applied extensively to biological systems [42,49]. Exact inference is in general difficult due to the intractability of the likelihood function, which has led to likelihood-free methods such as Approximate Bayesian Computation (ABC) [44,48]. [22] computes the probability that an underlying stochastic system satisfies a given property using data produced by the system and leveraging system's models. Along this line of work, the integration of verification of parameterised discrete-time Markov chains and Bayesian inference is considered in [38], with an extension to Markov decision processes in [39]. Both [38,39] work with small finite-state models with

© Springer Nature Switzerland AG 2020
A. Abate et al. (Eds.): CMSB 2020, LNBI 12314, pp. 255–279, 2020.
https://doi.org/10.1007/978-3-030-60327-4_14

fully observable traces, which allows the posterior probability distribution to be calculated analytically and parameters to be synthesised symbolically. On the contrary, here we work with partially observed data and stochastic models with intractable likelihoods, and must rely on likelihood-free methods and statistical parameter synthesis procedures. Building on previous work [36], which allowed for likelhood-free Bayesian Verification of systems, the presented framework is applicable to a wider variety of stochastic models.

Both probabilistic and statistical model checking have been applied to biological models [31, 32, 51], with tools for parameter synthesis [11, 12]. Although the parameter synthesis approach in [11] rigorously calculates the satisfaction probability over the whole parameter space, it suffers from scalability issues. A Bayesian approach to statistical model checking is considered in [27] and partly inspires this work. Parametric verification has been considered from a statistical approach underpinned by Gaussian Processes: smoothed Model checking [5] provides an estimate of the satisfaction probability with uncertainty estimates, and has been used for parameter estimation from Boolean observations [7] and for parameter synthesis [8]. [2] proposes a methodology that, given a reachability specification, computes a related probability distribution on the parameter space, and an automaton-based adaptation of the ABC method is introduced to estimate it.

Approach. Our framework is as follows (Sect. 3). Given a property of interest, a class of parametrised models and data from the underlying system, we simultaneously infer parameters and perform model-based statistical model checking. We then use a supervised machine learning method to determine regions of the parameter space that relate to models verifying the given property of interest. We integrate the generated posterior over these synthesised parameter regions, to quantify a probability (or credibility calculation) on whether or not the system satisfies the given property. We apply this framework to Chemical Reaction Networks (CRNs) [23, 49] (Sect. 4), representing the data-generating biological system, which can be modelled by parametrised continuous-time Markov Chains [28]. We argue that the alternative use of CRN data for black-box statistical model checking would be infeasible.

2   Background

2.1   Parametric Continuous-Time Markov Chains

Although our methodology can be applied to a number of parametrised stochastic models, in view of the applications of interest we work with discrete-state, continuous-time Markov chains [28].

Definition 1 (Continuous-time Markov Chain). *A continuous-time Markov chain (CTMC) $\mathcal{M}$ is a tuple (S, R, AP, L), where;*

- S is a finite, non-empty set of states,
- s_0 is the initial state of the CTMC,
- $R : S \times S \rightarrow \mathbb{R}_{\geq 0}$ is the transition rate matrix, where $R(s, s')$ is the rate of transition from state s to state s',
- $L : S \rightarrow 2^{AP}$ is a labelling function mapping each state, $s \in S$, to the set $L(s) \subseteq AP$ of atomic propositions AP, that hold true in s.

For the models in this paper, we assume s_0 is unique and deterministically given. The transition rate matrix R governs the dynamics of the overall model.

Definition 2 (Path of a CTMC). *Let* $\mathcal{M} = (S, R, AP, L)$ *be a CTMC. An infinite path of a CTMC* $\mathcal{M}$ *is a non-empty sequence* $s_0 t_0 s_1 t_1 \ldots$ *where* $R(s_i, s_{i+1}) > 0$ *and* $t_i \in \mathbb{R}_{>0}$ *for all* $i \geq 0$. *A finite path is a sequence* $s_0 t_0 s_1 t_1 \ldots s_{k-1} t_{k-1} s_k$ *such that* s_k *is absorbing. The value* t_i *represents the amount of time spent in the state* s_i *before jumping to the next state in the chain, namely state* s_{i+1}. *We denote by* $Path^{\mathcal{M}}(s)$ *the set of all (infinite or finite) paths of the CTMC* $\mathcal{M}$ *starting in state* s. *A trace of a CTMC is the mapping of a path through the labelling function* L.

Parametric CTMCs extend the notion of CTMC by allowing transition rates to depend on a vector of model parameters, $\theta \in \mathbb{R}^k$. The domain of each parameter θ_i is given by a closed bounded real interval describing the range of possible values, $[\theta_i^\perp, \theta_i^\top]$. The parameter space Θ is defined as the Cartesian product of the individual intervals, $\Theta = \times_{i \in \{1, \ldots, k\}} [\theta_i^\perp, \theta_i^\top]$, so that Θ is a hyperrectangle.

Definition 3 (Parametric CTMC). *Let* Θ *be a parameter space. A parametric Continuous-time Markov Chain (pCTMC) over* θ *is a tuple* (S, R_θ, AP, L):

- S, s_0, AP *and* L *are as in Definition 1, and*
- $\theta = (\theta_1, \ldots, \theta_k)$ *is the vector of parameters, taking values in a compact hyperrectangle* $\Theta \subset \mathbb{R}_{>0}^k$,
- $R_\theta : S \times S \rightarrow \mathbb{R}[\theta]$ *is the parametric rate matrix, where* $\mathbb{R}[\theta]$ *denotes a set of polynomials over* $\mathbb{R}^+$ *with variables* θ_k, $\theta \in \Theta$.

Given a pCTMC and a parameter space Θ, we denote with $\mathcal{M}_\Theta$ the set $\{\mathcal{M}_\theta, \theta \in \Theta\}$ where $\mathcal{M}_\theta = (S, R_\theta, AP, L)$ is the instantiated CTMC obtained by replacing the parameters in R with their valuation in θ. So a standard CTMC is induced by selecting a specific parameter $\theta \in \Theta$: the sampled paths of an instantiated pCTMC $\mathcal{M}_\theta$ are defined similarly to ω. In this work we deal with Chemical Reaction Networks (CRNs), which have dynamics that can be modelled by CTMCs.

Definition 4 (Chemical Reaction Network). *A Chemical Reaction Network (CRN)* $\mathcal{C}$ *is a tuple* $(M, X, W, \mathcal{R}, v)$, *where*

- $M = \{m_1, \ldots, m_n\}$ *is a set of* n *species,*
- $X(t) = (X_1(t), \ldots, X_n(t))$ *is a vector where each* X_i *represents the number of molecules of each species* i *at time* t. $X \subseteq \mathbb{N}^N$ *the state space,*

- $\mathcal{R} = \{r_1, \ldots, r_k\}$ is the set of chemical reactions, each of the form $r_j = (v_j, \alpha_j)$, with v_j the stoichiometry vector of size n and $\alpha_j = \alpha_j(X, v_j)$ is the propensity or rate function,
- $v = (v_1, \ldots, v_k)$ is the vector of (kinetic) parameters, taking values in a compact hyperrectangle $\Upsilon \subset \mathbb{R}^k$.

Each reaction j of the CRN is represented as $r_j : \sum_{i=1}^{n} u_{j,i} m_i \xrightarrow{\alpha_j} \sum_{i=1}^{n} u'_{j,i} m_i$, where $u_{j,i}$ ($u'_{j,i}$) is the amount of species m_i consumed (produced) by reaction r_j. CRNs are used to model many biological processes and can be modelled by CTMCs if we consider each state of the pCTMC to be a unique combination of the number of species, taking a given species count X_0 to be the initial state of the pCTMC, $s_0 = X_0$. Parametrising the reaction rates within a CRN results in a parametric CRN (pCRN), which can be modelled as a pCTMC. For the rest of this paper, with a slight abuse in notation, we will let $\mathcal{M}_\theta$ be the pCTMC that represents a pCRN, where θ are the kinetic rates.

2.2   Properties - Continuous Stochastic Logic

We wish to verify properties over CRNs and their pCTMC models. We employ a time-bounded fragment of *continuous stochastic logic* (CSL) [1,31].

Definition 5. *Let ϕ be a CSL formula interpreted over states $s \in S$ of a parametrised model $\mathcal{M}_\theta$, and φ be a formula over its paths. Its syntax is*

$$\phi := true \mid a \mid \neg\phi \mid \phi \wedge \phi \mid \phi \vee \phi \mid P_{\sim\zeta}[\varphi] \,,$$

$$\varphi := X^{[t,t']}\phi \mid \phi_1 U^{[t,t']}\phi_2 \,,$$

where $a \in AP$, $\sim \in \{<, \leq, \geq, >\}$, $\zeta \in [0,1]$, and $t, t' \in \mathbb{R}_{\geq 0}$.

$P_{\sim\zeta}[\varphi]$ holds if the probability of the path formula φ being satisfied from a given state meets $\sim \zeta$. Path formulas are defined by combining state formulas through temporal operators: $X^I \phi$ is true if ϕ holds if the next state of the Markov chain is reached at time $\tau \in I = [t, t']$, while $\phi_1 U^I \phi_2$ is true if ϕ_2 is satisfied at some $\tau \in I$ and ϕ_1 holds at all preceding time instants [31].

   We define a *satisfaction function* to capture how the satisfaction probability of a given property over a model paths relates to its parameters and initial state.

Definition 6 (Satisfaction Function). *Let ϕ be a CSL formula, $\mathcal{M}_\theta$ be a parametrised model over a space Θ, s_0 is the initial state, and $Path^{\mathcal{M}_\theta}(s_0)$ is the set of all paths generated by $\mathcal{M}_\theta$ with initial state s_0. Denote by $\Lambda_\phi : \theta \to [0,1]$ the satisfaction function such that*

$$\Lambda_\phi(\theta) = P\left(\{\omega \in Path^{\mathcal{M}_\theta}(s_0) \models \varphi\} \mid \omega(0) = s_0\right), \tag{1}$$

where a path $\omega \models \varphi$ if its associated trace satisfies the path formula φ corresponding to the CSL formula ϕ. That is, $\Lambda_\phi(\theta)$ is the probability that the set of paths from a given pCMTC $\mathcal{M}_\theta$ satisfies a property φ. If $\Lambda_\phi(\theta) \sim \zeta$, then we say that $\mathcal{M}_\theta \models \phi$.

2.3 Bayesian Inference

Given a set of observations or data, y_{obs}, a parametrised model (either stochastic or deterministic), $\mathcal{M}_\theta$, and prior information, the task of Bayesian inference is to learn the true model parameter via its probability distribution. Prior beliefs about the model parameters, expressed through a probability distribution $\pi(\theta)$, are updated via y_{obs}, where assumptions on the model's dynamics are encoded into the likelihood function $p(y_{obs}|\theta)$. Using Bayes' theorem, the posterior distribution is obtained as $\pi(\theta|y_{obs}) = p(y_{obs}|\theta)\pi(\theta)/\pi(y_{obs})$. When likelihood functions are intractable one can resort to likelihood-free methods, such as Approximate Bayesian Computation (ABC) [44], to approximate this posterior as $\pi_{ABC}(\theta|y_{obs}) \approx \pi(\theta|y_{obs})$.

Intractable Likelihoods for CRNs. We discuss next why we resort to likelihood-free methods for inferring parameters of CRN networks from noisy data observed at discrete points in time. A biochemical reaction network model is a discrete-state, continuous-time Markov process, which can be described by the chemical master equation (CME) [19],

$$\frac{d\mathcal{P}(x,t|x_0)}{dt} = \sum_{j=1}^{M} \alpha_j(x,\theta_j)\mathcal{P}(x - v_j, t|x_0) - \mathcal{P}(x, t|x_0)\sum_{j=1}^{M}\alpha_j(x,\theta_j), \quad (2)$$

where $\mathcal{P}(x,t|x_0)$ is the probability that the state of the Markov chain at time t is $X(t) = x$, given $X(0) = x_0$, $v_{j,i} = u'_{j,i} - u_{j,i}$ with v_j being the jth column of the stoichiometric matrix $v_{j,i}$ and θ is the vector of kinetic parameters in the CRN. The solution of the CME characterises the exact probability that the model is in any state at any time. Unfortunately, analytic solutions to the CME are only known for very special (and restrictive) CRNs. Precise traces $y \sim p(y|\theta)$ from the CRN of interest $\mathcal{M}_\theta$ can be generated by the stochastic simulation algorithm [18]. Furthermore, often $X(t)$ cannot be observed directly. Instead, an observation of the state vector sample path is observed, $Y(t) = g(X(t))$, where g is an arbitrary observation function on $X(t)$.

To illustrate why we work with likelihood-free inference, we consider the simplest case that $Y(t) = X(t)$, that is, the entire trace can be perfectly observed at time t: in this simpler instance, the likelihood is given by

$$p(Y|\theta) = \prod_{i=1}^{W} \mathcal{P}(Y(t_i), \theta, t_i - t_{i-1}|Y(t_{i-1})), \quad (3)$$

where $\mathcal{P}$ is the solution to the chemical master equation which we note is dependent on the kinetic parameters θ, $t_0 = 0$ and W is the number of observations taken. So even in this simple case the likelihood depends on the solution to the CME, which is analytically intractable for many cases. As a result, the Bayesian posterior will not be analytically tractable, hence we resort to likelihood-free methods, such as ABC.

Approximate Bayesian Computation. ABC methods [44] produce an approximation to the posterior probability distribution when the likelihood $p(y|\theta)$ is intractable. The likelihood is approximated by matching simulated data $y \sim p(y|\theta)$ with the observed data y_{obs}, according to some function of the distance $\|y - y_{obs}\|$ or correspondingly over summary statistics of the simulated and observed data, namely $\|s - s_{obs}\|$.

Ideally, the observations y_{obs} are directly mapped to the variables of the model, which is endowed with sufficient statistics y. However, in many real world settings, model variables cannot be fully observed, and moreover outputs y can be perturbed by noise due to measurement error. Since it is in general hard to identify a finite-dimensional set of sufficient statistics, it is common and computationally advantageous to use (insufficient) summary statistics $s = S(y)$, where function S performs a simplification of the signals y (e.g., averaging, smoothing, or sampling), and which ideally is so that $\pi(\theta|y_{obs}) = \pi(\theta|s_{obs})$ [40].

The procedure is as follows: first samples are generated by $\theta^* \sim \pi(\theta)$, each of which is used to generate simulated data $y \sim p(y|\theta)$, where the proposed sample θ^* is accepted if $\|y - y_{obs}\| \le h$ for some $h \ge 0$, $h \in \mathbb{R}^+$, and rejected if $\|y - y_{obs}\| > h$. This procedure is equivalent to drawing a sample (θ, y) from the joint distribution

$$\pi_{ABC}(\theta, y|y_{obs}) \propto K_h(\|y - y_{obs}\|)p(y|\theta)\pi(\theta), \tag{4}$$

where $K_h(u)$ is a standard smoothing kernel function [43], which depends on a predetermined distance h and on $u = \|y - y_{obs}\|$. A standard choice we use for the smoothing kernel function is the indicator function, where $K_h(\|y - y_{obs}\|) = 1$ if $\|y - y_{obs}\| \le h$, and $K_h(\|y - y_{obs}\|) = 0$ otherwise. Accordingly, the ABC approximation to the true posterior distribution is

$$\pi_{ABC}(\theta|y_{obs}) = \int \pi_{ABC}(\theta, y|y_{obs})dy. \tag{5}$$

As $h \to 0$ samples from the true posterior distribution are obtained [44] as:

$$\lim_{h \to 0} \pi_{ABC}(\theta|y_{obs}) \propto \int \delta_{y_{obs}}(y)p(y|\theta)\pi(\theta)dy = p(y_{obs}|\theta)\pi(\theta),$$

where $\delta_{y_{obs}}(y)$ is the Dirac delta measure, where $\delta_x(A) = 1$ if $x \in A$ and $\delta_x(A) = 0$ otherwise. In practice, it is highly unlikely that $y \approx y_{obs}$ can be generated from $p(y|\theta)$, thus a non-trivial scale parameter h is needed. Furthermore, the full datasets y_{obs} and y are often replaced by summary statistics s_{obs} and s, respectively, leading to sampling from the posterior distribution $\pi_{ABC}(\theta|s_{obs})$. The ABC approximation to $\pi(\theta|s_{obs})$ is given by

$$\pi_{ABC}(\theta|s_{obs}) \propto \int K_h(\|s - s_{obs}\|)p(y|\theta)\pi(\theta)dy, \tag{6}$$

where, by slight abuse of notation, $K_h(\|s - s_{obs}\|)$ is defined as for y, y_{obs}.

Approximate Bayesian Computation - Sequential Monte Carlo. The major issue with standard ABC is that if the prior $\pi(\theta)$ differs from the posterior distribution, $p(\theta|y_{obs})$, then the acceptance rates, namely the rates at which sampled parameters are accepted, will be low, thus resulting in more proposed parameters and associated simulations, which leads to an increase in computational burden. Approximate Bayesian Computation - Sequential Monte Carlo (ABCSMC) [47] techniques are developed to mitigate this issue. ABC-SMC algorithms [46,47] (cf. Appendix A) are designed to overcome this burden by constructing a sequence of slowly-changing intermediate distributions, $f_m(\theta)$, $m = 0, \ldots, M$, where $f_0(\theta) = \pi(\theta)$ is the initial sampling distribution and $f_M(\theta) = f(\theta)$ is the target distribution of interest, namely the approximated posterior, $\pi_{ABC}(\theta|s_{obs})$. A population of particles or samples from generation m, $\theta_m^{(i)}$, where $i = 1, \ldots, N$ is the number of particles, are propagated between these distributions sequentially, so that these intermediary distributions act as an importance sampling scheme [44], which is a technique used to sample from a distribution that over-weights specific regions of interest. This technique attempts to bridge the gap between the prior $\pi(\theta)$ and the (unknown) posterior $\pi(\theta|s_{obs})$. In the ABCSMC framework, a natural choice for the sequence of intermediary distributions is

$$f_m(\theta) = \pi_{ABC}^{h_m}(\theta, s|s_{obs}) \propto K_{h_m}(\|s - s_{obs}\|)p(y|\theta)\pi(\theta), \qquad (7)$$

where $m = 0, \ldots, M$ and h_m is a monotonically decreasing sequence, namely such that $h_m > h_{m+1} \geq 0$. As above, K_{h_m} is the standard smoothing kernel, which now depends on the distance h_m. We expect that $\lim_{h_m \to 0} \pi_{ABC}^{h_m}(\theta|s_{obs}) = \pi(\theta|s_{obs})$ [44], and that the more samples N are generated, the more accurate the approximated quantity will become.

A key part of the ABCSMC scheme is the generation of samples θ^* and the setting of weights (which is typical for other importance sampling schemes). Sample θ^* is initially ($m = 0$) taken from the prior and subsequently ($m > 0$) sampled from the intermediary distributions $f_{m-1}(\theta)$ through its corresponding weights (see below), as parameter $\theta_{m-1}^{(j)}$. Afterwards, θ^* is perturbed into θ^{**} by a kernel, $F_m(\theta^{**}|\theta^*)$. For the perturbed parameter, θ^{**}, a number of B_t simulations y_b, and in turn s^b, are generated from $p(y|\theta^{**})$, and the quantity $b_t(\theta^{**}) = \sum_{b=1}^{B_t} K_{h_m}(\|s^b - s_{obs}\|)$ is calculated. If $b_t(\theta^{**}) = 0$, then θ^{**} is discarded and we resample θ^* again. Otherwise, the accepted θ^{**} results in the pair $\{\theta_m^{(i)}, w_m^{(i)}\}$, where the corresponding weights $w_m^{(i)}$ are set to

$$w_m^{(i)} = \begin{cases} b_t\left(\theta_m^{(i)}\right), & \text{if } m = 0 \\[2ex] \dfrac{\pi\left(\theta_m^{(i)}\right) b_t\left(\theta_m^{(i)}\right)}{\sum_{j=1}^{N} w_{m-1}^{(j)} F_m\left(\theta_m^{(i)}|\theta_{m-1}^{(j)}\right)}, & \text{if } m > 0 \end{cases} \qquad (8)$$

and later normalised, after re-sampling each ith particle, $i = 1, \ldots, N$. If B_t is large, the estimate of $\pi_{ABC}(\theta|s_{obs})$ is accurate, which implies the acceptance probability is accurate, but it might come at the cost of many Monte Carlo draws. Conversely, if B_t is small, the acceptance probability is cheaper to evaluate [4] but can become highly variable.

The algorithm controls the transitioning between the intermediary distributions $f_{m-1}(\theta)$ and $f_m(\theta)$, by setting a user-inputted rate v, at which the thresholds h_m reduce until the algorithm stops. Stopping rules for ABCSMC schemes vary: here, we have opted for terminating the algorithm after a predetermined number M of steps (a.k.a. epochs). The algorithm returns weighted samples,

$$\left\{\theta_M^{(i)}, w_M^{(i)}\right\} \sim \pi_{ABC}^{h_M}(\theta|s_{obs}) \propto \int K_{h_M}(\|s - s_{obs}\|)p(y|\theta)\pi(\theta)dy.$$

2.4 Statistical Model Checking with the Massart Algorithm

Statistical model checking (SMC) techniques are used to estimate the validity of quantitative properties of probabilistic systems by simulating traces from an executable model of the system [33]. Unlike precise (up to numerics) probabilisitic model checking, SMC results are typically attained with statistical precision and can come, in particular, with confidence bounds [13,35]. In this work, we require Monte Carlo simulations to estimate the probability of properties of interest with a user-defined degree of accuracy (denoted below as ϵ). This can be obtained via standard concentration inequalities, such as the Chernoff [13] or the Okamoto [37] bounds. We wish to estimate a probability $\hat{\Lambda}_\phi(\theta)$ that approximates the unknown $\Lambda_\phi(\theta)$ within an absolute error ϵ and with a $(1 - \delta)$ confidence lower bound, namely

$$P(|\hat{\Lambda}_\phi(\theta) - \Lambda_\phi(\theta)| > \epsilon) \leq \delta. \tag{9}$$

For instance, the Okamoto bound ensures that drawing $n \geq n_{\mathcal{O}} = \lceil \frac{1}{2\epsilon^2} \log \frac{2}{\delta} \rceil$ simulations, results in an estimate $\hat{\Lambda}_\phi(\theta)$ with a statistical guarantee as in (9), where $\delta = 2 \exp\left(-2n\epsilon^2\right)$.

In this work, we leverage the sharper Massart bounds [34]: we use the Sequential Massart algorithm [25,26] (described below), which progressively defines confidence intervals of the estimated probability and then applies the Massart bounds [34]. Massart bounds depend on the unknown probability $\Lambda_\phi(\theta)$ that we are estimating, which forces one to numerically evaluate with certainty an interval in which $\Lambda_\phi(\theta)$ evolves. Let us denote by $C(\Lambda_\phi(\theta), I)$ the coverage of $\Lambda_\phi(\theta)$ by a confidence interval I, i.e., the probability that $\Lambda_\phi(\theta) \in I$.

Theorem 1 (Absolute-Error Massart Bound with Coverage [26]). *Let $\hat{\Lambda}_\phi(\theta)$ be the probability estimated from n Monte Carlo simulations, ϵ be a given error, $\hat{\Lambda}_\phi^L(\theta)$ and $\hat{\Lambda}_\phi^U(\theta)$ be the lower and upper bounds of a confidence interval $I = [\hat{\Lambda}_\phi^L(\theta), \hat{\Lambda}_\phi^U(\theta)]$ and I^c be its complement within $[0, 1]$. The Massart bound is defined as*

$$P(|\hat{\Lambda}_\phi(\theta) - \Lambda_\phi(\theta)| > \epsilon) \leq 2 \exp\left(-n\epsilon^2 h_a(\Lambda_\phi(\theta), \epsilon)\right) + C(\Lambda_\phi(\theta), I^c), \tag{10}$$

$$where \ h_a(\Lambda_\phi(\theta), \epsilon) = \begin{cases} \frac{9}{2} \frac{1}{(3\Lambda_\phi(\theta)+\epsilon)(3(1-\Lambda_\phi(\theta))-\epsilon)}, & if \ 0 < \Lambda_\phi(\theta) < 1/2, \\ \frac{9}{2} \frac{1}{(3(1-\Lambda_\phi(\theta))+\epsilon)(3\Lambda_\phi(\theta)+\epsilon)}, & if \ 1/2 \leq \Lambda_\phi(\theta) < 1. \end{cases}$$

Notice that the above theorem requires the true satisfaction probability $\Lambda_\phi(\theta)$, which is not known. We can replace it with its estimate $\hat{\Lambda}_\phi(\theta)$, which can be conservatively set to $\hat{\Lambda}_\phi(\theta) = \hat{\Lambda}_\phi^U(\theta)$ if $\hat{\Lambda}_\phi^U(\theta) < 1/2$, $\hat{\Lambda}_\phi(\theta) = \hat{\Lambda}_\phi^L(\theta)$ if $\hat{\Lambda}_\phi^L(\theta) > 1/2$, and $\hat{\Lambda}_\phi(\theta) = 1/2$ if $1/2 \in I$. The following sample-size result follows:

Theorem 2 ([26]). *Let α be a coverage parameter chosen such that $\alpha < \delta$ and $C(\Lambda_\phi(\theta), I^c) < \alpha$. Under the conditions of Theorem 1, a Monte Carlo algorithm $\mathcal{A}$ that outputs an estimate $\hat{\Lambda}_\phi(\theta)$ fulfils the condition in (9) if $n > \lceil \frac{1}{h_a(\Lambda_\phi(\theta),\epsilon)\epsilon^2} \log \frac{2}{\delta-\alpha} \rceil$.*

The Sequential Massart Algorithm requires three inputs: an error parameter ϵ and two confidence parameters δ and α. Initially, $\hat{\Lambda}_\phi^L(\theta) = 0$, $\hat{\Lambda}_\phi^U(\theta) = 1$, $C(\Lambda_\phi(\theta), [0,1]^c) = 0$, and $\hat{\Lambda}_\phi(\theta) = 1/2$, which results in the Okamoto-like bound with $h_a(1/2, \epsilon) \approx 2$ when $\epsilon \to 0$: the quantity $n_\mathcal{O} = \lceil \frac{1}{2\epsilon^2} \log \frac{2}{\delta} \rceil$ thus represents an upper-bound on the number of simulations required for the statistical guarantees. After each sampled trace, we update both a Monte Carlo estimator and a $(1-\alpha)$-confidence interval for $\Lambda_\phi(\theta)$. The updated confidence interval is then used in the Massart function to compute an updated required sample size n satisfying Theorem 2. This process is repeated until the calculated sample size is lower than or equal to the current number of simulated traces.

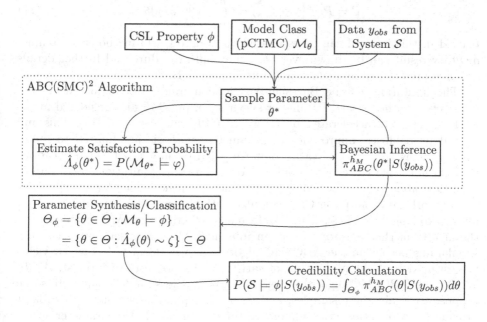

Fig. 1. Bayesian Verification via ABC(SMC)^2.

2.5 Bayesian Verification

In this work we extend the Bayesian Verification framework (cf. Fig. 4 in the Appendix) introduced in [36], which addresses the following problem. Consider a data generating stochastic system $\mathcal{S}$ (in this work, a CRN), where we denote the generated data as y_{obs}. We are interested in verifying a CSL property of interest ϕ over system $\mathcal{S}$ using sampled observations of the underlying system, y_{obs}, or a summary statistics $s_{obs} = S(y_{obs})$ thereof. We assume this goal cannot be reliably attained by means of statistical techniques directly applied on data y_{obs} (for instance, in the case studies later, we can access only very few observations). To tackle this problem, we integrate model-based techniques (formal verification) with the use of data (Bayesian inference). Suppose that we have sufficient knowledge to propose a parametric model that adequately describes the underlying system, $\mathcal{M}_\theta$. We employ Bayesian inference to learn the posterior probability distribution of the model, namely $\pi(\theta|s_{obs})$ from (possibly scarce) data s_{obs}. We also use this parametric model to formally verify the property of interest ϕ, as follows. We synthesise two complementary parameter regions, $\Theta_\phi = \{\theta \in \Theta : \mathcal{M}_\theta \models \phi\}$ and $\Theta_{\neg\phi} = \{\theta \in \Theta : \mathcal{M}_\theta \not\models \phi\}$. We then integrate the inferred posterior probability distribution over Θ_ϕ to obtain the credibility calculation, which represents the probability that the underlying system $\mathcal{S}$ satisfies the property:

$$C = P(\mathcal{S} \models \phi|s_{obs}) = \int_{\Theta_\phi} \pi(\theta|s_{obs})\mathrm{d}\theta. \tag{11}$$

If needed, this integral can be estimated via Monte Carlo methods. A complementary result can be drawn over $\Theta_{\neg\phi}$. The full procedure and further details are presented in [36] and summarised in Appendix B.

The limitations of the Bayesian Verification framework of [36] lie in the parameter synthesis part. Parameter synthesis of pCTMCs is considered in the work of [11], and accelerated by means of GPU processing in [12]. This and related probabilistic approaches to parameter synthesis are limited to finite-state systems that can be easily uniformised. In many practical applications they do not scale to realistic models. To address this limitation, we resort to statistical approaches (via SMC) for parameter synthesis, similar to [8]: whilst in this work we zoom in on CSL formulae with the provided semantics, it is of interest to look beyond this logic. We formally integrate the SMC technique into the algorithm that performs Bayesian inference. More precisely, we utilise the simulations needed in the ABCSMC algorithm to perform SMC, which yields the estimation of the probability of satisfying the property of interest, $\hat{\Lambda}_\phi(\theta)$. Whilst the ABCSMC algorithm rejects parts of the sampled parameters, we propose to retain these samples, and their corresponding simulations, to later provide a classification (via support vector machines) of the parameter space. With these statistically-estimated parameter regions, we complete the Bayesian Verification framework, as per (11). The new framework (detailed in the next section and presented in Fig. 1), which employs models to extract information from the observation data s_{obs}, is likelihood-free and entirely based on simulations, which makes it usable with models of different size and structure.

3   ABC(SMC)²: Approximate Bayesian Computation - Sequential Monte Carlo with Statistical Model Checking

We address the scalability limitations of our previous work [36], and specifically the parameter synthesis part: in [36] the synthesis was calculated symbolically, which practically limited the applicability to CTMCs with small state spaces and a few parameters. We incorporate here statistical model checking within the Bayesian inference framework and instead *estimate* parameter regions. We name the modified algorithm Approximate Bayesian Computation - Sequential Monte Carlo with Statistical Model Checking: ABC(SMC)² and present it in Fig. 1.

In the ABCSMC scheme (Algorithm 2 in Appendix A), a total of B_t simulations are performed for each sampled parameter θ^{**}, whether the sample is retained or not towards the approximate posterior $\pi_{ABC}^{h_M}(\theta|s_{obs})$: this leads to a considerable amount of wasted computational effort. We propose instead to statistically model check (SMC) each of the sampled parametrised models by means of the generated simulations, whilst parameter inference on the model is run (ABCSMC); we shall use the outcome of the SMC algorithm for the Bayesian Verification framework, by classifying the parameter synthesis regions using statistical approaches.

At any of the M iterations, for each sampled point $\theta^{**} \in \Theta$, we estimate the probability $\hat{\Lambda}_\phi(\theta^{**}) \approx \Lambda_\phi(\theta^{**})$, with statistical guarantees, that an instantiated model $\mathcal{M}_{\theta^{**}}$ satisfies a given property of interest ϕ, namely $P(\mathcal{M}_{\theta^{**}} \models \varphi) = \hat{\Lambda}_\phi(\theta^{**})$. We then proceed with the ABCSMC algorithm as normal, calculating whether the sampled parameter θ^{**} contributes to the approximate posterior (acceptance) or not (rejection). In addition to producing samples $\{\theta_{h_M}^{(i)}, w_{h_M}^{(i)}\}$, which allows one to construct an approximation of the posterior distribution $\pi_{ABC}^{h_M}(\theta|s_{obs})$, the algorithm outputs $\left\{\theta^{**}, \hat{\Lambda}_\phi(\theta^{**}), \hat{\Lambda}_\phi^L(\theta^{**}), \hat{\Lambda}_\phi^U(\theta^{**})\right\}$ for all the sampled parameters θ^{**} (whether accepted or not). These values are later used to train an SVM classifier to generate the parameter synthesis regions. We shall then integrate the approximate posterior over the classified parameter regions, to obtain a credibility calculation.

3.1   ABC(SMC)²

Recall that the output of the ABCSMC algorithm is a set of samples $\theta_M^{(i)}$ with their corresponding weights $w_M^{(i)}$, which satisfy the following:

$$\{\theta_M^{(i)}, w_M^{(i)}\} \sim \pi_{ABC}^{h_M}(\theta|s_{obs}) \propto \int K_{h_M}(\|s - s_{obs}\|) p(y|\theta)\pi(\theta)dy, \qquad (12)$$

where $i = 1, \ldots, N$ is the number of particles used to approximate the posterior. For each parameter θ^{**}, simulation data is generated from the model $y_b \sim p(y|\theta^{**})$ to calculate $s^b = S(y_b)$, for a total of B_t times, and this data is used to estimate $\hat{\Lambda}_\phi(\theta^{**}) \approx \Lambda_\phi(\theta^{**})$.

We utilise the sequential Massart algorithm [26] presented in the previous section for this SMC procedure. We replace the number of simulations for each sampled parameter, B_t, with the calculated minimum number of samples estimated in the sequential Massart algorithm [26], $B_t = n \leq n_{\mathcal{O}}$, to calculate an estimated probability $\hat{\Lambda}_\phi(\theta^{**})$ with accuracy and confidence. We sample θ^{**} a total of R times, whether or not these samples are accepted as samples from the posterior at any generation m. For these sampled parameters, $\theta^{(r)}$, $r = 1, \ldots, R$, we estimate the corresponding mean estimated probabilities $\hat{\Lambda}_\phi\left(\theta^{(r)}\right)$ and $(1-\delta)$ uncertainty bounds: $\left\{\theta^{(r)}, \hat{\Lambda}_\phi\left(\theta^{(r)}\right), \hat{\Lambda}_\phi^L\left(\theta^{(r)}\right), \hat{\Lambda}_\phi^U\left(\theta^{(r)}\right)\right\}$. Here R depends on the acceptance rate of the sampled parameters $\theta^{(r)}$, where $R \geq N \times M$, where N is the number of particles to sample and M is the total number of generations of the ABCSMC scheme. From this new algorithm, we obtain a set of weighted parameter vectors from the final generation M, $\{\theta_M^{(i)}, w_M^{(i)}\} \sim \pi_{ABC}^{h_M}(\theta|s_{obs})$ where $i = 1, \ldots, N$, as well as R sampled parameters and their corresponding estimated probabilities $\left\{\theta^{(r)}, \hat{\Lambda}_\phi\left(\theta^{(r)}\right), \hat{\Lambda}_\phi^L\left(\theta^{(r)}\right), \hat{\Lambda}_\phi^U\left(\theta^{(r)}\right)\right\}_{r=1}^R$.

We present the ABC(SMC)^2 scheme in Algorithm 1, with the $MASSART$ function corresponding to the Absolute-Error Massart Algorithm presented in Appendix C. The ABC(SMC)^2 algorithm takes as inputs a property of interest, ϕ, a prior probability distribution $\pi(\theta)$ an absolute-error tolerance ϵ as well as a coverage parameter α and confidence value δ. The estimated probabilities $\left\{\theta^{(r)}, \hat{\Lambda}_\phi(\theta)^{(r)}, \hat{\Lambda}_\phi^L(\theta)^{(r)}, \hat{\Lambda}_\phi^U(\theta)^{(r)}\right\}$, will be utilised for approximate parameter synthesis, which is discussed in the next section.

3.2   Approximate Parameter Synthesis via Statistical MC

The aim of parameter synthesis is to partition the parameter space Θ according to the satisfaction of the CSL property ϕ. Unlike the PMC-based synthesis in [36] (recalled in Sect. 2.5), we utilise a statistical approach to classify the parameter space, akin to [8]. So instead of employing the true satisfaction probability $\Lambda_\phi(\theta) \sim \zeta$ (where ζ is the probability bound contained in formula ϕ) to determine Θ_ϕ (and its complement), we use $\hat{\Lambda}_\phi(\theta^{(r)})$, a statistical approximation computed at each sampled parameter point $\theta^{(r)}$. Evidently, recalling the confidence parameter δ, we should compute $\hat{\Lambda}_\phi(\theta^{(r)}) \sim \zeta \pm \epsilon$ (where the sign $\pm$ depends on the direction of the inequality $\sim$).

In practice, we use the estimated lower $\hat{\Lambda}_\phi^L(\theta^{(r)})$ and upper bounds $\hat{\Lambda}_\phi^U(\theta^{(r)})$, such that $\Lambda_\phi(\theta^{(r)}) \in \left[\hat{\Lambda}_\phi^L(\theta^{(r)}), \hat{\Lambda}_\phi^U(\theta^{(r)})\right]$, to partition the parameter space as:

- $\Theta_\phi = \{\theta \in \Theta : \hat{\Lambda}_\phi^L(\theta) > \zeta\}$,
- $\Theta_{\neg\phi} = \{\theta \in \Theta : \hat{\Lambda}_\phi^U(\theta) < \zeta\}$,
- $\Theta_{\mathcal{U}} = \Theta \backslash (\Theta_\phi \cup \Theta_{\neg\phi})$.

Notice that these formulas are a function of $\theta \in \Theta$. Since in the ABC(SMC)^2 procedure we generate a finite number of parameter samples $\theta^{(r)}$, which are biased towards the sought posterior distribution, there might be areas of the parameter space Θ that are insufficiently covered. We thus resort to supervised

Algorithm 1. ABC(SMC)^2

Input:
- CSL specification ϕ
- Prior distribution $\pi(\theta)$ and data generating function $p(y|\theta)$
- A kernel function $K_h(u)$ and scale parameter $h > 0$ where $u = \|y - y_{obs}\|$
- $N > 0$, number of particles used to estimate posterior distributions
- Sequence of perturbation kernels $F_m(\theta|\theta^*)$, $m = 1, \ldots, M$
- A quantile $v \in [0, 1]$ to control the rate of decrease of thresholds h_m
- Summary statistic function $s = S(y)$
- Parameters for statistical MC: absolute-error value ϵ, confidence δ, coverage α

Output:
- Set of weighted parameter vectors $\left\{\theta_M^{(i)}, w_M^{(i)}\right\}_{i=1}^N$ drawn from $\pi_{ABC}(\theta|s_{obs}) \propto \int K_{h_M}(\|s - s_{obs}\|)p(s|\theta)\pi(\theta)ds$
- Set of parameters with corresponding estimated mean, $\hat{\Lambda}_\phi\left(\theta^{(r)}\right)$ and $(1 - \delta)$ confidence interval $\left[\hat{\Lambda}_\phi^L\left(\theta^{(r)}\right), \hat{\Lambda}_\phi^U\left(\theta^{(r)}\right)\right]$ of estimated probability to satisfy ϕ, $P\left(\mathcal{M}_{\theta^{(r)}} \models \varphi\right) = \hat{\Lambda}_\phi\left(\theta^{(r)}\right)$: $\left\{\theta^{(r)}, \hat{\Lambda}_\phi\left(\theta^{(r)}\right), \hat{\Lambda}_\phi^L\left(\theta^{(r)}\right), \hat{\Lambda}_\phi^U\left(\theta^{(r)}\right)\right\}$

1: Set $r = 0$
2: **for** $m = 0, \ldots, M$: **do**
3:    **for** $i = 0, \ldots, N$: **do**
4:      **if** $m = 0$ **then**
5:        Sample $\theta^{**} \sim \pi(\theta)$
6:      **else**
7:        Sample θ^* from the previous population $\{\theta_{m-1}^{(i)}\}$ with weights $\{w_{m-1}^{(i)}\}$ and perturb the particle to obtain $\theta^{**} \sim F_m(\theta|\theta^*)$
8:      **end if.**
9:      **if** $\pi(\theta^{**}) = 0$ **then**
10:       **goto** line 3
11:     **end if**
12:     Calculate $\left(\left\{\hat{\Lambda}_\phi(\theta^{**}), [\hat{\Lambda}_\phi^L(\theta^{**}), \hat{\Lambda}_\phi^U(\theta^{**})]\right\}, B_t, \sum_{b=1}^{B_t} K_{h_m}(\|s^b - s_{obs}\|), \bar{d}\right)$ from the modified Massart Algorithm: $MASSART\left(\epsilon, \delta, \alpha, h_m, \theta^{**}, s_{obs}\right)$
13:     Calculate $b_t(\theta^{**}) = \frac{1}{B_t}\sum_{b=1}^{B_t} K_{h_m}(\|s^b - s_{obs}\|)$
14:     Set $\left(\theta^{(r)}, \hat{\Lambda}_\phi\left(\theta^{(r)}\right), \hat{\Lambda}_\phi^L\left(\theta^{(r)}\right), \hat{\Lambda}_\phi^U\left(\theta^{(r)}\right)\right) = \left(\theta^{**}, \hat{\Lambda}_\phi(\theta^{**}), \hat{\Lambda}_\phi^L(\theta^{**}), \hat{\Lambda}_\phi^U(\theta^{**})\right)$
15:     $r \leftarrow r + 1$
16:     **if** $b_t(\theta^{**}) = 0$ **then**
17:       **goto** line 3
18:     **end if**
19:     Set $\theta_m^{(i)} = \theta^{**}$, $\bar{d}_m^{(i)} = \bar{d} = \frac{1}{B_t}\sum_{b=1}^{B_t}\|s^b - s_{obs}\|$ and calculate

20:
$$w_m^{(i)} = \begin{cases} b_t\left(\theta_m^{(i)}\right), & \text{if } m = 0 \\ \dfrac{\pi\left(\theta_m^{(i)}\right) b_t\left(\theta_m^{(i)}\right)}{\sum_{j=1}^N w_{m-1}^{(j)} F_m\left(\theta_m^{(i)}|\theta_{m-1}^{(j)}\right)}, & \text{if } m > 0 \end{cases}$$

21:   **end for**
22:   Normalise weights: $w_m^{(i)} \leftarrow w_m^{(i)} / \left(\sum_{i=1}^N w_m^{(i)}\right)$
23:   Set $h_{m+1} = (v/N)\sum_{i=1}^N \bar{d}_m^{(i)}$
24: **end for**
25: **return** $\left\{\left(\theta_M^{(i)}, w_M^{(i)}\right)\right\}_{i=1}^N, \left\{\theta^{(r)}, \hat{\Lambda}_\phi\left(\theta^{(r)}\right), \hat{\Lambda}_\phi^L\left(\theta^{(r)}\right), \hat{\Lambda}_\phi^U\left(\theta^{(r)}\right)\right\}_{r=1}^R$

learning techniques to classify parameter synthesis regions: we utilise support vector machines (SVMs) [14,45] as a classification technique. We train the SVM classifier on the data produced from the ABC(SMC)^2 algorithm, namely on the set $\{\theta^{(r)}, \hat{\Lambda}_\phi(\theta^{(r)}), \hat{\Lambda}_\phi^L(\theta^{(r)}), \hat{\Lambda}_\phi^U(\theta^{(r)})\}$ where $r = 1, \ldots, R$. The SVM which is trained on this data then provides a non-linear classifying function, $\xi_\phi(\theta)$, where $\xi_\phi(\theta) = 1$ if $\theta \in \Theta_\phi$, $\xi_\phi(\theta) = -1$ if $\theta \in \Theta_{\neg\phi}$ and $\xi_\phi(\theta) = 0$ if $\theta \in \Theta_\mathcal{U}$.

4   Experiments

Experimental Setup. All experiments have been run on an Intel(R) Xeon(R) CPU E5-1660 v3 @ 3.00 GHz, 16 cores with 16 GB memory. ABC(SMC)^2 is coded in C++, while Python is used for the SVM classifier. A comparison of the parameter synthesis technique via PRISM or via SMC and SVM can be seen in Appendix D.

SIR System and Parameterised Model. Towards an accessible explanation of the ABC(SMC)^2 algorithm, we consider the stochastic SIR epidemic model [29], which has the same structure (stoichiometry over species counts) as CRNs [12]. The model describes the dynamics of three epidemic types, a susceptible group (S), an infected group (I), and a recovered group of individuals (R) - here we let S, I and R evolve via the rules

$$S + I \xrightarrow{k_i} I + I, \quad I \xrightarrow{k_r} R.$$

This is governed by the rate parameters $\theta = (k_i, k_r)$, and each state of the pCTMC describes the combination of the number of each type (S, I, R) (this equates to molecule/species counts in CRNs). The initial state of the pCTMC is $s_0 = (S_0, I_0, R_0) = (95, 5, 0)$. We wish to verify the property $\phi = P_{>0.1}((I > 0)U^{[100,150]}(I = 0))$, i.e. whether, with a probability greater than 0.1, the infection dies out within a time interval between $t = 100$ and $t = 150$ s. We confine our parameters to the set $\Theta = [k_i^\perp, k_i^\top] \times [k_r^\perp, k_r^\top] = [5 \times 10^{-5}, 0.003] \times [0.005, 0.2]$. We generate observation data from the SIR model with three different parameter choices, corresponding to the CTMCs $\mathcal{M}_{\theta_\phi}$, $\mathcal{M}_{\theta_{\neg\phi}}$ and $\mathcal{M}_{\theta_\mathcal{U}}$, where $\theta_\phi = (0.002, 0.075)$, $\theta_{\neg\phi} = (0.001, 0.15)$ and $\theta_\mathcal{U} = (0.002, 0.125)$. From Fig. 5a (in Appendix D), we see that $\theta_\phi \in \Theta_\phi$, $\theta_{\neg\phi} \in \Theta_{\neg\phi}$, and finally $\theta_\mathcal{U}$ is near the borderline. These models will correspond to three "true" underlying stochastic systems $\mathcal{S}$, with associated observation data. For each instance, we work with observed data y_{obs} that is sampled at a finite number of time steps. The observed data consists of only 5 simulated traces, observed at 10 time points. The summary statistics $S(y_{obs}) = s_{obs}$ is the average of the 5 traces. It is worth emphasising that with so few observation traces, black-box SMC (directly based on observation traces, not on model-generated simulations) would be hopeless.

Application of ABC(SMC)^2 Algorithm. Our algorithm outputs samples from the approximated posterior and their corresponding weights, $\left\{\theta_M^{(i)}, w_M^{(i)}\right\} \sim \pi_{ABC}^{h_M}(\theta|s_{obs})$ where $i = 1, \ldots, N$. By the strong law of large numbers, letting $\bar{\theta}_M = \sum_{i=1}^N \theta_M^{(i)} w_M^{(i)}$, $P\left(\lim_{N\to\infty} \sum_{i=1}^N w_M^{(i)} \theta_M^{(i)} - \mathbb{E}[\bar{\theta}_M] = 0\right) = 1$.

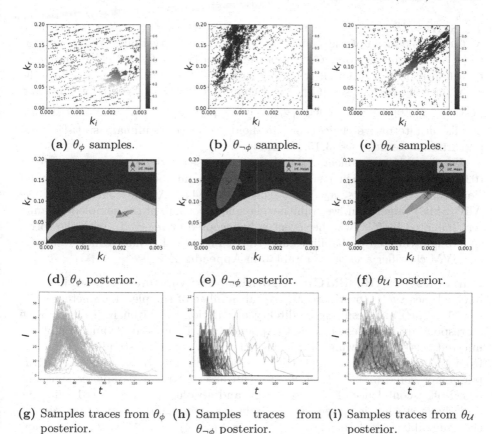

(a) θ_ϕ samples.

(b) $\theta_{\neg\phi}$ samples.

(c) $\theta_\mathcal{U}$ samples.

(d) θ_ϕ posterior.

(e) $\theta_{\neg\phi}$ posterior.

(f) $\theta_\mathcal{U}$ posterior.

(g) Samples traces from θ_ϕ posterior.

(h) Samples traces from $\theta_{\neg\phi}$ posterior.

(i) Samples traces from $\theta_\mathcal{U}$ posterior.

Fig. 2. Bayesian Verification results from ABC(SMC)² for Case θ_ϕ (2a and 2d), Case $\theta_{\neg\phi}$(2b and 2e), and Case $\theta_\mathcal{U}$ (2c and 2f). Sampled points θ with estimated probabilities $\hat{\Lambda}_\phi(\theta)$ (2a, 2b and 2c). Inferred posterior $\pi_{ABC}^{h_M}(\theta|s_{obs})$ and parameter regions (2d, 2e and 2f). Traces of I molecules simulated from $\mathcal{M}_\theta$, with θ sampled from the θ_ϕ posterior (2g), the $\theta_{\neg\phi}$ posterior (2h) and the $\theta_\mathcal{U}$ posterior (2i). The set Θ_ϕ, is shown in yellow, whereas $\Theta_{\neg\phi}$ is shown in blue. The undecided areas $\Theta_\mathcal{U}$ is shown in magenta. For Fig. (2g–2i) the traces colour represent which set the sampled parameters are members of. (Color figure online)

Thus we assume that the approximated posterior can be modelled by a multivariate Normal distribution, $\pi_{ABC}^{h_M}(\theta|s_{obs}) \approx \mathcal{N}(\bar{\theta}_M, \Sigma_M)$, where the mean is given by $\bar{\theta}_M$ and the elements of the empirical covariance matrix are defined as

$$\Sigma_{Mjk} = \frac{1}{1 - \sum_{i=1}^{N}(w_M^{(i)})^2} \sum_{i=1}^{N} w_M^{(i)} \left(\theta_M^{(i)} - \bar{\theta}_M\right)_j \left(\theta_M^{(i)} - \bar{\theta}_M\right)_k.$$

We choose the number of samples to be $N = 500$; the number of sequential steps to be $M = 20$; the kernel function $K_h(u)$ to be a simple indicator function, i.e. $K_h(u) = 1$ if $u < h$, $K_h(u) = 0$ otherwise; the rate at which the thresholds

h_m decrease to be $v = 0.5$; and the summary statistic $s = S(y)$ is chosen to be the sample mean of the simulations and of the observations. We choose $\pi(\theta)$ to be a uniform prior over Θ. The perturbation kernel $F_m(\theta^{**}|\theta^*)$ is chosen to be a multivariate Normal distribution, so that $\theta^{**} \sim \mathcal{N}(\theta^*, 2\Sigma_{m-1})$, where the covariance is twice the second moment computed over the accepted weights and particles at step $m - 1$, namely $\left\{ \theta_{m-1}^{(i)}, w_{m-1}^{(i)} \right\}$, where $i = 1, \dots, N$. For further details on alternative choices for threshold sequences, summary statistics and perturbation kernels, see [3,15,16,41,44].

For the SMC component of the algorithm, we select the parameters $(\epsilon, \delta, \alpha) = (0.01, 0.05, 0.001)$, which results in a maximum number of necessary simulations that equals $B_t \leq n_\mathcal{O} = \lceil \frac{1}{2\epsilon^2} \log \frac{2}{\delta} \rceil = 18445$. At the conclusion of the ABC(SMC)^2 algorithm, we train the classifier over half of the sampled parameters (denoted by $\theta^{(r)}$, whether eventually accepted or rejected), with the corresponding estimated probabilities and test it on the other half, which results in the SVM classifier accuracy in Table 3 in Appendix E.

Outcomes of ABC(SMC)^2 Algorithm. For the three case studies, the inferred mean $\bar{\theta}_M$, covariance Σ_M, total number of sampled parameters ($\theta^{(r)}$, $r = 1, \dots, R$) and resulting credibility calculation are given in Table 1, with corresponding runtimes in Table 4 (Appendix E). Figures 2d, 2e and 2f plot the inferred posterior, showing the mean (denoted by $\times$) and 2 standard deviations from the mean (corresponding ellipse around the mean), as well as the true parameter value ($\triangle$). In Case θ_ϕ, we can assert, with a parameter synthesis based off a confidence of $(1 - \delta) = 0.95$ and absolute-error $\epsilon = 0.01$, that the underlying stochastic system $\mathcal{S}$ does indeed satisfy the property of interest, as the credibility calculation gives $P(\mathcal{S} \models \phi | S(y_{obs})) = 1$. Case $\theta_{\neg\phi}$ has a low probability of satisfying the property of interest ($P(\mathcal{S} \models \phi | S(y_{obs})) = 0.0054$), whereas for Case $\theta_\mathcal{U}$ the inferred mean converges to the true mean that we would expect the estimated probability of satisfying the property to converge to, which is 0.5.

Table 3 and Fig. 3 suggest that simulation times are largely dependent on the estimated probabilities, $\hat{\Lambda}_\phi(\theta)$: the closer the estimated probabilities are to 0.5,

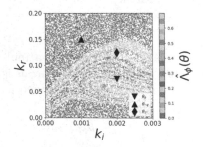

Fig. 3. True parameter values with corresponding estimated probabilities using SMC (15000 uniform samples).

Table 1. Number of SMC simulations used in ABC(SMC)^2.

Case	$\hat{\Lambda}_\phi(\theta)$	Total simulations
θ_ϕ	0.47254	18445
$\theta_{\neg\phi}$	0.00408719	2202
$\theta_\mathcal{U}$	0.100433	14775

the larger the number of simulations required. To improve the runtime of Case $\theta_{\mathcal{U}}$, we would need to reduce variance and improve the accuracy of the inferred parameters, for instance by increasing the number of observed data points y_{obs} or with an alternative choice of either the summary statistics chosen or of the perturbation kernels [16].

5   Discussion and Future Work

The new ABC(SMC)^2 framework allows the Bayesian Verification framework of [36] to be applied to a wide variety of models. In ABC(SMC)^2 we have newly utilised the simulations needed for the likelihood-free inference (also present in [36]) for statistical model checking of properties of interest and used the outputs of this procedure to allow for approximate parameter synthesis. The ABC(SMC)^2 framework presented here hinges largely on the ABCSMC scheme of [47] and is thus bound to its limitations: theoretically, there is nothing stopping the framework to be considered for models with a higher number of latent variables, but we would expect a higher runtime due to more proposals $F_m(\theta^{**}|\theta^*)$ being rejected due to the higher dimensionality and would thus need to employ alternative proposal distributions to those considered here. Another limitation of the framework is the dependence on a possibly large number of simulations for parameter synthesis (the SMC part), which however it is a strong alternative to the parameter synthesis technique of [12] that leverages GPU acceleration. To address possibly large simulation times, we can leverage ongoing research on approximation techniques to speed up simulations of CRNs [6,20,23,48]. Furthermore, the overall ABC(SMC)^2 scheme can easily be parallelised over its components, namely CRN simulations [50], ABCSMC inference [24] and the SMC [26] algorithm for verification.

We plan to apply the framework to different model classes, such as stochastic differential equations [17,21] and to incorporate the problem of Bayesian model selection [30,47].

Acknowledgements. Gareth W. Molyneux acknowledges funding from the University of Oxford and the EPSRC & BBSRC Centre for Doctoral Training in Synthetic Biology (grant EP/L016494/1).

A   Approximate Bayesian Computation - Sequential Monte Carlo (ABCSMC) Algorithm

---

Algorithm 2. ABCSMC

Input:

- Prior $\pi(\theta)$ and data-generating likelihood function $p(y_{obs}|\theta)$
- A kernel function $K_h(u)$ and scale parameter $h > 0$ where $u = \|y - y_{obs}\|$
- $N > 0$, number of particles used to estimate posterior distributions
- Sequence of perturbation kernels $F_m(\theta|\theta^*)$, $m = 1, \ldots, M$
- A quantile $v \in [0, 1]$ to control the rate of decrease of h_m
- Summary statistic function $s = S(y)$
- $B_t > 0$, number of simulations per sampled particle. For stochastic systems $B_t > 1$

Output:

- Set of weighted parameter vectors $\left\{\theta_M^{(i)}, w_M^{(i)}\right\}_{i=1}^N$ drawn from $\pi_{ABC}(\theta|s_{obs}) \propto \int K_{h_M}(\|s - s_{obs}\|)p(y|\theta)\pi(\theta)ds$

1: **for** $m = 0, \ldots, M$: **do**
2:    **for** $i = 0, \ldots, N$: **do**
3:       **if** $m = 0$ **then**
4:          Generate $\theta^{**} \sim \pi(\theta)$
5:       **else**
6:          Generate θ^* from the previous population $\{\theta_{m-1}^{(i)}\}$ with weights $\{w_{m-1}^{(i)}\}$ and perturb the particle to obtain $\theta^{**} \sim F_m(\theta|\theta^*)$
7:       **end if**
8:       **if** $\pi(\theta^{**}) = 0$ **then**
9:          **goto** line 3
10:      **end if**
11:      **for** $b = 1, \ldots, B_t$: **do**
12:         Generate $y_b \sim p(y|\theta^{**})$
13:         Calculate $s^b = S(y_b)$
14:      **end for**
15:      Calculate $b_t(\theta^{**}) = \sum_{b=1}^{B_t} K_{h_m}(\|s^b - s_{obs}\|)$
16:      **if** $b_t(\theta^{**}) = 0$ **then**
17:         **goto** line 3
18:      **end if**
19:      Set $\theta_m^{(i)} = \theta^{**}$, $\bar{d}_m^{(i)} = \frac{1}{B_t}\sum_{b=1}^{B_t}\|s^b - s_{obs}\|$ and calculate

20:

$$
w_m^{(i)} = \begin{cases} b_t\left(\theta_m^{(i)}\right), & \text{if } t = 0 \\[2mm] \dfrac{\pi\left(\theta_m^{(i)}\right) b_t\left(\theta_m^{(i)}\right)}{\sum_{j=1}^N w_{m-1}^{(j)} F_m\left(\theta_m^{(i)}|\theta_{m-1}^{(j)}\right)}, & \text{if } t > 0 \end{cases}
$$

21:   **end for**
22:   Normalise weights: $w_m^{(i)} \leftarrow w_m^{(i)}/\left(\sum_{i=1}^N w_m^{(i)}\right)$
23:   Set $h_{m+1} = (v/N)\sum_{i=1}^N \bar{d}_m^{(i)}$
24: **end for**
25: **return** $\left\{\left(\theta_M^{(i)}, w_M^{(i)}\right)\right\}_{i=1}^N$

B   Bayesian Verification Framework

There are 3 aspects to the Bayesian Verification framework. The Bayesian inference, parameter synthesis and probability or credibility calculation. The inference technique we use has been covered in the main text and here we focus on the parameter synthesis and the probability calculation.

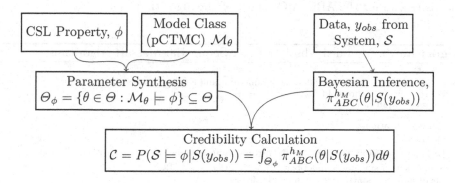

Fig. 4. Bayesian Verification Framework of [36].

B.1   Credibility Calculation

In the final phase of the approach, a probability estimate is computed corresponding to the satisfaction of a CSL specification formula ϕ by a system of interest such that $S \models \phi$, which we denote as the credibility. To calculate the credibility that the system satisfies the specified property, we integrate the posterior distribution $\pi(\theta|y_{obs})$ over the feasible set of parameters Θ_ϕ:

Definition 7. *Given a CSL specification ϕ and observed data y_{obs} and $s_{obs} = S(y_{obs})$ from the system S, the probability that $S \models \phi$ is given by*

$$C = P(S \models \phi|s_{obs}) = \int_{\Theta_\phi} \pi(\theta|s_{obs})\,d\theta, \tag{13}$$

where Θ_ϕ denotes the feasible set of parameters.

C   Absolute-Error Massart Algorithm

Here we present the slightly modified Sequential Massart Algorithm with Absolute Error (Algorithm 3). The outputs of Algorithm 3 are $\hat{\Lambda}_\phi(\theta)$, the total number of simulation undertaken B_t, the sum of the kernel smoothing functions $\sum_{b=1}^{B_t} K_{h_m}(\|s^b - s_{obs}\|)$ and the mean summary statistic produced from n simulations, $\bar{d}$. The algorithm is slightly modified to consider the distance function that is crucial for the ABCSMC aspect of the algorithm.

Algorithm 3. Modified Absolute-Error Sequential Massart Algorithm

---

Input:
- Absolute-error value ϵ, a confidence parameter δ and coverage parameter α.
- Current distance threshold h_m.
- Sampled parameter θ^{**}.
- True data s_{obs}
- CSL specification ϕ

Output:
- Estimated probability $\hat{\Lambda}_\phi(\theta^{**})$ with corresponding bounds $[\Lambda_\phi^L(\theta^{**}), \Lambda_\phi^U(\theta^{**})]$.
- Sum of kernel smoothing functions $\sum_{b=1}^{B_t} K_{h_m}(\|s^b - s_{obs}\|)$.
- Mean summary statistic from B_t simulations $\bar{d}$.

Set Initial number of successes, $l = 0$, and initial iteration $k = 0$.
Set $B_t = n_O$, where $n_O = \lceil \frac{1}{2\epsilon^2} \log \frac{2}{\delta} \rceil$ is the Okamoto bound and the initial confidence interval
$I_0 = [a_0, b_0] = [0, 1]$ in which $\Lambda_\phi(\theta^{**})$ belongs to.
while $k < B_t$ **do**
    $k \leftarrow k + 1$
    Generate trace $y^{(k)} \sim p(y|\theta^{**})$ and calculate $s^k = S(y^{(k)})$.
    Calculate $K_{h_m}(\|s^k - s_{obs}\|)$
    $z(y^{(k)}) = \mathbb{1}(y^{(k)} \models \phi)$
    $l \leftarrow l + z(y^{(k)})$
    $I_k = [a_k, b_k] \leftarrow \text{CONFINT}(l, k, \alpha)$
    if $1/2 \in I_k$ **then**
        $B_t = n_O$
    else if $b_k < 1/2$ **then**
        $B_t = \lceil \frac{2}{h_a(b_k, \epsilon)\epsilon^2} \log \frac{2}{\delta - \alpha} \rceil$
    else
        $B_t = \lceil \frac{2}{h_a(a_k, \epsilon)\epsilon^2} \log \frac{2}{\delta - \alpha} \rceil$
    end if
    $B_t \leftarrow \min(B_t, n_O)$
end while
Calculate $\bar{d} = (1/B_t) \sum_{b=1}^{B_t} s^b$.
Calculate $\sum_{b=1}^{B_t} K_{h_m}(\|s^b - s_{obs}\|)$.
Set $a_k = \hat{\Lambda}_\phi^L(\theta^{**})$, $b_k = \hat{\Lambda}_\phi^U(\theta^{**})$.
return $\hat{\Lambda}_\phi(\theta^{**}) = l/B_t, \sum_{b=1}^{B_t} K_{h_m}(\|s^b - s_{obs}\|), \bar{d}, [\hat{\Lambda}_\phi^L(\theta^{**}), \hat{\Lambda}_\phi^U(\theta^{**})]$.

D   Parameter Synthesis: A Motivating Comparison

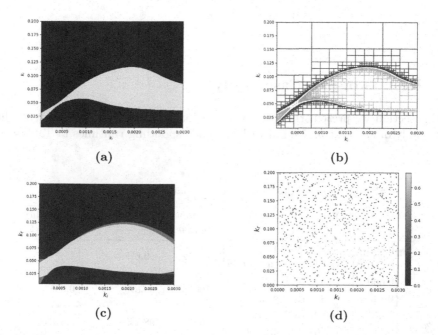

(a)

(b)

(c)

(d)

Fig. 5. The set Θ_ϕ, is shown in yellow (lighter colour), meanwhile $\Theta_{\neg\phi}$, is shown in blue (darker colour) $\Theta_{\neg\phi}$. The undecided areas, $\Theta_{\mathcal{U}}$ (if any) are shown in magenta. (5a) Parameter regions synthesised by GPU-Accelerated PRISM [12]. (5b) Gridding scheme. (5c) Parameter regions from SVM classification using 1000 samples from a uniform distribution. (5d) Estimated probabilities $\Lambda_\phi(\theta^*)$. (Color figure online)

The PRISM-based parameter synthesis technique dissects the parameter space into 14413 grid regions (cf. Fig. 5b), which results in calculating the satisfaction probability at 57652 points.

Instead, we consider sampling 1000 points from a Uniform distribution over the parameter space. We run the Massart algorithm at each point to obtain an estimated probability with corresponding $(1 - \delta)$ confidence bounds, where $\delta = 0.05$. With these samples and probabilities, we classify parameter regions with an SVM, which results in Fig. 5c, with corresponding estimated probabilities in Fig. 5d. The runtimes presented in Table 2 suggest that we obtain a good approximation of the parameter synthesis region in half the time of the GPU-accelerated PRISM tool, which could be further improved if we parallelised the computation [26]. These considerations have led us to embed the statistical parameter synthesis in the parameter inference algorithm.

Table 2. Parameter synthesis runtimes.

Parameter synth	Times [seconds]
PRISM-GPU	3096
SVM & SMC	1653.8

E　Results of SIR Case Study

In this section we present the inferred posterior from the SIR case study in the main body of the text in Table 3 with the corresponding runtimes in Table 4.

Table 3. Inferred posterior and Bayesian Verification Results.

Case	$\bar{\theta}_M$	Σ_M	Sampled Pars θ^{**}	SVM Accuracy Accuracy	Credibility Calculation
θ_ϕ	$\begin{bmatrix} 0.00215 \\ 0.07050 \end{bmatrix}$	$\begin{bmatrix} 1.46 \cdot 10^{-8} & 4.24 \cdot 10^{-7} \\ 4.24 \cdot 10^{-7} & 1.97 \cdot 10^{-5} \end{bmatrix}$	10952	99.6%	1
$\theta_{\neg\phi}$	$\begin{bmatrix} 0.00072 \\ 0.14519 \end{bmatrix}$	$\begin{bmatrix} 2.47 \cdot 10^{-8} & 3.41 \cdot 10^{-6} \\ 3.41 \cdot 10^{-6} & 9.22 \cdot 10^{-4} \end{bmatrix}$	10069	99.8%	0.0054
$\theta_\mathcal{U}$	$\begin{bmatrix} 0.00193 \\ 0.11337 \end{bmatrix}$	$\begin{bmatrix} 8.89 \cdot 10^{-8} & 5.86 \cdot 10^{-6} \\ 5.86 \cdot 10^{-6} & 4.21 \cdot 10^{-4} \end{bmatrix}$	10807	98.7%	0.6784

Table 4. Runtimes for algorithms.

Case	Times [seconds]		
	ABC(SMC)^2	SVM Optimisation	SVM Classification
θ_ϕ	64790	168	3.98
$\theta_{\neg\phi}$	8014	82	4.25
$\theta_\mathcal{U}$	35833	2166	5.12

References

1. Aziz, A., Sanwal, K., Singhal, V., Brayton, R.: Verifying continuous time Markov chains. In: Alur, R., Henzinger, T.A. (eds.) CAV 1996. LNCS, vol. 1102, pp. 269–276. Springer, Heidelberg (1996). https://doi.org/10.1007/3-540-61474-5_75

2. Bentriou, M., Ballarini, P., Cournède, P.-H.: Reachability design through approximate Bayesian computation. In: Bortolussi, L., Sanguinetti, G. (eds.) CMSB 2019. LNCS, vol. 11773, pp. 207–223. Springer, Cham (2019). https://doi.org/10.1007/978-3-030-31304-3_11

3. Bonassi, F.V., West, M., et al.: Sequential Monte Carlo with adaptive weights for approximate Bayesian computation. Bayesian Anal. **10**(1), 171–187 (2015)

4. Bornn, L., Pillai, N.S., Smith, A., Woodard, D.: The use of a single pseudo-sample in approximate Bayesian computation. Stat. Comput. **27**(3), 583–590 (2017)

5. Bortolussi, L., Milios, D., Sanguinetti, G.: Smoothed model checking for uncertain continuous-time Markov chains. Inf. Comput. 247(C), 235–253 (2016)

6. Bortolussi, L., Palmieri, L.: Deep abstractions of chemical reaction networks. In: Češka, M., Šafránek, D. (eds.) CMSB 2018. LNCS, vol. 11095, pp. 21–38. Springer, Cham (2018). https://doi.org/10.1007/978-3-319-99429-1_2

7. Bortolussi, L., Sanguinetti, G.: Learning and designing stochastic processes from logical constraints. In: Joshi, K., Siegle, M., Stoelinga, M., D'Argenio, P.R. (eds.) QEST 2013. LNCS, vol. 8054, pp. 89–105. Springer, Heidelberg (2013). https://doi.org/10.1007/978-3-642-40196-1_7

8. Bortolussi, L., Silvetti, S.: Bayesian statistical parameter synthesis for linear temporal properties of stochastic models. In: Beyer, D., Huisman, M. (eds.) TACAS 2018. LNCS, vol. 10806, pp. 396–413. Springer, Cham (2018). https://doi.org/10.1007/978-3-319-89963-3_23

9. Box, G., Tiao, G.: Bayesian Inference in Statistical Analysis. Wiley Classics Library, Wiley (1973)

10. Broemeling, L.: Bayesian Inference for Stochastic Processes. CRC Press, Cambridge (2017)

11. Ceska, M., Dannenberg, F., Paoletti, N., Kwiatkowska, M., Brim, L.: Precise parameter synthesis for stochastic biochemical systems. Acta Inf. **54**(6), 589–623 (2014)

12. Ceska, M., Pilar, P., Paoletti, N., Brim, L., Kwiatkowska, M.Z.: PRISM-PSY: precise GPU-accelerated parameter synthesis for stochastic systems. In: Tools and Algorithms for the Construction and Analysis of Systems - 22nd International Conference, TACAS 2016, Held as Part of the European Joint Conferences on Theory and Practice of Software, ETAPS 2016, Eindhoven, The Netherlands, 2–8 April 2016, Proceedings. pp. 367–384 (2016)

13. Chernoff, H., et al.: A measure of asymptotic efficiency for tests of a hypothesis based on the sum of observations. Ann. Math. Stat. **23**(4), 493–507 (1952)

14. Cortes, C., Vapnik, V.: Support-vector networks. Mach. Learn. **20**(3), 273–297 (1995)

15. Del Moral, P., Doucet, A., Jasra, A.: An adaptive sequential Monte Carlo method for approximate Bayesian computation. Stat. Comput. **22**(5), 1009–1020 (2012)

16. Filippi, S., Barnes, C.P., Cornebise, J., Stumpf, M.P.: On optimality of kernels for approximate Bayesian computation using sequential Monte Carlo. Stat. Appl. Genetics Molecular Biol. **12**(1), 87–107 (2013)

17. Gardiner, C.: Stochastic Methods: A Handbook for the Natural and Social Sciences, 4 edn, vol. 13, Springer, Heidelberg (2009)

18. Gillespie, D.T.: Exact stochastic simulation of coupled chemical reactions. J. Phys. Chem. **81**(25), 2340–2361 (1977)

19. Gillespie, D.T.: A rigorous derivation of the chemical master equation. Phys. A: Stat. Mech. Appl. **188**(1), 404–425 (1992)

20. Gillespie, D.T.: Approximate accelerated stochastic simulation of chemically reacting systems. J. Chem. Phys. **115**(4), 1716–1733 (2001)

21. Gillespie, D.T., Gillespie, D.T.: The chemical Langevin equation. J. Chem. Phys. **297**, 2000 (2000)
22. Haesaert, S., Van den Hof, P.M., Abate, A.: Data-driven and model-based verification via Bayesian identification and reachability analysis. Automatica **79**, 115–126 (2017)
23. Higham, D.J.: Modeling and simulating chemical reactions. SIAM Rev. **50**(2), 347–368 (2008)
24. Jagiella, N., Rickert, D., Theis, F.J., Hasenauer, J.: Parallelization and high-performance computing enables automated statistical inference of multi-scale models. Cell Syst. **4**(2), 194–206 (2017)
25. Jegourel, C., Sun, J., Dong, J.S.: Sequential schemes for frequentist estimation of properties in statistical model checking. In: Bertrand, N., Bortolussi, L. (eds.) QEST 2017. LNCS, vol. 10503, pp. 333–350. Springer, Cham (2017). https://doi.org/10.1007/978-3-319-66335-7_23
26. Jegourel, C., Sun, J., Dong, J.S.: On the sequential Massart algorithm for statistical model checking. In: Margaria, T., Steffen, B. (eds.) ISoLA 2018. LNCS, vol. 11245, pp. 287–304. Springer, Cham (2018). https://doi.org/10.1007/978-3-030-03421-4_19
27. Jha, S.K., Clarke, E.M., Langmead, C.J., Legay, A., Platzer, A., Zuliani, P.: A Bayesian approach to model checking biological systems. In: Degano, P., Gorrieri, R. (eds.) CMSB 2009. LNCS, vol. 5688, pp. 218–234. Springer, Heidelberg (2009). https://doi.org/10.1007/978-3-642-03845-7_15
28. Karlin, S., Taylor, H., Taylor, H., Taylor, H., Collection, K.M.R.: A First Course in Stochastic Processes. No. vol. 1, Elsevier Science (1975)
29. Kermack, W.: A contribution to the mathematical theory of epidemics. Proc. Royal Soc. London A: Math. Phys. Eng. Sci. **115**(772), 700–721 (1927)
30. Kirk, P., Thorne, T., Stumpf, M.P.: Model selection in systems and synthetic biology. Current Opinion Biotechnol. **24**(4), 767–774 (2013)
31. Kwiatkowska, M., Norman, G., Parker, D.: Stochastic model checking. In: Bernardo, M., Hillston, J. (eds.) SFM 2007. LNCS, vol. 4486, pp. 220–270. Springer, Heidelberg (2007). https://doi.org/10.1007/978-3-540-72522-0_6
32. Kwiatkowska, M., Thachuk, C.: Probabilistic model checking for biology. In: Software Safety and Security. NATO Science for Peace and Security Series - D: Information and Communication Security, IOS Press (2014)
33. Legay, A., Delahaye, B., Bensalem, S.: Statistical model checking: an overview. In: Barringer, H., Falcone, Y., Finkbeiner, B., Havelund, K., Lee, I., Pace, G., Roşu, G., Sokolsky, O., Tillmann, N. (eds.) RV 2010. LNCS, vol. 6418, pp. 122–135. Springer, Heidelberg (2010). https://doi.org/10.1007/978-3-642-16612-9_11
34. Massart, P.: The tight constant in the Dvoretzky-Kiefer-Wolfowitz inequality. The annals of Probability pp. 1269–1283 (1990)
35. Metropolis, N., Ulam, S.: The Monte Carlo method. J. Am. Stat. Assoc. **44**(247), 335–341 (1949)
36. Molyneux, G.W., Wijesuriya, V.B., Abate, A.: Bayesian verification of chemical reaction networks. In: Sekerinski, E., et al. (eds.) Formal Methods. FM 2019 International Workshops. LNCS, vol. 12233, pp. 461–479. Springer, Cham (2020)
37. Okamoto, M.: Some inequalities relating to the partial sum of binomial probabilities. Ann. Inst. Stat. Math. **10**(1), 29–35 (1959)
38. Polgreen, E., Wijesuriya, V.B., Haesaert, S., Abate, A.: Data-efficient Bayesian verification of parametric Markov chains. In: Quantitative Evaluation of Systems - 13th International Conference, QEST 2016, Quebec City, QC, Canada, 23–25 August 2016, Proceedings. pp. 35–51 (2016)

39. Polgreen, E., Wijesuriya, V.B., Haesaert, S., Abate, A.: Automated experiment design for data-efficient verification of parametric Markov decision processes. In: Quantitative Evaluation of Systems - 14th International Conference, QEST 2017, Berlin, Germany, 5–7 September 2017, Proceedings. pp. 259–274 (2017)
40. Prangle, D.: Summary statistics in approximate Bayesian computation. arXiv preprint arXiv:1512.05633 (2015)
41. Prangle, D., et al.: Adapting the ABC distance function. Bayesian Anal. **12**(1), 289–309 (2017)
42. Schnoerr, D., Sanguinetti, G., Grima, R.: Approximation and inference methods for stochastic biochemical kinetics: a tutorial review. J. Phys. A: Math. Theor. **50**(9), 093001 (2017)
43. Sisson, S., Fan, Y., Beaumont, M.: Overview of abc. Handbook of Approximate Bayesian Computation pp. 3–54 (2018)
44. Sisson, S.A., Fan, Y., Beaumont, M.: Handbook of Approximate Bayesian Computation. Chapman and Hall/CRC, Cambridge (2018)
45. Smola, A.J., Schölkopf, B.: A tutorial on support vector regression. Stat. Comput. **14**(3), 199–222 (2004)
46. Toni, T., Stumpf, M.P.: Simulation-based model selection for dynamical systems in systems and population biology. Bioinformatics **26**(1), 104–110 (2010)
47. Toni, T., Welch, D., Strelkowa, N., Ipsen, A., Stumpf, M.P.: Approximate Bayesian computation scheme for parameter inference and model selection in dynamical systems. J. Royal Soc. Interface **6**(31), 187–202 (2008)
48. Warne, D.J., Baker, R.E., Simpson, M.J.: Simulation and inference algorithms for stochastic biochemical reaction networks: from basic concepts to state-of-the-art. J. Royal Soc. Interface **16**(151), 20180943 (2019)
49. Wilkinson, D.: Stochastic Modelling for Systems Biology, Second Edition. Chapman & Hall/CRC Mathematical and Computational Biology, Taylor & Francis (2011)
50. Zhou, Y., Liepe, J., Sheng, X., Stumpf, M.P., Barnes, C.: GPU accelerated biochemical network simulation. Bioinformatics **27**(6), 874–876 (2011)
51. Zuliani, P.: Statistical model checking for biological applications. Int. J. Softw. Tools Technol. Transfer **17**(4), 527–536 (2015)

Parallel Parameter Synthesis
for Multi-affine Hybrid Systems
from Hybrid CTL Specifications

Eva Šmijáková, Samuel Pastva, David Šafránek$^{(\boxtimes)}$, and Luboš Brim

Faculty of Informatics, Masaryk University, Brno, Czech Republic
{xsmijak1,xpastva,safranek,brim}@fi.muni.cz

Abstract. We consider the parameter synthesis problem for multi-affine hybrid systems and properties specified using a hybrid extension of CTL (HCTL). The goal is to determine the sets of parameter valuations for which the given hybrid system satisfies the desired HCTL property. As our main contribution, we propose a shared-memory parallel algorithm which efficiently computes such parameter valuation sets. We combine a rectangular discretisation of the continuous dynamics with the discrete transitions of the hybrid system to obtain a single over-approximating semi-symbolic transition system. Such system can be then analysed using a fixed-point parameter synthesis algorithm to obtain all satisfying parametrisations. We evaluate the scalability of the method and demonstrate its applicability in a biological case study.

Keywords: Hybrid systems · Parameter synthesis · Rectangular abstraction · Semi-symbolic · Hybrid CTL

1 Introduction

In real-world dynamical systems, one encounters a complex interplay of both *continuous* and *discrete* dynamics. This type of behaviour appears in cyber-physical systems, biochemical or biophysical systems (systems biology), economic and social interaction models, or in the infectious disease control (epidemic systems). In many cases, the continuous part reflects the natural phenomena and the discrete part arises due to some (not necessarily digital) embedded control mechanism.

Such systems are formalised by means of *hybrid systems* (also *hybrid automata* (HA)). These typically consist of several *modes*, each describing the continuous evolution of the system using ordinary differential equations (ODE). The modes are then connected using conditional discrete *jumps*. *Parameters* often need to be introduced into the continuous ODE flow or the conditions

L. Brim–This work has been supported by the Czech Science Foundation grant No. 18-00178S.

A. Abate et al. (Eds.): CMSB 2020, LNBI 12314, pp. 280–297, 2020.
https://doi.org/10.1007/978-3-030-60327-4_15

of the discrete jumps to represent an unknown or uncertain behaviour of the system.

We focus on multi-affine hybrid systems, which have a multi-affine vector field in every mode. Such systems have a large set of applications including infectious disease models [37], altitude and velocity control systems [5], models of gene regulation [32], or models of other biological systems with mixed continuous and discrete variables [18,34].

In typical scenarios, hybrid systems have too many mutually dependent variables and parameters to be studied *analytically*. To examine a proposed hypothesis, one commonly relies on *computational* methods, as these exhibit better scalability and often do not require expert knowledge.

To rigorously represent a hypothesis about some abstract observable sequence of events (or event branching) in the behaviour of a hybrid system, we use temporal logic. We employ an expressive *hybrid extension* of the computation tree logic (HCTL) [2]. HCTL extends CTL with first-order quantifiers as well as specialised operators (*bind* ↓ and *at* @) for reasoning about properties of states.

Given an HCTL formula and a parametrised hybrid system, the goal of the *parameter synthesis problem* is then to determine the set of parametrisations for which the system satisfies the given HCTL specification.

Paper Contributions. We introduce a novel approach to the parameter synthesis of multi-affine hybrid systems that addresses the state space explosion problem using parallelism and symbolic parameter representation.

- As a specification language, we utilise the expressive HCTL logic. HCTL enables properties such as un-reachability, general oscillation and stability.
- We consider a wide family of multi-affine hybrid automata (MHA) with parameters in the continuous flow as well as mode invariants and jumps.
- For such an MHA, we construct a compound parametrised Kripke structure that over-approximates its behaviour. The continuous modes of the automaton are discretised using rectangular abstraction [4]. To efficiently represent this structure, we extend the semi-symbolic approach (i.e. an explicit state space with symbolic parameter sets) proposed in [8].
- We propose a parallel, shared-memory parameter synthesis algorithm based on [10]. Given a parametrised hybrid automaton H and an HCTL property φ, we compute the set of parametrisations of H for which $H \models \varphi$.
- We evaluate the scalability of the method and demonstrate its applicability in a case study based on a complex biological system.

In general, in this paper we give a significant extension of our existing framework for piece-wise multi-affine continuous-time systems [8,10,11] by making it working with a more general class of systems—multi-affine hybrid automata. First, we adapt the rectangular abstraction to correctly capture MHA. Second, we provide novel algorithms working with this class of hybrid systems.

Related Work. Hybrid systems are rather ubiquitous in systems biology. A comprehensive overview appears in [14,39]. The problem of parameter identification for HA has also been targeted from several different perspectives:

The closest work considering *multi-affine hybrid automata* is implemented in the tool Hydentify [12]. Hydentify considers parameters only in the continuous flow function (we allow parametrised jump guards and invariants as well). The most significant difference is the abstraction—Hydentify employs several abstractions that simultaneously over-approximate and under-approximate the MHA by *linear hybrid automata* (LHA). In our case, rectangular abstraction is employed to explicitly discretise the vector field.

On the one hand, LHA abstraction has the advantage of preserving the timing information, and it also enables the use of efficient symbolic reachability analysis algorithms, such as in [26]. On the other hand, Hydentify has to iteratively decompose the parameter space by repeating the (non-parametrised) reachability task. In our case, a coarser (rectangular) abstraction is not limited to reachability (we use HCTL as the specification language) and the analysis of parameter space can be performed symbolically in a single iteration of the parameter synthesis algorithm, without the need for explicit decomposition.

Rectangular abstraction [6] for parameter synthesis of non-linear (piece-wise multi-affine) dynamical systems has been originally used in [4] and implemented in the tool RoVerGeNe (for LTL specifications). The extension to MHA provided by Hydentify has significantly improved scalability and precision of RoVerGeNe, but it restricts specifications to reachability and safety questions. Our work allows more flexibility in the parametrisation of the MHA and extends the set of supported specification formalisms.

Counterexample-guided abstraction refinement (CEGAR) [19] is also applicable to parameter synthesis of LHAs [25]. In this case, the counterexamples are paths to the bad states, and the model is refined by restricting the domains of parameters. The main advantage of CEGAR is efficiency. Compared to the standard reachability analysis, it is often faster and requires smaller state space.

Breach [22,23] uses simulation-based techniques to analyse hybrid systems. It performs parameter synthesis of general, non-linear HA with respect to properties in signal temporal logic. A similar approach is used in Biocham [17] for LTL [38] and CTL [24]. Simulation-based methods use numerical solvers, and therefore their precision significantly relies on the quality of parameter space sampling (unlike abstraction based approaches, there are no global formal guarantees). Breach minimises this parameter space sampling error using sensitivity analysis. Meanwhile, U-Check [13] combines statistics and machine learning to address the problem of good sampling with statistical guarantees [3,15]. These approaches are limited to time-bounded behaviour, whereas our approach allows time-unbounded analysis.

If some parametrisations with the desired behaviour are already known, it is possible to synthesise the whole set of parametrisations for which the system preserves the same behaviour using the inverse method [1]. Tool HYMITATOR implements this approach for LHAs [27].

Parameter synthesis problem can also be encoded into first-order logic formulae and solved using δ-complete decision procedures [28]. In particular, a formula is used to describe the states reachable within a finite number of steps. Parameter synthesis is then reduced to computing a satisfying valuation of the parameters for such formula [33]. While mostly limited to time-bounded analysis, this framework was successfully used to obtain parameter ranges indicating disorders in cardiac cells models [31,35].

2   Parameter Synthesis of Hybrid Automata

In this section, we first define the studied class of hybrid automata with parameters, and then show how such automata can be transformed into discrete parametrised Kripke structures (PKS). The construction proceeds in two steps. First, each continuous mode of the hybrid automaton is translated into a partial PKS. Then, a compound PKS is created as a combination of the partial PKSs with jump and reset conditions of the hybrid automaton. We argue that this Kripke structure over-approximates the behaviour of the original automaton. We then define hybrid CTL and its semantics over Kripke structures and finally, parameter synthesis problem for hybrid automata and hybrid CTL.

Preliminaries. We use $\mathbb{R}$ and $\mathbb{N}$ to denote the set of real and natural numbers respectively. $\mathcal{P}(A)$ denotes the power set (set of all subsets) of A. In general, we write A^B to denote the set of all possible functions $B \rightarrow A$. When B is a set of variables or parameters of the system, such function is often referred to as *valuation* (and A^B is thus a set of all valuations).

To describe the semantics of a discrete system with parameters, we use the notion of *parametrised Kripke structure* [16]:

Definition 1. *Let AP be a set of atomic propositions. A* parametrised Kripke structure *(PKS) over AP is a tuple $K = (P, S, I, \rightarrow, L)$ where:*

- *P is a finite set of* parametrisations;
- *S is a finite set of* states;
- *$I \subseteq S$ is the set of* initial states;
- *$L : S \rightarrow \mathcal{P}(AP)$ is a* labelling *of states with atomic propositions;*
- *$\rightarrow \subseteq S \times P \times S$ is a* parametrised transition relation.

We write $s \xrightarrow{p} t$ instead of $(s, p, t) \in \rightarrow$. We assume that the PKS is total, i.e. for all $s \in S$ and $p \in P$, there exists at least one t such that $s \xrightarrow{p} t$. By fixing a parametrisation $p \in P$ we obtain an ordinary (i.e. non-parametrised) Kripke structure (KS) [20] $K_p = (S, I, \rightarrow_p, L)$. Here, S, I and L are the same as in K and $\rightarrow_p = \{(s, t) \mid (s, p, t) \in \rightarrow\}$.

A *path* in a non-parametrised KS, K_p, is an infinite sequence of states $\sigma : \mathbb{N} \rightarrow S$, such that for all $i \in \mathbb{N}, (\sigma(i-1), \sigma(i)) \in \rightarrow_p$. We use σ_i instead of $\sigma(i)$ to denote the i-th element on the path. The set of all paths in a KS starting at $s \in S$ is denoted $\Sigma_s = \{\sigma \mid \sigma \text{ is a path in the KS and } \sigma_0 = s\}$.

2.1 Parametrised Hybrid Automata

Hybrid systems combine the behaviour of discrete and continuous systems. The typical example of such a system is a physical system whose behaviour depends on a discrete controller. Hybrid systems can be modelled as *hybrid automata* [30].

In practice, some parts of a hybrid system might be unknown or customizable. To describe such systems, we consider *parametrised hybrid automata* which allow both parameters in the continuous flow functions as well as parameters influencing the switching of modes [23,27].

Definition 2. *A parametrised hybrid automaton (PHA) H is an ordered tuple* $H = (\Pi, Q, X, F, J, D, E, G, R)$ *where:*

- *Π is a finite set of real-valued* parameters;
- *Q is a finite set of* discrete modes;
- *X is a finite set of real-valued* variables;
- *$F : Q \times \mathbb{R}^X \times \mathbb{R}^\Pi \to \mathbb{R}^X$ is a* parametrised vector field *defining the local continuous flow via a differential equation* $\dot{x} = F(q, x, \pi)$;
- *$J \subseteq Q \times \mathbb{R}^X$ is a set of* initial states *of the PHA;*
- *$D : Q \times \mathbb{R}^\Pi \to \mathcal{P}(\mathbb{R}^X)$ is a parametrised domain of a mode, sometimes also called an* invariant *of a mode;*
- *$E \subseteq Q \times Q$ is a set of edges (transitions,* jumps*) between modes;*
- *$G : E \times \mathbb{R}^\Pi \to \mathcal{P}(\mathbb{R}^X)$ gives a* parametrised *guard condition for each jump;*
- *$R : E \to \mathcal{P}(\mathbb{R}^X \times \mathbb{R}^X)$ specifies for each jump a* reset relation *which describes possible variable valuations before and after a jump.*

In this paper, we further restrict PHA to represent a so called piece-wise multi-affine automaton. Specifically, we assume F to be piece-wise multi-affine in both variables and parameters. Furthermore, the invariants, jump conditions and reset conditions of the automaton have to be described using a Boolean combination of inequalities which are affine in both variables and parameters as well.

Finally, we assume the automaton to be bounded. *That is, the domain of every real-valued parameter $p \in \Pi$ is an interval $[p_{min}, p_{max}]$ and for every variable $x \in X$, we also have an interval $[x_{min}, x_{max}]$ such that it covers all values satisfying the related invariant conditions.*

The requirement of multi-affinity and boundedness is necessary in order to enable efficient abstraction techniques for the continuous flow of the automaton and efficient manipulation of invariants and jumps in general. If one can supply such operations for a broader class of PHAs, the restrictions can be lifted. We will further discuss such possible extensions of our method where appropriate.

A *state* of a PHA is a pair (q, x) where $q \in Q$ is the current discrete mode and $x \in \mathbb{R}^X$ is the current valuation of X. The state is *valid* in a PHA for a parameter valuation $\pi \in \mathbb{R}^\Pi$ if it fulfils the invariant condition, i.e. $x \in D(q, \pi)$.

There are two types of flows in a PHA. The first type is given by the trajectories of the continuous vector field. This flow is called *local* and is relevant as long as the evolved state fulfils the invariant. The second type of flow is the *jump* flow which corresponds to transitions between the individual discrete modes. Jump $j \in E$ is allowed between (q, x) and (q', x') for a parameter valuation π only if:

- both (q, x) and (q', x') are valid (fulfil the corresponding mode invariants);
- the guard condition of j is satisfied, i.e., $x \in G(j, \pi)$;
- the reset relation of j is such that x resets to x', i.e., $(x, x') \in R(j)$.

Note that in many practical cases, the reset relation is simply an identity, in which case one naturally considers only jumps where $x = x'$ and the third condition thus becomes unnecessary.

It is important to emphasise that the parameters can be present at three different places in a PHA: flow function parameters, predicates defining an invariant of a mode, and predicates defining a guard condition of a jump. By fixing a parameter valuation $\pi \in \mathbb{R}^{\Pi}$ of a PHA H, we obtain an automaton H_π which has the same semantics as a standard non-parametrised hybrid automaton specified in [30].

2.2 Rectangular Abstraction of Parametrised Hybrid Automata

In order to interpret hybrid CTL formulae over a PHA, we first need to transform the PHA into a PKS. In our case, the PKS over-approximates the behaviour of the original multi-affine PHA. The construction proceeds in two steps. First, a partial PKS of the continuous (local) flow is constructed for each discrete mode of the PHA. Second, the partial PKSs are merged into a single compound PKS that also incorporates the jumps between individual modes.

Local Mode Abstraction. In order to construct a PKS describing the continuous behaviour of a single discrete mode, we employ *rectangular abstraction* [4] for piece-wise multi-affine continuous models. This abstraction was chosen since it can work with parametrised systems and in the past has been successfully used for parameter synthesis in ODE models [8]. The choice of abstraction method greatly influences the class of automata the approach can handle. If a more general abstraction method is available, the restriction to piece-wise multi-affine systems can be lifted. Here, we first give a high-level overview of the abstraction method and then a brief technical summary. For full technical explanation, the reader is referred to [4].

The abstraction divides the continuous state space into a set of n-dimensional rectangles. Transitions are only introduced between adjacent rectangles, i.e. rectangles that share an $(n-1)$-dimensional facet. A transition is introduced between two rectangles, r_1 and r_2, if there exists a continuous trajectory which flows from r_1 to r_2 through their connecting facet (or more precisely, when the absence of such trajectory cannot be decided). A self-loop on a rectangle is introduced whenever it cannot be decided that eventually every trajectory escapes the rectangle. As argued in [4], this over-approximates the behaviour of the original continuous flow. In our case, we further extend the method by removing the discrete states (rectangles) which do not contain any points satisfying the invariant conditions of the PHA. This does not influence the over-approximation, since these states would not appear in the original PHA neither.

When dealing with parameters, instead of a yes-no answer, the abstraction procedure computes for each transition a description of the parameter valuation set for which it is enabled (i.e. the above mentioned conditions are met), thus yielding a parametrised Kripke structure. In our case, these parameter valuation sets can be described using combinations of affine inequalities (due to the chosen abstraction method). In case of a more general class of hybrid automata, a more general representation such as semi-algebraic sets or SMT formulae are necessary.

The procedure assumes a PHA H as defined above and a fixed discrete mode $q \in Q$. As a result, we obtain a PKS $K^q = (P^q, S^q, I^q, \to^q, L^q)$ which describes the local behaviour of H in mode q.

Furthermore, for each continuous variable $v \in X$ we assume a sequence of thresholds $\{\theta_1^v, \ldots, \theta_{n_v}^v\} \subset \mathbb{R}$ ordered such that $\theta_1^v \leq \theta_2^v \leq \cdots \leq \theta_{n_v}^v$ and $\theta_1^v = v_{min}$ and $\theta_{n_v}^v = v_{max}$. These thresholds partition the continuous state space of the automaton into n-dimensional intervals $[\theta_{j_1}^{v_1}, \theta_{j_1+1}^{v_1}] \times \cdots \times [\theta_{j_n}^{v_n}, \theta_{j_n+1}^{v_n}]$ (here, v_1 through v_n are the variables of the PHA). These intervals are referred to as rectangles. Each rectangle is uniquely identified via an n-tuple of indices: $\square(j_1, \ldots, j_n) = [\theta_{j_1}^{v_1}, \theta_{j_1+1}^{v_1}] \times \cdots \times [\theta_{j_n}^{v_n}, \theta_{j_n+1}^{v_n}]$, where the range of each j_i is $\{1, \ldots, n_{v_i} - 1\}$. In each rectangle, the flow F of the PHA must be multi-affine. Notice that for such rectangle, we can easily perform basic set operations ($\in, \subseteq$, $\cap, \ldots$) since it is essentially a set of real valued tuples (each specifying a single valuation of variables of H).

Using this rectangular partitioning, we can construct K^q as follows:

- The parameter valuations of K^q are given by all possible valuations of parameters of H, i.e., $P^q := \mathbb{R}^\Pi$.
- The state space of the PKS is created by taking all n-dimensional rectangles that contain at least one valid state of H (that is, $\exists p \in P^q : D(q,p) \cap \square(j_1, \ldots, j_n) \neq \emptyset$) and extending them with q to indicate the mode which the rectangle belongs to:

$$S^q := \{(q, r = \square(j_1, \ldots, j_n)) \mid \exists p \in P^q, \exists x \in r : x \in D(q,p)\}$$

  Note that the validity check for each rectangle can be performed due to the imposed restrictions on the guards of the PHA.
- Initial states are given by the rectangles that contain at least one initial state of the original PHA, i.e., $I^q := \{(q,r) \in S^q \mid \exists x \in r : (q,x) \in J\}$.
- The labelling function assigns a proposition $a \in AP$ to a rectangle r if at least one point in r satisfies the proposition in H. Formally, $L^q((q,r)) := \{a \in AP \mid (q,r) \in S^q \wedge \exists x \in r : x \models a\}$. Here, a is usually some inequality (or a Boolean combination of inequalities) defined over variables, e.g., $v_1 \geq 3$.
- Finally, the procedure for constructing the transition relation $\to^q$ is described in [4]—since it only requires the knowledge of S^q and the differential equations F assumed in mode q, it remains largely the same as for general ODE models.

The main difference is that here, some rectangles may not be included in the state space since they do not contain any valid states of H. Such rectangles are excluded from the abstraction.

Compound Abstraction. Assuming the PKS K^q for each mode $q \in Q$ is available, we can construct the *compound* PKS (CPKS) that extends the model with jump transitions between individual modes. CPKS $C = (P, S, I, \rightarrow, L)$ is constructed as follows:

- $P := \mathbb{R}^\Pi$—all K^q share the same set of parameter valuations;
- $S := \bigcup_{q \in Q} S^q$—individual states already contain the mode label and thus the sets of states of local PKSs are disjoint;
- $I := \bigcup_{q \in Q} I^q$;
- Proposition labelling function follows individual mode labellings as well. Additionally, we introduce artificial proposition $\mathcal{A}(q)$ that allows us to explicitly reference all states of a mode q:

$$L((q, r)) := L^q((q, r)) \cup \{\mathcal{A}(q)\}$$

- The transition relation $\rightarrow$ consists of two parts, $\rightarrow_l$ (local) and $\rightarrow_j$ (jump), defined as follows:

$$\rightarrow_l := \bigcup_{q \in Q} \rightarrow^q$$

$$\rightarrow_j := \bigcup_{e = (q_1, q_2) \in E} \{((q_1, s), p, (q_2, t)) \in S \times P \times S \mid$$

$$\exists x \in s, x' \in t : (x, x') \in R(e) \wedge x \in G(e, p)\}$$

As we can see, a jump transition is created for every pair of rectangles where some combination of continuous values satisfies the guard and reset conditions of some original jump in H. Due to the restrictions imposed on jump and reset conditions, we can explicitly compute the set of parametrisations for which this relation holds.

The CPKS C created using this procedure over-approximates the behaviour of the original multi-affine PHA. The over-approximation of the continuous flows follows from the correctness of the rectangular abstraction. The discrete jumps of the automaton are then conservatively re-created in the Kripke structure as described above. That is, whenever a jump is possible from a rectangle (jump guard is satisfied in some subset of the rectangle), a transition in the CPKS is created.

Figure 1 illustrates an example of a rectangular abstraction of a hybrid system. On the left, an original hybrid automaton with three modes is depicted. The black areas do not fulfil the modes invariant conditions. The grey areas represent guard and reset relations of individual jumps. The dashed arrows represent the mapping of a jump. Here, first jump is mapped to a singular point whereas the second jump does not override the coordinates of the variables $(x' = x)$. The solid arrows depict the vector fields. On the right side, the resulting rectangular abstraction of the hybrid automaton is shown. The semantics of black and grey rectangles are the same as on the left side. The small arrows represent local (flow) transitions and dashed arrows represent jump transitions.

2.3 Hybrid CTL

We use the *hybrid extension* of the computational tree logic (HCTL) [2] to reason about properties of interest at the level of CPKS. HCTL allows to express time-unbounded properties at a very general level. For example, reachability of a single stable state (or a cycle) can be expressed without addressing a concrete state (or states on a concrete cycle). This is achieved by state variable quantification and state binding operators. Branching operators of CTL are used to fully reflect the non-determinism present in the CPKS.

The formulae of HCTL are defined using the following abstract syntax:

$$\varphi ::= true \mid q \mid \neg\varphi \mid \varphi_1 \wedge \varphi_2 \mid \mathbf{EX}\,\varphi \mid \mathbf{E}\,(\varphi_1\,\mathbf{U}\,\varphi_2) \mid \mathbf{AF}\,\varphi \mid\, \downarrow x.\varphi \mid @x.\varphi \mid \exists x.\varphi$$

Here, q ranges over AP and x ranges over the set of state variables V.

In the following, we adapt the semantics of HCTL previously defined in [7]. In particular, we remove action-labeled operators that were needed for the purpose of discrete bifurcation analysis.

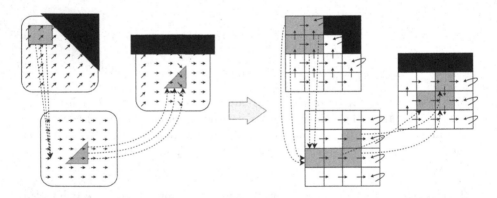

Fig. 1. (left) A schematic depiction of a hybrid automaton (without parameters) and (right) its corresponding CPKS. Black areas represent parts of the state space that do not satisfy the mode invariant conditions. Grey areas connected by dashed arrows depict the possible discrete jumps between individual modes based on their guard and reset relations. As we can see, to preserve over-approximation, the abstract grey rectangles cover larger area than the grey triangles in the original system (i.e. spurious behaviour is introduced in the abstract system).

First, we extend the KS model with a valuation of the state variables $h :$ $V \to S$. We use $h[x \mapsto s]$ to denote a valuation which maps the variable x to state s and is otherwise defined as the valuation h. Formally, $h[x \mapsto s](x) = s$, $h[x \mapsto s](y) = h(y)$ for all $y \neq x$. Second, the definition of the satisfaction of an HCTL formula in a state of the KS is stated in the following way.

Definition 3. *Let K be a KS and $h : V \to S$ be a valuation of state variables. The satisfaction relation for states and paths of K w.r.t. HCTL formulae is defined as follows:*

$$(K, h, s) \models true$$
$$(K, h, s) \models p \iff p \in L(s)$$
$$(K, h, s) \models \neg\varphi \iff (K, h, s) \not\models \varphi$$
$$(K, h, s) \models \varphi_1 \wedge \varphi_2 \iff (K, h, s) \models \varphi_1 \text{ and } (K, h, s) \models \varphi_2$$
$$(K, h, s) \models \mathbf{EX}\,\varphi \iff \exists\sigma \in \Sigma_s : (K, h, \sigma_1) \models \varphi$$

$$(K, h, s) \models \mathbf{E}\,(\varphi_1\,\mathbf{U}\,\varphi_2) \iff \exists\sigma \in \Sigma_s \text{ and } \exists i \in \mathbb{N} : (K, h, \sigma_i) \models \varphi_2$$
$$\text{and } \forall j < i : (K, h, \sigma_j) \models \varphi_1$$
$$(K, h, s) \models \mathbf{AF}\,\varphi \iff \forall\sigma \in \Sigma_s, \exists i \in \mathbb{N} : (K, \sigma_i) \models \varphi$$
$$(K, h, s) \models \,\downarrow x.\varphi \iff (K, h[x \mapsto s], s) \models \varphi$$
$$(K, h, s) \models \exists x.\varphi \iff \exists s' \in S : (K, h[x \mapsto s'], s) \models \varphi$$
$$(K, h, s) \models @x.\varphi \iff (K, h, h(x)) \models \varphi$$

Usually, we are interested in formulae without free variables. In such case we write $(M, s) \models \varphi$ instead of $(M, h, s) \models \varphi$ as then the choice of h is not relevant. We also use standard syntactic extensions of HCTL such as universal quantification $\forall x.\varphi$ meaning $\neg\exists x.\neg\varphi$ and CTL operators $\mathbf{EF}$ (existential finally), $\mathbf{AG}$ (for all globally), and $\mathbf{AX}$ (for all successors).

The following three operators make the core of the hybrid extension: $\exists x$ (*exists*), $\downarrow x$ (*bind*), and $@x$ (*at*). The $\exists x$ operator has the same meaning of existential quantification as in the first-order logic. The $\downarrow x$ operator provides a more specialised alternative to *exists* as it allows to assign the current state to the variable x. Finally, the $@x$ operator points the subformula to the state that has been stored in the variable x.

Some examples of properties of CPKSs expressible in HCTL:

- Reachability of a mode q: $\mathbf{EF}\,\mathcal{A}(q)$
- Unreachability of a mode q: $\neg\mathbf{EF}\,\mathcal{A}(q)$
- Stable steady state (sink): $\downarrow s.\mathbf{AX}\,s$
- Cycle: $\downarrow s.\mathbf{EX}\,\mathbf{EF}\,s$

As it has been discussed in Sect. 2.2, paths of CPKS over-approximate the (uncountable) set of all hybrid trajectories of the abstracted multi-affine PHA. As a consequence, at the level of PHA we can interpret only those properties the validity of which cannot be violated by the abstraction. For example, if a universally quantified formula **AF** φ is satisfied at state s in the CPKS then it essentially refers to *all* paths starting in s. These paths over-approximate all PHA trajectories starting at any point of the rectangle represented by s (there is also a universal quantification at the level of points covered by the rectangle). In fact, these paths may include some spurious path—a path that do not correspond to the behaviour of the PHA. In this (completely universally quantified) situation there is no problem—the corresponding property can be considered valid also at the level of the abstracted PHA. However, if a formula **EF** φ is checked true at some state s of the CPKS then it might be (existentially) true just for some spurious path and therefore it cannot be guaranteed to be satisfied in the PHA. As a consequence, only properties containing universal quantification (or negation of existential quantification) can be correctly interpreted at the level of PHA (this is the case of all the example properties mentioned above).

2.4   Parameter Synthesis Problem

The goal of parameter synthesis is to find parameter valuations for which the required properties hold. Given a PHA H, we first transform it to a CPKS C. Then the constraints on C using HCTL are specified, obtaining a HCTL formula φ. Afterwards, we solve a parameter synthesis problem for C and φ by computing the function $\mathcal{F}_\varphi^C : S \to \mathcal{P}(P)$ such that $\mathcal{F}_\varphi^C(s) = \{p \in P \mid (C_p, s) \models \varphi\}$. Finally we can map this result back to the semantics of PHA H.

For example, the HCTL formula for a reachability of a PHA mode q_f is **EF** q_f. From the result of a parameter synthesis, we can tell for every rectangular region of the PHA, for what parameter valuations there exists a path from this region to a state with mode q_f.

3   The Algorithmics

Our method relies on the parallel HCTL parameter synthesis algorithm described in [10]. Conceptually, the algorithm follows the idea of *coloured model checking* for CTL [16]. The examined PKS is represented semi-symbolically: while the state space is explicit, the parameter space is handled symbolically.

This approach allows efficient parallelisation of the time-consuming reachability procedures by partitioning the state space of the PKS between available processors. Meanwhile, the parameter space is still represented using compact symbolic data structures (intervals, polytopes, SMT formulae [8], etc.) which allows us to reasonably handle large number of parametrisations. The important consequence is that the parametrisations are not processed individually (such as during a parameter scan), but in sets. This allows significant performance benefits when similar parametrisations lead to similar behaviour, as these similar parametrisations are typically all processed together in a single symbolic set.

For basic CTL, the algorithm [16] recursively follows the structure of the formula, synthesising parametrisations for individual states and sub-formulae. For HCTL, the procedure needs to be extended with valuations of the free variables. Furthermore, the relationships between individual sub-formulae can be much more intricate. For example, to avoid duplicate computation, one has to consider more complex conditions than simple syntactic equality (e.g. $\downarrow x : \mathbf{AX} x$ and $\downarrow y : \mathbf{AX} y$ being effectively the same formula).

In [10], a *parametrised dependency graph* is constructed on-the-fly, based on the structure of the HCTL property and the structure of the PKS state space. Such graph is then lazily evaluated, yielding the parameter synthesis results for individual states such that the $\mathcal{F}_\varphi^C$ mapping can be constructed. Another advantage of the on-the-fly dependency graph approach is the fact that partial results can be immediately discarded once they are no longer needed.

We demonstrate the approach in Fig. 2. The input for the algorithm is a CPKS, a *solver* which handles symbolic operations over parametrisation sets, a communication *channel* for transferring partial results between parallel workers and a collection of HCTL properties. The dependency graph is then constructed on the fly and evaluated with concrete parametrisation sets. Once the root nodes of the graph are evaluated, the parameter synthesis result is available.

Our implementation of this procedure is developed with the help of existing algorithms and data-structures provided by Pithya [9]. In Pithya, this parameter synthesis algorithm has been successfully applied to standard ODE models before [7,21]. As discussed in Sect. 2.2, we compute a rectangular abstraction of the individual modes of the hybrid system and connect them using discrete jump transitions into a single CPKS. Such CPKS can be then handled using existing parameter synthesis procedure provided by Pithya. The prototype implementation is available at http://github.com/sybila/hybrid-generator.

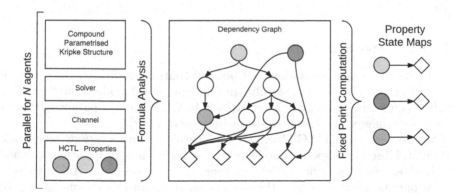

Fig. 2. Workflow scheme of the parallel HCTL parameter synthesis algorithm. First, the inputs are translated into a dependency graph of individual sub-problems, merging any duplicates in the process (in this scheme, diamonds indicate computed values, circles indicate sub-problems to be evaluated). The graph is then iteratively evaluated, yielding the parametrisation mapping $S \rightarrow \mathcal{P}(P)$ for every property.

4 Experimental Evaluation

Our experimental evaluation consists of two parts. We demonstrate the applicability of our approach to real-world biological systems by analysing a known diauxic shift model [36]. We then also explore scalability of our parallel approach with increasing number of available processors.

4.1 Diauxic Shift Model

We consider a hybrid model from [36], representing a *diauxic shift* in an organism with a carbon catabolite repression. Such an organism prefers a specific carbon source (C_1) and grows rapidly while the source is available. With C_1 depleted, the organism switches to another source (C_2) and regulates its growth aggressively.

The model has two regulated proteins, RP and T_2. The presence of RP is regulated based on the carbon source C_1; if the amount of C_1 falls below the threshold γ, RP is suppressed. Similarly, T_2 is regulated by RP; if RP falls below the threshold α, T_2 is suppressed. The model thus consists of four hybrid modes—the combinations of RP and T_2 being suppressed or not. We use [on, on], [on, off], [off, on], and [off, off] to denote these hybrid modes. The values indicate the status of RP and T_2 respectively (on is active, off is suppressed).

In its original form, the model is too simplistic for our experiments. The carbon sources in the model are never replenished. Eventually, all carbon is exhausted and all activity dies out. This is sufficient for short term simulations but not for long-term analysis using temporal properties. We thus extend the model with artificial sources of C_1 and RP (in_{C_1} and in_{RP}) considered as parameters. Complete model is shown in Fig. 3. We pre-process the model as described in [29], approximating the non-linear dynamics with piece-wise multi-affine functions.

4.2 Parameter Synthesis

Based on a few simple simulations, we observe that, depending on the values of in_{C_1} and in_{RP}, the model can stabilise in different hybrid modes as well as oscillate between modes indefinitely. Our goal will be to identify the parametrisations where these different types of long-term behaviour occur.

We use the formula **AG** $\mathcal{A}(mode)$ to identify parametrisations where the system stabilises in a specific mode (here, $\mathcal{A}(mode)$ is a proposition satisfied in all states of the specific *mode*). Furthermore, we use $\downarrow x :$ **AX** x to synthesise parametrisations where the system stabilises in exactly one state. To detect oscillations between two modes, we use the formula **AG** (**AF** ($\mathcal{A}(m_1) \wedge$ **A** ($\mathcal{A}(m_1)$ **U** ($\mathcal{A}(m_2) \wedge$ **A** ($\mathcal{A}(m_2)$ **U** $\mathcal{A}(m_1)$))))) where m_1 and m_2 are different modes. The formula can be extended by adding additional modes (and until operators) to describe more complex oscillations.

$$\dot{C}_1 = -k_{cat_1} T_1 \frac{C_1}{K_{C_1} + C_1} + in_{C_1}$$

$$\dot{C}_2 = -k_{cat_2} T_2 \frac{C_2}{K_{C_2} + C_2}$$

$$\dot{RP} = \mathbf{k_{RP} R} \frac{\mathbf{M}}{\mathbf{K_{RP} + M}} - kd_{RP} RP + in_{RP}$$

$$\dot{T}_1 = k_{T_1} R \frac{M}{K_{T_1} + M} - kd_{T_1} T_1$$

$$\dot{T}_2 = \mathbf{k_{T_2} R} \frac{\mathbf{M}}{\mathbf{K_{T_2} + M}} - kd_{T_2} T_2$$

$$\dot{R} = k_R R \frac{M}{K_R + M} - kd_R R$$

$$\dot{M} = k_{cat_1} T_1 \frac{C_1}{K_{C_1} + C_1} + k_{cat_2} T_2 \frac{C_2}{K_{C_2} + C_2}$$

$$- \mathbf{k_{RP} R} \frac{\mathbf{M}}{\mathbf{K_{RP} + M}} - k_{T_1} R \frac{M}{K_{T_1} + M}$$

$$- \mathbf{k_{T_2} R} \frac{\mathbf{M}}{\mathbf{K_{T_2} + M}} - k_R R \frac{M}{K_R + M}$$

$$k_{cat_1} = 0.3$$
$$k_{cat_2} = 0.2$$
$$k_R = 0.03$$
$$k_{RP} = 0.05$$
$$k_{T_1} = k_{T_2} = 0.1$$
$$K_{C_1} = K_{C_2} = 1.0$$
$$K_{T_1} = K_{T_2} = 1.0$$
$$K_R = K_{RP} = 1.0$$
$$kd_{T_1} = kd_{T_2} = 0.1$$
$$kd_R = 0.001$$
$$kd_{RP} = 0.1$$
$$\alpha = \gamma = 1.0$$
$$in_{C_1} \in [0, 2]$$
$$in_{RP} \in [0, 0.4]$$

Fig. 3. Differential equations of the extended diauxic shift model. The bold expressions are only present in the corresponding hybrid modes. That is, $k_{RP} R \frac{M}{K_{RP} + M}$ is removed in modes where RP is suppressed. The same happens for $k_{T_2} R \frac{M}{K_{T_2} + M}$ when T_2 is suppressed.

The abstraction procedure considers $C_1 \in [0, 50]$, $C_2 \in [0, 35]$, $M \in [0, 20]$, $RP \in [0, 3.5]$, $T_1 \in [0, 8]$, $T_2 \in [0, 2.5]$, $R \in [0, 40]$ (bounds obtained by simulations and sampling in the parameter space) and roughly 8 thresholds per variable, producing ≈ 2 million valid discrete states.

The results of the analysis are shown in Fig. 4. Notice that the system cannot stabilise in the mode [on, on] and there is no possibility of oscillation between three different modes. Also observe that not all behaviour stabilising in a specific mode reaches a sink state. These only appear for the [off, off] mode.

4.3 Scalability

To evaluate the scalability of our approach, we again use the diauxic shift model together with an HCTL property specifying that the model oscillates between all four discrete modes (green areas in Fig. 4). We chose four variants of the model with varying number of continuous parameters and size of the discrete state space (depending on the number of thresholds selected during abstraction).

We conducted all measurements using a machine equipped with AMD Ryzen Threadripper 2990WX 32-Core Processor and 64 GB of memory. The runtime of each experiment is shown in Table 1.

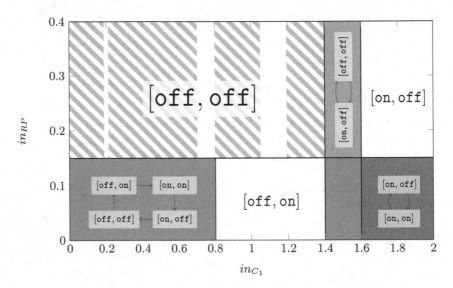

Fig. 4. Parameter synthesis results for the extended diauxic shift model. The white areas depict stable regions (**AG** *mode*) with striped areas indicating the presence of sinks ($\downarrow x : \textbf{AX}\ x$). Coloured areas depict different types of oscillation between modes indicated in that area. (Color figure online)

Table 1. Scalability results for several variants of the diaux shift model.

Parameter count	State count	1 cpu	2 cpu	4 cpu	8 cpu	16 cpu	32 cpu
1	≈1.9e6	123s	95s	70s	55s	43s	38s
	≈4.8e4	5.1s	3.9s	2.7s	1.8s	1.5s	1.3s
2	≈1.9e6	145s	111s	83s	66s	50s	43s
	≈4.8e4	5.9s	4.3s	3.2s	2.2s	1.6s	1.4s

5   Conclusion

In this work, we have presented a novel approach towards parameter synthesis of a significant class of hybrid systems with an expressive logic HCTL. Our approach uses rectangular abstraction to transfer the problem to the domain of state transition systems with a parametrised transition relation. For such an abstract system, we show how to compute the set of parametrisations for which the given HCTL specification holds. We evaluate our approach on a non-trivial real-world biological model.

The first advantage of our approach is the possibility of addressing time-unbounded properties of the systems dynamics. The second advantage is the variety of parameters (flow dynamics, jump guards, invariants) our approach can accommodate. Finally, our algorithm relies on the ideas of parallel parameter synthesis for CTL, thus enabling full utilisation of high-performance hardware.

The disadvantage of the method appears to be the employed abstraction. While very fast and useful in many cases, its exact error (due to the over-approximation) is not precisely quantifiable. Additionally, this approach still suffers from state space explosion with respect to the number of system variables.

As future work, we aim to refine this proof of concept implementation into a stable tool. Moreover, the tool can benefit from future improvements in Pithya, such as performance and scalability optimisations as well as advanced data structures for manipulating symbolic parametrisation sets, thus allowing even wider range of hybrid automata.

References

1. André, É.: IMITATOR: a tool for synthesizing constraints on timing bounds of timed automata. In: Leucker, M., Morgan, C. (eds.) ICTAC 2009. LNCS, vol. 5684, pp. 336–342. Springer, Heidelberg (2009). https://doi.org/10.1007/978-3-642-03466-4_22

2. Arellano, G., et al.: "Antelope": a hybrid-logic model checker for branching-time Boolean GRN analysis. BMC Bioinform. 12(1), 490 (2011)

3. Bartocci, E., Bortolussi, L., Nenzi, L., Sanguinetti, G.: On the robustness of temporal properties for stochastic models. In: Dang, T., Piazza, C. (eds.) Proceedings Second International Workshop on Hybrid Systems and Biology, HSB 2013, Taormina, Italy, 2nd September 2013. EPTCS, vol. 125, pp. 3–19 (2013)

4. Batt, G., Belta, C., Weiss, R.: Model checking genetic regulatory networks with parameter uncertainty. In: Bemporad, A., Bicchi, A., Buttazzo, G. (eds.) HSCC 2007. LNCS, vol. 4416, pp. 61–75. Springer, Heidelberg (2007). https://doi.org/10.1007/978-3-540-71493-4_8

5. Belta, C.: On controlling aircraft and underwater vehicles. In: Proceedings of the IEEE International Conference on Robotics and Automation, vol. 5, pp. 4905–4910 (2004)

6. Belta, C., Habets, L.: Controlling a class of nonlinear systems on rectangles. IEEE Trans. Autom. Control 51(11), 1749–1759 (2006)

7. Beneš, N., Brim, L., Demko, M., Pastva, S., Šafránek, D.: A model checking approach to discrete bifurcation analysis. In: Fitzgerald, J., Heitmeyer, C., Gnesi, S., Philippou, A. (eds.) FM 2016. LNCS, vol. 9995, pp. 85–101. Springer, Cham (2016). https://doi.org/10.1007/978-3-319-48989-6_6

8. Beneš, N., Brim, L., Demko, M., Pastva, S., Šafránek, D.: Parallel SMT-based parameter synthesis with application to piecewise multi-affine systems. In: Artho, C., Legay, A., Peled, D. (eds.) ATVA 2016. LNCS, vol. 9938, pp. 192–208. Springer, Cham (2016). https://doi.org/10.1007/978-3-319-46520-3_13

9. Beneš, N., Brim, L., Demko, M., Pastva, S., Šafránek, D.: Pithya: a parallel tool for parameter synthesis of piecewise multi-affine dynamical systems. In: Majumdar, R., Kunčak, V. (eds.) CAV 2017. LNCS, vol. 10426, pp. 591–598. Springer, Cham (2017). https://doi.org/10.1007/978-3-319-63387-9_29

10. Beneš, N., Brim, L., Pastva, S., Šafránek, D.: Parallel parameter synthesis algorithm for hybrid CTL. Sci. Comput. Program. 185, 102321 (2019)

11. Beneš, N., Brim, L., Demko, M., Pastva, S., Šafránek, D.: Pithya: a parallel tool for parameter synthesis of piecewise multi-affine dynamical systems. Int. J. Struct. Stab. Dyn. (2017)

12. Bogomolov, S., Schilling, C., Bartocci, E., Batt, G., Kong, H., Grosu, R.: Abstraction-based parameter synthesis for multiaffine systems. Hardw. Softw.: Verif. Test. **11**, 19–35 (2015)
13. Bortolussi, L., Milios, D., Sanguinetti, G.: U-Check: model checking and parameter synthesis under uncertainty. In: Campos, J., Haverkort, B.R. (eds.) QEST 2015. LNCS, vol. 9259, pp. 89–104. Springer, Cham (2015). https://doi.org/10.1007/978-3-319-22264-6_6
14. Bortolussi, L., Policriti, A.: Hybrid systems and biology. In: Bernardo, M., Degano, P., Zavattaro, G. (eds.) SFM 2008. LNCS, vol. 5016, pp. 424–448. Springer, Heidelberg (2008). https://doi.org/10.1007/978-3-540-68894-5_12
15. Bortolussi, L., Sanguinetti, G.: Smoothed model checking for uncertain continuous time Markov chains. CoRR abs/1402.1450 (2014)
16. Brim, L., Češka, M., Demko, M., Pastva, S., Šafránek, D.: Parameter synthesis by parallel coloured CTL model checking. In: Roux, O., Bourdon, J. (eds.) CMSB 2015. LNCS, vol. 9308, pp. 251–263. Springer, Cham (2015). https://doi.org/10.1007/978-3-319-23401-4_21
17. Calzone, L., Fages, F., Soliman, S.: BIOCHAM: an environment for modeling biological systems and formalizing experimental knowledge. Bioinformatics **22**(14), 1805–1807 (2006)
18. Chiang, H.K., Fages, F., Jiang, J.R., Soliman, S.: Hybrid simulations of heterogeneous biochemical models in SBML. ACM Trans. Model. Comput. Simul. **25**(2), 14:1–14:22 (2015)
19. Clarke, E., Grumberg, O., Jha, S., Lu, Y., Veith, H.: Counterexample-guided abstraction refinement. In: Emerson, E.A., Sistla, A.P. (eds.) CAV 2000. LNCS, vol. 1855, pp. 154–169. Springer, Heidelberg (2000). https://doi.org/10.1007/10722167_15
20. Clarke Jr., E.M., Grumberg, O., Peled, D.A.: Model Checking. MIT Press, Cambridge (1999)
21. Demko, M., Beneš, N., Brim, L., Pastva, S., Šafránek, D.: High-performance symbolic parameter synthesis of biological models: a case study. In: Bartocci, E., Lio, P., Paoletti, N. (eds.) CMSB 2016. LNCS, vol. 9859, pp. 82–97. Springer, Cham (2016). https://doi.org/10.1007/978-3-319-45177-0_6
22. Donzé, A.: Breach, a toolbox for verification and parameter synthesis of hybrid systems. In: Touili, T., Cook, B., Jackson, P. (eds.) CAV 2010. LNCS, vol. 6174, pp. 167–170. Springer, Heidelberg (2010). https://doi.org/10.1007/978-3-642-14295-6_17
23. Donzé, A., Krogh, B., Rajhans, A.: Parameter synthesis for hybrid systems with an application to Simulink models. In: Majumdar, R., Tabuada, P. (eds.) HSCC 2009. LNCS, vol. 5469, pp. 165–179. Springer, Heidelberg (2009). https://doi.org/10.1007/978-3-642-00602-9_12
24. Fages, F., Rizk, A.: From model-checking to temporal logic constraint solving. In: Gent, I.P. (ed.) CP 2009. LNCS, vol. 5732, pp. 319–334. Springer, Heidelberg (2009). https://doi.org/10.1007/978-3-642-04244-7_26
25. Frehse, G., Jha, S.K., Krogh, B.H.: A counterexample-guided approach to parameter synthesis for linear hybrid automata. In: Egerstedt, M., Mishra, B. (eds.) HSCC 2008. LNCS, vol. 4981, pp. 187–200. Springer, Heidelberg (2008). https://doi.org/10.1007/978-3-540-78929-1_14
26. Frehse, G., et al.: SpaceEx: scalable verification of hybrid systems. In: Gopalakrishnan, G., Qadeer, S. (eds.) CAV 2011. LNCS, vol. 6806, pp. 379–395. Springer, Heidelberg (2011). https://doi.org/10.1007/978-3-642-22110-1_30

27. Fribourg, L., Kühne, U.: Parametric verification and test coverage for hybrid automata using the inverse method. In: Delzanno, G., Potapov, I. (eds.) RP 2011. LNCS, vol. 6945, pp. 191–204. Springer, Heidelberg (2011). https://doi.org/10.1007/978-3-642-24288-5_17

28. Gao, S., Kong, S., Clarke, E.M.: dReal: an SMT solver for nonlinear theories over the reals. In: Automated Deduction - CADE-24, pp. 208–214 (2013)

29. Grosu, R., et al.: From cardiac cells to genetic regulatory networks. In: Gopalakrishnan, G., Qadeer, S. (eds.) CAV 2011. LNCS, vol. 6806, pp. 396–411. Springer, Heidelberg (2011). https://doi.org/10.1007/978-3-642-22110-1_31

30. Henzinger, T.A.: The theory of hybrid automata. In: Proceedings 11th Annual IEEE Symposium on Logic in Computer Science, pp. 278–292, July 1996

31. Islam, M.A., et al.: Bifurcation analysis of cardiac alternans using δ-decidability. In: Bartocci, E., Lio, P., Paoletti, N. (eds.) CMSB 2016. LNCS, vol. 9859, pp. 132–146. Springer, Cham (2016). https://doi.org/10.1007/978-3-319-45177-0_9

32. de Jong, H.: Modeling and simulation of genetic regulatory systems: a literature review. J. Comput. Biol. **9**(1), 67–103 (2002)

33. Kong, S., Gao, S., Chen, W., Clarke, E.: dReach: δ-reachability analysis for hybrid systems. In: Baier, C., Tinelli, C. (eds.) TACAS 2015. LNCS, vol. 9035, pp. 200–205. Springer, Heidelberg (2015). https://doi.org/10.1007/978-3-662-46681-0_15

34. Lincoln, P., Tiwari, A.: Symbolic systems biology: hybrid modeling and analysis of biological networks. In: Alur, R., Pappas, G.J. (eds.) HSCC 2004. LNCS, vol. 2993, pp. 660–672. Springer, Heidelberg (2004). https://doi.org/10.1007/978-3-540-24743-2_44

35. Liu, B., Kong, S., Gao, S., Zuliani, P., Clarke, E.M.: Parameter synthesis for cardiac cell hybrid models using δ–decisions. In: Mendes, P., Dada, J.O., Smallbone, K. (eds.) CMSB 2014. LNCS, vol. 8859, pp. 99–113. Springer, Cham (2014). https://doi.org/10.1007/978-3-319-12982-2_8

36. Liu, L., Bockmayr, A.: Formalizing metabolic-regulatory networks by hybrid automata. bioRxiv (2019)

37. Liu, X., Stechlinski, P.: Infectious Disease Modeling. Springer, Cham (2020). https://doi.org/10.1007/978-3-319-53208-0

38. Rizk, A., Batt, G., Fages, F., Soliman, S.: Continuous valuations of temporal logic specifications with applications to parameter optimization and robustness measures. Theor. Comput. Sci. **412**(26), 2827–2839 (2011). Foundations of Formal Reconstruction of Biochemical Networks

39. Stéphanou, A., Volpert, V.: Hybrid modelling in biology: a classification review. Math. Model. Nat. Phenom. **11**(1), 37–48 (2016)

Core Models of Receptor Reactions to Evaluate Basic Pathway Designs Enabling Heterogeneous Commitments to Apoptosis

Marielle Péré[1,2](✉), Madalena Chaves[1](✉), and Jérémie Roux[1,2](✉)

[1] Université Côte d'Azur, Inria, INRAE, CNRS, Sorbonne Université, Biocore Team, Sophia Antipolis, France
{marielle.pere,madalena.chaves}@inria.fr, jeremie.roux@univ-cotedazur.fr
[2] Université Côte d'Azur, CNRS UMR 7284, Inserm U 1081, Institut de Recherche sur le Cancer et le Vieillissement de Nice, Centre Antoine Lacassagne, 06107 Nice, France

Abstract. Isogenic cells can respond differently to cytotoxic drugs, such as the tumor necrosis factor-related apoptosis inducing ligand (TRAIL), with only a fraction committing to apoptosis. Since non-genetic transient resistance to TRAIL has been shown to dependent on caspase-8 dynamics at the receptor level *in vitro*, here we investigate the core reactions leading to caspase-8 activation, based on mass-action kinetics models, to evaluate the basic mechanisms giving rise to the observed heterogeneous response. In this work, we fit our models to single-cell trajectories of time-resolved caspase-8 activation measured in clonal cells after treatment with TRAIL. Then, we analyse our results to assess the relevance of each model and evaluate how well it captures the extent of biological heterogeneity observed *in vitro*. Particularly, we focus on a positive feedback loop on caspase-8, the impacts of initial condition variations and the relevance of the caspase-8 degradation.

Keywords: ODE · Mass-action kinetics · Parameter identification · Apoptosis · Fractional killing · TRAIL · Caspase-8

1 Introduction

Apoptosis plays a key role in human tissue homeostasis. Its disruption causes well-known diseases such as Alzheimer, Parkinson (excessive apoptosis), or auto-immune disorders and cancers (lack of apoptosis).

To induce cell death in tumor cells, many treatments have been designed and tested so far, such as TRAIL-receptor ligands, which present the advantage of sparing healthy cells. TRAIL binds the death receptors (DR4/5) of the cancer cell, initiating the extrinsic apoptosis pathway. Then, a Death-Inducing Signaling Complex (DISC) is formed in the cytoplasm with adaptor-proteins

© Springer Nature Switzerland AG 2020
A. Abate et al. (Eds.): CMSB 2020, LNBI 12314, pp. 298–320, 2020.
https://doi.org/10.1007/978-3-030-60327-4_16

such as FADD (Fas-Associated protein with Death Domain). This association allows the recruitment of the pro-caspase 8 and 10 (hereafter pC8 and pC10) and other proteins. These pro-caspases compete at the DISC level with c-FLIP [8], an anti-apoptotic protein, to activate the initiator caspase 8 (C8) [31] via dimerization (or even trimerization) and self-cleavage of pC8 [19]. In many cell types, once activated, C8 triggers cell death by mediating Bid cleavage causing the mitochondrial outer membrane permeabilization (MOMP, [4]) which induces the activation of the effective caspases 3 and 7 (C3 and C7), or "executioner caspase", leading to DNA fragmentation and cell death [21,30].

Although TRAIL has been a very promising drug thanks to its ability to target cancer cells specifically, it showed only limited success in the clinic due to a lack of efficiency. In fact, single-cell studies revealed that cells from the same clonal population commit differently to cell death when treated with TRAIL (or other pro-apoptotic drugs), with an important variability in the time of death for the sensitive cells and with a fraction of cells evading apoptosis entirely. When the remaining resistant cells are retreated a second time with cancer drugs (even saturating doses), fractional killing is once again observed. [32,33].

A number of studies and mathematical modeling efforts have evaluated the origins of drug response heterogeneity, proposing mechanisms such as the random fixation of TRAIL on the DR4/5 [1,3], the presence of decoy receptors (which impair the formation of a functional DISC after ligand binding [2]) or the p53 gene effects on TRAIL efficiency [24,25]. The gene CD-95 has also an impact as it regulates FADD, an essential protein for the pC8 binding to the DISC [5,26–28]. c-FLIP antagonist role has been revealed as well, and gives a better understanding of how it "competes" with pC8 at DISC level to trigger (or not) apoptosis [5–7,9], (even if C8 and FLIP seem to bind the DISC on different sites, pC8 favors c-FLIP recruitment [8]). The action of C10 is less well identified. It may be an anti-apoptotic factor in some cases [10], as some members of Bcl-2 family that competes for activating MOMP downstream [4,21,30]. But C10 has also a pro-death role [11,12], it can trigger apoptosis in absence of C8 [13,14] and favor anti-tumorigenesis [15]. Finally, C8 activation has been defined as a determining factor in cell death decision [16], by showing a threshold in rate and timing for C8 activation that distinguishes resistant and sensitive cells [17].

These studies lead to the conclusion that cell decision happens before MOMP and the effector caspase cascade.

Here, taking these insights into consideration with C8 threshold as the main determinant of cell fate, we aim to identify within the core reactions, basic pathway designs that capture cell response heterogeneity to TRAIL, and features of C8 dynamics. Once identified, the next goal is to characterize these regulatory events, to understand how and to what extent, some proteins may influence the C8 dynamic and determine how their variation is correlated to the cell-to-cell variability.

In that aim, we especially focus on three points: (i) FADD role and its capacity for regulating C8, (ii) the relevance of caspase clusters composed of C8 and C10, and (iii) the regulatory effect of the effector caspases on C8 which depends on a positive feedback loop. To investigate the effect of these interactions and their relative timing on apoptosis, we then propose four alternative minimal ODE models. Next, based on the results of Roux et al. [17], these models are calibrated from single-cell data and the distributions of the different parameters are analysed to find links between the models, the C8 dynamic and the cell fates. Finally, we study the feedback loop action, quantify the influence of FADD and C10 and validate our models, explaining the special distribution of C8 degradation.

2 Modeling the Main Processes of Extrinsic Apoptosis Initiation

The first goal is to establish the mechanisms responsible for the main pathway dynamics, and their impact on the C8 activation threshold distinguishing between TRAIL resistant and sensitive cells. The second aim is to understand how these mechanistic models can reproduce cell response heterogeneity.

To this end, this study focuses on three different regulation points: the FADD protein and its capacity for regulation of C8, the importance of C8/C10 cluster in C8 activation [10] and the possible presence of a dowstream regulatory effect of C8 [21, 30], symbolized here by a positive feedback loop from the effector caspase cascade on the C8. In each case, our analyses aim to understand the effect of a given mechanism on the C8 dynamics main features and in which measure this process is a source of heterogeneity or, at least, source of extrinsic noise.

2.1 Models' Assumptions

To capture the extrinsic apoptosis core reactions, our models are thus constructed with a minimal number of components and steps: the TRAIL binding on the death-receptor DR4/5, the recruitment of the FADD protein and the initiator pC8 to form the DISC, the pC8 dimerization, and finally the activation of C8. (c-FLIP is considered to be in very small quantities and so has a lower impact on C8 recruitment.)

TRAIL is denoted by T, the DR4/5 receptors become a single component named R (for Receptor), the pC8 and C8 are grouped to form a unique protein C8. Instead of the recruitment of a single pC8, our models assume two molecules simultaneously bind to DISC, since only dimerization or trimerization of pC8 can trigger apoptosis. F_D denotes the FADD protein and Z_0 the complex TRAIL-receptors. The downstream caspase cascade, the MOMP and cell death are grouped into the component D, with a intermediary complex Z_1.

2.2   Extrinsic Apoptosis Initiation Core Models (EAICM)

Four extrinsic apoptosis initiation core models (EAICM) are proposed, corresponding to the four possible combinations of presence or not of a feedback loop on C8 conjugated with either the adaptor protein or C8/C10 binding.

The feedback loop is represented by the red links on Fig. 1. Two models focus on C10/C8 coupling, where the C8 dimerization happens before the C10 binding (models -cf and -c) to understand how C10 interacts with C8, and finally two others, where only the FADD reaction and the C8 dimerization are taken into account (models -af and -a) to examine the importance of the adaptor protein FADD, especially its regulatory capacity of pC8 recruitment.

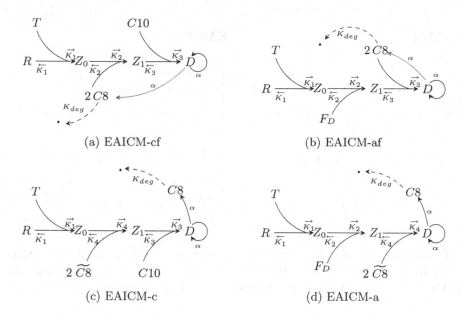

(a) EAICM-cf   (b) EAICM-af

(c) EAICM-c   (d) EAICM-a

Fig. 1. Extrinsic apoptosic initiation core models (EAICM) schemes (Color figure online)

In models without feedback loop, $\widetilde{C8}$ is a constant parameter representing available pC8.

To model the different reactions, we apply the mass-action kinetics and obtain four models of the form $dX/dt = f_{P_r}(X)$, with $f : \mathbb{R}^7 \rightarrow \mathbb{R}^7$ depending on the time-independent reaction rate vector $P_r = (\overrightarrow{K_1}, \overleftarrow{K_1}, ..., \alpha)$, and the initial conditions:

$$\begin{cases} X_0^c = (T_0, R_0, C8_0, C10_0, Z_{0,0}, Z_{1,0}, D_0) \\ X_0^a = (T_0, R_0, C8_0, F_{D,0}, Z_{0,0}, Z_{1,0}, D_0) \end{cases}$$

EAICM-cf:

$$\begin{cases}
\dfrac{dT}{dt} = -\vec{K}_1\, TR + \overleftarrow{K}_1\, Z_0, \\[2mm]
\dfrac{dR}{dt} = -\vec{K}_1\, TR + \overleftarrow{K}_1\, Z_0, \\[2mm]
\dfrac{dZ_0}{dt} = \vec{K}_1\, TR - \overleftarrow{K}_1\, Z_0 - \vec{K}_2\, Z_0 C8^2 + \overleftarrow{K}_2\, Z_1, \\[2mm]
\dfrac{dC8}{dt} = -2\vec{K}_2\, Z_0 C8^2 + 2\overleftarrow{K}_2\, Z_1 + \alpha\, D - K_{deg}\, C8, \\[2mm]
\dfrac{dZ_1}{dt} = \vec{K}_2\, Z_0 C8^2 - \overleftarrow{K}_2\, Z_1 - \vec{K}_3\, Z_1 C10 + \overleftarrow{K}_3\, D, \\[2mm]
\dfrac{dC10}{dt} = -\vec{K}_3\, C10 Z_1 + \overleftarrow{K}_3\, D, \\[2mm]
\dfrac{dD}{dt} = \vec{K}_3\, Z_1 C10 - \overleftarrow{K}_3\, D.
\end{cases} \tag{1}$$

EAICM-af:

$$\begin{cases}
\dfrac{dT}{dt} = -\vec{K}_1\, TR + \overleftarrow{K}_1\, Z_0, \\[2mm]
\dfrac{dR}{dt} = -\vec{K}_1\, TR + \overleftarrow{K}_1\, Z_0, \\[2mm]
\dfrac{dZ_0}{dt} = \vec{K}_1\, TR - \overleftarrow{K}_1\, Z_0 - \vec{K}_2\, Z_0 F_D + \overleftarrow{K}_2\, Z_1, \\[2mm]
\dfrac{dF_D}{dt} = -\vec{K}_2\, F_D Z_0 + \overleftarrow{K}_2\, Z_1, \\[2mm]
\dfrac{dZ_1}{dt} = \vec{K}_2\, Z_0 F_D - \overleftarrow{K}_2\, Z_1 - \vec{K}_3\, Z_1 C8^2 + \overleftarrow{K}_3\, D, \\[2mm]
\dfrac{dC8}{dt} = -2\vec{K}_3\, Z_1 C8^2 + 2\overleftarrow{K}_3\, D + \alpha\, D - K_{deg}\, C8, \\[2mm]
\dfrac{dD}{dt} = \vec{K}_3\, Z_1 C8^2 - \overleftarrow{K}_3\, D.
\end{cases} \tag{2}$$

EAICM-c:

$$\begin{cases}
\dfrac{dT}{dt} = -\vec{K}_1\, TR + \overleftarrow{K}_1\, Z_0, \\[2mm]
\dfrac{dR}{dt} = -\vec{K}_1\, TR + \overleftarrow{K}_1\, Z_0, \\[2mm]
\dfrac{dZ_0}{dt} = \vec{K}_1\, TR - \overleftarrow{K}_1\, Z_0 - \vec{K}_4\, Z_0 \widetilde{C8}^2 + \overleftarrow{K}_4\, Z_1, \\[2mm]
\dfrac{dC8}{dt} = \alpha\, D - K_{deg}\, C8, \\[2mm]
\dfrac{dZ_1}{dt} = \vec{K}_4\, Z_0 \widetilde{C8}^2 - \overleftarrow{K}_4\, Z_1 - \vec{K}_3\, Z_1 C10 + \overleftarrow{K}_3\, D, \\[2mm]
\dfrac{dC10}{dt} = -\vec{K}_3\, C10 Z_1 + \overleftarrow{K}_3\, D, \\[2mm]
\dfrac{dD}{dt} = \vec{K}_3\, Z_1 C10 - \overleftarrow{K}_3\, D.
\end{cases} \tag{3}$$

EAICM-a:

$$\begin{cases}
\dfrac{dT}{dt} = -\vec{K}_1\, TR + \overleftarrow{K}_1\, Z_0, \\[2mm]
\dfrac{dR}{dt} = -\vec{K}_1\, TR + \overleftarrow{K}_1\, Z_0, \\[2mm]
\dfrac{dZ_0}{dt} = \vec{K}_1\, TR - \overleftarrow{K}_1\, Z_0 - \vec{K}_2\, Z_0 C10 + \overleftarrow{K}_2\, Z_1, \\[2mm]
\dfrac{dF_D}{dt} = -\vec{K}_2\, F_D Z_0 + \overleftarrow{K}_2\, Z_1, \\[2mm]
\dfrac{dZ_1}{dt} = \vec{K}_2\, F_D Z_0 - \overleftarrow{K}_2\, Z_1 - \vec{K}_4\, Z_1 \widetilde{C8}^2 + \overleftarrow{K}_4\, D, \\[2mm]
\dfrac{dC8}{dt} = \alpha\, D - K_{deg}\, C8, \\[2mm]
\dfrac{dD}{dt} = \vec{K}_4\, Z_1 \widetilde{C8}^2 - \overleftarrow{K}_4\, D.
\end{cases} \tag{4}$$

Comparing these four alternatives to experimental measurements is then necessary to investigate which of the mechanisms more faithfully reproduces the data and is capable of better generating the single-cell dynamic properties.

3   Single Cell Model Calibration

Our models are calibrated using single cell data from Roux et al. [17]. The data measure the C8 activity before MOMP happens for 414 single cells (114 resistant and 300 sensitive) treated only with 50 ng/mL of TRAIL (and not with cyclohex-imide contrary to [21,30]), for 10 h. These data were obtained using the Initiator Caspase-Reporter Protein (IC-RP [21]), a FRET pair of fluorescent proteins that are linked by the peptide sequence of Bid, cleaved by C8. (FRET therefore decreases once IC-RP molecules are cleaved by C8.) In the same time, Bid is cleaved in tBid, which regulates MOMP in extrinsic apoptosis. As there is no degradation of IC-RP, contrary to tBid, it accumulates leading to the FRET stabilization at the end of the experiment for resistant cells that corresponds to the tBid degradation.

The four EAIC models are fitted to each single cell traces separately, as opposed to fitted to one averaged trace [41,42]. This approach is meant to study each single cell's heterogeneous features and it allows to obtain the parameter distribution without any assumption.

One model topology is used for both resistant and sensitive cells, since the clonal cells are genetically homogeneous. (The main differences between the two populations are attributed to the protein expression levels.)

As only data on the evolution of FRET ratio in time is available, and because the models do not take into account the FRET activation, we assume that the FRET creation corresponds only to a re-scale of C8, *ie* that the FRET dynamic is obtained from the C8 dynamic by changing the amplitude of the C8 curve and the activation time with a supplementary delay, and so the method compares directly the implemented C8 concentration to the real cleaved C8, with great attention to the slope as the FRET slope is a major indicator of the C8 activation speed.

3.1 From Qualitative Criteria to Quantitative Reference Values

To evaluate and compare the four models, it is essential to define a set of criteria to determine how closely each model approaches the real data. This involves translating the main qualitative properties of the C8 curves into quantitative values that can be calculated from the model's solutions. Three fundamental properties are relevant in C8 dynamic and can be evaluated as reference values, as follows (see Fig. 2): (i) the time delay before activation of C8 is triggered; (ii) the mean slope during the C8 activation phase; and (iii) the C8 concentration reaches a stabilization value, over the last 300 min (especially for resistant cells). These properties can be turned into reference values by defining:

– T_{100000} evaluates the initial delay by $C8(T_{100000}) = 100000$ molecules;
– S is the C8 activation slope, as the maximum of the derivative of $C8(t)$ between 25 and 275 min, computed using the Matlab function *sgolayfilt*;

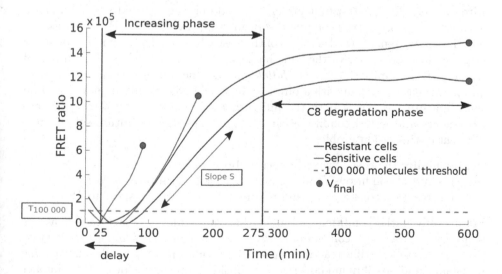

Fig. 2. Reference values and C8 features scheme

– V_{final} gives the final stabilization value, *ie* C8(600), or the value of C8 at death time, for sensitive cells.

It must be noticed that the initial decreasing phase isn't taken into account. It is due to the photoactivation of the FRET and doesn't depend on the apoptosis initiation and as a result, of our models.

3.2 Distinguishing the Effects of Initial Conditions and Rate Parameters on the System Dynamics

Here, we use a nonlinear least-squares method to determine the parameters $P = (P_r, P_i^j, j \in \{c, a\})$ of our models $dX/dt = f_{P_r}(X)$, $P_i^j \in X_0$, where $P_r = (\overleftarrow{K_1}, \overrightarrow{K_1}, ..., K_{deg}, \alpha)$ represents the reaction rates and $P_i^c = (R_0, C8_0, C10_0)$ and $P_i^a = (R_0, C8_0, F_{D,0})$ represent the initial conditions to be evaluated during the model fit, of models EAICM-cf or EAICM-af, respectively. The other initial conditions are fixed with values from literature [21].

An euclidean norm is used to compute the cost, given by the differences between the measurements, denoted by $C8_{t_i}$, $t_i \in \mathbb{T} = \{5, 10, ..., 600\}$ and the computed solution $C8^c$ of the chosen model taken every 5 min. To take into account the slope and the final C8 concentration relevance, the cost is weighted from the 25th min (approximately the beginning time of the increasing phase) until the end with heavier weight ω between the 25th and the 275th (for the slope calculated during the increasing phase). For instance, $\omega = 1000$ between min 25 and min 275. After 280 min, $\omega = 500$. Finally, denoted T_d, the cell death time, the cost $\mathcal{C}$ is given by:

$$\mathcal{C}^2 = \sum_{t_i \in \{5,...,\min(20,T_d)\}} \left(C8_{t_i} - C8_{t_i}^c\right)^2 + \sum_{t_i \in \{25,...,\min(275,T_d)\}} \omega \times \left(C8_{t_i} - C8_{t_i}^c\right)^2 + \sum_{t_i \in \{280,...,\min(600,T_d)\}} \frac{\omega}{2} \times \left(C8_{t_i} - C8_{t_i}^c\right)^2.$$

(5)

Alternatively, for the resistant population, adding the squared slope difference between the data and the computed solution, improves the fit. For the sensitive population, we remove the last parts of the cost when the death time T_d is smaller than the first boundary of the time interval for each one of the three terms of the sum. To minimize $\mathcal{C}$, we used *Matlab* and its function *fminsearchbnb*, to solve an optimization problem with a physiologically significant initial guess based on the literature. To access both the individual and joint effects of reaction rate parameters and initial conditions on the dynamics, the algorithm solves three different optimization problems,

F1. Minimize the cost C with respect to both P_i and P_r;
F2. Fix initial conditions P_i and minimize cost C with respect to P_r;
F3. Fix reaction constants P_r and minimize cost C with respect to P_i.

Fitting only initial conditions, assumes that the model is "exact" and that the response heterogeneity comes from environmental conditions and extrinsic noise only. Conversely, fitting reaction rates only, means that the models have some variability and possibly unknown or not considered reactions or proteins impact the behaviour of C8.

It may be expected that the heterogeneity factors are a mix of the two explanations and so the fit obtained on both initial conditions and reaction rates is the best but, in this case, the results are less straightforward to interpret.

4 Analysing Mechanisms for Generating Heterogeneity

To simulate the models, we set the initial conditions for TRAIL at $T_0 = 1500$ (from [21]), and the intermediary complexes $Z_{0,0}$, $Z_{1,0}$ and D_0 equal to 0. Simulations are performed with *ode23* for 600 min with a weight $\omega = 1000$ for C. For the parameter set and the other initial conditions, when they aren't estimated by the algorithm, values obtained during a first manual fit on a median real cell are used.

4.1 Comparison of the Four Core Apoptosis Models

The first point is to elucidate which of the reactions, binding of the receptor complex to F_D or to C10, best reproduces the behaviour heterogeneity of C8. To determine which of the models of type 1 or 2 best captures the extrinsic apoptosis dynamics, the norm C and the reference values are computed for 114 resistant cells and 300 sensitive ones. Then, for each type of fit F1 to F3, we confront the four models by computing, for each cell and each model, the absolute value of the difference between the data slope and the $C8^c$ slope (that is to say $|S_{EAICM,i} - S_{data,i}|$, $i \in \{1, ..., 414\}$). Then, comparing the four results for each cell, the number of cells for which each model gives the lowest result is counted. The model with the highest score (*i.d.* the largest number of cell for which the given model gives the lowest result comparing the four models) is considered to have the best performance, as summarized in Table 1. In Appendix A, tables for the cost C, the C8 final value and the delay are given.

Table 1. Number of cell best approached per model and type of fits according to the slope

	fate	EAICM-cf	EAICM-c	EAICM-af	EAICM-a	Best model
F1	S. cells	120	78	57	45	EAICM-cf
	R. cells	59	11	32	12	EAICM-cf
F2	S. cells	108	79	71	42	EAICM-cf
	R. cells	75	12	26	1	EAICM-cf
F3	S. cells	269	8	20	3	EAICM-cf
	R. cells	51	23	31	9	EAICM-cf

Table 1 shows clearly that EAICM-cf performs better, suggesting that the caspase cluster and the feedback loop are the main mechanisms necessary to reproduce the variability in C8 slope and general cell response heterogeneity. The same results are obtained for the delay criteria. Moreover, the feedback loop seems essential to capture cell C8 dynamics, because none of the models without feedback loop accurately reproduces the three C8 properties. This result agrees with the findings of Schwarzer et al. [36] in which they demonstrate *in vivo*, the downstream inducing apoptosis effectors' effects on caspase 8. These outcomes also reveal that the clusterization of C8/C10, and so the recruitment and the activation of C8, is more important to C8 dynamics than the presence of F_D in pC8 fixation on DISC. Tummers et al. showed that caspase-8 mediates inflammasome activation independently of FADD in epithelial cells [38], further evidence that FADD isn't mandatory for caspase 8 activity. Future work would expand the study of this cluster reaction, perhaps adding more variables to take into account the effects of other proteins since the reactions around pC8 recruitment (especially its interactions with pC10 and c-FLIP) are still unclear.

Another hypothesis could also be made in this case, assuming that in EAICM-cf, the F_D action is not present in the equations but indeed taken into account since C8 is still recruited at the DISC level.

4.2   The Feedback Loop Mechanism

The second question to address in this Section concerns the effects of the positive feedback loop on C8 to understand its importance on C8 dynamics.

To evaluate the feedback loop impacts on the C8 dynamic, we use the parameters obtained from fit F1, on both initial conditions and reaction rates. Figure 3 and Fig. 4 (a) and (c) compare the FRET ratio and the $C8^c$ curve corresponding to the models 1 with and without feedback for selected resistant and sensitive cells from the cell populations in [17]. It seems clear that the model without feedback fails to reproduce the initial delay before C8 activation. In a second plot, Fig. 3 and Fig. 4 (b) and (d) compare the relative weights of the different terms that contribute to C8 activation. This is a method developed by Casagranda et al. in [34] and consists in representing the absolute values curve of each term that composes the C8 equation, divided by the sum of all absolute values, to normalize. For instance, if we consider the following C8 equation of EAICM-cf:

$$\frac{dC8}{dt} = -2\overrightarrow{K_2}\,Z_0 C8^2 + 2\overleftarrow{K_2}\,Z_1 + \alpha\,D - K_{deg}\,C8, \tag{6}$$

then the plotted curves are:

$$
\left\{
\begin{array}{l}
\dfrac{|K_{deg}\,C8|}{|K_{deg}\,C8| + |\alpha\,D| + |2\overleftarrow{K_2}\,Z_1| + |2\overrightarrow{K_2}\,Z_0C8^2|}, \\[2.5ex]
\dfrac{|\alpha\,D|}{|K_{deg}\,C8| + |\alpha\,D| + |2\overleftarrow{K_2}\,Z_1| + |2\overrightarrow{K_2}\,Z_0C8^2|}, \\[2.5ex]
\dfrac{|2\overleftarrow{K_2}\,Z_1|}{|K_{deg}\,C8| + |\alpha\,D| + |2\overleftarrow{K_2}\,Z_1| + |2\overrightarrow{K_2}\,Z_0C8^2|}, \\[2.5ex]
\dfrac{|2\overrightarrow{K_2}\,Z_0C8^2|}{|K_{deg}\,C8| + |\alpha\,D| + |2\overleftarrow{K_2}\,Z_1| + |2\overrightarrow{K_2}\,Z_0C8^2|}.
\end{array}
\right.
\tag{7}
$$

Similar plots for the EAICM-af and EAICM-a models can be found in Appendix B. First, comparing Fig. 3 and Fig. 4, notice that there are essentially no differences between resistant and non resistant cells in the component-wise analysis. However, there is no activation delay in C8 curve for the models without a feedback loop. Then, focusing on the $|\alpha D|$ variation (corresponding to the feedback loop effect), one can observe that $|\alpha D|$ reaches its maximum and $|K_{deg}C8|$ its minimum at approximately the same moment, which also coincides with the moment when C8 starts increasing. Recall that αD drives all the effective caspase cascade and the feedback loop, so the coincidence between maximum of αD and

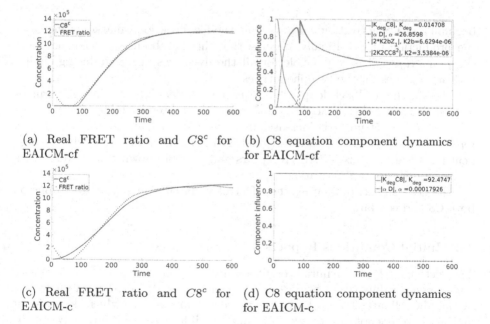

(a) Real FRET ratio and $C8^c$ for EAICM-cf

(b) C8 equation component dynamics for EAICM-cf

(c) Real FRET ratio and $C8^c$ for EAICM-c

(d) C8 equation component dynamics for EAICM-c

Fig. 3. Comparison of C8 dynamics and main properties for models EAICM-cf (a), (b) and EAICM-c (c), (d), for the resistant cell n. 10

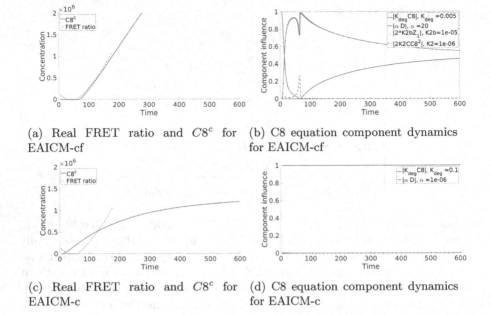

(a) Real FRET ratio and $C8^c$ for EAICM-cf

(b) C8 equation component dynamics for EAICM-cf

(c) Real FRET ratio and $C8^c$ for EAICM-c

(d) C8 equation component dynamics for EAICM-c

Fig. 4. Comparison of C8 dynamics and main properties for models EAICM-cf (a), (b) and EAICM-c (c), (d) for the sensitive cell n. 121 - *simulations were performed for 600 min for comparison needs*

beginning of C8 activation suggests that the feedback loop markedly increases the production of C8. Finally, observe that, in the absence of feedback loop, the term $|K_{deg}C8|$ is responsible for all the dynamics of C8, inducing similar activation slopes for the two phenotypes.

Overall, the feedback loop helps to refine cell decision, by improving modulation of the activation slope, as illustrated by the term αD: for the sensitive cell, in the first 50 min αD increases in a much steeper manner. The feedback represents a supplementary set of regulatory mechanisms that is surely independent from the complex TRAIL/receptors and possibly downstream, yet with a decisive impact on C8 activation.

The next step is evaluating the effect of variability in initial conditions on both C8 and cell fate.

4.3 Initial Conditions Impacts on Slope Values

This section analyses the initial conditions distributions and compares them with our reference values, to identify some mathematical patterns that can help predicting the cell fate. The goal is to find those distributions for which the resistant and sensitive phenotypes present a significant difference, or a link between the initial conditions and C8 dynamics.

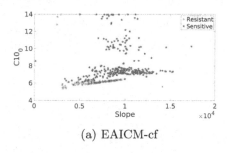

(a) EAICM-cf

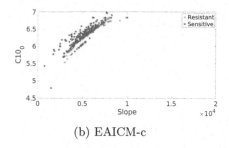

(b) EAICM-c

Fig. 5. Scatter plot of $C10_0$ values according to the slope, depending on cell fate, for the EAICM-cf and EAICM-c

To represent the data obtained after model's fitting, the bar chart of the cell density after model fitting according to their parameter distribution and the scatter plot of the initial condition distribution in logscale according to our reference values (for example, the slope) are used. For each type of graph, resistant and sensitive cells are differenciated to find specific behaviours.

The parameters used for comparison are those obtained from fit F3 (only on the initial conditions), to evaluate the environmental impacts. A clear difference for $C10_0$ between resistant and sensitive cells is observed on the logscale scatter plots in Fig. 5, with a linear correlation between the slope and the initial protein value with highly clustered points for the two types of cells. This is also the case for the F_D distribution that can be found in Appendix C. To understand how these two initial conditions, as well as R_0 variation, affect the C8 dynamics, Fig. 6 shows the evolution of the $C8^c$ curves for each model, as two of the initial conditions are fixed and the third is given by the third is giving by the median value obtained with the fit on all the parameters for resistant cells (given in Appendix D, in black dash dots on Fig. 6) multiplied by $m \in \{0, 0.2, 0.4, 0.6, 0.8, 1.2, 1.4, 1.6, 1.8, 2, 4, 10\}$.

First of all, observe that an increase in the receptor number enhances the slope of C8 and so it speeds up the C8 production and delays the C8 degradation since the stabilization happens later but it doesn't influence the total C8 production (C8 stabilization at the same value). Hence, R_0 is likely to contribute to determination of the C8 activation threshold.

A saturation effect is observed in every model, for the recruited C8, that can't exceed a certain threshold in the total C8 production. This is in agreement with single cell traces since, independently of the TRAIL dose, even at saturated concentration with all the receptors occupied, not every cell commits to apoptosis. An improvement in our models may be necessary to take into account the necessary receptors trimerization that leads to DISC formation [16].

Another observation is that larger $C10_0$ induce larger values for C8 stabilization. An increase in $C10_0$ enhances the C8 production speed but doesn't impact the degradation beginning time. Observe that $C10_0$ also plays a significant role in feedback loop-free models. This effect of $C10_0$ on C8 behaviour

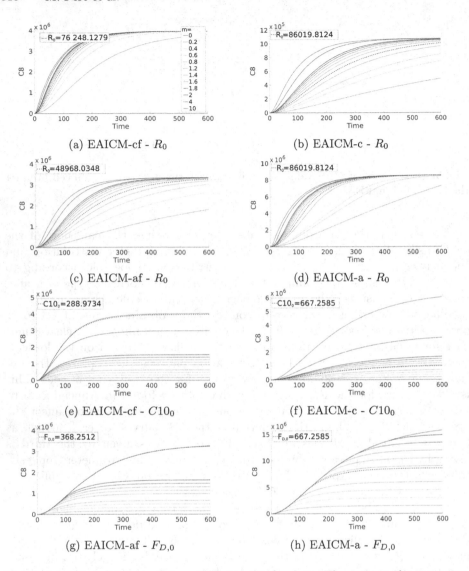

Fig. 6. Initial condition variation effects on C8 dynamic. The estimated parameters P_i are indicated at the top left corners and used as reference values to vary the initial condition, in the range $[0, 10X_0]$, where $X_0 = R_0$ in (a)–(d); $X_0 = C10_0$ in (e)–(f); and $X_0 = F_{D,0}$ in (g)–(h).

confirms the essential role of caspase cluster to trigger cell-death, as shown in Dickens et al. [16].

Finally, increasing $F_{D,0}$ delays C8 degradation and improves C8 production or recruitment, but doesn't speed up the C8 production since the activation slope doesn't show much variation. Furthermore, increasing $F_{D,0}$ leads to an increase in C8, thus making it possible to exceed the C8 threshold responsible for cell death and confirming that FADD is necessary to trigger the extrinsic cell death as demonstrated by Kuang et al. in [18]. Similarly to $C10_0$, F_D also has more influence on the model without feedback loop, suggesting that the feedback loop has a saturation effect on C8 dynamic.

4.4 Model Validation and Degradation Specificity

Comparison of the reaction rates distributions, singles out C8 degradation rate which exhibits a large discrepancy between resistant and sensitive populations, with values related by a factor $K_{deg}^r \approx 10K_{deg}^s$.

As seen in Sect. 4.2, degradation is the process that counteracts C8 activation and, when the term $K_{deg}C8$ becomes sufficiently high, the stabilization phase sets in. Decreasing the degradation rate constant should lead to higher activation slopes and effectively "switch" cells from the resistant to the sensitive populations.

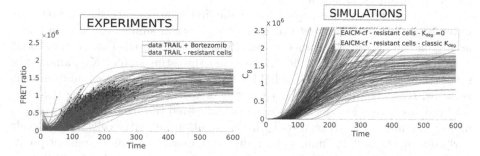

(a) Comparison of FRET ratio between resistant cells treated only with TRAIL and cells treated with TRAIL and Bortezomib.

(b) Comparison of $C8^c$ from EAICM-cf for resistant cells treated only with TRAIL, with $K_{deg} = 0$ and with classic degradation.

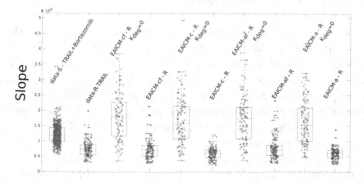

(c) Slope distribution according to the model used and the C8 degradation rate.

Fig. 7. Degradation study

Data from [17] includes a second group of 563 cells treated with 50 ng/mL of TRAIL and 100 ng/mL of Bortezomib, a proteasome inhibitor drug that blocks C8 degradation and drives the cell to commit apoptosis. To validate our models, our hypothesis is that, setting K_{deg} to zero in model EAICM-cf (while keeping other parameters as estimated for each resistant cell), will elicit the same response as Bortezomib, thus transforming the resistant population into sensitive. Figure 7(a) shows the FRET ratio of the two groups of cells: the resistant population of 100 cells treated only with TRAIL and the second group of 563 cells treated with TRAIL and Bortezomib. These experimental results are to be compared with Fig. 7(b), that represents the $C8^c$ EAICM-cf model curves for our original resistant population, with all the corresponding estimated parameters except for K_{deg}, which is set to 0.

Observe that the model predictions in Fig. 7(b) and (c) are quite similar to the experimental data. Figure 7 shows that, imposing a null degradation for our model, allows to reproduce a large heterogeneity range and the main features (delay and bigger slope) of C8 dynamic of the population treated with Bortezomib, thus validating our hypothesis.

Why does the sensitive population of the first group of cells show a markedly lower K_{deg} constant? Perhaps a (negative) feedback or similar mechanism is also acting on the degradation process, annulling its effect in the case of a steep C8 activation. However, it might be the case that the estimation of K_{deg} among the sensitive population is not fully reliable: indeed, recall that the degradation term is linear, $K_{deg}C8$, and that active caspase 8 is absent at the beginning (C8(0) = 0), implying a very low degradation when compared to terms of the form K_2Z_1 or αD which are proportional to T_0R_0. In addition, sensitive cells die relatively early during the first 150 min, so that there are much fewer measurement points available than for resistant cells. New modeling steps are needed to further study the C8 degradation process.

5   Discussion and Conclusion

This paper studies the role and the relevance of several components of the extrinsic apoptosis initiation pathway in cell response heterogeneity. Four minimal ODE models are proposed, taking into account the major steps of the extrinsic pathway: the TRAIL/receptors association, the DISC formation with the recruitment of pro-caspase 8 and, either a focus on the FADD action, or a particular attention to the cluster formation of pC8 and pC10. These models also represent the C8 activation with (or without) a positive feedback loop on C8 to integrate a supplementary regulation of C8 downstream. Finally, as cell decision to commit apoptosis seems to happen before effective caspase activation and MOMP, all the downstream apoptosis steps were combined in a single variable.

The models were calibrated to single cell data from a cloned population treated with death ligand TRAIL. The corresponding initial conditions and/or parameters were analysed to search for correlations between molecular factors and/or network interactions, and the resulting cell fates.

Our analysis selects two mechanisms that significantly contribute to cell response heterogeneity: the clusterization of the caspases C8/C10 and subsequent C8 activation and, to a larger extent, the positive feedback loop. The formation of C8/C10 clusters accelerates C8 activation by increasing C8 production as well as the slope of the curve (see the effect of $C10_0$), while the F_D reaction does not greatly affect the slope but delays the stabilization time. Therefore, caspase clusterization has a greater capacity to generate variability in cell response.

The positive feedback is important in the timing of C8 dynamics, particularly in reproducing the initial delay observed in C8 activation. Studying the components of the C8 equation shows that activation of C8 is triggered when the feedback loop has a maximum effect on C8 and degradation is still negligible. Conversely, when the degradation and the feedback loop terms reach similar levels C8 leaves the high slope phase, revealing that the balance between feedback loop and C8 degradation plays a major role in cell fate.

Another role of the feedback loop is to introduce a saturation on the maximum level of C8 induced by variability in initial conditions: indeed, for our two models with positive feedback, increasing the initial numbers of molecules leads to an increase in the maximum C8 levels, but this maximum value has an upper-bound independent of the initial numbers. This reveals a large robustness of the feedback models with respect to variations in initial amounts of molecules.

Finally, our models faithfully reproduce the experiments involving Bortezomib, a drug that blocks C8 degradation. In our models, application of Bortezomib is represented by setting $K_{deg} = 0$, and the corresponding effect is to increase all activation slopes into the range observed for the sensitive population. Based on the mechanisms and interactions selected by our methods, future work includes the development of a more detailed model to answer further questions such as the need for trimerization of the death receptor, understand the process of caspase degradation during the first hours of C8 activation, or adding new variables to investigate the impact of the anti-apoptotic component c-FLIP.

Appendices

A   Comparison Models Tables

See Tables 2, 3, 4

Table 2. Number of cell best approached per model and type of fits, comparing $\mathcal{C}$ value

Fit ╲ Model	fate	EAICM-cf	EAICM-c	EAICM-af	EAICM-a	Best model
F1	S. cells	177	20	95	8	EAICM-cf
	R. cells	51	3	52	8	EAICM-cf/EAICM-af
F2	S. cells	0	20	0	280	EAICM-a
	R. cells	0	102	0	12	EAICM-c
F3	S. cells	0	63	1	236	EAICM-a
	R. cells	2	95	0	17	EAICM-c

Table 3. Number of cell best approached per model and type of fits comparing the delay, ie $|T_{100000,EAICM,i} - T_{100000,data,i}|$, $i \in \{1, ..., 414\}$

Fit ╲ Model	fate	EAICM-cf	EAICM-c	EAICM-af	EAICM-a	Best model
F1	S. cells	132	98	20	50	EAICM-cf
	R. cells	46	43	9	16	EAICM-cf
F2	S. cells	130	103	8	59	EAICM-cf
	R. cells	55	48	1	10	EAICM-cf
F3	S. cells	222	20	17	41	EAICM-cf
	R. cells	64	10	33	7	EAICM-cf

Table 4. Number of cell best approached per model and type of fits according to C8 final value, *ie* comparing $|V_{final,EAICM,i} - V_{final,data,i}|$, $i \in \{1, ..., 414\}$

Fit \ Model	fate	EAICM-cf	EAICM-c	EAICM-af	EAICM-a	Best model
F1	S. cells	91	108	46	55	EAICM-c
	R. cells	45	16	35	18	EAICM-cf
F2	S. cells	68	111	59	62	EAICM-c
	R. cells	51	10	44	9	EAICM-cf
F3	S. cells	263	0	23	14	EAICM-cf
	R. cells	9	26	62	17	EAICM-af

B   Feedback Loop Effects for EAICM-af and EAICM-a

See Figs. 8 and 9

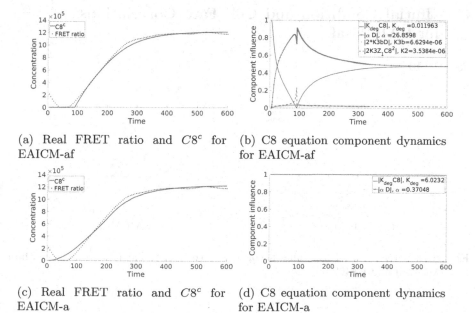

(a) Real FRET ratio and $C8^c$ for EAICM-af

(b) C8 equation component dynamics for EAICM-af

(c) Real FRET ratio and $C8^c$ for EAICM-a

(d) C8 equation component dynamics for EAICM-a

Fig. 8. Comparison of C8 main features with the dynamic of each C8 equation component of EAICM-af (a), (b) and EAICM-a (c), (d) for the resistant cell n. 10

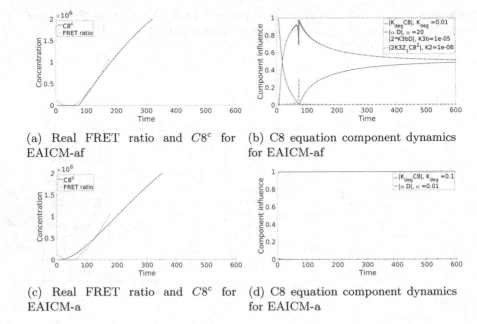

(a) Real FRET ratio and $C8^c$ for EAICM-af

(b) C8 equation component dynamics for EAICM-af

(c) Real FRET ratio and $C8^c$ for EAICM-a

(d) C8 equation component dynamics for EAICM-a

Fig. 9. Comparison of C8 main features with the dynamic of each C8 equation component of EAICM-af (a), (b) and EAICM-a (c), (d) for the sensitive cell n. 121 - *simulations were performed for 600 min for comparison needs*

C   Initial Condition and Cell Fate Correlations for EAICM-af

See Fig. 10

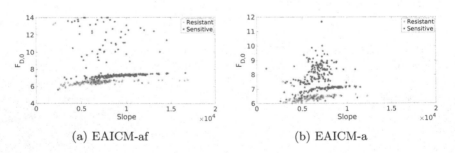

(a) EAICM-af

(b) EAICM-a

Fig. 10. Scatter plot of $F_{D,0}$ values according to the slope, depending on the cell fate for EAICM-af and EAICM-a

D Median Parameter Values from the Fit on Both Initial Conditions and Reaction Rates Used in Fig. 6

See Table 5

Table 5. Median reaction rates and initial conditions for all models determined with the fit on both initial conditions and reactions rates

	EAICM-cf		EAICM-c		EAICM-af		EAICM-a	
	R. cells	S. cells	R. cells	S. cells	R. cells	S. cells	R. cells	S. cells
$\overrightarrow{K_1}$	4.3955e-07	2.7388e-07	6.5892e-08	1.2254e-07	1.6320e-07	4.0018e-07	6.5892e-08	1.2254e-07
$\overleftarrow{K_1}$	0.0052	0.01129	1.1176	1.7906	3.4358e-04	0.0011	1.1177	1.7907
$\overrightarrow{K_2}$	1.5590e-05	2.4304e-05			0.0525	0.0649	25.5081	14.8725
$\overleftarrow{K_2}$	2.9114e-04	9.6920e-04			3.6929e-06	1.2142e-05	2.3934	2.1489
$\overrightarrow{K_3}$	0.0012	0.002792	0.002045	0.002206	4.5915e-05	2.1800e-04		
$\overleftarrow{K_3}$	0.0273	0.1607	20.1550	26.9179	2.0294	9.2971		
$\overrightarrow{K_4}$			16.5523	25.6951			16.5524	25.6952
$\overleftarrow{K_4}$			2.6201	2.6749			2.6202	2.6750
K_{deg}	0.0133	0.004012	0.0001165	0.001765	0.0122	0.0108	0.0117	0.0018
α	36.2215	48.8287	1.3188	279.2583	27.3338	76.6511	131.8895	279.2584
R_0	7.6248e+04	6.7850e+04	8.6019e+04	4.0387e+04	4.8968e+04	5.6593e+04	8.6020e+04	4.0388e+04
$C_{8,0}$	288.9734	905.9665	667.2585	337.2630	368.2512	663.5847	667.2585	337.2631
$C_{10,0}$	2.2325e+03	3.1050e+04	761.9486	1.0829e+04				
$F_{D,0}$					2.9291e+03	5.3681e+04	761.9486	1.0830e+04

E Operation of the Parameter Model and Reference Value tables

In addition of this article, we provide all the parameters tables and the reference values tables obtained with our 3 types of fit for the 414 cells treated with TRAIL only. A line corresponds to one parameter in that order $(\mathcal{C}, \overrightarrow{K_1}, \overleftarrow{K_1}, \overrightarrow{K_2}, \overleftarrow{K_2}, \overrightarrow{K_3}, \overleftarrow{K_3}, \overrightarrow{K_4}, \overleftarrow{K_4}, \alpha, K_{deg}, R_0, C8_0, C10_0$ or $F_{D,0})$ and without $\overrightarrow{K_4}$ and $\overleftarrow{K_4}$ for models with feedback loop.

Parameters_EAICM-cf_NON_resist_fit_Pr_only.mat
Parameters_EAICM-c_NON_resist_fit_Pr_only.mat
Parameters_EAICM-af_NON_resist_fit_Pr_only.mat
Parameters_EAICM-a_NON_resist_fit_Pr_only.mat

12 × 300 table that gives the 8 reactions rates (10 for models without feedback loop) in the first lines and the 3 initial conditions obtained from the fit only on reaction rates for the 300 sensitive cells for each model in the last lines.

Parameters_EAICM-cf_resist_fit_Pr_only.mat
Parameters_EAICM-c_resist_fit_Pr_only.mat
Parameters_EAICM-af_resist_fit_Pr_only.mat
Parameters_EAICM-a_resist_fit_Pr_only.mat

12 × 300 table that gives the 8 reactions rates (10 for models without feedback loop) in the first lines and the 3 initial conditions obtained from the fit only on reaction rates for the 114 resistant cells for each model in the last lines.

Parameters_EAICM-cf_NON_resist_fit_Pi_only.mat
Parameters_EAICM-c_NON_resist_fit_Pi_only.mat
Parameters_EAICM-af_NON_resist_fit_Pi_only.mat
Parameters_EAICM-a_NON_resist_fit_Pi_only.mat
} 12 × 300 table that gives the 8 reactions rates (10 for models without feedback loop) in the first lines and the 3 initial conditions obtained from the fit only on intial conditions for the 300 sensitive cells for each model in the last lines.

Parameters_EAICM-cf_resist_fit_Pi_only.mat
Parameters_EAICM-c_resist_fit_Pi_only.mat
Parameters_EAICM-af_resist_fit_Pi_only.mat
Parameters_EAICM-a_resist_fit_Pi_only.mat
} 12 × 300 table that gives the 8 reactions rates (10 for models without feedback loop) in the first lines and the 3 initial conditions obtained from the fit only on initial conditions for the 114 resistant cells for each model in the last lines.

Parameters_EAICM-cf_NON_resist_fit_Pi_Pr.mat
Parameters_EAICM-c_NON_resist_fit_Pi_Pr.mat
Parameters_EAICM-af_NON_resist_fit_Pi_Pr.mat
Parameters_EAICM-a_NON_resist_fit_Pi_Pr.mat
} 12 × 300 table that gives the 8 reactions rates (10 for models without feedback loop) in the first lines and the 3 initial conditions obtained from the fit on both reaction rates and initial conditions for the 300 sensitive cells for each model in the last lines.

Parameters_EAICM-cf_resist_fit_Pi_Pr.mat
Parameters_EAICM-c_resist_fit_Pi_Pr.mat
Parameters_EAICM-af_resist_fit_Pi_Pr.mat
Parameters_EAICM-a_resist_fit_Pi_Pr.mat
} 12 × 300 table that gives the 8 reactions rates (10 for models without feedback loop) in the first lines and the 3 initial conditions obtained from the fit on both reaction rates and intial conditions for the 114 resistant cells for each model in the last lines.

With the same classification, the files that begin by "Reference_value" followed by the model's name, the cell fate ("resist" or "NON_resist") and the type of fit ("Pr_only", "Pi_only", "Pi_Pr"), contained 3 lines that gives the value of the slope, the C8 final value and the delay with T_{100000} in this order with as many columns as cells.

References

1. Matveeva, A., et al.: Heterogeneous responses to low level death receptor activation are explained by random molecular assembly of the Caspase-8 activation platform. PLoS Comput. Biol. **15**(9), e1007374 (2019)
2. Bouralexis, S., Findlay, D.M., Evdokiou, A.: Death to the bad guys: targeting cancer via Apo2L/TRAIL. Apoptosis **10**(1), 35–51 (2005). https://doi.org/10.1007/s10495-005-6060-0
3. Shlyakhtina, Y., Pavet, V., Gronemeyer, H.: Dual role of DR5 in death and survival signaling leads to TRAIL resistance in cancer cells. Cell Death Dis. **8**(8), e3025 (2017)
4. Eskes, R., Desagher, S., Antonsson, B., Martinou, J.C.: Bid induces the oligomerization and insertion of Bax into the outer mitochondrial membrane. Mol. Cell. Biol. **20**(3), 929–935 (2000)
5. Fricker, N., Beaudouin, J., Richter, P., Eils, R., Krammer, P.H., Lavrik, I.N.: Model-based dissection of CD95 signaling dynamics reveals both a pro-and anti-apoptotic role of c-FLIPL. J. Cell Biol. **190**(3), 377–389 (2010)
6. Han, L., Zhao, Y., Jia, X.: Mathematical modeling identified c-FLIP as an apoptotic switch in death receptor induced apoptosis. Apoptosis **13**(10), 1198–1204 (2008). https://doi.org/10.1007/s10495-008-0252-3
7. Tsuchiya, Y., Nakabayashi, O., Nakano, H.: FLIP the Switch: Regulation of Apoptosis and Necroptosis by cFLIP. Int. J. Mol. Sci. **16**(12), 30321–30341 (2015)

8. Hughes, M. A., Powley, I.R., Jukes-Jones, R., Horn, S., Feoktistova, M., Fairall, L. Schwabe, J. WR., Leverkus, M., Cain, K. and MacFarlane, M. : Co-operative and hierarchical binding of c-FLIP and caspase-8: a unified model defines how c-FLIP isoforms differentially control cell fate. In Molecular cell, vol. 61, n. 6, p. 834–849. Elsevier (2016)

9. Hillert, L.K., et al.: Long and short isoforms of c-FLIP act as control checkpoints of DED filament assembly. Oncogene **39**(8), 1756–1772 (2020)

10. Horn, S., et al.: Caspase-10 negatively regulates caspase-8-mediated cell death, switching the response to CD95L in favor of NF-κB activation and cell survival. Cell Rep. **19**(4), 785–797 (2017)

11. Wang, J., Chun, H.J., Wong, W., Spencer, D.M., Lenardo, M.J.: Caspase-10 is an initiator caspase in death receptor signaling. Proc. Natl. Acad. Sci. **98**(24), 13884–13888 (2001)

12. Wachmann, K., et al.: Activation and specificity of human caspase-10. Biochemistry **49**(38), 8307–8315 (2010)

13. Kischkel, F.C., et al.: Death receptor recruitment of endogenous caspase-10 and apoptosis initiation in the absence of caspase-8. J. Biol. Chem. **276**(49), 46639–46646 (2001)

14. Raulf, N., et al.: Differential response of head and neck cancer cell lines to TRAIL or SMAC mimetics is associated with the cellular levels and activity of caspase-8 and caspase-10. Br. J Cancer **111**(10), 1955–1964 (2014)

15. Kumari, R., Deshmukh, R.S., Das, S.: Caspase-10 inhibits ATP-citrate lyase-mediated metabolic and epigenetic reprogramming to suppress tumorigenesis. Nat. Commun. **10**(1), 1–15 (2019)

16. Dickens, L.S., et al.: A death effector domain chain DISC model reveals a crucial role for caspase-8 chain assembly in mediating apoptotic cell death. Molecular Cell **47**(2), 291–305 (2012)

17. Roux, J., et al.: Fractional killing arises from cell-to-cell variability in overcoming a caspase activity threshold. Molecular Syst. Biol. **11**(5) (2015)

18. Kuang, A.A., Diehl, G.E., Zhang, J., Winoto, A.: FADD is required for DR4-and DR5-mediated apoptosis LACK of TRAIL-induced apoptosis in FADD-deficient mouse embryonic fibroblasts. J. Biol. Chem. **275**(33), 25065–25068 (2000)

19. Chang, D.W., Xing, Z., Capacio, V.L., Peter, M.E., Yang, X.: Interdimer processing mechanism of procaspase-8 activation. EMBO J. **22**(16), 4132–4142 (2003)

20. Schleich, K., et al.: Molecular architecture of the DED chains at the DISC: regulation of procaspase-8 activation by short DED proteins c-FLIP and procaspase-8 prodomain. Cell Death Differ. **23**(4), 681 (2016)

21. Albeck, J.G., Burke, J.M., Spencer, S.L., Lauffenburger, D.A., Sorger, P.K.: Modeling a snap-action, variable-delay switch controlling extrinsic cell death. PLoS Biology **6**(12), e299 (2008)

22. Lederman, E. E.Hope, J. M., King, M R.: Mass action kinetic model of apoptosis by TRAIL-functionalized leukocytes. Front. Oncol. **8** (2018)

23. Bertaux, F., Stoma, S., Drasdo, D., Batt, G.: Modeling dynamics of cell-to-cell variability in TRAIL-induced apoptosis explains fractional killing and predicts reversible resistance. PLoS Comput. Biol. **10**(10), e1003893 (2014)

24. Chong, K.H., Samarasinghe, S., Kulasiri, D., Zheng, J.: Mathematical modelling of core regulatory mechanism in p53 protein that activates apoptotic switch. J. Theor. Biol. **462**, 134–147 (2019)

25. Ballweg, R., Paek, A.L., Zhang, T.: A dynamical framework for complex fractional killing. Sci. Rep. **7**(1), 8002 (2017)

26. Buchbinder, J.H., Pischel, D., Sundmacher, K., Flassig, R.J., Lavrik, I.N.: Quantitative single cell analysis uncovers the life/death decision in CD95 network. PLoS Comput. Bbiol. **14**(9), e1006368 (2018)

27. Bentele, M., et al.: Mathematical modeling reveals threshold mechanism in CD95-induced apoptosis. J. Cell Biol. **166**(6), 839–851 (2004)

28. Neumann, L., et al.: Dynamics within the CD95 death-inducing signaling complex decide life and death of cells. Molecular Syst. Biol. vol. 6(1) (2010)

29. Paek, A.L., Liu, J.C., Loewer, A., Forrester, W.C., Lahav, G.: Cell-to-cell variation in p53 dynamics leads to fractional killing. Cell **165**(3), 631–642 (2016)

30. Rehm, M., Huber, H.J., Dussmann, H., Prehn, J.H.M.: Systems analysis of effector caspase activation and its control by X-linked inhibitor of apoptosis protein. EMBO J. **25**(18), 4338–4349 (2006)

31. Martin, D.A., Siegel, R.M., Zheng, L., Lenardo, M.J.: Membrane oligomerization and cleavage activates the caspase-8 (FLICE/MACHα1) death signal. J. Biol. Chem. **273**(8), 4345–4349 (1998)

32. Fallahi-Sichani, M., Honarnejad, S., Heiser, L.M., Gray, J.W., Sorger, P.K.: Metrics other than potency reveal systematic variation in responses to cancer drugs. Nat. Chem. Biol. **9**(11), 708 (2013)

33. Flusberg, D.A., Roux, J., Spencer, S.L., Sorger, P.K.: Cells surviving fractional killing by TRAIL exhibit transient but sustainable resistance and inflammatory phenotypes. Molecular Biol. Cell **24**(14), 2186–2200 (2013)

34. Casagranda, S., Touzeau, S., Ropers, D., Gouzé, J.L.: Principal process analysis of biological models. BMC Syst. Biol. **12**(1), 68 (2018). https://doi.org/10.1186/s12918-018-0586-6

35. Hillert, L.K., et al.: Dissecting DISC regulation via pharmacological targeting of caspase-8/c-FLIP L heterodimer. Cell Death Diff. 1–14 (2020)

36. Schwarzer, R., Jiao, H., Wachsmuth, L., Tresch, A., Pasparakis, M.: FADD and caspase-8 regulate gut homeostasis and inflammation by controlling MLKL-and GSDMD-mediated death of intestinal epithelial cells. Immunity (2020)

37. Strasser, A., Vaux, D.L.: Cell death in the origin and treatment of cancer. Molecular Cell (2020)

38. Tummers, B., et al.: Caspase-8-dependent inflammatory responses are controlled by its adaptor, FADD, and Necroptosis. Immunity (2020)

39. Amaral, M.P., Bortoluci, K.R.: Caspase-8 and FADD: where cell death and inflammation collide. Immunity **52**(6), 890–892 (2020)

40. Chaudhry, M.Z., et al.: Cytomegalovirus inhibition of extrinsic apoptosis determines fitness and resistance to cytotoxic CD8 T cells. Proc. Natl. Acad. Sci. **117**(23), 12961–12968 (2020)

41. Llamosi, A., et al.: What population reveals about individual cell identity: single-cell parameter estimation of models of gene expression in yeast. PLoS Comput. Biol. **12**(2), e1004706 (2016)

42. Pereira, L.C.G.: Thesis : Modeling cell response heterogeneity to pro-apoptotic ligands. COMUE Université Côte d'Azur (2015–2019)

Drawing the Line: Basin Boundaries
in Safe Petri Nets

Stefan Haar[1(✉)], Loïc Paulevé[2], and Stefan Schwoon[1]

[1] Inria and LSV, CNRS & ENS Paris-Saclay, Université Paris-Saclay, Gif-sur-Yvette,
France
`stefan.haar@inria.fr`
[2] Univ. Bordeaux, Bordeaux INP, CNRS, LaBRI, UMR5800, Talence, France

Abstract. Attractors of network dynamics represent the long-term
behaviours of the modelled system. Understanding the basin of an attractor, comprising all those states from which the evolution will eventually lead into that attractor, is therefore crucial for understanding the
response and differentiation capabilities of a dynamical system. Building on our previous results [2] allowing to find attractors via Petri net
Unfoldings, we exploit further the unfolding technique for a backward
exploration of the state space, starting from a known attractor, and show
how all strong or weak basins of attractions can be explicitly computed.

Keywords: Dynamical systems · Qualitative models · Attractors ·
Concurrency · Biological networks · Epigenetics · Reprogramming

1 Introduction

Multistability is a central feature of models for biological systems. It is implied by
many fundamental biological processes, such as cellular differentiation, cellular
reprogramming, and cell fate decision.

In qualitative models such as Boolean and multivalued networks, multistability is tied to the notion of *attractors* and to their *basins*. Attractors are usually
defined as the smallest subsets of states from which the system cannot escape.
Then, basins refer to the states of the system which can reach a given attractor.
One can distinguish two kinds of basins for an attractor A: the *weak* basin of
A which gathers all the states that can reach A, but possibly others; and the
strong basin of A which is the subset of the weak basin which cannot reach other
attractors than A. The strong basin includes the attractor itself, and possibly
other preceding states [15].

Understanding how the system switches from a weak to a strong basin is a
recurrent question when analysing models of signalling and gene regulatory networks [5,19]. In [10], the authors provide a method for identifying the states in
which one transition leads to losing the reachability of a given attractor (bifurcation transitions). Whereas the approach can still enumerate only the bifurcation
transitions, it then loses the precious information of the contexts in which the

© Springer Nature Switzerland AG 2020
A. Abate et al. (Eds.): CMSB 2020, LNBI 12314, pp. 321–336, 2020.
https://doi.org/10.1007/978-3-030-60327-4_17

transitions make the system bifurcate from the attractor. Thus, besides listing of the states on the boundary of a strong basins, the challenge resides in identifying the specific contexts and sequences of transitions leading to a strong basin.

Let us illustrate on the small automata network example showing bifurcation transitions reproduced in Fig. 1. The model gathers 3 automata a, b, and c with respectively 3, 2, and 3 local states. The automata start in the state in gray, then can undergo transitions (fully) asynchronously. Transitions labeled with the local states of other automata can only be taken whenever the referenced automata are in these states. This results in the transition graph of Fig. 2. Two attractors are reachable: the fixpoint (i.e. singleton attractor) $\langle a_2, b_1, c_0 \rangle$, and the cyclic attractor where c is fixed to 2, and a and b oscillate between 0 and 1. The states in gray in Fig. 2 form the weak basin of the fixpoint$\langle a_2, b_1, c_0 \rangle$. In specific states of this basin, firing the transition t_8, which is the transition of c from 1 to 2 makes the system lose the capability to reach the fixpoint, by entering in (the strong basin of) the cyclic attractor.

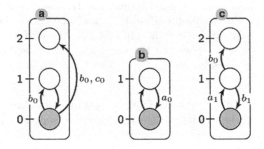

Fig. 1. Running example: Automata network, from [10]; grey-shaded states are initial

In this paper, we address the computation of strong basins that are reachable from a fixed initial state. We rely on concurrency theory to obtain compact representations of reachable sequences of transitions by means of *safe Petri nets* unfoldings, which avoid the explicit exploration of all interleavings. They provide a compact and insightful representation of strong basins, by focusing on the causality and context of transitions.

Safe (or 1-bounded) Petri nets [20] are close to Boolean and multivalued networks [3], although they enable a more fine-grained specification of the conditions for triggering value changes. Focussing on safe PNs entails no limitation of generality of the model, as two-way behaviour-preserving translations between boolean and multivalued models exist (see [3] and the appendix of [2] for discussion). Indeed, instead of a function whose evaluation gives the next value for each node of the network, Petri nets explicitly specify the nodes that actually enable each value change, which are typically very few. *Unfoldings* [8] of Petri nets, which are essentially event structures in the sense of Winskel et al. [21] with additional information about *states* that are crucial in our work here, bring an acyclic representation of the possible sequences of transitions, akin

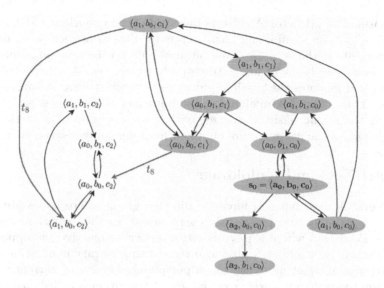

Fig. 2. Transition graph for the automata network of Fig. 1, from [10]. Attractors are $\{\langle a_2, b_1, c_0\rangle$ and $\{\langle a_1, b_1, c_2\rangle, \langle a_0, b_1, c_2\rangle, \langle a_0, b_0, c_2\rangle, \langle a_1, b_0, c_2\rangle\}$.

to Mazurkiewicz traces [7][1] but enriched with branching information. From an unfolding, all reachable states and even attractors [2] can be extracted. Like the authors of [11] did for transition systems, we will exploit both forward and reverse dynamics, by constructing reverse unfoldings to explore co-reachable states (i.e. states *from which* a given set can be reached).

Here again, the restriction to 1-safe nets is a technical convenience for unfoldings but not a strict necessity; the use of finite complete prefixes and application of our techniques is possible for general *bounded*[2] nets via careful model translations, not discussed here due to lack of space. Also, there is a variety of other formal approaches that use concurrency to avoid state space explosion, such as partial order reduction in reachability-related tasks following Godefroid [12], or dihomotopy in the sense of Goubault [13]; their angle of attack is on the level of transition systems, whereas our approach focusses on local causal relations.

Outline. After providing the necessary definitions for Petri nets and their unfoldings in Sect. 2, we will turn our attention to attractors in Sect. 3. The algorithm that we had developed in [2] for *finding* the complete list of attractors in a safe Petri net, will be extended here. The extended algorithm ATTMAP below provides, in addition to the attractors, information about which system state, called *marking* for Petri nets, allows to reach which attractor. The output of ATTMAP includes a function $\mathsf{Sig}(\bullet)$ that assigns to every marking the set of attractors in whose basin of attraction the marking lies. Inversely, the

---

[1] Which Cousot et al. [6] have recently use as theoretical foundation for capturing which entity of a program is *responsible* of a given behavior.

[2] Note that infinite-state Petri nets do not have finite complete prefixes in our sense.

strong basin of an attractor **A** consists precisely of those markings M for which $\mathrm{Sig}(M) = \{\mathbf{A}\}$. The second, and final, step is then taken in Sect. 5: we develop two algorithmic methods (on-line and off-line) built on the same principle; they delimit the strong basin of any attractor of the net, via a reverse Petri net unfolding that explores the possible processes that might have led into a given attractor. Truncating this unfolding at the boundary of the strong basin allows to exhibit the strong basin as the set of *interior configurations* in the sense defined below, of the net structure obtained from unfolding. Sect. 6 concludes.

2   Petri Nets and Unfoldings

Petri Nets. A Petri net is a bipartite directed graph where nodes are either *places* or *transitions*, and places may carry *tokens*. In this paper, we consider only *safe* Petri nets where a place is either active or inactive (as opposed to general Petri nets, where each place can receive an arbitrary number of tokens, safe Petri nets allow at most one token per place). The set of currently active places form the state, or *marking*, of the net. A transition is called enabled in a marking if all its input places are active, and no output place is active unless it is also an input place. The *firing* of a transition modifies the current marking of the net by rendering the input places inactive and output places active.

Formally, a *net* is a tuple $N = (P, T, F)$, where T is a set of *transitions*, P a set of *places*, and $F \subseteq (P \times T) \cup (T \times P)$ is a *flow relation* whose elements are called *arcs*. A subset $M \subseteq P$ of the places is called a *marking*. A *Petri net* is a tuple $\mathcal{N} = \langle N, M_0 \rangle$, with $M_0 \subseteq P$ an *initial marking*.

In figures, places are represented by circles and the transitions by boxes (each one with a label identifying it). The arrows represent the arcs. The initial marking is represented by dots (or tokens) in the marked places. The *reverse net* of $\mathcal{N}$ is $\overleftarrow{\mathcal{N}} \overset{def}{=} (P, T, F^{-1})$. For any node $x \in P \cup T$, we call *pre-set* of x the set $^\bullet x = \{y \in P \cup T \mid (y, x) \in F\}$ and *post-set* of x the set $x^\bullet = \{y \in P \cup T \mid (x, y) \in F\}$. A transition $t \in T$ is *enabled* at a marking M, denoted $M \overset{t}{\rightarrow}$, if and only (i) $^\bullet t \subseteq M$, and (ii) $(M \cap t^\bullet) \subseteq {}^\bullet t$. Note that the second requirement is usually not made in the Petri net literature; however in the class of safe nets (see below), condition (ii) is true whenever (i) is true. An enabled transition t can *fire*, leading to the new marking $M' = (M \setminus {}^\bullet t) \cup t^\bullet$; in that case write $M \overset{t}{\rightarrow} M'$. A *firing sequence* is a (finite or infinite) word $w = t_1 t_2 t_3 \ldots$ over T such that there exist markings $M_1, M_2, \ldots$ such that $M_0 \overset{t_1}{\rightarrow} M_1 \overset{t_2}{\rightarrow} M_2 \overset{t_3}{\rightarrow} \ldots$. All markings in such a firing sequence are called *reachable* from the initial marking M_0. We denote the set of markings reachable from some marking M in $\mathcal{N}$ by $\mathbf{R}_{\mathcal{N}}(M)$ (dropping the subscript $\mathcal{N}$ if no confusion can arise).

A marking M reachable in $(\overleftarrow{N}, M_0)$ is said *co-reachable* from M_0 in N; denote as $\overleftarrow{\mathbf{R}}_N(M_0)$ the set of co-reachable markings for M_0 and $\mathcal{N}$.

Semantics of Safe Petri Nets. As an important restriction, we require all Petri nets arising in this paper to be *safe*, that is, in any reachable marking M and any transition t such that $^\bullet t \subseteq M$, one has $(M \cap t^\bullet) \subseteq {}^\bullet t$. For an easier

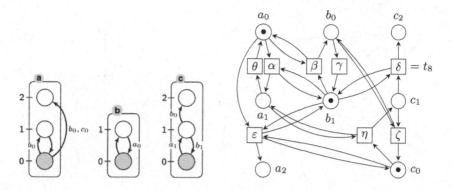

Fig. 3. Right hand side: A Petri net version of the automata network from Fig. 1 (initial state shifted to $< a_0, b_1, c_0 >$), reproduced on the left hand side.

formalization of the following concepts, it is possible to introduce *complementary places*: a place $\overline{p}$ is a complement of place p iff

1. ${}^\bullet p = \overline{p}^\bullet$ and $p^\bullet = {}^\bullet\overline{p}$, and
2. $[M_0 \cap \{p, \overline{p}\}| = 1$.

A Petri net $\mathcal{N} = (P, T, F, M_0)$ is called *complete* iff every place in P has at least one complement in P. Complete Petri nets are safe by construction, since the number of tokens on a pair of complementary places is an invariant of the transition firing. Moreover, the reverse net of a complete net is also complete, and thus safe. For a net $N = (P, T, F)$, the *completion* of N is $\hat{N} = (P \uplus \overline{P}, T, F \uplus \overline{F})$, where $\overline{P} = \{\overline{p} \mid p \in P\}$ is disjoint from P, and

$$\overline{F} \stackrel{def}{=} \{(\overline{p}, t) \mid (t, p) \in F \cap (T \times P)\} \cup \{(t, \overline{p}) \mid (p, t) \in F \cap (P \times T)\}$$

From an initial marking of the net, one can recursively derive all possible transitions and reachable markings, resulting in the *marking graph* (Definition 1). The marking graph is always finite in the case of safe Petri nets. The attractors reachable from some initial marking of the net are the terminal strongly connected components of the associated reachability graph.

Definition 1. *Let $N = (P, T, F)$ be a net and $\mathcal{M}$ a set of markings. The marking graph induced by $\mathcal{M}$ is a directed graph $(\mathcal{M}, \mathcal{E})$ such that $\mathcal{E} \subseteq \mathcal{M} \times \mathcal{M}$ contains (M, M') if and only if $M \xrightarrow{t} M'$ for some $t \in T$; the arc (M, M') is then labeled by t. The reachability graph of a Petri net $(\mathcal{N}, M_0)$ is the marking graph $\mathcal{G}(N, M_0)$ induced by $\mathbf{R}_\mathcal{N}(M_0)$. The transition system of a complete net $N = (P \uplus \overline{P}, T, F \uplus \overline{F})$ is the marking graph $TS(N)$ induced by $\{M \cup \overline{P \setminus M} \mid M \subseteq 2^P\}$.*

Note that complementary places are uniquely defined. To keep the presentation legible and compact, we will henceforth drop all complement places from both notations and figures.

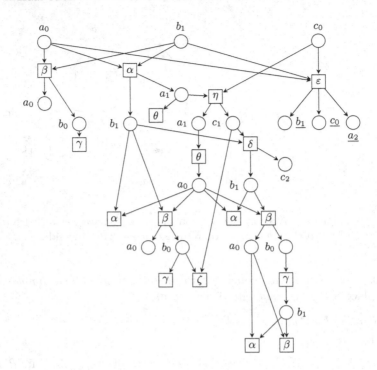

Fig. 4. A finite complete prefix of the unfolding of the Petri net of Fig. 3; underlined names indicate fixed conditions, events without outgoing arcs are cut-offs.

Unfoldings. Let us now recall the basics of Petri net unfoldings and how to use them in finding attractors, following [2]. Roughly speaking (a more extensive treatment can be found, e.g., in [8]), the unfolding of a Petri net $\mathcal{N}$ is an acyclic Petri net $\mathcal{U}$ that has the same behaviours as $\mathcal{N}$ (modulo homomorphism). In general, $\mathcal{U}$ is an infinite net, but if $\mathcal{N}$ is safe, then it is possible to compute a finite prefix $\preceq$ of $\mathcal{U}$ that is "complete" in the sense that every reachable marking of $\mathcal{N}$ has a reachable counterpart in $\preceq$, and vice versa [8,18].

Definition 2 (Causality, conflict, concurrency). *Let $N = \langle P, T, F \rangle$ be a net and $x, y \in P \cup T$ two nodes of N. We say that x is a* causal predecessor *of y, noted $x < y$, if there exists a non-empty path of arcs from x to y. We note $x \leq y$ if $x < y$ or $x = y$. If $x \leq y$ or $y \leq x$, then x and y are said to be* causally related. *x and y are in* conflict, *noted $x \# y$, if there exist $u, v \in T$ such that $u \neq v$, $u \leq x$, $v \leq y$, and $^\bullet u \cap {}^\bullet v \neq \emptyset$. We call x and y* concurrent, *noted x co y, if they are neither causally related nor in conflict.*

Definition 3 (Occurrence net). *Let $\mathcal{N} = \langle P, T, F, M_0 \rangle$ be a Petri net. We say that $\mathcal{N}$ is an* occurrence net *if it satisfies the following properties:*

1. *The causality relation $<$ is acyclic;*
2. *$|^\bullet p| \leq 1$ for all places p, and $p \in M_0$ iff $|^\bullet p| = 0$;*

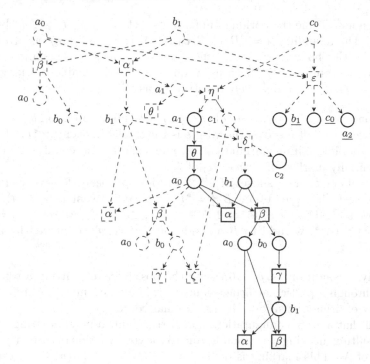

Fig. 5. The attractors for the Petri net of Fig. 3, represented in the unfolding prefix of Fig. 4.

3. *For every transition t, $t \# t$ does not hold, and $[x] \overset{\text{def}}{=} \{x \mid x \leq t\}$ is finite.*

We say that $\mathcal{N}$ is a reverse occurrence net *iff $\overleftarrow{\mathcal{N}}$ is an occurrence net.*

As we said before, an *unfolding* is an "acyclic" version of a safe Petri net $\mathcal{N}$. This notion of acyclicity is captured by Definition 3.

As is convention in the unfolding literature, we shall refer to the places of an occurrence net as *conditions* and to its transitions as *events*. Due to the structural constraints, the firing sequences of occurrence nets have special properties: if some condition c is marked during a run, then the token on c was either present initially or produced by one particular event (the single event in $^\bullet c$); moreover, once the token on c is consumed, it can never be replaced by another token, due to the acyclicity constraint on $<$.

Definition 4 (Configuration, cut). *Let $\mathcal{N} = \langle B, E, G, \mathbf{c_0} \rangle$ be an occurrence net. A set $C \subseteq E$ is called* configuration *of $\mathcal{N}$ if (i) C is causally closed, i.e. $e' < e$ and $e \in E$ imply $e' \in E$; and (ii) C is conflict-free, i.e. if $e, e' \in C$, then $\neg(e \# e')$. The* cut *of C, denoted $\mathbf{cut}(C)$, is the set of conditions $(\mathbf{c_0} \cup C^\bullet) \setminus {}^\bullet C$.*

Intuitively, a configuration is a set of events that can fire during a firing sequence of $\mathcal{N}$, and its cut is the set of conditions marked after that firing sequence. Note that $\emptyset$ is a configuration, and that $\mathbf{c_0}$ is its cut.

We can now define the notion of unfoldings. Let $\mathcal{N} = \langle P, T, F, M_0 \rangle$ be a safe Petri net. The unfolding $\mathcal{U} = \langle B, E, G, \mathbf{c}_0 \rangle$ of $\mathcal{N}$ is an (infinite) occurrence net (equipped with a homomorphism h) such that the firing sequences and reachable markings of $\mathcal{U}$ are exactly the firing sequences and reachable markings of $\mathcal{N}$ (modulo h). $\mathcal{U}$ can be inductively constructed as follows:

1. The condition set B is a subset of $(E \cup \{\bot\}) \times P$. For a condition $b = \langle e, p \rangle$, we will have $e = \bot$ iff $b \in \mathbf{c}_0$; otherwise e is the singleton event in ${}^\bullet b$. Moreover, $h(b) = p$. The initial marking $\mathbf{c}_0$ contains exactly one condition $\langle \bot, p \rangle$ for each initially marked place $p \in M_0$ of $\mathcal{N}$.
2. The events of E are a subset of $2^B \times T$. More precisely, for every cut $\mathbf{c}$ and $B' \subseteq \mathbf{c}$ such that $\{ h(b) \mid b \in B' \} = {}^\bullet t$, we have an event $e = \langle B', t \rangle$. In this case, we add edges $\langle b, e \rangle$ for each $b \in B'$ (i.e. ${}^\bullet e = B'$), we set $h(e) = t$, and for each $p \in t^\bullet$, we add to B a condition $b = \langle e, p \rangle$ connected by an edge $\langle e, b \rangle$.

Intuitively, a condition $\langle e, p \rangle$ represents the possibility of putting a token onto place p through a particular firing sequence, while an event $\langle B', e \rangle$ represents a possibility of firing transition e in a particular context.

Recall that a finite configuration C of $\mathcal{U}$ represents a possible firing sequence whose resulting marking corresponds, due to the construction of $\mathcal{U}$, to a reachable marking of $\mathcal{N}$. This marking is defined as $Mark(C) := \{ h(b) \mid b \in \mathbf{cut}(C) \}$. Since $\mathcal{U}$ is infinite in general, we are interested in computing an initial portion of it (a *prefix*) that completely characterizes the behaviour of $\mathcal{N}$.

Definition 5 (complete prefix). *Let $\mathcal{N} = \langle P, T, F, M_0 \rangle$ be a safe Petri net and $\mathcal{U} = \langle B, E, G, \mathbf{c}_0 \rangle$ its unfolding. A finite occurrence net $\preceq = \langle B', E', G', \mathbf{c}_0 \rangle$ is said to be a* prefix *of $\mathcal{U}$ if $E' \subseteq E$ is causally closed, $B' = \mathbf{c}_0 \cup E'^\bullet$, and G' is the restriction of G to B' and E'. A prefix $\preceq$ is said to be* complete *if for every reachable marking M of $\mathcal{N}$ there exists a configuration C of $\preceq$ such that (i) $Mark(C) = M$, and (ii) for each transition $t \in T$ enabled in M, there is an event $\langle B'', t \rangle \in E'$ enabled in $\mathbf{cut}(C)$ (Fig. 5).*

We shall write $\Pi_0(\mathcal{N}, M)$ to denote an arbitrary complete prefix of $\mathcal{N}$ from initial marking M. It is known [9,18] that the construction of such a complete prefix is indeed possible, and efficient tools [14,22] exist for this purpose. The precise details of this construction are out of scope for this paper; for what follows it suffices to know that it essentially follows the construction of $\mathcal{U}$ outlined above but that certain events are flagged as *cut-offs* when they do not "contribute any new reachable markings". The construction then does not continue beyond any such cut-off event.

3   Attractors

Definitions and Fundamental Properties. The notion of *attractor* denotes, informally, a set of states from which the system cannot 'escape', i.e. from any state of an attractor, only states inside the same attractor are reachable; that is, the attractors are exactly the terminal SCCs of the transition system. The *strong basin* of an attractor **A** collects those states from which the system eventually enters **A** (never to leave it again), and its *weak basin* those states from which the system may enter **A**.

Definition 6. *Let* $\mathcal{N} = (P, T, F)$ *be a net. An* attractor $\mathbf{A} \subseteq 2^P$ *is a bottom (terminal) strongly connected component (SCC) of* $TS(\mathcal{N})$. *Denote the set of attractors (of* $\mathcal{N}$) *by* $\mathcal{A}$, *and the set of attractors reachable from a marking* M *by*

$$\mathsf{Sig}(M) \stackrel{def}{=} \{\mathbf{A} \in \mathcal{A} : \ \mathbf{A} \cap \mathbf{R}(M) \neq \emptyset\}.$$

In particular, an attractor **A** *is a fixed point iff there is* $M \in \mathcal{M}$ *such that* $\mathbf{A} = \{M\}$, *and for any* $t \in T$, $M \stackrel{t}{\rightarrow} M'$ *implies* $M = M'$.

It is important to stress the fact that the SCCs to be considered as attractors have to be terminal. In the example, the system may (though this is not likely) cycle forever in the set of states in which neither a_2 nor c_2 holds; however, this set is not an attractor. When we wish to give a dynamic characterization of attraction, and in particular of *basins of attraction*, it is not enough to require the existence of infinite runs that stay inside a given state set; we need to restrict to those runs that 'eventually explore all accessible branches'. This intuition can be captured by the notion of *fairness*: any transition that is *enabled* infinitely often, must also eventually occur :

Definition 7. *In* $\mathcal{N}$ *as above, an infinite firing sequence* $M_0 \stackrel{t_1}{\rightarrow} M_1 \stackrel{t_2}{\rightarrow} \ldots$ *is* fair *iff for all* $t \in T$:

$$\left| \left\{ i \in \mathbb{N}; \ M_i \stackrel{t}{\rightarrow} \right\} \right| = \infty \Rightarrow \{j \in \mathbb{N}; \ t_j = t\} \neq \emptyset \tag{1}$$

Note that this notion corresponds to *weak* fairness in the sense of [23], which is sufficient for our purposes (see also Abadi et al. [1]); we thus speak only of fairness here. Any such fair sequence will eventually leave any spurious SCC, and, the state space of the net is finite, sooner or later enter a bottom SCC, which, of course, it cannot leave anymore. We are thus ready to define:

Definition 8. *Let* $\mathcal{A}' \subseteq \mathcal{A}$. *The* strong basin $\mathcal{B}_{\mathcal{A}'} \subseteq 2^P$ *of* $\mathcal{A}'$ *is the set of markings from which every fair firing sequence leads eventually into some* $\mathbf{A} \in \mathcal{A}'$; *the* weak basin $\mathcal{W}_{\mathcal{A}'} \subseteq 2^P$ *of* $\mathcal{A}'$ *is the set of markings from which some* $\mathbf{A} \in \mathcal{A}'$ *is reachable. By abuse of notation, we will write* $\mathcal{B}_{\mathbf{A}}$ $(\mathcal{W}_{\mathbf{A}})$ *for* $\mathcal{B}_{\mathcal{A}'}$ $(\mathcal{W}_{\mathcal{A}'})$ *when* $\mathcal{A}' = \{\mathbf{A}\}$.

Note that for all attractors **A**, we have $\mathbf{A} \subseteq \mathcal{B}_{\mathbf{A}} \subseteq \mathcal{W}_{\mathbf{A}}$. Also, two distinct attractors **A**, **A**' must be disjoint, and their *strong* basins too. However, two *weak* basins $\mathcal{B}_{\mathbf{A}}, \mathcal{B}_{\mathbf{A}'}$ are *never* disjoint, as each contains at least M_0.

Signatures for Configurations. Lifting the notion of attraction to the level of *configurations*, and by abuse of notation, we set, for any configuration C, Hence,

$$\mathsf{Sig}(C) \overset{def}{=} \mathsf{Sig}(Mark(C)).$$

Note that in general, for any M there will be several C such that $Mark(C) = M$.

4   Extracting Attractors from Unfoldings

In this section, we present a new method that identifies, for a given Petri net $\mathcal{N}$, both its attractors and their basins, based on the unfolding of $\mathcal{N}$. We first recall the method from our previous work [2] that identifies attractors, and then present the new algorithm.

Representation of Attractors as Finite Complete Prefixes. The method from [2] uses unfoldings in two ways: first to find a set of markings which intersects all the attractors, and secondly to output the attractors as a set of finite complete prefixes.

Every attractor **A** can be compactly represented as a finite complete prefix of the unfolding of the Petri net $\mathcal{N}$ initialized at some marking $M \in \mathbf{A}$. Let us denote this prefix $\mathcal{U}_M$: the markings associated to the configurations of $\mathcal{U}_M$ are precisely those of the attractor, moreover the prefix shows the dynamics of the net while in the attractor. Lastly, the size of $\mathcal{U}_M$ (as number of non cut-off events) can be up to exponentially smaller (in case of highly concurrent behaviour) than the number of markings in the attractor and never exceeds it.

Maximal Configurations and Attractors. Let us recall from [2]:

Property 1. Let $\mathcal{N}$ be a Petri net and $\mathcal{U}$ a finite complete prefix of its unfolding. For every attractor **A** of $\mathcal{N}$, there exists (at least) one maximal configuration of $\mathcal{U}$ whose associated marking belongs to **A**.

The Attractor Map. The following algorithm generates a 'map' of attractors, i.e. the set of these attractors together with the information which marking obtained from a maximal configuration of the first complete prefix leads into which attractor. It is an extension of the algorithm from [2] for *finding* attractors.

Algorithm ATTMAP. **Initialize.** $\rightsquigarrow \overset{def}{=} \emptyset$, $\mathcal{A}^* \overset{def}{=} \emptyset$ and $\hat{\mathcal{A}} \overset{def}{=} \emptyset$.

1. Compute a finite complete prefix Π_0 of the unfolding of $\mathcal{N}$; initialise $\Pi \overset{def}{=} \Pi_0$.
2. Initialize $\mathcal{M}$ to the set $\mathcal{M}_{\max}$ of markings corresponding to maximal configurations of Π_0.
3. Loop: for M in $\mathcal{M}$ do
   – Compute a finite complete prefix Π_M of the Petri net $\mathcal{N} = (N, M)$. Grow Π by appending a copy of Π_M to every configuration C_M of Π such that $Mark(C_M) = M$.

- Compute the set $\mathbf{next}_M \stackrel{def}{=} \{M' \in \mathcal{M} \backslash \{M\} : M' \in \mathbf{R}(M)\}$ of markings in $\mathcal{M}$ that are reachable from M (reachability check done using Π_M).
  - If $\mathbf{next}_M \neq \emptyset$ then update $\leadsto := \leadsto \cup (\{M\} \times \mathbf{next}_M)$;
  - If $\mathbf{next}_M = \emptyset$ then update $\hat{\mathcal{A}} := \hat{\mathcal{A}} \cup \{\Pi_M\}$ and $\mathcal{A}^* := \mathcal{A}^* \cup \{M\}$.
4. Output the attractor candidates, i.e. marking set $\mathcal{A}^*$ and the set $\hat{\mathcal{A}}$ of unfolding prefixes, and the transitive closure $\leadsto \stackrel{def}{=} (\leadsto)^*$ of $\leadsto$.
5. Define the equivalence relation $\equiv$ on $\mathcal{A}^*$ by $M \equiv M' \stackrel{\triangle}{\leftrightarrow} M \leadsto M' \wedge M' \leadsto M$
6. In the quotient of $\mathcal{A}^*$ under $\equiv$, one obtains the set of root markings of attractors as the set $\underline{\mathcal{A}^*}$ of $\leadsto$-maximal elements; the set $\mathcal{A}$ of attractors is the set of prefixes rooted in some marking from $\underline{\mathcal{A}^*}$.
7. Compute $\mathsf{Sig}(\bullet)$:
   - For every marking $M \in \underline{\mathcal{A}^*}$, $\mathsf{Sig}(M) \stackrel{def}{=} [M]_\equiv$.
   - For other $M \in \mathcal{M}$: $\mathsf{Sig}(M) \stackrel{def}{=} \{[M']_\equiv : M' \in \hat{\mathcal{A}} \wedge M \leadsto M'\}$.
   - For all families of configurations $(C_i)_{i \in \mathcal{I}}$,
     - $\mathsf{Sig}(\bigcap_{i \in \mathcal{I}} C_i) = \bigcup_{i \in \mathcal{I}} \mathsf{Sig}(C_i)$, and
     - if $C_\mathcal{I} \stackrel{def}{=} \bigcup_{i \in \mathcal{I}} C_i \in \mathcal{C}$, then $\mathsf{Sig}(C_\mathcal{I}) = \bigcap_{i \in \mathcal{I}} \mathsf{Sig}(C_i)$.

Note that every attractor can be represented as an occurrence net $\mathbf{A}_M \stackrel{def}{=} \mathcal{U}_M$ rooted at some attractor marking M, and that no M can be the root of (or even belong to the marking set of) two distinct attractors. Moreover, the sets $\mathcal{C}(M)$ contain full information about which marking from $\mathcal{M}$ allows to reach which attractors, via the relation $\leadsto \subseteq \mathcal{M}^2$.

Comparison with the Original Algorithm From [2]

- ATTMAP explores, for every $M \in \mathcal{M}$, *all* markings that are reachable from M, whereas in [2] it was sufficient to detect *existence* of some such markings; the worst-case complexity is thus increased by one exponential factor.
- The information about how some attractor was reached is stored during the procedure, which induces only a bounded increase of computational and storage effort.
- A more fundamental difference is that attractors come out of the [2] algorithm as individual markings, rather than equivalence classes as in ATTMAP. The reason is that the [2] algorithm discards every marking from which an attractor representative marking is reachable; this reduction is not available in the above since all reachability information between the candidate markings is to be stored. In general, this will include mutual reachabilities; the quotient with respect to $\equiv$ contracts these strongly connected components. Since, by a classical result from order theory, the preorder $\leadsto$ is collapsed into a partial order by the quotient operation, one retrieves indeed all attractors from the maximal nodes of $\leadsto /_\equiv$.

5   Basins and Their Boundaries

In this section, we will present methods for computing an unfolding-based representation of the strong basin of a given attractor. The two methods represent different tradeoffs w.r.t. time vs space requirements. We will present a so-called *on-line* method first and the *off-line* method later.

On-line Method for Computing Strong Flow Basins. Let N be a net and $\mathbf{A}$ an attractor of N, i.e. a terminal SCC of $TS(N)$. The following procedure gives a method to compute $\mathcal{B}(\mathbf{A})$.

1. Fix any marking $M \in \mathbf{A}$, and compute $\Phi(M) \overset{def}{=} \Pi_0(\overleftarrow{N}, M) = (B, E, F, \mathbf{c}_0)$. Clearly, all $M' \in \mathbf{A}$, and even all $M'' \in \mathcal{W}(\mathbf{A})$, are represented in $\Phi(M)$.
2. Let $\mathcal{M}^{\leftarrow}$ be the set of markings of maximal configurations of $\Phi(M)$.
3. Set $\mathcal{C}_{\mathbf{A}} := \emptyset$. For all $M' \in \mathcal{M}^{\leftarrow}$, do the following:
   (a) Apply the algorithm ATTMAP on the net (N, M'); this computes, among other things, a complete prefix $\Pi(M')$ as well as $\mathrm{Sig}(M'')$ for all markings M'' reachable from M'.
   (b) For all minimal configurations C' of $\Pi(M')$ satisfying $\mathrm{Sig}(C') = \{\mathbf{A}\}$, find a configuration C of $\Phi(M)$ such that $Mark(C) = Mark(C')$, and add C to $\mathcal{C}_{\mathbf{A}}$.
4. With $E' := \bigcup_{C \in \mathcal{C}_{\mathbf{A}}} C$ and $B' = {}^{\bullet}E' \cup E'^{\bullet} \cup \mathbf{c}_0$, let $\Psi(\mathbf{A}) = (B', E', F, \mathbf{c}_0)$, i.e. $\Psi(\mathbf{A})$ is the restriction of $\Phi(M)$ to the events of configurations in $\mathcal{C}_{\mathbf{A}}$.

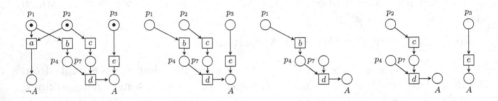

Fig. 6. Illustration of the techniques and concepts used in the computation of $\Psi(\mathbf{A})$. The attractors of the Petri net on the left hand side are the fixed points, i.e. singleton markings, $\mathbf{A} = \{\{A\}\}$ and $\overline{\mathbf{A}} = \{\{\neg A\}\}$. From the initial marking $M_0 = \{p_1, p_2\}$ both attractors are reachable. Reverse exploration from $\mathbf{A}$ yields $\mathcal{M}^{\leftarrow} = \{\{p_1, p_2\}, \{p_3\}\}$. Computing $\Phi(\{A\})$ yields the reverse occurrence net second-from-left. The three interior configurations of $\Psi(\mathbf{A})$ are shown in the figures from center to right; the strong basin of $\mathbf{A}$ is the collection of these configurations and their suffixes.

$\mathcal{B}(\mathbf{A})$ is now represented by $\Psi(\mathbf{A})$ and $\mathcal{C}_{\mathbf{A}}$ in the sense that $M_1 \in \mathcal{B}(\mathbf{A})$ iff there exists a configuration $C' \subseteq C$ with $Mark(C') = M_1$ and $C \in \mathcal{C}$. We shall call $\mathcal{C}_{\mathbf{A}}$ the *interior configurations* of $\mathbf{A}$, motivated by the following two results:

Lemma 1. *Let $C' \subseteq C$ be a configuration of $\Psi(\mathbf{A})$ such that $C \in \mathcal{C}_{\mathbf{A}}$. Then $Mark(C') \in \mathcal{B}(\mathbf{A})$.*

Proof: Let $M_1 = Mark(C')$ and $M'' = Mark(C)$. Since $C' \subseteq C$, we have that M'' is reachable from M_1 in $\overleftarrow{N}$, and hence M_1 is reachable from M'' in

N. Since $\mathrm{Sig}(M'') = \{\mathbf{A}\}$, the only attractor reachable from M_1 can be $\mathbf{A}$, too. Moreover, $\mathbf{A}$ is indeed reachable from M_1 because C' is a configuration of $\Phi(M)$, and therefore reachable in $\overleftarrow{N}$ from M, where $M \in \mathbf{A}$.   □

Lemma 2. *For every state $M_1 \in \mathcal{B}(\mathbf{A})$, there is a configuration C' of $\psi(\mathbf{A})$ such that $Mark(C') = M_1$ and $C' \subseteq C$ for some $C \in \mathcal{C}_{\mathbf{A}}$.*

Proof: In step 1 of the algorithm, $\Phi(M)$ represents all markings, including M_1, that can reach $\mathbf{A}$ in N and hence M. Therefore, in $\Phi(M)$ there exists some configuration whose marking is M_1 and which reaches some maximal configuration of $\Phi(M)$. Thus, among the maximal markings in $\mathcal{M}^{\leftarrow}$, there must be some M' such that M' reaches M through M_1. On each such path there must be a first marking M'' with $\mathrm{Sig}(M'') = \{\mathbf{A}\}$, i.e. M'' can reach only $\mathbf{A}$ and no other attractor. Since $\Pi(M')$ contains all markings reachable from M', step 3(b) of the algorithm is bound to find some configuration C' whose marking is such an M''. Since M'' is also reachable from M in $\overleftarrow{N}$, it is represented in $\Psi(M)$, and therefore a configuration C with $Mark(C) = M''$ will be added to $\mathcal{C}_{\mathbf{A}}$.   □

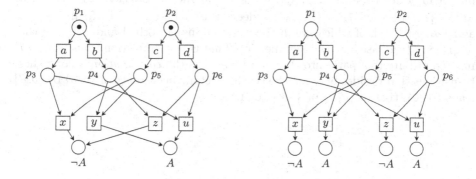

Fig. 7. A net (left) and its complete unfolding (right) with a bifurcation between two attractors, $\{A\}$ and $\{\neg A\}$ that requires coordination of two choices. From markings $\{p_3, p_5\}$ and $\{p_4, p_6\}$, it is inevitable to reach $\{\neg A\}$. That is, the two concurrent choices - between a and b on the one hand, and c and d on the other - must be coordinated to ensure that, e.g., $\{A\}$ is eventually reached; no local choice can achieve this alone.

Note that in general the set of configurations of net $\Psi(\mathbf{A})$ overapproximates the basin, as there may be non-interior configurations spanned by $\Psi(\mathbf{A})$. The example in Fig. 6 illustrates this point, see the discussion in the caption.

Off-Line Computation. In step 3 of the above on-line method, the algorithm ATTMAP is called at runtime, every time some marking M' from $\mathcal{M}^{\leftarrow}$ is inspected. An alternative procedure, which we call *off-line*, would consist in computing, before any basin is inspected, the signature for all states of the transition system of N; then, the value of $\mathrm{Sig}(C)$ would be found, when needed, by a lookup. Indeed, this computation can be implemented by applying ATTMAP to a modified complete net with places $P \cup \overline{P}$: add a new place p_0 to P and its complement place $\overline{p_0}$ to $\overline{P}$. Make $M_0' := P \cup \{p_0\}$ the initial marking, and for each place $p \in P$, add a transition t_p with ${}^\bullet t_p = \{p_0, p\}$ and $t_p{}^\bullet = \{p_0, \overline{p}\}$. Add a further transition t_0 from p_0 to $\overline{p_0}$, i.e. ${}^\bullet t_0 = \{p_0\}$ and $t_0{}^\bullet = \{\overline{p_0}\}$, and add $\overline{p_0}$ to the presets and postsets of all the original transitions of N. In this way, the modified net can first decide to move to any marking $M \subseteq P$ before firing t_0 and then behaving just like $\mathcal{N}$ would, if starting at M. The advantage of the off-line method is that it avoids computing the same signatures several times, in the case of overlaps, and that it produces a very rich set of information about the dynamics of a net, for any further use. Moreover, the computation of $\Phi(M)$ can be limited and treat events e with $Mark([e]) \neq \{\mathbf{A}\}$ as cut-off points. Its drawback is in the potentially very big data structures to be explored; the on-line method may be preferred if the system exhibits many small basins, limiting the number of signatures that are actually required. Which method is to be preferred, will need to be decided for every net individually.

Example. It is worth noting that the entry into a strong basin need *not* be linked to a unique transition; depending on the context, the same transition may lead either towards or away from some attractor. Consider Fig. 7, letting $M_1 \stackrel{def}{=} \{p_3, p_5\}$ and $M_2 \stackrel{def}{=} \{p_4, p_6\}$. Indeed, the predecessors of M_1 are $\{p_1, p_5\}$ and $\{p_3, p_2\}$, both of which are in the weak basins of both A and $\neg A$, and thus not in $\mathcal{B}_{\neg A}$ (the case for M_2 is symmetric). That is, any transition from $\{a, b, c, d\}$ may contribute to a path into A, or into $\neg A$; it is the *coordination* among these transitions that decides between the two attractors: occurrence of $\{a, c\}$ or $\{b, d\}$ leads to $\neg A$, that of $\{a, d\}$ or $\{b, c\}$ to A.

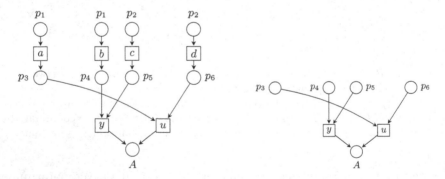

Fig. 8. Continuation of the example from Fig. 7. Left: backward unfolding from A to obtain $\Phi(\{A\})$; one has $\mathcal{M}^{\leftharpoonup} = \{\{p_1\}, \{p_2\}\}$. Right: $\Psi(\{A\})$, allowing two maximal and interior configurations.

Boundaries. We note that the above two cases exhibit two very different behaviours at the boundaries of their strong basins. Let us define a *boundary configuration* to be a configuration C that is not interior, but such that there exists an 'immediate predecessor' interior configuration $C' \subseteq$ with $C \backslash C'$ consisting in a single event. Then, in the example of Fig. 6, we have a unique boundary configuration $C = \{b, c, d\}$ with immediate predecessors $C_1 = \{b, d\}$ and $C_2 = \{c, d\}$, and we observe that C is obtained by a sort of closure operation from the interior configurations, in the sense that $C = C_1 \cup C_2$. By contrast, the example of Figs. 7 and 8, the boundary configurations are $C^1 = \{a, u\}$, $C^2 = \{d, u\}$, $C^3 = \{b, y\}$, and $C^4 = \{c, y\}$, and the immediate interior successors $C_1 = y$ and $C_2 = \{u\}$. No obvious combination of C_1 and C_2 can produce any C^i. Further classification and study of these (and potentially other) boundary types is left to future work.

6   Conclusion

We have developped Petri net-represented structures that allow to identify completely the strong basins of attraction for all attractors present in a finite safe Petri net. Future work will investigate further the different types of boundaries encountered here, and aim at refining and evaluating *robustness* of attractors and *reprogramming strategies* [16,17] in the context of concurrency. Finally, in regard to benchmarks in prior work relying on Petri net unfoldings [2,4], the time and space consumption of the proposed algorithms allows to envisage their application to networks with two-digit numbers of nodes. In future work, we will investigate the implementation of the on-line and off-line algorithms and their tractability on real-world models of biological systems.

Acknowledgments. This research was supported by Agence Nationale de la Recherche (ANR) with the ANR-FNR project AlgoReCell (ANR-16-CE12-0034); Labex DigiCosme (project ANR-11-LABEX-0045-DIGICOSME) operated by ANR as part of the program "Investissement d'Avenir" Idex Paris-Saclay (ANR-11-IDEX-0003-02).

References

1. Abadi, M., Lamport, L.: The existence of refinement mappings. Theor. Comput. Sci. **82**(2), 253–284 (1991)
2. Chatain, T., Haar, S., Jezequel, L., Paulevé, L., Schwoon, S.: Characterization of reachable attractors using Petri Net Unfoldings. In: Mendes, P., Dada, J.O., Smallbone, K. (eds.) CMSB 2014. LNCS, vol. 8859, pp. 129–142. Springer, Cham (2014). https://doi.org/10.1007/978-3-319-12982-2_10
3. Chatain, T., Haar, S., Kolcák, J., Paulevé, L., Thakkar, A.: Concurrency in Boolean networks. Natural Comput. (2019, to appear)
4. Thomas Chatain and Loïc Paulevé. Goal-driven unfolding of petri nets. In Roland Meyer and Uwe Nestmann, editors, 28th International Conference on Concurrency Theory, CONCUR 2017, 5–8 September 2017, Berlin, Germany, LIPIcs, vol. 85, pp. 18:1–18:16. Schloss Dagstuhl - Leibniz-Zentrum für Informatik (2017)
5. Cohen, D.P.A., Martignetti, L., Robine, S., Barillot, E., Zinovyev, A., Calzone, L.: Mathematical modelling of molecular pathways enabling tumour cell invasion and migration. PLoS Comput. Biol. **11**(11), e1004571 (2015)

6. Deng, C., Cousot, P.: Responsibility analysis by abstract interpretation. In: Chang, B.-Y.E. (ed.) SAS 2019. LNCS, vol. 11822, pp. 368–388. Springer, Cham (2019). https://doi.org/10.1007/978-3-030-32304-2_18
7. Diekert, V., Rozenberg, G. (eds.): The Book of Traces. World Scientific (1995)
8. Esparza, J., Heljanko, K.: Unfoldings - A Partial-Order Approach to Model Checking. Springer, Heidelberg (2008). https://doi.org/10.1007/978-3-540-77426-6
9. Esparza, J., Römer, S., Vogler, W.: An improvement of McMillan's unfolding algorithm. FMSD **20**, 285–310 (2002)
10. Fitime, L.F., Roux, O., Guziolowski, C., Paulevé, L.: Identification of bifurcation transitions in biological regulatory networks using Answer-Set Programming. Algorithm Molecular Biol. **12**(1), 19 (2017)
11. Fueyo, S., Monteiro, P.T., Naldi, A., Dorier, J., Remy, É, Chaouiya, C.: Reversed dynamics to uncover basins of attraction of asynchronous logical models. F1000Research **30**(6) (2017)
12. Godefroid, P. (ed.): Partial-Order Methods for the Verification of Concurrent Systems. LNCS, vol. 1032. Springer, Heidelberg (1996). https://doi.org/10.1007/3-540-60761-7
13. Goubault, É., Raussen, M.: Dihomotopy as a tool in state space analysis tutorial. In: Rajsbaum, S. (ed.) LATIN 2002. LNCS, vol. 2286, pp. 16–37. Springer, Heidelberg (2002). https://doi.org/10.1007/3-540-45995-2_8
14. Khomenk, V.: Punf. http://homepages.cs.ncl.ac.uk/victor.khomenko/tools/punf/
15. Klarner, H., Siebert, H., Nee, S., Heinitz, F.: Basins of attraction, commitment sets and phenotypes of Boolean networks. IEEE/ACM Trans. Comput. Biol. Bioinform. **17**, 115–1124 (2018)
16. Mandon, H., Su, C., Haar, S., Pang, J., Paulevé, L.: Sequential reprogramming of Boolean networks made practical. In: Bortolussi, L., Sanguinetti, G. (eds.) CMSB 2019. LNCS, vol. 11773, pp. 3–19. Springer, Cham (2019). https://doi.org/10.1007/978-3-030-31304-3_1
17. Mandon, H., Su, C., Pang, J., Paul, S., Haar, S., Paulevé, L.: Algorithms for the sequential reprogramming of Boolean networks. IEEE/ACM Trans. Computat. Biol. Bioinform. (2019, to appear)
18. McMillan, K.L.: Using unfoldings to avoid the state explosion problem in the verification of asynchronous circuits. In: CAV, pp. 164–177 (1992)
19. Mendes, N.D., Henriques, R., Remy, E., Carneiro, J., Monteiro, P.T., Chaouiya, C.: Estimating attractor reachability in asynchronous logical models. Front. Physiol. **9** (2018)
20. Murata, T.: Petri nets: properties, analysis and applications. Proc. IEEE **77**(4), 541–580 (1989)
21. Nielsen, M., Plotkin, G.D., Winskel, G.: Petri nets, event structures and domains, Part I. Theor. Comput. Sci. **13**, 85–108 (1981)
22. Schwoon, S.: Mole. http://www.lsv.ens-cachan.fr/~schwoon/tools/mole/
23. Vogler, W.: Fairness and partial order semantics. Inf. Process. Lett. **55**(1), 33–39 (1995)

Tools

ModRev - Model Revision Tool for Boolean Logical Models of Biological Regulatory Networks

Filipe Gouveia$^{(\boxtimes)}$ [iD], Inês Lynce [iD], and Pedro T. Monteiro [iD]

Department of Computer Science and Engineering, INESC-ID/Instituto Superior
Técnico, Universidade de Lisboa, Lisbon, Portugal
{filipe.gouveia,ines.lynce,pedro.tiago.monteiro}@tecnico.ulisboa.pt

Abstract. Biological regulatory networks can be represented by computational models, which allow the study and the analysis of biological behaviours, therefore providing a better understanding of a given biological process. However, as new information is acquired, biological models may need to be revised, in order to also account for this new information. Here, we present a model revision tool, capable of repairing inconsistent Boolean biological models. Moreover, the tool is able to confront the models, both with steady state observations, as well as time-series data, considering both synchronous and asynchronous update schemes. The tool was tested with a well-known biological model that was corrupted with different random changes. The presented tool was able to successfully repair the majority of the corrupted models.

1 Introduction

Computational models of biological regulatory networks are of great interest in Systems Biology [7]. These models, representing complex biological processes, allow to study and analyse such processes and the corresponding biological behaviours. Such computational models accommodate the test of hypotheses, the identification of predictions *in silico*, and the identification of network properties of biological regulatory networks.

As new experimental data become available, computational models may become inconsistent, i.e., models may not be able to reproduce the new information acquired. In this case, models need to be revised [8]. However, this model revision process is mainly a manual task, performed by a modeler, and therefore prone to error. Moreover, repairing an inconsistent model is not an easy task, due to the inherent combinatorial problem associated to all the possible changes that can be made to render a model consistent. Furthermore, the construction of biological models is also typically a manual task, thus accentuating the importance of the model revision process.

This work was supported by national funds through Fundação para a Ciência e a Tecnologia (FCT) with reference SFRH/BD/130253/2017 (PhD grant) and UIDB/50021/2020 (INESC-ID multi-annual funding).

A. Abate et al. (Eds.): CMSB 2020, LNBI 12314, pp. 339–348, 2020.
https://doi.org/10.1007/978-3-030-60327-4_18

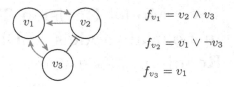

$$f_{v_1} = v_2 \wedge v_3$$

$$f_{v_2} = v_1 \vee \neg v_3$$

$$f_{v_3} = v_1$$

Fig. 1. Example of a Boolean logical model.

This paper presents a new tool, MODREV, that combines and implements the methods for model revision from previously published works [5,6]. MODREV is capable of assessing whether a Boolean logical model of a biological regulatory network is consistent with new experimental observations. In case of inconsistency, the tool repairs the model to render it consistent.

MODREV is able to consider steady state observations or time-series observations as experimental data. Moreover, both synchronous and asynchronous update schemes are supported when considering time-series observations.

The paper is organised as follows. Section 2 presents the background and related work. The MODREV tool is presented in Sect. 3. Experimental results are shown in Sect. 4. We discuss the tool features and prospects in Sect. 5.

2   Preliminaries

Biological regulatory networks are composed of biological compounds, and the corresponding interactions, representing complex biological processes. Different formalisms can be use to build computational models of regulatory networks, such as Ordinary Differential Equations (ODE) [7], Piecewise Linear Differential Equations (PLDE) [2], Logical Formalism [14], Sign Consistency Model (SCM) [13], among others [2]. Here, we consider the Boolean logical formalism [14], which has proven useful to study and analyse biological behaviours.

A Boolean logical model is usually represented by a *regulatory graph* and a set of *regulatory functions*. A *regulatory graph* is defined as a tuple (V, E) where V is a set of nodes representing the biological compounds, and E is the set of directed edges representing interactions between biological compounds. Each node is associated with a Boolean variable, representing whether the corresponding compound is present or absent. Edges are associated with a sign, representing positive interactions (activations) or negative interactions (inhibitions). If there is an edge from node v_1 to node v_2 we say that v_1 is a *regulator* of v_2. Each node is also associated with a *regulatory function*, which is a Boolean function that given the value of that node regulators determines its next value.

Figure 1 shows an example of a regulatory graph and corresponding regulatory functions, where green pointed arrows represent positive interactions, and red blunt arrows represent negative interactions.

2.1 Related Work

Few approaches of model revision processes have been proposed. A first approach of model revision over Thomas' logical formalism [14] considers that a model is inconsistent if it is over-constrained [10]. The revision process removes constraints until the model becomes consistent, probably leading to under-constrained models, not representing correctly the real biological process. Some approaches to model revision consider the Sign Consistency Model formalism [3,13]. This formalism, although similar to the logical formalism, relies on the sign algebra for the sign of the regulatory functions. Therefore, this type of models lacks in expressiveness in the definition of regulatory functions when compared to the logical models. In [9], properties found in literature, called rule of thumbs, are considered to repair inconsistent models. However, this approach has limitations regarding the repair operations that can be performed, the definition of regulatory functions, and the generation of the networks' dynamics. Recently, a model revision approach was proposed for Boolean logical models, with more expressiveness regarding the definition of regulatory functions [8]. However, it does not take into account the impact of the regulatory function on the networks' dynamics. Also, it does not consider adding a missing regulator in the model as a possible repair operation.

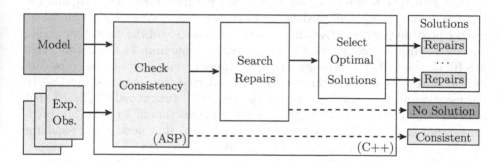

Fig. 2. Tool architecture.

3 ModRev Tool

MODREV is a freely available model revision tool for Boolean logical models of biological regulatory networks[1]. This paper presents a tool that implements the model revision methods presented in [5] to repair inconsistent models under steady state, and implements the method presented in [6] for time-series observations (see [5,6] for a detailed description of the methods).

Considering a Boolean logical model and a set of experimental observations, MODREV determines whether the model is consistent with the observations. In case of inconsistency, it determines the minimum set of nodes that must be

---
[1] https://filipegouveia.github.io/ModelRevisionASP/.

repaired. Four possible repair operations are considered: change a regulatory function; change the type of interaction (from activation to inhibition and vice-versa); remove a regulator; and add a regulator.

In order to repair an inconsistent model, the following lexicographic optimisation criteria is defined to minimise the number of operations of: 1) add/remove regulator; 2) change interaction type; 3) function change. These criteria allows to give preference to function changes over changes in the structure of the network.

Figure 2 illustrates the tool architecture. Dashed arrows represent alternative flows, where it is not possible to repair a model, or the model is already consistent and no repair is needed.

3.1   Input and Output

Regulatory functions supported by MODREV are monotone non-degenerate Boolean functions. Biologically, a monotone function means that each regulator only has one role, either an activator, or an inhibitor, but not both. A non-degenerate function means that each regulator influences the output of the regulatory function. Otherwise, it should not be a regulator. This model revision process requires the regulatory functions to be represented in Blake Canonical Form[1], which is a disjunction of all the prime implicants of the function [5,6].

The MODREV tool is based on Answer Set Programming (ASP) [4], and the input is defined using ASP predicates. To represent a Boolean logical model we use the predicate vertex(V), to indicate that V is a node of the regulatory graph, and the predicate edge(V1,V2,S) to represent an edge from V1 to V2 with a sign $S \in \{0,1\}$, where 0 (1) represents a negative (positive) interaction. The predicate vertex may be omitted if the node can be inferred from edge predicates. To represent regulatory functions, we use the predicate functionOr(V,1..N) that indicates that the regulatory function of V is a disjunction of N terms. The predicate functionAnd(V,T,R) is then used to represent that node R is a regulator of V and is present in the term T of the regulatory function.

MODREV is able to confront a model with a set of experimental observations, either in steady state, or a time-series data. To represent the set of experimental observations, the predicate exp(E) is used to identify an experimental observation E. To represent the observed values of an experiment E, we use the predicate obs_vlabel(E,V,S), which means that node V in experiment E has an observed value $S \in \{0,1\}$ considering steady state observations. If time-series observations are to be considered instead, a similar predicate (obs_vlabel(E,T,V,S)) is used, where T represents the time-step of the observed value.

If MODREV identifies that a given model is not consistent with a set of observations, it produces all the optimum solutions (repairs) that render the model consistent. Considering the optimisation criteria defined above, the optimum set of repair operations are produced.

4   Experimental Evaluation

We tested our tool using a Boolean logical model of the segment polarity (SP) network which plays a role in the fly embryo segmentation [12]. We corrupted the model using four probability parameters (in percentage): F, the probability of changing a regulatory function; E, the probability of changing the sign of an edge; R, the probability of removing a regulator; and A, the probability of adding a regulator. Table 1 shows 24 combinations of these parameters that have been considered. For each parameter configuration, 100 corrupted instances were generated. Also, five time-series observations with twenty time-steps were considered.

Given a corrupted model and a set of experimental observations, our tool is able to repair most of the models under a time limit of one hour. Figure 3 shows the median solving times for each configuration. Considering the synchronous update scheme, it is possible to observe that, for the configurations with added or removed regulators, a greater repair time is needed. This is due to the change in the dimension of the regulatory function, which has a big impact on the tool performance. Considering the asynchronous update scheme, we can verify that the tool repairs the corrupted models in less than 2 s. This difference between the

Table 1. Percentage values of F, E, R, and A parameters, of the 24 configurations.

Config.	1	2	3	4	5	6	7	8	9	10	11	12	13	14	15	16	17	18	19	20	21	22	23	24
F	5	25	50	100	0	0	0	0	0	0	0	0	0	0	0	0	0	0	0	25	50	100	5	10
E	0	0	0	0	5	10	15	20	25	50	75	0	0	0	0	0	0	0	0	5	25	50	25	10
R	0	0	0	0	0	0	0	0	0	0	0	1	5	10	15	0	0	0	0	0	0	0	5	5
A	0	0	0	0	0	0	0	0	0	0	0	0	0	0	0	1	5	10	15	0	0	0	5	5

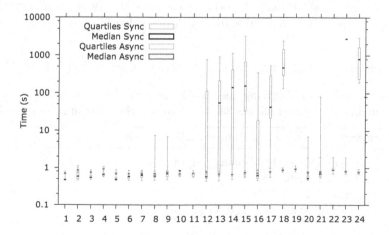

Fig. 3. Median time in seconds of solved instances for each corruption configuration, under synchronous and asynchronous update schemes.

two update schemes relies on the fact that, in the asynchronous update, only one regulatory function is updated at each time step. Therefore, fewer constraints must be verified when looking for possible repair operations.

5   Discussion

Currently, the interaction (input/output) with the MODREV tool is based on ASP predicates. To be able to facilitate the future interoperability with the qualitative modelling community, we plan to implement an import/export facility to be integrated into the BioLQM library [11]. Additionally, we are currently improving the comparison with time-series data through the implementation of the fully asynchronous update scheme. This will allow to be more permissive on the generated dynamics.

A   Tutorial

The model shown in Fig. 1 is represented by the following listing:

```
vertex(v1). vertex(v2). vertex(v3).
edge(v1,v2,1). edge(v1,v3,1). edge(v2,v1,1).
edge(v3,v1,1). edge(v3,v2,0).
functionOr(v1,1..1).
functionAnd(v1,1,v2). functionAnd(v1,1,v3).
functionOr(v2,1..2).
functionAnd(v2,1,v1). functionAnd(v2,2,v3).
functionOr(v3,1..1).
functionAnd(v3,1,v1).
```

Now let us consider that we want to define a steady state observation in which v_1 has value 0, v_2 has value 0, and v_3 has value 1, as following:

```
exp(p1).
obs_vlabel(p1,v1,0). obs_vlabel(p1,v2,0). obs_vlabel(p1,v3,1).
```

Using MODREV tool, giving the model defined above (as a file model.lp) and the steady state (as a file obsSS.lp), execute the following command:

```
$ ./modrev -m model.lp -obs obsSS.lp -ss
```

```
### Found solution with 1 repair operation.
    Inconsistent node v3.
        Repair #1:
            Flip sign of edge (v1,v3).
```

This output means that the model in Fig. 1 can be repaired by changing the interaction type between v_1 and v_3. If we repair the model and execute the above command again, the result will be:

This network is consistent!

Now let us assume that the user knows that the interaction between v_1 and v_3 is correct, and wants to prevent repairs over it. The predicate `fixed(v1,v3).` can be used to define that the edge between these nodes can not be changed or removed. Adding this predicate to the model and running the command above, we obtain the following result:

```
### Found solution with 2 repair operations.
    Inconsistent node v3.
        Repair #1:
            Change function of v3 to (v1) || (v3)
            Add edge (v3,v3) with sign 1.
```

A different set of repair operations is obtained that does not change the fixed edge. Now assume that the user wants to prevent any repair over the node v_3. The predicate `fixed(v3).` can be used to prevent that node to be inconsistent. However, in this example, if we prevent any change to node v_3, considering its regulatory function, and that v_1 has value 0 and v_3 has value 1, and we are in the presence of a steady state, it becomes impossible to repair the network. In this case, when the model is over-constrained, using the same command as before, the tool produces the following message:

```
It is not possible to repair this network.
```

Consider now that we have, for the same model in Fig. 1, a time-series data as shown in Table 2. Consider that this experimental observation with three time-steps (0, 1 and 2) is considering a synchronous update scheme.

Table 2. Synchronous time-series data

		Time		
		0	1	2
Node	v_1	0	1	0
	v_2	0	0	0
	v_3	1	0	0

We can represent the time-series data using the following listing:

```
#const t = 2.
exp(p2).
obs_vlabel(p2,0,v1,0). obs_vlabel(p2,0,v2,0).
obs_vlabel(p2,0,v3,1).
obs_vlabel(p2,1,v1,1). obs_vlabel(p2,1,v2,0).
obs_vlabel(p2,1,v3,0).
obs_vlabel(p2,2,v1,0). obs_vlabel(p2,2,v2,0).
obs_vlabel(p2,2,v3,0).
```

Note that we start the file indicating the maximum value of time step with `#const t = 2`.

Using MODREV to verify whether the model is consistent, while considering the above time-series data (as a file `obsTS01.lp`) under a synchronous update scheme, execute the following command:

```
$ ./modrev -m model.lp -obs obsTS01.lp -up s
```

This will produce the following result:

```
### Found solution with 5 repair operations.
    Inconsistent node v1.
        Repair #1:
            Change function of v1 to (v2) || (v3)
    Inconsistent node v2.
        Repair #1:
            Change function of v2 to (v1 && v3)
            Flip sign of edge (v1,v2).
        Repair #2:
            Change function of v2 to (v1 && v3)
            Flip sign of edge (v3,v2).
    Inconsistent node v3.
        Repair #1:
            Change function of v3 to (v1 && v2)
            Add edge (v2,v3) with sign 1.
        Repair #2:
            Change function of v3 to (v1 && v3)
            Add edge (v3,v3) with sign 1.
```

Note that now we have multiple choices to render the model consistent. To repair node v_2, for example, one can apply the operations in `Repair #1` or in `Repair #2`. The same applies to repair node v_3.

If instead of a time-series data under a synchronous update scheme, we are under an asynchronous update scheme, the previous command would change from `-up s` to `-up a`. The option `-up` indicates the update scheme to be considered, with argument `s` for synchronous and `a` for asynchronous.

MODREV also supports incomplete time-series data. Assume that we have the experimental observation shown in Table 3, where node v_3 was not observed, and a value of v_1 was also not observed.

Consider the following representation of an incomplete time-series data:

```
#const t = 2.
exp(p3).
obs_vlabel(p3,0,v1,0). obs_vlabel(p3,0,v2,1).
obs_vlabel(p3,1,v2,0).
obs_vlabel(p3,2,v1,1). obs_vlabel(p3,2,v2,0).
```

Table 3. Incomplete synchronous time-series data

		Time		
		0	1	2
Node	v_1	0		1
	v_2	1	0	0
	v_3			

Executing the following command, while considering the above experimental observation (as a file obsTS02.lp) under synchronous update scheme, produces the result below.

```
$ ./modrev -m model.lp -obs obsTS02.lp -up s
```

```
### Found solution with 3 repair operations.
    Inconsistent node v1.
        Repair #1:
            Change function of v1 to (v2) || (v3)
            Flip sign of edge (v2,v1).
    Inconsistent node v2.
        Repair #1:
            Change function of v2 to (v1 && v3)
```

MODREV tool also supports confronting a model with multiple experimental observations at the same time. For example, we could confront the model of Fig. 1 with the two time-series data above, using the command:

```
$ ./modrev -m model.lp -obs obsTS01.lp obsTS02.lp -up s
```

Note that the directive #const t = 2 must only be defined once.

References

1. Crama, Y., Hammer, P.L.: Boolean Functions: Theory, Algorithms, and Applications. Cambridge University Press, Cambridge (2011)
2. De Jong, H.: Modeling and simulation of genetic regulatory systems: a literature review. J. Comput. Biol. **9**(1), 67–103 (2002)
3. Gebser, M., et al.: Repair and prediction (under inconsistency) in large biological networks with answer set programming. In: Lin, F., Sattler, U., Truszczynski, M. (eds.) Principles of Knowledge Representation and Reasoning: Proceedings of the Twelfth International Conference, KR 2010. AAAI Press (2010)
4. Gebser, M., Kaminski, R., Kaufmann, B., Schaub, T.: Answer set solving in practice. In: Synthesis Lectures on Artificial Intelligence and Machine Learning, vol. 6, no. 3, pp. 1–238 (2012)
5. Gouveia, F., Lynce, I., Monteiro, P.T.: Revision of Boolean models of regulatory networks using stable state observations. J. Comput. Biol. **27**(2), 144–155 (2020)

6. Gouveia, F., Lynce, I., Monteiro, P.T.: Semi-automatic model revision of Boolean regulatory networks: confronting time-series observations with (a)synchronous dynamics. bioRxiv preprint (2020) https://doi.org/10.1101/2020.05.10.086900
7. Karlebach, G., Shamir, R.: Modelling and analysis of gene regulatory networks. Nat. Rev. Mol. Cell Biol. **9**(10), 770 (2008)
8. Lemos, A., Lynce, I., Monteiro, P.T.: Repairing Boolean logical models from time-series data using Answer Set Programming. Algorithms Molecular Biol. **14**(1), 9 (2019)
9. Merhej, E., Schockaert, S., Cock, M.D.: Repairing inconsistent answer set programs using rules of thumb: a gene regulatory networks case study. Int. J. Approximate Reason. **83**, 243–264 (2017)
10. Mobilia, N., Rocca, A., Chorlton, S., Fanchon, E., Trilling, L.: Logical modeling and analysis of regulatory genetic networks in a non monotonic framework. In: Ortuño, F., Rojas, I. (eds.) IWBBIO 2015. LNCS, vol. 9043, pp. 599–612. Springer, Cham (2015). https://doi.org/10.1007/978-3-319-16483-0_58
11. Naldi, A.: BioLQM: A java toolkit for the manipulation and conversion of logical qualitative models of biological networks. Front. Physiol. **9** (2018)
12. Sánchez, L., Chaouiya, C., Thieffry, D.: Segmenting the fly embryo: logical analysis of the role of the segment polarity cross-regulatory module. Int. J. Dev. Biol. **52**(8), 1059–1075 (2002)
13. Siegel, A., Radulescu, O., Le Borgne, M., Veber, P., Ouy, J., Lagarrigue, S.: Qualitative analysis of the relation between DNA microarray data and behavioral models of regulation networks. Biosystems **84**(2), 153–174 (2006)
14. Thomas, R.: Boolean formalization of genetic control circuits. J. Theor. Biol. **42**(3), 563–585 (1973)

fnyzer: A Python Package
for the Analysis of Flexible Nets

Jorge Júlvez[1]([✉])[iD] and Stephen G. Oliver[2,3][iD]

[1] Department of Computer Science and Systems Engineering, University of Zaragoza,
Zaragoza, Spain
`julvez@unizar.es`
[2] Cambridge Systems Biology Centre, University of Cambridge, Cambridge, UK
`sgo24@cam.ac.uk`
[3] Department of Biochemistry, University of Cambridge, Cambridge, UK

Abstract. This paper introduces *fnyzer*, a Python package for the analysis of Flexible Nets (FNs). FNs is a modelling formalism for dynamical
systems that can accommodate a number of uncertain parameters, and
that is particularly well suited to model the different types of networks
arising in systems biology. *fnyzer* offers different types of analysis, can
handle nonlinear dynamics, and can transform models expressed in Systems Biology Markup Language (SBML) into FN format.

1 Overview

Flexible Nets (FNs) [5] is a modelling formalism for dynamical systems, inspired
by Petri [10] nets, that can handle uncertain parameters and that offer different
analysis possibilities. FNs have four types of vertices: *places, transitions, event
handlers,* and *intensity handlers. Places* are represented as circles and model
state variables, e.g. metabolites. The value of the state variable, e.g. metabolite
concentration, modelled by a place is called marking. *Transitions* are represented
as rectangles and model processes that can modify the marking, e.g. reactions.

In addition to places and transitions, FNs incorporate: a) *event handlers,*
represented as dots, that model how the transitions modify the marking; and b)
intensity handlers, also represented as dots, that model how the marking modifies
the speed (or intensity) of the transitions. This way, places and transitions are not
connected directly but only through handlers. This connection can be established
either by means of *arcs,* which model consumption/production of marking or
intensity, or *edges,* which model the use of marking or intensity. In order to
account for the relationships "process–marking change" and "marking–speed
change", both, event and intensity handlers, are associated with sets of equalities
and inequalities.

As an example, the FN in Fig. 1(a) is composed of 3 places A, B, and C
(with initial markings 6, 4, and 0 respectively) that are connected through the
event handler v by means of arcs. The equalities associated with v determine
the stoichiometry of the reaction modelled by the net, namely $a{=}2b$ establishes

A. Abate et al. (Eds.): CMSB 2020, LNBI 12314, pp. 349–355, 2020.
https://doi.org/10.1007/978-3-030-60327-4_19

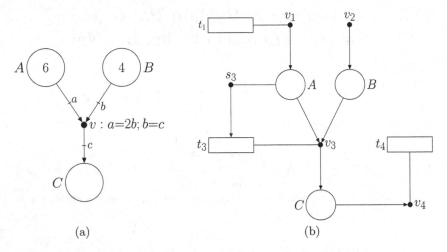

Fig. 1. FNs modelling the stoichiometry of reaction $R : 2A + B \to C$ (a) and; exchange reactions with partially known dynamics (b).

that two units of A are consumed per each unit of B that is consumed; and $b=c$ establishes that one unit of B is consumed per each unit of C that is produced. This way, the FN models the reaction $R : 2A + B \to C$, and it does not specify a speed as there is no transition in the net. FNs are defined in $f$nyzer by Python dictionaries, e.g. the Python dictionary that defines the FN in Fig. 1(a) is:

```
stonet = { # FN modelling reaction R: 2A + B -> C
  'name': 'stonet',
  'solver': 'glpk',
  'places': {'A': {'m0': 6}, 'B': {'m0': 4}, 'C': {'m0': 0}},
  'vhandlers': {                          # Event handler
    'v': [{'a': ('A','v'), 'b': ('B','v'), 'c': ('v','C')},
          'a == 2*b', 'b == c']},         # Stoichiometry
  'obj': {'f': "m['C']", 'sense': 'max'},# Objective function
  'options': {'antype': 'un'}            # Untimed analysis
}
```

Thus, in addition to the net structure determined by the keys `places` and `vhandlers` the dictionary contains information about the objective function and the type of analysis to be carried out by *fnyzer*.

Figure 2 sketches (in FN fashion) the main tasks performed by *fnyzer* in order to analyse an FN. First, a set of mathematical constraints is derived from the FN definition and the desired type of analysis. This set of constraints, which represent necessary reachability conditions, together with an objective function are used to set up a programming problem by using the package Pyomo [4]. This programming problem is solved by a state-of-the-art solver (current supported solvers are CPLEX [1], Gurobi [3] and GLPK [9]) and the obtained solution is saved in a spreadsheet and plotted.

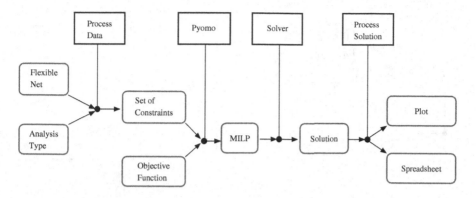

Fig. 2. Main pipeline of *fnyzer* from the input data to the results.

2 Installation and Use

fnyzer is an open **Python** package that can be installed with **pip** the standard tool for installing **Python** packages:

```
$ pip install fnyzer
```

The online documentation of **fnyzer** detailing all the available options can be found at https://fnyzer.readthedocs.io, the source code is available at https://bitbucket.org/Julvez/fnyzer, and the file **nets/fnexamples.py** in that repository contains a number of FN examples, e.g. the above dictionary **stonet** is in that file.

Assuming that the mentioned file **fnexamples.py** is in your working directory, the execution of:

```
$ fnyzer fnexamples.py stonet
```

produces: a) the spreadsheet **stonet.xls** with the optimization results and CPU times; and b) the file **stonet.pkl** with the pickled FN object (see "Accessing saved objects" in the next section). For the proposed objective function, $max\ m[C]$ (i.e. maximize the final concentration of C), the value obtained is 3 (the final concentrations obtained for A and B are 0 and 1 respectively).

3 Main Features

Handling Uncertain Parameters. Assume that the initial concentration of A in Fig. 1(a) is uncertain, but known to be in the interval $[4, 7]$. This can be captured in the **stonet**dictionary above by setting **'A': {'m0': None}** and including a keyword modelling such constraint:

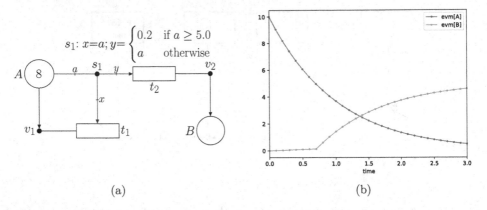

(a)   (b)

Fig. 3. (a) Guarded FN modelling the activation of reaction $R_2 : \emptyset \rightarrow B$ when A goes below 5.0; (b) Time trajectories of the concentrations of A and B as plotted by *fnyzer*.

```
'm0cons':  ["4 <= m0['A']", "m0['A'] <= 7"]
```

The optimization of the resulting net (execute "fnyzer fnexamples.py unstonet') yields 3.5 as the maximum final concentration for C.

Uncertain stoichiometric weights can be incorporated in a similar way. Assume that the reaction is $R : nA+B \rightarrow C$ where n is only partially known, e.g. $n \in [19, 21]$. Such a reaction can be modelled by substituting in the dictionary above 'a == 2*b' by two inequalities '19*b <= a', 'a <= 21*b'.

The FN in Fig. 1(b) models a small reaction network composed of the reactions reported in Table 1. Each reaction R_i is modelled by the event handler v_i (also the intensity handler s_3 in the case of R_3) and the net elements connected to it. The uncertain rates of R_1 and R_4 are accounted for by the constraints:

```
'10cons':  ["1 <= 10['t1']", "10['t1'] <= 4", "10['t4'] <= 3"]
```

that are included in the dictionary (see **excnet** in the file **fnexamples.py** that defines the net. The completely unknown rate of R_2 is modelled by absence of transitions connected to v_2. The known rate of R_3 is modelled by s_3 and t_3.

Table 1. Reactions modelled by the FN in Fig. 1(b).

Reaction	Modelled by	Rate
$R_1 : \emptyset \rightarrow A$	v_1	In the interval $[1, 4]$
$R_2 : \emptyset \rightarrow B$	v_2	Unknown
$R_3 : A + B \rightarrow C$	v_3 and s_3	Equal to the concentration of A
$R_4 : C \rightarrow \emptyset$	v_4	In the interval $[0, 3]$

Handling Nonlinear Dynamics. The rates of the reactions of the FN in Fig. 1(b) are either constant or depend linearly on the concentration of metabolites. In order to account for the complex and nonlinear dynamics exhibited by biological systems, FNs can associate piecewise linear functions with their intensity handlers. Consider the FN in Fig. 3(a) which models the reactions $R_1 : A \rightarrow \emptyset$ and $R_2 : \emptyset \rightarrow B$. The intensity handler s_1 has associated: a) a linear function $x = a$ which determines that the rate of t_1 is equal to the concentration of A; and b) a piecewise linear function $y = \begin{cases} 0.2 & \text{if } a \geq 5.0 \\ a & \text{otherwise} \end{cases}$ which establishes that the rate of R_2 is constant and equal to 0.2 if the concentration of A is greater than or equal to 5.0, and equal to the concentration of A otherwise. In the **Python** dictionary that describes the FN, this is defined by means of two regions (provided by the key `'regs'`), and linear functions associated with them:

```
'regs': {'off': ["m['A'] >= 5"], 'on': ["m['A'] <= 5"]},
'shandlers': {
    's1': [{'a':('A','s1'), 'x':('s1','t1'), 'y':('s1','t2')},
           'x == a',
           {'off': ['y == 0.2'], 'on': ['y == a']}]
},
```

Types of Analysis. FNs can be analysed by **fnyzer** under 4 interpretations: untimed [6], transient state [5], Model Predictive Control (MPC) [7], and steady state [8]. Assume that it is desired to compute the maximum concentration of A of the FN in Fig. 1(b) in the steady state. This can be achieved by setting the analysis type to `'antype': 'st'`, and the objective function to:

```
'obj': {'f': "avm['A']", 'sense': 'max'}
```

where **avm** denotes average marking. The value obtained by *fnyzer* is 3.0. At this concentration of A, the flux of all the reactions is 3.0. If the minimum concentration is desired instead, then the `'sense'` of the objective function must be set to `'min'`. The resulting concentration is 1.0. These values were obtained by executing "fnyzer fnexamples.py excnet".

As an example of MPC, time trajectories of the concentrations of A and B can be obtained by setting the analysis type to `'antype': 'mpc'`. In the trajectories shown in Fig. 3(b), which were generated by "fnyzer fnexamples .py guardnet", 30 time intervals of length 0.1 were considered.

Importing SBML Models. In order to facilitate the manipulation of existing models, **fnyzer** offers the possibility of translating **COBRA** [2] models to FNs. Assume that a Systems Biology Markup Language (SBML) model, MODEL000.xml, is available in the working directory, then the lines:

```
>>> from fnyzer import optimize, cobra2fn
>>> import cobra
>>> cobra_model = cobra.io.read_sbml_model('MODEL000.xml')
>>> fndic = cobra2fn(cobra_model)
```

convert the model into the dictionary fndic that defines the corresponding FN and that can be extended, modified and analysed.

Accessing Saved Objects. fnyzer saves the analysis results in a file that can be easily accessed. For instance, the following lines, read the file guardnet.pkl generated by "fnyzer fnexamples.py guardnet", save the results in a different spreadsheet, plot the trajectories, and write the concentration of A over time:

```
>>> import pickle
>>> datafile = open("guardnet.pkl", 'rb')
>>> fn = pickle.load(datafile)
>>> datafile.close()
>>> fn.writexls("new_guardnet.xls")
>>> fn.plotres()
>>> [net.places['A'].m for net in fn.lnets]
```

In the above lines, the object fn provides access to all the values of the variables in the FN (see the online documentation for details).

Acknowledgments. This work was supported by the Spanish Ministry of Science, Innovation and Universities [ref. Medrese-RTI2018-098543-B-I00], by the Biotechnology & Biological Sciences Research Council (UK) grant no. BB/N02348X/1 as part of the IBiotech Program, and by the Industrial Biotechnology Catalyst (Innovate UK, BBSRC, EPSRC) to support the translation, development and commercialisation of innovative Industrial Biotechnology processes.

References

1. IBM ILOG CPLEX Optimizer (2010). https://www.ibm.com/analytics/cplex-optimizer
2. Ebrahim, A., Lerman, J.A., Palsson, B.O., Hyduke, D.R.: COBRApy: COnstraints-Based Reconstruction and Analysis for Python. BMC Syst. Biol. **7**(1), 74 (2013). https://doi.org/10.1186/1752-0509-7-74
3. Gurobi Optimization Inc: Gurobi optimizer reference manual (2015). http://www.gurobi.com
4. Hart, W.E., et al.: Pyomo-Optimization Modeling in Python, vol. 67, 2nd edn. Springer, Cham (2017). https://doi.org/10.1007/978-3-319-58821-6
5. Júlvez, J., Dikicioglu, D., Oliver, S.G.: Handling variability and incompleteness of biological data by flexible nets: a case study for Wilson disease. NPJ Syst. Biol. Appl. **4**(1), 7 (2018). https://doi.org/10.1038/s41540-017-0044-x
6. Júlvez, J., Oliver, S.G.: Flexible Nets: a modeling formalism for dynamic systems with uncertain parameters. Discrete Event Dyn. Syst. **29**(3), 367–392 (2019). https://doi.org/10.1007/s10626-019-00287-9
7. Júlvez, J., Oliver, S.G.: Modeling, analyzing and controlling hybrid systems by Guarded Flexible Nets. Nonlinear Anal. Hybrid Syst. **32**, 131–146 (2019). https://doi.org/10.1016/j.nahs.2018.11.004
8. Júlvez, J., Oliver, S.G.: Steady State Analysis of Flexible Nets. IEEE Trans. Autom. Control 1 (2019). https://doi.org/10.1109/TAC.2019.2931836

9. Makhorin, A.: GLPK (gnu linear programming kit) (2012). http://www.gnu.org/software/glpk/glpk.html

10. Murata, T.: Petri Nets: properties, analysis and applications. Procs. IEEE **77**(4), 541–580 (1989)

eBCSgen: A Software Tool
for Biochemical Space Language

Matej Troják[(✉)], David Šafránek, Lukrécia Mertová, and Luboš Brim

Systems Biology Laboratory, Masaryk University, Brno, Czech Republic
xtrojak@fi.muni.cz

Abstract. eBCSgen is a tool for development and analysis of models written in Biochemical Space Language (BCSL). BCSL is a rule-based language for biological systems designed to combine compact description with a specific level of abstraction which makes it accessible to users from life sciences. Currently, eBCSgen represents the only tool completely supporting BCSL. It has the form of a command line interface which is integrated into Galaxy – a web-based bioinformatics platform automating data-driven and model-based analysis pipelines.

1 Introduction

Rule-based modelling is a promising approach in systems biology which can be used to write mechanistic models of complex reaction systems. Compared to traditional mechanistic or mathematical approaches such as reaction-based modelling or ordinary differential equations (ODEs), the rule-based approach provides a compact form of model description that scales well with the size and complexity of the modelled system.

Key features of rule-based languages, such as structures binding [3,10], regulatory interactions [17], modularity [16], or spatial aspects [12], combined with a language-specific level of abstraction, require the development and analysis of rule-based models to be supported by software tools. Moreover, to appropriately reflect the needs of the biological domain, it is necessary to enable analysis of models with incomplete information (e.g., unknown kinetic parameters).

Several existing rule-based languages are provided with well-established software support. Kappa [6] is supported by the Kappa platform [3], providing a model editor, stochastic simulation, several static analysis procedures accompanied by graphical visualisation, and a generator of ODE models. BioNetGen package [10] provides the tool RuleBender [20] for construction, debugging, analysis (e.g. simulation, parameter scan), and visualisation of models (e.g. influence graph, contact map). The software environment BioCham [4] with its custom language [5] supports multiple semantics and allows, for example, checking of temporal properties expressed in CTL, analysing models with respect to FO-LTL properties (measuring the robustness, parameter sensitivity), and simulating models. Some other languages employ embedding to an existing programming language (e.g. Chromar [12], PySB [14]).

© Springer Nature Switzerland AG 2020
A. Abate et al. (Eds.): CMSB 2020, LNBI 12314, pp. 356–361, 2020.
https://doi.org/10.1007/978-3-030-60327-4_20

Our experiences with using these languages in direct collaboration with biologists have shown that in most cases, it is difficult to train users outside computer science to use these languages directly. Although some of the tools are quite intuitive, they either do not support various useful features or work with a very detailed level of abstraction that makes models hard to understand, maintain, and re-use. To that end, we have introduced Biochemical Space Language (BCSL) [9,18], a high-level rule-based language that combines several features of rule-based frameworks in a single formalism, recently extended by quantitative aspects [19].

In this paper, we present eBCSgen, a tool for the development and analysis of models written in BCSL. The tool is integrated into Galaxy [1], a web-based platform for data-intensive biomedical research. It provides a convenient way to use eBCSgen by the target group of users due to the extensive popularity of Galaxy in biology-oriented community. Our tool provides interactive model editor, model simulation, analysis of the model with respect to PCTL [11] properties, useful static analysis methods, and interactive data visualisation. We demonstrate the tool usability on a case study describing circadian rhythms in cyanobacteria. eBCSgen is available online[1] within Galaxy. It is accompanied with a short tutorial[2].

2   Biochemical Space Language

In this section, we briefly show the primary features of BCSL on several examples. For more details, we recommend [18] for formal definition and [19] for formal semantics and analysis methods.

A BCSL model is given by a set of *rules*, an *initial state*, and a set of *definitions* (optional). A rule describes a pattern how *agents* (structured objects) can interact. Examples of the rules are provided below. The initial state defines the number of individual agents in the initial solution of the system. The definitions assign particular values to parameters. Additionally, there can be fourth part defining complex aliases (see below). In the following expression

A(act { off })::cyt + B( )::cyt ⇒ A(act { on }).B( )::cyt @ k1×[A( )::cyt]×[B( )::cyt]

there is a rule describing the interaction of agent A( ) with agent B( ), creating a complex A( ).B( ); moreover, the agent A( ) changes the state of its feature act from off to on, meaning its activity was turned on. This interaction takes place inside cyt (cytosol) physical compartment. The *rate* of the rule is given by a mass action kinetic law with no restriction on the particular state of the agent A( ), parameterised by a parameter k1.

The example above demonstrated the basic features of the language – the formation of a complex and a state change. Additional features are for example complex dissociation (the rule above with opposite direction), agent formation (the left-hand side of the rule is empty), and degradation (no right-hand side).

---

[1] https://biodivine-vm.fi.muni.cz/galaxy/.
[2] https://biodivine.fi.muni.cz/galaxy/eBCSgen/tutorial.

Among these basic constructs, the language offers several syntactic features which make BCSL models more readable and compact. *Nesting* allows "zooming" inside of individual agents to emphasise the particular part of the agent. For example, the rule

```
act{off}:A():A().B()::cell ⇒ act{on}:A():A().B()::cell @ k2×[A().B()::cell]
```

describes the state change of feature `act` *inside* (indicated by single colon symbol) of agent `A()` as a part of complex `A().B()` (localised in compartment `cell`). Note that agent `ffA(acto).B()::cell` represents an equivalent form to that on the left-hand side of the rule without the usage of the nesting operator.

The syntax using nesting is particularly useful in combination with *complex aliases* and *variables*. The complex alias allows defining a short name for a particular complex (for example, `AB = A().B()`, `AC = A().C()`, `AD = A().D()`). The variable can substitute multiple agents on a particular position in the rule, providing a compact way of aggregating repeating patterns. The rule

```
act{off}:A():?::cell ⇒ act{on}:A():?::cell @ k2×[?::cell]
```

followed by `? = AB, AC, AD` describes the previous rule and two additional rules compactly – individual complex aliases are substituted on the position declared by question mark.

The semantics of the model is given by transitive rewriting of the rules starting with the initial state. The rule rewriting is defined by *match–replace* relation, which first selects the suitable candidates from the state (they have to satisfy the left-hand side of the rule) and then they are replaced according to the right-hand side of the rule. During this process, the rate is evaluated as a function of the state. Finally, all outgoing rates from the state are normalised to obtain the probability of the transition. Following the idea of using approximate models with discrete-time semantics [2], the obtained transition system is a Discrete Time Markov Chain (DTMC) or a parametric Markov Chain (pMC) [7,13], depending on whether there are some parameters used in rule rates which do not have defined value in *definitions* section of the model.

3   Implementation

The tool eBCSgen is implemented in Python programming language. It is developed as a command-line tool with a GUI provided by integration into Galaxy [1], a web-based scientific analysis platform used to analyse large datasets. With its three primary features – accessibility, reproducibility, and communication – it is a very convenient and practical alternative to an individual GUI development.

The Galaxy interface of eBCSgen offers interactive *model editor*, which can be used to create and edit BCSL models. The interactivity is ensured by automatic syntax highlighting and real-time code validation. Any errors in the model code are immediately highlighted.

The probabilistic behaviour of BCSL model can be analysed with respect to PCTL [11] properties. PCTL is an extension of computation tree logic (CTL) which allows for probabilistic quantification of described properties. The given property is checked using `Storm` model checker [8] after the corresponding transition system is generated. The tool allows checking whether a given probability threshold is satisfied or to find the probability of satisfaction for given path formula. In the case of the parameterised model, Storm is used to solve *parameter synthesis*. If the formula has defined probability threshold, then Storm computes the partitioning of the given parameter space (defined by the user) to regions which satisfy (resp. violate) the property. If the threshold is not given, a probability function of parameters is computed instead, which evaluates to the probability of satisfaction for particular parameterisation.

In addition to these analysis methods, the tool provides *stochastic simulation* and several static analysis techniques to improve the scalability issues of exhaustive computational methods. In particular, we have developed a method to *detect* (potentially) *redundant* rules in the model. The absence of a redundant rule in the model does not change the behaviour of the model since a more general rule already exists in the model. This can be useful in large models to detect potentially conflicting rules and to make the model more compact. Another analysis technique is used to *reduce the context* of the model to the minimal level in order to produce a smaller and more abstract model. The resulting model still preserves some properties while making the analysis of the model computationally simpler. Finally, the static analysis of *unreachability* can be used to check whether an agent is unreachable without the need to enumerate the transition system. This analysis is based on the idea that in order to reach an agent, there must be a rule which either produces the agent or its more abstract form.

An important part of the presentation of data produced by eBCSgen is visualisation. The result for both types of simulation can be visualised in an interactive chart and the result of parameter synthesis can be displayed in a visualisation which shows slices of 2D parameter space projections (Fig. 1 left). Moreover, it is possible to visualise the generated transition system and the result of he sampling of the probability function of parameters produced by parameter synthesis.

4   Experimental Results

For the purpose of evaluation, we consider the model Miyoshi et al. [15] describing circadian rhythms in cyanobacteria formed by three proteins controlled by repeated phosphorylation and dephosphorylation of key protein and complex formation with other proteins. A simplified version of the model composed of 10 rules in BCSL (in contrast to 27 explicit reactions) has been analysed using a prototype version of eBCSgen in [19].

In this paper, we consider a full version of the model[3] with 9 rules (note the number of rules is lower, but the rules are more detailed, representing almost

---

[3] The model files and computed analysis results are available here:
https://biodivine.fi.muni.cz/galaxy/eBCSgen/case-studies/cmsb-2020.
The results for simplified analysis are also publicly available:
https://biodivine.fi.muni.cz/galaxy/eBCSgen/case-studies/nfm-2020.

700 explicit reactions) using the abstract syntax described above. Analysis of this model provided more detailed results compared to the simplified model. Similarly to the case study in [19], we analysed phosphorylation and dephosphorylation phases separately using PCTL parameter synthesis. The more detailed analyses showed that the oscillatory behaviour is dependent on particular values of parameters responsible for the dephosphorylation phase (Fig. 1 left). The original results showed much higher robustness with respect to parameter values (Fig. 1 right).

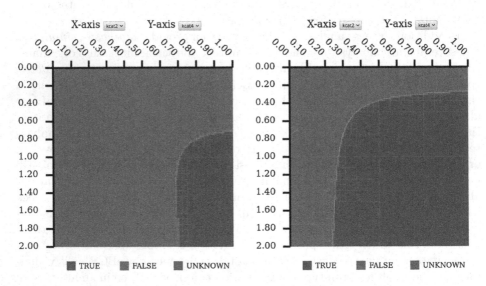

Fig. 1. Visualisation of the partitioning of the parameter space as a result of parameter synthesis for Miyoshi et al. model. The results are computed for the dephosphorylation phase with respect to the property of reaching a fully dephosphorylated target protein complex with a probability higher than 0.99. The *left* figure shows original results from [19] for a simplified version of the model. The *right* figure shows results for the full version of the model.

5   Conclusion

We presented the tool eBCSgen with its primary features and capabilities. The tool serves as a base for development and analysis of models written in BCSL. It focuses on user-accessibility of its features. Since the most of the computational analysis is performed by external tools (e.g. PCTL property checking in Storm, simulation in Python package `scipy`), we focused on the language description, tool capabilities, visualisation as important factors for the user experience, and briefly explained experimental results from the biological domain. Regarding the performance of eBCSgen, the present bottleneck is the generation of explicit transition system. Our future steps are to focus on the scalability issues trying to completely avoid this step using symbolic approaches or on-the-fly techniques.

References

1. Afgan, E., et al.: The Galaxy platform for accessible, reproducible and collaborative biomedical analyses: 2018 update. Nucleic Acids Res. **46**(W1), W537–W544 (2018)
2. Barbuti, R., et al.: An intermediate language for the stochastic simulation of biological systems. TCS **410**(33–34), 3085–3109 (2009)
3. Boutillier, P., et al.: The Kappa platform for rule-based modeling. Bioinformatics **34**(13), i583–i592 (2018)
4. Calzone, L., et al.: BIOCHAM: an environment for modeling biological systems and formalizing experimental knowledge. Bioinformatics **22**(14), 1805–1807 (2006)
5. Chabrier-Rivier, N., Fages, F., Soliman, S.: The biochemical abstract machine BIOCHAM. In: Danos, V., Schachter, V. (eds.) CMSB 2004. LNCS, vol. 3082, pp. 172–191. Springer, Heidelberg (2005). https://doi.org/10.1007/978-3-540-25974-9_14
6. Danos, V., Laneve, C.: Formal molecular biology. Theor. Comput. Sci. **325**, 69–110 (2004)
7. Daws, C.: Symbolic and parametric model checking of discrete-time Markov chains. In: Liu, Z., Araki, K. (eds.) ICTAC 2004. LNCS, vol. 3407, pp. 280–294. Springer, Heidelberg (2005). https://doi.org/10.1007/978-3-540-31862-0_21
8. Dehnert, C., Junges, S., Katoen, J.P., Volk, M.: A **STORM** is coming: a modern probabilistic model checker. In: Majumdar, R., Kunčak, V. (eds.) CAV 2017. LNCS, vol. 10427, pp. 592–600. Springer, Cham (2017). https://doi.org/10.1007/978-3-319-63390-9_31
9. Děd, T., et al.: Formal biochemical space with semantics in Kappa and BNGL. Electr. Notes Theor. Comput. Sci. **326**, 27–49 (2016)
10. Harris, L.A., et al.: BioNetGen 2.2: advances in rule-based modeling. Bioinformatics **32**(21), 3366–3368 (2016)
11. Hasson, H., Jonsson, B.: A logic for reasoning about time and probability. FAOC **6**, 512–535 (1994)
12. Honorato-Zimmer, R., et al.: Chromar, a language of parameterised agents. Theor. Comput. Sci. **765**, 97–119 (2019)
13. Lanotte, R., et al.: Parametric probabilistic transition systems for system design and analysis. FAOC **19**(1), 93–109 (2007)
14. Lopez, C.F., et al.: Programming biological models in Python using PySB. Mol. Syst. Biol. **9**(1), 646 (2013)
15. Miyoshi, F., et al.: A mathematical model for the Kai-protein-based chemical oscillator and clock gene expression rhythms in Cyanobacteria. J. Biol. Rhythms **22**(1), 69–80 (2007)
16. Pedersen, M., et al.: A high-level language for rule-based modelling. PloS One **10**(6), e0114296 (2015)
17. Romers, J.C., Krantz, M.: rxncon 2.0: a language for executable molecular systems biology. bioRxiv (2017)
18. Troják, M., et al.: Executable biochemical space for specification and analysis of biochemical systems. arXiv 2002.00731 (2020)
19. Troják, M., et al.: Parameter synthesis and robustness analysis of rule-based models. In: NASA Formal Methods Symposium (2020, published). https://doi.org/10.1007/978-3-030-55754-6_3
20. Xu, W., et al.: RuleBender: a visual interface for rule-based modeling. Bioinformatics **27**(12), 1721–1722 (2011)

What is a Cell Cycle Checkpoint? The TotemBioNet Answer

Déborah Boyenval[(✉)], Gilles Bernot, Hélène Collavizza, and Jean-Paul Comet

University Côte d'Azur, I3S Laboratory, UMR CNRS 7271, CS 40121,
06903 Sophia Antipolis Cedex, France
{deborah.boyenval,gilles.bernot,helene.collavizza,
jean-paul.comet}@univ-cotedazur.fr

Abstract. TotemBioNet is a new software platform to assist the design of qualitative regulatory network models by combining "genetically modified Hoare logic", temporal logic model checking and optimized enumeration techniques. TotemBioNet is particularly efficient to manage parameter identification, the most critical step of formal modelling. It is also remarkably flexible and efficient to check properties in order to explore biological assumptions. To illustrate this efficacy, we address the classical example of the cell cycle, where the passage from one phase to the next one, called *checkpoint*, is crucial but is usually a rather fuzzy informal concept. The cyclic behaviour of the cell cycle is specified by temporal logic and the order of individual events inside each phase is explored thanks to quantifiers introduced in Hoare logic. This way, TotemBioNet rapidly suggests a sensible formalization of the notion of checkpoint.

Keywords: Regulatory network · Discrete modelling · Parameter identification · Hoare logic · Temporal logic · Model-checking · Cell cycle

1 Formal Methods for Thomas Regulatory Networks

In the 70's, *qualitative models* based on discrete mathematics [10,17] have proved useful to understand the main causalities that govern observed phenotypes [18,19], and the multivalued framework of René Thomas and Houssine Snoussi has become a classic for biological regulatory networks. It gained new power with the introduction of formal methods in the early 2000s [4], concomitantly with [5] for signalling networks.

A qualitative model is an influence graph where the important actors for the biological question, as well as their interactions, have been inventoried on the basis of biological knowledge. Formally, the graph covers a huge set of different discrete models because the strengths of combined activations and inhibitions are unknown. They are encoded by a set of discrete *parameters* [15], which we need to identify: Any biological knowledge reduces the number of possible parameter values, by rejecting the parameterizations that do not comply. Parameter identification is the most difficult part of the modelling activity.

© Springer Nature Switzerland AG 2020
A. Abate et al. (Eds.): CMSB 2020, LNBI 12314, pp. 362–372, 2020.
https://doi.org/10.1007/978-3-030-60327-4_21

Simulations rapidly reach their limit, because of the non-determinism of trajectories and of a high number of parameterizations: A finite number of simulations cannot establish general properties. Formal methods solve this difficulty. There are software platforms based on Thomas' semantics: some study the *invariants* of a given model using Petri net tools [12] and some others perform *model checking* algorithms inspired from program verification techniques [1,11]. *Temporal logics* allow to check very general properties, including those universally quantified on traces, and SMBioNet [4], based on CTL, has made possible to *exhaustively* treat *sets of models* (sets of parameterizations) by optimized enumeration algorithms. Nevertheless, to find the *exhaustive* set of possible parameterizations, enumeration asks for exponentially growing computation time w.r.t. the size and connectivity of the regulatory network.

On another note, biological experiments directly request a set of traces in the model, and the "genetically modified Hoare logic" [3] is much more effective than temporal logic for this task. Instead of commonly enumerating parameterizations as in many approaches [14,16] and [9], it produces a set of constraints on the parameters that characterizes those in which these traces exist. The program Hoare-fol [13] efficiently computes these constraints.

TotemBioNet stands out by combining all these approaches in such a way that modellers converge *rapidly* toward the exhaustive set of parameterizations satisfying all *formalized* biological knowledge.

Finding the set of parameterizations compatible with current knowledge does not end the modelling activity. It remains to use the model to study the *biological question...* which, most of the time, is *not yet formalized*. Each possible explanation constitutes an hypothesis, and formalizing the latter is necessary to, at least, check its *consistency*: if the set of compatible parameterizations is empty then the hypothesis is inconsistent. TotemBioNet's ability to repeatedly question the model about its diverse properties, and to obtain quick answers, allows for a fast convergence towards a formalisation of the biological question.

2   The Platform TotemBioNet

TotemBioNet supports two variants of temporal logics: CTL and a dedicated *fair-path* CTL, needed for certain reachability properties in the Thomas framework. Indeed, the quantifiers A and E in CTL often induce artifactual results because they consider *unfair* paths that cross an infinite number of times a given state but never fire a possible transition from this state. So, universally quantified CTL formulas to study, for instance, attraction basins become unfairly false. Fair-path CTL quantifiers A and E simply ignore unfair paths [13]. TotemBioNet automatically translates fair-path CTL formulas into (more complex) CTL formulas, allowing it to benefit from usual CTL model checking algorithms.

Besides, TotemBioNet Hoare triples contain: a pre-condition describing the possible initial states of a given biological experiment, a path, and an observed post-condition. The path abstracts the curves obtained from experimental observations: according to thresholds setting, a threshold crossing of a variable v along the curve is written $v+$ if v increases, or $v-$ if v decreases. When experimental conditions are not precise enough to know which variable passes its threshold first, *existential quantifiers* can express this uncertainty: $\exists(v_1+; v_2-$, $v_2-; v_1+)$ means that v_1 has increased and v_2 has decreased, but in an unknown order. Also, universal quantifiers permit to abstract together a collection of similar experimental observations. Genetically modified Hoare logic extends classical Hoare logic by formalising under which conditions on the parameters each $v+$ or $v-$ of the path can occur. Then, the usual *weakest pre-condition* is the constraint on the parameters that makes the abstract path possible [3].

The inputs of `TotemBioNet` are: an influence graph, any knowledge on the parameter values, and properties on the dynamics of the system expressed using CTL, fair-path CTL, or Hoare triples. `TotemBioNet` integrates *Hoare-fol* [8] and an extended version of *SMBioNet* [4]. First, *Hoare-fol* computes and simplifies the weakest pre-condition wp w.r.t. genetically modified Hoare logic [3]. Then, the enumeration process of `TotemBioNet` is based on that of *SMBioNet*, which exploits self-influences and the *Snoussi constraint* (more resources cannot reduce the expression level) to greatly reduce the enumeration complexity.

`TotemBioNet` enumerates all parameterizations that satisfy wp: *i)* if $wp \equiv$ *False*, the enumeration process stops, *ii)* if wp is a conjunction of atoms of the form $(K_{v_i} \leq s_i)$ or $\neg(K_{v_i} \leq s_i)$ where K_{v_i} is a parameter and s_i a threshold for variable v_i, then the enumeration domains of K_{v_i} are reduced, and *iii)* if wp contains disjunctions, the validity of wp is checked on the fly. This considerably reduces the search space of all possible parameterizations and `TotemBioNet` generates, for each remaining parameterization, one input file for the model checker *NuSMV* [6]. This file contains the conjunction of temporal formulas and an automaton which encodes the state transition graph for the current parameterization. `TotemBioNet` also offers *environment variables* used to freeze some variables according to an experimental environment (by the way, it also reduces the number of parameters).

`TotemBioNet`, see https://gitlab.com/totembionet/totembionet, comes with many examples that illustrate the combination of CTL properties, fair-path CTL properties and Hoare triples. `TotemBioNet` allows one to describe the influence graph with *yEd* graph editor (https://www.yworks.com/products/yed). A typical session consists in building the influence graph using *yEd*, in automatically generating the corresponding text file and then in adding temporal properties and Hoare triples in concrete syntax. `TotemBioNet` generates an output file (possibly in `csv` format) which contains all parameterizations, labeled with "OK" when the dynamic properties are verified, and if not with all the properties which are not satisfied.

The global `TotemBioNet` process is illustrated in Fig. 1.

3   `TotemBioNet` Use Case: A Simplified Cell Cycle Model

The cell cycle is a series of events leading to correct duplication of DNA of a cell (synthesis or S phase) and its division into two genetically identical daughter cells

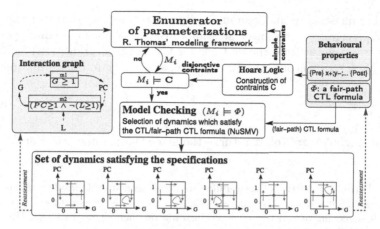

Fig. 1. `TotemBioNet`' processing flow

(mitosis or M phase). Gap phases G1 and G2 lie respectively before S and M. Progression through the cell cycle is driven by Cyclins/Cyclin-dependant kinases complexes (Cyc/Cdks) and their inhibitors. A 5-variables cell-cycle model has been designed in [2] where the variables sk, a and b are the main Cyc/Cdks involved in the mammalian cell cycle and en and ep their inhibitors. The interaction graph and its variables are detailed in Appendix A. Moreover, the succession of phases, G1, S, G2 and M, has been described in [2] *via* the Hoare

triple $H_{init} : \left\{ G1_{init} \right\} \begin{array}{c} sk+; sk+; en-; \\ a+; sk-; sk-; b+; \\ a-; ep+; \\ en+; b-; ep-; \end{array} \left\{ G1_{init} \right\}$ where the pre- and post-condition

$G1_{init}$ specifies the state $sk = 0, ep = 0, a = 0, b = 0, en = 1$, G1 is the blue subsequence, S the red one, G2 the dark gray one and M the green one.

Our main question is: *Is this small model powerful enough to represent checkpoints?* First, notice that the model assumes constant infusion of growth factors and consequently its dynamics must *always* be cyclic. The ability of `TotemBioNet` to mix several formal approaches allows us to combine H_{init} and this property specified using fair-path CTL: $\varphi_{cyclic} \equiv G1_{init} \Rightarrow AX(AF(G1_{init}))$ (where $AXAF$ means a strict future). `TotemBioNet` results are synthetized in the first line of Table 1: from the 100800 parameterizations satisfying Snoussi constraint, 676 of them satisfy the weakest precondition calculated from H_{init}. Then 609 out of the 676 also validate φ_{cyclic}. Notice the great efficiency of Hoare Logic: The use of the equivalent CTL formula φ_{init} (see Appendix B) instead of the H_{init} Hoare triple in the first experiment of the table would drastically increase the computation time: from 6.1 s to 18.5 min for the same result.

Another question to understand checkpoint is the order of transitions inside phases. [2] suggests that some transitions inside a phase may admit permutations: all transitions except the first one for G1 and S, all transitions for G2 and none for M. H_{forall} encodes the 12 possible paths owing to the *Forall* quantifier,

$$\mathbf{H_{forall}} : \left\{ G1_{init} \right\} \begin{array}{c} sk+; \quad Forall((sk+;en-),(en-;sk+)); \\ a+; \quad Forall((sk-;sk-;b+),(sk-;b+;sk-),(b+;sk-;sk-)); \\ Forall((ep+;a-),(a-;ep+)); \\ en+;b-;ep-; \end{array} \left\{ G1_{init} \right\}$$

and surprisingly `TotemBioNet` returns the same 609 parameterizations!

Table 1. Formal properties of a simplified 5-variables cell cycle model: H_m is the set of models satisfying Hoare and Snoussi constraints. S_m is the set of selected models after model-checking of a temporal logic formula on each element of H_m. (Performed on an Intel Core i7-8650U processor, 1.90 GHz, 8 cores.)

Exp	Hoare triple	$|H_m|$	Temporal logic formula	$|S_m|$	Computation time (s)
1	H_{init}	676	φ_{cyclic}	609	6.1
2	H_{forall}	676	φ_{cyclic}	609	6.1
3	H_{perm}	0	φ_{cyclic}	0	0.24
4	H_{permG1}	260	φ_{cyclic}	240	2.4
5	H_{permG1}	260	$\varphi_{cyclic} \wedge \varphi_{G2/M} \wedge \varphi_{G1/S}$	**28**	2.9

This suggests that a *phase* could be simply a bag of transitions that can be performed in arbitrary order. We check this idea with the Hoare triple H_{perm} (Appendix C) and `TotemBioNet` returns a unsatisfiable weakest precondition. So, let us check individually for G1, S and M. H_{permG1} asks for all permutations in G1, whereas S, G2 and M are the same as in H_{forall} (Appendix D): `TotemBioNet` returns 240 parameterizations. For S and for M no parameterization is selected. We conclude that the order of transitions suggested in [2] is constrained within S and M but not within G1 and G2.

Now, having a better idea of what goes on within a phase, it appears that a *checkpoint* between two phases, p_1 and p_2 should ensure that none of the possible *first* transitions of p_2 can be performed before one of the transitions of p_1. The most biologically important and the most studied checkpoints are G2/M and G1/S. They can be formalized using Hoare logic but the CTL formula is simpler if we remark that the first state of a phase is unique, whatever the order of the previous phases:

$$\varphi_{G2/M} \equiv \begin{pmatrix} sk = 0 \\ \wedge\ ep = 0 \\ \wedge\ a = 1 \\ \wedge\ b = 1 \\ \wedge\ en = 0 \end{pmatrix} \Rightarrow \neg \begin{pmatrix} EX(\mathbf{en} = 1 \wedge EX(a = 0 \wedge EX(ep = 1))) \\ \vee\ EX(\mathbf{en} = 1 \wedge EX(ep = 1 \wedge EX(a = 0))) \\ \vee\ EX(a = 0 \wedge EX(\mathbf{en} = 1 \wedge EX(ep = 1))) \\ \vee\ EX(ep = 1 \wedge EX(\mathbf{en} = 1 \wedge EX(a = 0))) \end{pmatrix}.$$

Similarly $\varphi_{G1/S}$ is given in Appendix E, and TotemBioNet returns 28 parameterizations satisfying $\varphi_{G2/M}$ and $\varphi_{G1/S}$ in addition to H_{permG1} and φ_{cyclic}. Thus, *checkpoints can be captured in a purely discrete framework*. The biologically less studied checkpoints, S/G2 and M/G1, have been also formalized (Appendix F). No parameterisation is selected suggesting that the current model is not detailed enough to satisfy S/G2 or M/G1 checkpoint, as defined.

4   Conclusion

TotemBioNet combines in an optimized manner Hoare logic and (different variants of) temporal logic. Two of our current works are to facilitate an incremental analysis of models, as well as to provide a more user-friendly interface with jupyter notebook as BioCHAM [7] and CoLoMoTo [11] do. TotemBioNet aims at offering a growing palette of formal methods to the modellers, so that each biological knowledge can be formalized according to the most suited one. Thanks to the versatility and efficacy of TotemBioNet, the general properties of qualitative Thomas models can be rapidly checked during their design. We showed on the small cell cycle model initially specified with a Hoare triple in [2] how the main properties of the phases can be explored, leading to a proper formalization of the notions of phase and checkpoint.

Acknowledgements. We are grateful to all contributors/users: M. Folschette (Hoarefol), S. Ndèye and E. Gallésio (**antlr4** parser and installation scripts), L. Gibart (beta tests on big models). We are also indebted to A. Richard for *SMBioNet* and the constructive proof of translation from *fair-path CTL* to *CTL*. This work also benefited from fruitful collaborations and discussions with J. Behaegel and F. Delaunay.

Appendix A: Static Description of the Cell Cycle Model

See Fig. 2.

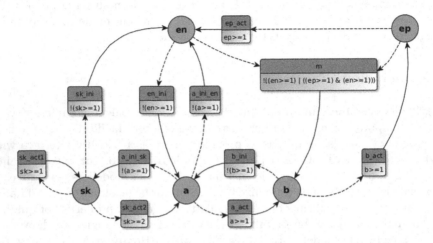

Fig. 2. A 5-variable interaction graph of the mammalian cell cycle, from [2].
Progression through the cell cycle is driven by 2 types of genetic entities: complexes of
Cyclins/Cyclin-dependant kinases (Cyc/Cdks) and their inhibitors known as *ennemies*.
The 5 variables of the graph represent these entities, in orange. *sk* is the abstraction
of both complexes CycE/Cdk2 and CycH/Cdk7, known as *starting kinases*. *a* and
b respectively represent CycA/Cdk1 and CycB/Cdk1. *en* is the abstraction of the
main Cyc/Cdks ennemies: the anaphase-promoting complex APC/Cdh1, cyclin-kinase
inhibitors *p21* and *p27*, and Wee1 protein. The variable *ep* is the anaphase-promoting
complex APC/Cdc20, which is a Cyc/Cdks ennemy involved in mitosis exit and so-
called *exit protein*. Regulations between variables are described in [2]. This interaction
graph was designed using the tool *yEd* (www.yworks.com/products/yed). (Color figure
online)

Appendix B: Equivalent Specification of H_{init} using a Fair CTL Formula

In the first experiment, the cell cycle is specified by the H_{init} Hoare triple. Here,
the cell cycle is specified by the φ_{init} CTL formula depicting the H_{init} path.

$\varphi_{init} \equiv$

$$
\begin{pmatrix}
(sk = 0 \wedge ep = 0 \wedge a = 0 \wedge b = 0 \wedge en = 1) -> EX(\mathbf{sk} = \mathbf{1} \wedge ep = 0 \wedge a = 0 \wedge b = 0 \wedge en = 1) \\
\wedge (EX(\mathbf{sk} = \mathbf{2} \wedge ep = 0 \wedge a = 0 \wedge b = 0 \wedge en = 1) \wedge (EX(sk = 2 \wedge ep = 0 \wedge a = 0 \wedge b = 0 \wedge \mathbf{en} = \mathbf{0}) \\
\wedge (EX(sk = 2 \wedge ep = 0 \wedge \mathbf{a} = \mathbf{1} \wedge b = 0 \wedge en = 0) \wedge (EX(\mathbf{sk} = \mathbf{1} \wedge ep = 0 \wedge a = 1 \wedge b = 0 \wedge en = 0) \\
\wedge (EX(\mathbf{sk} = \mathbf{0} \wedge ep = 0 \wedge a = 1 \wedge b = 0 \wedge en = 0) \wedge (EX(sk = 0 \wedge ep = 0 \wedge a = 1 \wedge \mathbf{b} = \mathbf{1} \wedge en = 0) \\
\wedge (EX(sk = 0 \wedge ep = 0 \wedge \mathbf{a} = \mathbf{0} \wedge b = 1 \wedge en = 0) \wedge (EX(sk = 0 \wedge \mathbf{ep} = \mathbf{1} \wedge a = 0 \wedge b = 1 \wedge en = 0) \\
\wedge (EX(sk = 0 \wedge ep = 1 \wedge a = 0 \wedge b = 1 \wedge \mathbf{en} = \mathbf{1}) \wedge (EX(sk = 0 \wedge ep = 1 \wedge a = 0 \wedge \mathbf{b} = \mathbf{0} \wedge en = 1) \\
\wedge (EX(sk = 0 \wedge \mathbf{ep} = \mathbf{0} \wedge a = 0 \wedge b = 0 \wedge en = 1))))))))))))
\end{pmatrix}
$$

Appendix C: Specification of H_{perm}

This Hoare triple encodes a cell cycle in which phases are described by all permutations of their respective transitions. G1 is specified in blue, S in red, G2 in grey and M in green.

$$
\mathbf{H_{perm}} : \left\{ G1_{init} \right\}
\begin{array}{c}
Forall((sk+; sk+; en-), (sk+; en-; sk+), (en-; sk+; sk+)); \\
Forall((a+; sk-; sk-; b+), (a+; sk-; b+; sk-), \\
(a+; b+; sk-; sk-), (sk-; a+; sk-; b+), \\
(sk-; a+; b+; sk-), (b+; a+; sk-; sk-), \\
(sk-; sk-; a+; b+), (sk-; b+; a+; sk-), \\
(b+; sk-; a+; sk-), (sk-; sk-; b+; a+), \\
(sk-; b+; sk-; a+), (b+; sk-; sk-; a+)) \\
Forall[(ep+; a-), (a-; ep+)]; \\
Forall((en+; b-; ep-), (en+; ep-; b-), \\
(ep-; b-; en+), (ep-; en+; b-), \\
(b-; en+; ep-), (b-; ep-; en+));
\end{array}
\left\{ G1_{init} \right\}
$$

Appendix D: Specification of H_{permG1}

This Hoare triple describes the cell cycle in which G1 in addition to G2 allows all permutations of its transitions.

$$
\mathbf{H_{permG1}} : \left\{ G1_{init} \right\}
\begin{array}{c}
Forall((sk+; sk+; en-), \\
(sk+; en-; sk+), (en-; sk+; sk+)); \\
a+; \\
Forall((sk-; sk-; b+), (sk-; b+; sk-), \\
(b+; sk-; sk-)); \\
Forall((ep+; a-), (a-; ep+)); \\
en+; b-; ep-;
\end{array}
\left\{ G1_{init} \right\}
$$

Appendix E: Specification of $\varphi_{G1/S}$ with CTL

The premise $G1_{init}$ of the formula $\varphi_{G1/S}$ is the precondition of the Hoare triple H_{init} defined in [2]. It defines the initial state of G1. The first transition of S, $a+$, must not occur before any G1 transition. Thus 9 paths must not exist starting from the first G1 state encoded in premise.

The notation $EX(a = 1 \wedge EX(sk = 1 \wedge EX(en = 0 \wedge EX(sk = 2)))$ is a CTL version of the Hoare path: $a+; sk+; en-; sk+$.

$$
\varphi_{G1/S} \equiv \left(G1_{init} \right) \Rightarrow \neg
\begin{pmatrix}
EX(\mathbf{a{=}1} \wedge EX(sk = 1 \wedge EX(en = 0 \wedge EX(sk = 2)))) \\
\vee\ EX(sk = 1 \wedge EX(\mathbf{a{=}1} \wedge EX(en = 0 \wedge EX(sk = 2)))) \\
\vee\ EX(sk = 1 \wedge EX(en = 0 \wedge EX(\mathbf{a{=}1} \wedge EX(sk = 2)))) \\
\vee\ EX(\mathbf{a{=}1} \wedge EX(en = 0 \wedge EX(sk = 1 \wedge EX(sk = 2)))) \\
\vee\ EX(sk = 1 \wedge EX(\mathbf{a{=}1} \wedge EX(sk = 1 \wedge EX(en = 0)))) \\
\vee\ EX(sk = 1 \wedge EX(sk = 2 \wedge EX(\mathbf{a{=}1} \wedge EX(en = 0)))) \\
\vee\ EX(\mathbf{a{=}1} \wedge EX(sk = 1 \wedge EX(sk = 2 \wedge EX(en = 0)))) \\
\vee\ EX(en = 0 \wedge EX(\mathbf{a{=}1} \wedge EX(sk = 1 \wedge EX(sk = 2)))) \\
\vee\ EX(en = 0 \wedge EX(sk = 1 \wedge EX(\mathbf{a{=}1} \wedge EX(sk = 2))))
\end{pmatrix}
$$

Appendix F: Specification and Checking of S/G2 and M/G1 Checkpoints with CTL

The premise of $\varphi_{S/G2}$ formula (see below) encodes the first state of S. $a-$ and $ep+$ are the 2 possible first events of G2 according to H_{permG1}. They must not occur before completion of S events. Thus 21 paths must not exist starting from the state in premise. $\varphi_{S/G2}$ is then defined as:

$$
\begin{pmatrix} sk=2 \\ \wedge\ ep=0 \\ \wedge\ a=0 \\ \wedge\ b=0 \\ \wedge\ en=0 \end{pmatrix} \Rightarrow \neg
\begin{pmatrix}
EX(a=1 \wedge EX(sk=1 \wedge EX(sk=0 \wedge EX(\mathbf{a=0} \wedge EX(b=1 \wedge EX(ep=1)))))) \\
\vee\ EX(a=1 \wedge EX(sk=1 \wedge EX(\mathbf{a=0} \wedge EX(sk=0 \wedge EX(b=1 \wedge EX(ep=1)))))) \\
\vee\ EX(a=1 \wedge EX(\mathbf{a=0} \wedge EX(sk=1 \wedge EX(sk=0 \wedge EX(b=1 \wedge EX(ep=1)))))) \\
\vee\ EX(a=1 \wedge EX(b=1 \wedge EX(sk=1 \wedge EX(\mathbf{a=0} \wedge EX(sk=0 \wedge EX(ep=1)))))) \\
\vee\ EX(a=1 \wedge EX(b=1 \wedge EX(\mathbf{a=0} \wedge EX(sk=1 \wedge EX(sk=0 \wedge EX(ep=1)))))) \\
\vee\ EX(a=1 \wedge EX(\mathbf{a=0} \wedge EX(b=1 \wedge EX(sk=1 \wedge EX(sk=0 \wedge EX(ep=1)))))) \\
\vee\ EX(a=1 \wedge EX(sk=1 \wedge EX(b=1 \wedge EX(\mathbf{a=0} \wedge EX(sk=0 \wedge EX(ep=1)))))) \\
\vee\ EX(a=1 \wedge EX(sk=1 \wedge EX(\mathbf{a=0} \wedge EX(b=1 \wedge EX(sk=0 \wedge EX(ep=1)))))) \\
\vee\ EX(a=1 \wedge EX(\mathbf{a=0} \wedge EX(sk=1 \wedge EX(b=1 \wedge EX(sk=0 \wedge EX(ep=1)))))) \\
\vee\ EX(a=1 \wedge EX(sk=1 \wedge EX(sk=0 \wedge EX(\mathbf{ep=1} \wedge EX(b=1 \wedge EX(a=0)))))) \\
\vee\ EX(a=1 \wedge EX(sk=1 \wedge EX(\mathbf{ep=1} \wedge EX(sk=0 \wedge EX(b=1 \wedge EX(a=0)))))) \\
\vee\ EX(a=1 \wedge EX(\mathbf{ep=1} \wedge EX(sk=1 \wedge EX(sk=0 \wedge EX(b=1 \wedge EX(a=0)))))) \\
\vee\ EX(\mathbf{ep=1} \wedge EX(a=1 \wedge EX(sk=1 \wedge EX(sk=0 \wedge EX(b=1 \wedge EX(a=0)))))) \\
\vee\ EX(a=1 \wedge EX(b=1 \wedge EX(sk=1 \wedge EX(\mathbf{ep=1} \wedge EX(sk=0 \wedge EX(a=0)))))) \\
\vee\ EX(a=1 \wedge EX(b=1 \wedge EX(\mathbf{ep=1} \wedge EX(sk=1 \wedge EX(sk=0 \wedge EX(a=0)))))) \\
\vee\ EX(a=1 \wedge EX(\mathbf{ep=1} \wedge EX(b=1 \wedge EX(sk=1 \wedge EX(sk=0 \wedge EX(a=0)))))) \\
\vee\ EX(\mathbf{ep=1} \wedge EX(a=1 \wedge EX(b=1 \wedge EX(sk=1 \wedge EX(sk=0 \wedge EX(a=0)))))) \\
\vee\ EX(a=1 \wedge EX(sk=1 \wedge EX(b=1 \wedge EX(\mathbf{ep=1} \wedge EX(sk=0 \wedge EX(a=0)))))) \\
\vee\ EX(a=1 \wedge EX(sk=1 \wedge EX(\mathbf{ep=1} \wedge EX(b=1 \wedge EX(sk=0 \wedge EX(a=0)))))) \\
\vee\ EX(a=1 \wedge EX(\mathbf{ep=1} \wedge EX(sk=1 \wedge EX(b=1 \wedge EX(sk=0 \wedge EX(a=0)))))) \\
\vee\ EX(\mathbf{ep=1} \wedge EX(a=1 \wedge EX(sk=1 \wedge EX(b=1 \wedge EX(sk=0 \wedge EX(a=0))))))
\end{pmatrix}
$$

Similarly, the premise of $\varphi_{M/G1}$ formula (see below) encodes the first state of M. $sk+$ and $en-$ are the 2 possible first events of G1 according to H_{permG1}. Thus the 8 paths enabling these events to occur before completion of M events must not exist, starting from the state in premise. $\varphi_{M/G1}$ is then defined as:

$$
\begin{pmatrix} sk=0 \\ \wedge\ ep=1 \\ \wedge\ a=0 \\ \wedge\ b=1 \\ \wedge\ en=0 \end{pmatrix} \Rightarrow \neg
\begin{pmatrix}
EX(en=1 \wedge EX(b=0 \wedge EX(\mathbf{sk=1} \wedge EX(ep=0 \wedge EX(sk=2 \wedge EX(en=0)))))) \\
\vee\ EX(en=1 \wedge EX(\mathbf{sk=1} \wedge EX(b=0 \wedge EX(ep=0 \wedge EX(sk=2 \wedge EX(en=0)))))) \\
\vee\ EX(\mathbf{sk=1} \wedge EX(en=1 \wedge EX(b=0 \wedge EX(ep=0 \wedge EX(sk=2 \wedge EX(en=0)))))) \\
\vee\ EX(en=1 \wedge EX(b=0 \wedge EX(\mathbf{sk=1} \wedge EX(ep=0 \wedge EX(en=0 \wedge EX(sk=2)))))) \\
\vee\ EX(en=1 \wedge EX(\mathbf{sk=1} \wedge EX(b=0 \wedge EX(ep=0 \wedge EX(en=0 \wedge EX(sk=2)))))) \\
\vee\ EX(\mathbf{sk=1} \wedge EX(en=1 \wedge EX(b=0 \wedge EX(ep=0 \wedge EX(en=0 \wedge EX(sk=2)))))) \\
\vee\ EX(en=1 \wedge EX(b=0 \wedge EX(\mathbf{en=0} \wedge EX(ep=0 \wedge EX(sk=1 \wedge EX(sk=2)))))) \\
\vee\ EX(en=1 \wedge EX(\mathbf{en=0} \wedge EX(b=0 \wedge EX(ep=0 \wedge EX(sk=1 \wedge EX(sk=2))))))
\end{pmatrix}
$$

`TotemBioNet` extracts no model (Table 2) for each of these checkpoints, from which we conclude that the model is not precise enough to capture them.

Table 2. Verification of S/G2 and M/G1 checkpoints. H_m is the set of models satisfying Hoare and Snoussi constraints. S_m is the set of selected models after model-checking of a temporal logic formula on each element of H_m.

Exp	Hoare triple	$|H_m|$	Temporal logic formula	$|S_m|$	Computation time (s)
6	H_{permG1}	260	$\varphi_{cyclic} \wedge \varphi_{G2/M} \wedge \varphi_{G1/S} \wedge \varphi_{S/G2}$	0	3.5
7	H_{permG1}	260	$\varphi_{cyclic} \wedge \varphi_{G2/M} \wedge \varphi_{G1/S} \wedge \varphi_{M/G1}$	0	3.2

References

1. Batt, G., Bergamini, D., de Jong, H., Garavel, H., Mateescu, R.: Model checking genetic regulatory networks using GNA and CADP. In: Graf, S., Mounier, L. (eds.) SPIN 2004. LNCS, vol. 2989, pp. 158–163. Springer, Heidelberg (2004). https://doi.org/10.1007/978-3-540-24732-6_12
2. Behaegel, J., Comet, J.P., Bernot, G., Cornillon, E., Delaunay, F.: A hybrid model of cell cycle in mammals. J. Bioinform. Comput. Biol. **14**(1), 1640001 (2016)
3. Bernot, G., Comet, J.P., Khalis, Z., Richard, A., Roux, O.F.: A genetically modified Hoare logic. Theor. Comput. Sci. **765**, 145–157 (2019)
4. Bernot, G., Comet, J.P., Richard, A., Guespin, J.: Application of formal methods to biological regulatory networks: extending Thomas' asynchronous logical approach with temporal logic. J. Theor. Biol. **229**(3), 339–347 (2004)
5. Chabrier, N., Fages, F.: Symbolic model checking of biochemical networks. In: Priami, C. (ed.) CMSB 2003. LNCS, vol. 2602, pp. 149–162. Springer, Heidelberg (2003). https://doi.org/10.1007/3-540-36481-1_13
6. Cimatti, A., et al.: NuSMV 2: an OpenSource tool for symbolic model checking. In: Brinksma, E., Larsen, K.G. (eds.) CAV 2002. LNCS, vol. 2404, pp. 359–364. Springer, Heidelberg (2002). https://doi.org/10.1007/3-540-45657-0_29
7. Fages, F., Soliman, S.: On robustness computation and optimization in BIOCHAM-4. In: Češvka, M., Šafránek, D. (eds.) CMSB 2018. LNCS, vol. 11095, pp. 292–299. Springer, Cham (2018). https://doi.org/10.1007/978-3-319-99429-1_18
8. Folschette, M.: The Hoare-fol tool. Technical report, Univ. Lille & CNRS UMR 9189 (2019). https://hal.archives-ouvertes.fr/hal-02409801
9. Guziolowski, C., et al.: Exhaustively characterizing feasible logic models of a signaling network using answer set programming. Bioinformatics **30**, 1942 (2013)
10. Kauffman, S.A.: Metabolic stability and epigenesis in randomly constructed genetic nets. J. Theor. Biol. **22**(3), 437–467 (1969)
11. Naldi, A., et al.: The CoLoMoTo interactive notebook. Front. Physiol. **9**, 680 (2018)
12. Remy, E., Ruet, P., Mendoza, L., Thieffry, D., Chaouiya, C.: From logical regulatory graphs to standard Petri nets: dynamical roles and functionality of feedback circuits. In: Transactions on Computational Systems Biology VII, pp. 56–72 (2006)
13. Richard, A.: Fair Paths in CTL (2008). https://gitlab.com/totembionet/totembionet
14. Schwab, J., Kühlwein, S., Ikonomi, N., Kühl, M., Kestler, H.: Concepts in Boolean network modeling: what do they all mean? Comput. Struct. Biotechnol. J. **18**, 571-582 (2020)

15. Snoussi, E.: Qualitative dynamics of a piecewise-linear differential equations: a discrete mapping approach. Dyn. Stab. Syst. **4**, 189–207 (1989)
16. Streck, A., Thobe, K., Siebert, H.: Comparative statistical analysis of qualitative parametrization set, September 2015
17. Thomas, R.: Boolean formalization of genetic control circuits. J. Theor. Biol. **42**(3), 563–585 (1973)
18. Thomas, R.: Logical analysis of systems comprising feedback loops. J. Theor. Biol. **73**(4), 631–56 (1978)
19. Thomas, R., Gathoye, A., Lambert, L.: A complex control circuit. Regulation of immunity in temperate bacteriophages. Eur. J. Biochem. **71**(1), 211–227 (1976)

Kaemika App: Integrating Protocols and Chemical Simulation

Luca Cardelli[✉]

University of Oxford, Oxford, UK
luca.a.cardelli@gmail.com

Abstract. Kaemika is an app available on the four major app stores. It provides deterministic and stochastic simulation, supporting natural chemical notation enhanced with recursive and conditional generation of chemical reaction networks. It has a liquid-handling protocol sublanguage compiled to a virtual digital microfluidic device. Chemical and microfluidic simulations can be interleaved for full experimental-cycle modeling. A novel and unambiguous representation of directed multigraphs is used to lay out chemical reaction networks in graphical form.

Keywords: Molecular programming · Digital microfluidics

1 Introduction

Kaemika is a chemical reaction simulator, including a modern graphical user interface and a functional programming language for platform independent (command line free) operation. It provides basic deterministic and stochastic simulation functionality, supporting natural chemical notation and enhancing it with the recursive and conditional generation of chemical reaction networks. It innovates primarily in the integration of liquid-handling protocols with chemical kinetics, providing a unified semantics for laboratory procedures and the evolution of multiple chemicals samples. Based on a previously presented protocol language [1], the app demonstrates its potential by compiling its geometry-free descriptions to a virtual digital microfluidic device that interleaves droplet routing simulation with chemical simulation, for full experimental-cycle modeling. Another contribution is a regular and compact representation of directed multigraphs, which includes a new representation of Petri nets but is further specialized for presenting chemical reaction networks in graphical form.

2 Simulation of Chemical Reaction Networks

Kaemika[1] offers deterministic and stochastic simulation of chemical reaction networks, aiming for uniformity of techniques over all expressible reaction networks. Mass action kinetics is used by default, but Hill, Arrhenius, and other kinetics can be expressed via common algebraic and elementary transcendental functions. This includes supplying continuous and discontinuous input waveforms.

---

[1] /'kimika/, a homophone of the Italian word for chemistry.

© Springer Nature Switzerland AG 2020
A. Abate et al. (Eds.): CMSB 2020, LNBI 12314, pp. 373–379, 2020.
https://doi.org/10.1007/978-3-030-60327-4_22

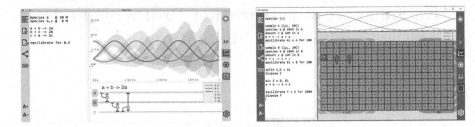

Fig. 1. Graphical user interface (macOS left, Windows UWP right). Script editor (left), plot (top), reaction score (Mac, bottom right), microfluidics (Windows, bottom right). Menus and buttons on the sides. Press play to simulate.

Stochasticity is supported via the Linear Noise Approximation (LNA [8]). Numerical LNA simulations can produce displays for standard deviation, variance, coefficient of variation, Fano factor, and (variance, etc., of) linear combinations of species. LNA numerical simulations can be applied to all expressible kinetics, which would be hard to do with other stochastic techniques (e.g., the Gillespie algorithm needs to be adapted for non mass action kinetics). LNA is supported also symbolically, providing formal derivatives for the covariance of any pair of species for all expressible kinetics (as long as the kinetic functions are differentiable), which can then be externally studied analytically.

The focus on the LNA technique is due to its general and uniform applicability, and to its relative speed and single-shot operation. The LNA is an approximation of the chemical master equation, and we should complement it with other techniques whenever possible. But the great convenience of the LNA makes it, in my opinion, the default every-day solution, especially in the context of dealing with any (multimolecular, Hill, etc.) reactions that a user may write.

While these simulation techniques are not particularly novel, they are applied in a uniform and consistent way to facilitate experimentation, so that if a user can write down a chemical model, then the tool can in fact simulate it at the click of a button (within the numerical bounds of an ODE solver). An example is the extension of the LNA semantics to liquid-handling simulations (which is novel), where the noise present in a compartment is correctly propagated when the compartment is split or merged with other compartments.

All this functionality is packaged in the interface as a single "play" button for simulation, plus a toggle for the LNA, and a corresponding "stop" button.

3   Programmatic Generation of Networks and Protocols

Despite their transparency and simplicity, chemical reaction networks become awkward when they contain many reactions, many repeated subsystems, and many parameters. This is a classical abstraction problem that has been identified and addressed long ago [10] and more recently [12]. Kaemika originated from the desire to build "programmable" (arbitrarily parameterizable) reaction networks

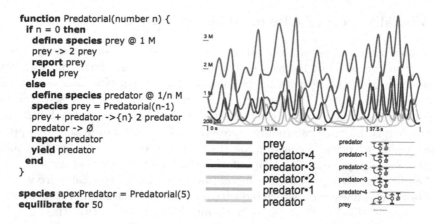

```
function Predatorial(number n) {
  if n = 0 then
    define species prey @ 1 M
    prey -> 2 prey
    report prey
    yield prey
  else
    define species predator @ 1/n M
    species prey = Predatorial(n-1)
    prey + predator ->{n} 2 predator
    predator -> Ø
    report predator
    yield predator
  end
}

species apexPredator = Predatorial(5)
equilibrate for 50
```

Fig. 2. Predator-eat-predator. Left: a program generating a variable-size chemical network, reactions in red. Top: simulation plot for n = 5. Bottom center: legend. Bottom right: graphical representation of the generated reaction network. (Color figure online)

with natural chemical notation, e.g. to study their algorithmic capabilities, and from frustration with existing tools that did not seem to quite meet that need. (Network generation and repeated triggering of simulations are needed particularly for the protocol subsystem. The alternative use of general programming languages leads to loosing the notational convenience of chemical reactions.)

Kaemika adopts modern concepts from functional programming to solve this problem. First, there is functional programming itself for complete, higher order, abstraction ("Can a species, or a network, be a parameter to a network?"). We then use *nominal* semantics [6] to deal with the generation and lexical binding of an unbounded number of unique chemical species ("If I create new species inside a loop, can I plot them?"). Finally we use an *output monad* [11], which is a somewhat grandiose but systematic scheme for generating a network of chemical reactions from a functional computation ("Can I produce a network whose size is determined by conditional execution?"). All answers are "yes!".

A short example will have to suffice here. The "Predatorial" function in Fig. 2 creates a stack of predator-prey relationships in Lotka-Volterra style, and returns the apex predator. To note: (1) the function is recursive; it internally creates new species ('prey', 'predator'), initializes them ('@'), and returns them ('yield'), (2) the new species are 'reported' as they are created, so that they can be plotted, (3) chemical notation (in red) is freely intermixed with flow control, (4) 'equilibrate' runs a simulation and plots it, combining all the reports. The 'equilibrate' statement can be repeatedly invoked. Through some variations of the 'report' statement one can also capture simulation timecourses, recombine them within other simulations, and export them as data.

4   Visualization of Chemical Reaction Networks

Automated layout of reaction networks (multigraphs) is usually highly unsatisfactory in the sense of hiding the symmetries of the network, and awkward in the sense of requiring constant panning and zooming. Kaemika uses a new graphical representation of directed multigraphs with multiplicities, which are those needed to unambiguously represent chemical reactions. In first instance, the problem is the same as visually representing Petri nets; even here we appear to be making an original contribution. In addition, catalysts are given a more compact visual representation that extends the basic one for Petri nets.

We call this new representation a *reaction score*. Like a musical score it has a set of horizontal lines, each associated with a chemical species rather than a pitch. Reactions are added to the score in horizontally-bounded vertical tiles. Neither the horizontal nor vertical orders are important (unlike in musical notation), and it is useful to be able to manually or automatically reorder species and reactions to cluster them in different ways. Each reaction $A \to B$ is first recast in the form $C, A' \to B'$ where for each species s if $n * s$ occurs in A and $m * s$ occurs in B, then $min(n, m) * s$ are moved into C, and the rest are left in A' or B' (not both). The reaction $A' \to B'$ is laid out as a Petri net transition and interconnected. (The Petri net places are stretched out as horizontal lines. The transition "bars" are placed vertically, handling multimolecular reactions with repeated connections, or are omitted in 1-input/1-output cases such as all the ones in Fig. 3). Additional catalytic connections, using a different visual style, are introduced between the species in C and the *stem* (transition) of $A' \to B'$.

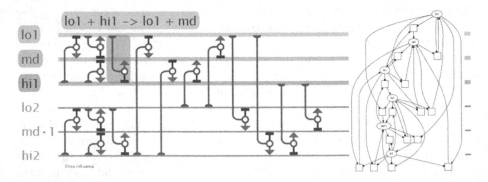

Fig. 3. Reaction score. Horizontal lines are species, vertical tiles are reactions. Blue/blunt are reagents, red/sharp are products, green/circle are catalysts. Note some evident substructures and symmetries. On the right, for comparison, is the same multigraph rendered by GraphViz, where the symmetries become hidden. (Color figure online)

This representation is complete (any reaction network can be automatically laid out) and unambiguous (the original reaction network can be recovered from it, except for the reaction rates and initial conditions).

5   Protocols and Digital Microfluidics

The Kaemika system provides a virtual liquid handling device for the simulation and visualization of protocols (Fig. 1). We focus on digital microfluidics because of its generality, simplicity, and programmability, in that a single device can execute all the basic liquid-handling protocols [2,13], and support automated observation of the samples [9].

A Kaemika protocol contains information about the kinetics of the reactions that naturally occur within samples, and also about laboratory manipulations performed on samples [1]. The two are linked because lab operations affect concentrations, volumes, and temperatures, which affect kinetics. Correspondingly, the execution of a Kaemika protocol intertwines the simulation of individual reaction networks with the microfluidic manipulation of the samples, including intertwining the plotting of simulations and the visualization of liquid handling. The state of a sample at the end of a chemical simulation is propagated to the following liquid handling operation, and conversely.

A typical digital microfluidics device has a rectangular array of electrically controlled pads, and some means of adding and removing liquid droplets over its surface. Injection and extraction may by done by hand, or by extruding standard-size droplets from larger on-device reservoirs, or by pumps at the device's periphery. The standard droplet size is around 1 μL. Droplets can be moved by changing the electrical properties of adjacent pads, and multiple droplets can be moved in parallel. Droplets can be merged by causing one to move over the pad of another, and split by electrically pulling them in opposite directions. An overhead camera or an on-surface sensing apparatus may provide feedback about the position of the droplets.

In a Kaemika droplet simulation, each "sample" (a container for species and reactions) is represented by a droplet on the device. Mixing, splitting, and disposing of samples is handled by appropriate routing of the droplets over the device surface: this is automatic, and does not require geometric instructions.

Some physical assumptions are needed for timing, for observation, and for the handling of temperatures and volumes. We assume that a region of the device is maintained at a *cool* temperature. All the staging and mixing operation are executed in this region, because chemical reactions are assumed not to be happening during liquid handling: cool temperature and quick execution can approximate those conditions. We also assume that another region of the device is maintained at a *hot* temperature, and an intermediate region is at *warm*, ambient temperature. Times passes, logically, only during "equilibrate" operations, which move droplets into one of the warm or hot regions, according to need, hold them there for the prescribed time, and then move them back to the cool region. Observation capabilities (and subsequent feedback into protocols) are highly hardware dependent [9]: we provide in the language general observability of concentrations, but this will have to be matched to physical device capabilities.

6   Implementation and Deployment

The main audience for Keamika is research and higher education environments, although we have tried to widen its adoption potential by supporting mobile and dual keyboard/touch devices, and by deploying it to app stores.

Kaemika is written in C# using the Visual Studio/Xamarin IDE, and is available on four app stores: Windows UWP, macOS, iOS, and Android. A single Visual Studio solution is used for all platforms, with shared application logic, compiled under either Windows or macOS; the source code is on GitHub [4]. The language syntax is based on the Gold LALR parser generator [5]. The ODE solver is OSLO [7]. The basic simulation functionality, including LNA, is common with many other tools, e.g. [3,10], which otherwise focus on other modeling aspects. The main Windows and macOS GUI interfaces consist of two similar separate forms; a separate touch-optimized GUI is used for mobile displays, with Xamarin providing a unified interface to Android/iOS. Low-level graphics (lines, splines, fonts, etc.) is shared across Windows/iOS/Android via Skia graphics, but separate from CoreGraphics for macOS. XAML, which subverts lexical scoping, typing, error accountability, and reliability, is painstakingly circumvented.

In practice, supporting multiple platforms is not hard, and software changes propagate easily across them. Rather, the challenge is navigating the parkour-like registration, provisioning, and app submission procedures of each app store. Still, I strongly advise this path since it has huge benefits for users in terms of tool installation, and also of usability (flawed GUIs are store-rejected). In the end it has huge benefits for developers too, in terms of removing variability of user configurations and all the related distribution and support issues, which I found even more challenging than app store approvals.

References

1. Abate, A., Cardelli, L., Kwiatkowska, M., Laurenti, L., Yordanov, B.: Experimental biological protocols with formal semantics. In: Češka, M., Šafránek, D. (eds.) CMSB 2018. LNCS, vol. 11095, pp. 165–182. Springer, Cham (2018). https://doi.org/10.1007/978-3-319-99429-1_10
2. Alistar, M., Gaudenz, U.: OpenDrop: an integrated do-it-yourself platform for personal use of biochips. Bioengineering (Basel) **4**(2), 45 (2017)
3. Cardelli, L., Tribastone, M., Tschaikowski, M., Vandin, A.: ERODE: a tool for the evaluation and reduction of ordinary differential equations. In: Legay, A., Margaria, T. (eds.) TACAS 2017. LNCS, vol. 10206, pp. 310–328. Springer, Heidelberg (2017). https://doi.org/10.1007/978-3-662-54580-5_19
4. Cardelli, L.: https://github.com/luca-cardelli/KaemikaXM
5. Cook, D.: Design and development of a grammar oriented parsing system. MSc Project, California State University Sacramento (2004)
6. Crole, R., Nebel, F.: Nominal lambda calculus: an internal language for FM-Cartesian closed categories. ENTCS **298**, 93–117 (2013)
7. Dalchau, N.: Open solving library for ODEs. https://www.microsoft.com/en-us/research/project/open-solving-library-for-odes/
8. Ethier, S., Kurtz, T.: Markov Processes. Wiley, Hoboken (2009)

9. Freire, S.: Perspectives on digital microfluidics. Sens. Actuators A: Phys. **250**, 15–28 (2016)

10. Pedersen, M., Phillips, A.: Towards programming languages for genetic engineering of living cells. J. R. Soc. Interface **6**, S437–S450 (2009)

11. Petricek, T.: What we talk about when we talk about monads. Art Sci. Eng. Program. **2**(2), 12 (2018)

12. Vasic, M., Soloveichik, D., Khurshid, S.: CRN++: molecular programming language. Nat. Comput. **19**(12), 1–17 (2020)

13. Willsey, M., et al.: Puddle: a dynamic, error-correcting, full-stack microfluidics platform. In: ASPLOS (2019)

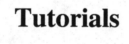
Tutorials

Tutorial: The CoLoMoTo Interactive Notebook, Accessible and Reproducible Computational Analyses for Qualitative Biological Networks

Loïc Paulevé[(⊠)]

University of Bordeaux, Bordeaux INP, CNRS, LaBRI, UMR5800, 33400 Talence,
France
loic.pauleve@labri.fr

Analysing models of biological networks typically relies on workflows in which different software tools with sensitive parameters are chained together, many times with additional manual steps. The accessibility and reproducibility of such workflows is challenging, as publications often overlook analysis details, and because some of these tools may be difficult to install, and/or have a steep learning curve.

The CoLoMoTo Interactive Notebook [1][1] provides a unified environment to edit, execute, share, and reproduce analyses of qualitative models of biological networks. This framework combines the power of different technologies to ensure repeatability and to reduce users' learning curve of these technologies. The framework is distributed as a Docker image with the tools ready to be run without any installation step besides Docker, and is available on Linux, macOS, and Microsoft Windows. The embedded computational workflows are edited with a Jupyter web interface, enabling the inclusion of textual annotations, along with the explicit code to execute, as well as the visualization of the results. Resulting notebook files can then be shared and re-executed in the same environment.

To date, the CoLoMoTo Interactive Notebook provides access to the software tools listed in Table 1 for the modeling and analysis of Boolean and multi-valued networks. More tools will be included in the future. We developed a Python interface for each of these tools to offer a seamless integration in the Jupyter web interface and ease the chaining of complementary analyses. An *executable paper* entirely edited within the CoLoMoTo Interactive Notebook accessible and demonstrating its main features is available at http://doi.org/10.3389/fphys.2018.00787.

The prime aim of the CoLoMoTo Interactive Notebook is to foster the production of accessible and reproducible computational analysis of biological models, with a focus on qualitative models. The CoLoMoTo Docker image can be easily extend to include additional tools, and be used by standard workflow systems, such as SnakeMake, to lighten the burden of installing the different software tools and make them accessible on different operating systems.

Loïc Paulevé—LP was supported by ANR-FNR project "AlgoReCell" (ANR-16-CE12-0034).

[1] http://colomoto.org/notebook.

A. Abate et al. (Eds.): CMSB 2020, LNBI 12314, pp. 383–385, 2020.
https://doi.org/10.1007/978-3-030-60327-4

The CoLoMoTo Interactive Notebook is also relevant for teaching purposes. With Jupyter, students can straightforwardly execute, modify, and extend a template notebook to learn methods for analysing models of biological networks.

Table 1. Software tools distributed to date in the CoLoMoTo Notebook

Software tool	Description
bioLQM	Logical Qualitative Modelling toolkit
bioLQM	Logical Qualitative Modelling toolkit
CellCollective	Model repository and knowledge base
GINsim	Boolean and multi-valued network modelling
MaBoSS	Markovian Boolean Stochastic Simulator
mpbn	Most Permissive Boolean Networks
NuSMV	Symbolic model-checker
Pint	Static analyzer for dynamics of Automata Networks
R-BoolNet	Analysis and reconstruction of Boolean networks dynamics

Objective of the Ttutorial. Give an introduction to different tools related to qualitative modelling and analysis of biological networks, and promote frameworks to ease accessibility and reproducibility of computational analyses.

Program. After a brief introduction to the qualitative modelling of biological network, the tutorial session will demonstrate how to:

1. create a Boolean network from scratch
2. load an existing model of biological system (e.g.., hosted on GINsim repository or cellcollective.org)
3. compute stable states with bioLQM and perform model-checking with NuSMV
4. perform stochastic simulations with MaBoSS
5. perform control prediction and validation using Pint and MaBoSS
6. overview of other included and coming tools and features
7. extend the environment with your own tool

Importantly, the tutorial will emphasize how to use notebooks to improve reproducible research, with good practices for editing and sharing them, and how to re-execute and adapt them.

Requirements. The CoLoMoTo Notebook requires Docker and optionally Python. Then, the setup and launch is a matter of a single command. For people who do not want or fail to install Docker, they will still be able to follow the tutorial without requiring any specific installation using tmpnb.colomoto.org.

Reference

1. Naldi, A., et al.: The CoLoMoTo interactive notebook: accessible and reproducible computational analyses for qualitative biological networks. Front. Physiol. **9**, 680 (2018). https://doi.org/10.3389/fphys.2018.00680

Integrating Experimental Pharmacology and Systems Biology for GPCR Drug Discovery

Susanne Roth, Yaroslav Nikolaev, Mirjam Zimmermann, Nadine Dobberstein, Maria Waldhoer, and Aurélien Rizk[✉]

InterAx Biotech, Villigen, Switzerland
rizk@interaxbiotech.com

Abstract. G protein-coupled receptors (GPCRs) regulate many physiological and pathophysiological processes and constitute a major class of drug targets. Our ability to design drugs that control the cellular responses via GPCRs relies on our understanding of how receptors encode and transfer information. Both, i) receptor conformational changes and ii) activation dynamics of GPCRs are crucial in governing downstream signaling events, hence, a deeper understanding of how ligands activate and regulate these events are crucial for drug development and optimization. At InterAx we use Systems Biology as a computational tool to integrate theoretical knowledge with experimental data of GPCR mediated signaling and trafficking events. As a result, deeper mechanistic insights into the dynamic cellular signaling systems activated by these receptors are achievable. Importantly, such mathematical models allow to predict experimental outcomes and deliver novel insights into drug actions on GPCRs.

In this tutorial we will show how an ordinary differential equation (ODE) model can be developed to describe the trafficking and signaling events of the beta-2 adrenergic receptor (B2AR) in response to various agonists (= asthma drugs). We will present i) time-resolved data of receptor internalization and recycling and of cyclic AMP accumulation that can be used to parameterize such models, ii) quantification of parameter identifiability by profile likelihood estimation, iii) validation of the approach through comparison of predicted kinetic parameters with direct experimental estimation and iv) clustering of compounds based on predicted kinetic parameters and comparison with known in-vivo effects.

Keywords: Ordinary differential equations · G protein-coupled receptors · Drug discovery · Time-resolved signalling assays · Compound clustering

Author Index

Printed in the United States
By Bookmasters